AF324308

POPULATION DYNAMICS

Algebraic and Probabilistic Approach

POPULATION DYNAMICS

Algebraic and Probabilistic Approach

Utkir A. Rozikov

V. I. Romanovskiy Institute of Mathematics, Tashkent, Uzbekistan

World Scientific

NEW JERSEY · LONDON · SINGAPORE · BEIJING · SHANGHAI · HONG KONG · TAIPEI · CHENNAI · TOKYO

Published by

World Scientific Publishing Co. Pte. Ltd.

5 Toh Tuck Link, Singapore 596224

USA office: 27 Warren Street, Suite 401-402, Hackensack, NJ 07601

UK office: 57 Shelton Street, Covent Garden, London WC2H 9HE

Library of Congress Control Number: 2020001588

British Library Cataloguing-in-Publication Data
A catalogue record for this book is available from the British Library.

POPULATION DYNAMICS
Algebraic and Probabilistic Approach

ISBN 978-981-121-122-5

For any available supplementary material, please visit
https://www.worldscientific.com/worldscibooks/10.1142/11578#t=suppl

To my family

Contents

Probabilistic approach   225

Concrete populations dynamics   291

Preface

A population is a summation of all the organisms of the same group or species, which live in a particular geographical area, and have the capability of interbreeding. A state of a population is a distribution of probabilities of the different types of organisms. Type partition is called differentiation (for example, sex differentiation which defines a bisexual population).

The main mathematical problem for a given population is to examine the evolution (time dependent dynamics of states) of the population. The mathematical methods used in the study of this problem are based on probability theory, stochastic processes, dynamical systems, nonlinear analysis, differential and difference equations, and (non-)associative algebras.

In November of 2019, MathSciNet found more than 39300 entries for "population" in the entire database. By this database the first publication related to the population is [268], which was published in 1924. In just 2018 there were 1700 publications. Therefore, theory of populations is very popular topic in mathematics. Since there are a lot of literature devoted to population dynamics, it is impossible to collect all of them in one book.

The book systematically presents algebraic and probabilistic approaches in the theory (mainly since 2010) of population dynamics.

In the algebraic approach part several kinds of algebra (Evolution algebras of "self-reproduction", "free population", "bisexual population", "chicken population") related to population dynamics are discussed. Moreover, flows (in particular chains) of algebras are described.

The probabilistic approach part presents Markov processes of cubic stochastic (in a fixed sense) matrices (MPCSM), which are continuous-time dynamical systems, whose states are stochastic cubic matrices satisfying an analogue of the Kolmogorov-Chapman equation. MPCSM considered for two specially chosen notions of stochastic cubic matrices and two multiplications of such matrices (known as Maksimov's multiplications). Time-dependent behavior of such processes is given with applications to a population with the possibility of twin births.

The book also presents several dynamical systems of biological models: dynamical systems of free population; dynamics of sex linked population; dynamical systems generated by a gonosomal evolution operator; dynamical system and an evolution algebra of mosquito population; ocean ecosystems.

The long list of literature devoted to population dynamics makes it difficult for a beginner to start learning the theory. The main aim of this book is to facilitate the reader's in-depth understanding by giving a systematic review of the theory of population dynamics which has wide applications in biology, mathematics, medicine, and physics.

MathSciNet found 46 books devoted to population dynamics, see for example [13], [14], [23], [98], [99], [110], [111], [114], [155], [153], [183], [186], [207], [244], [246], [262], [269]. But in this book, I used modern papers and internet sources which were not previously used in textbooks on theory of population dynamics. I am sure that the book will be very helpful for readers interested in dynamical systems and mathematical biology.

The book is based on materials collected by the author during several visits to the University of Cambridge, Newton Institute and University of Leeds, UK; ICTP, Trieste, Italy; "La Sapienza" University, Rome, Italy; IHES, Bures-sur-Yvette, France; Universite du Sud Toulon Var, CPT and IMéRA, Marseille, France; University Paris-Est, France; University Santiago de Compostela and University Granada, Spain; University of Bonn and Ruhr University, Bochum, Germany.

Acknowledgements. I thank all the above-mentioned institutions for their warm hospitality and excellent working conditions. I am very grateful to my coauthors and collaborators A. T. Absalamov, J. M. Casas, A. Dzhumadil'daev, N. N. Ganikhodjaev, R. N. Ganikhodzhaev, A. J. M. Hardin, A. Yu. Khamraev, A. Labra, M. Ladra, B. J. Mamurov, F. M. Mukhamedov, Sh. N. Murodov, S. Nazir, B. A. Omirov, N. B. Shamsiddinov, S. K. Shoyimardonov, J. P. Tian, R. Varro, M. V. Velasco, A. Zada, U. U. Zhamilov.

I also thank World Scientific Publishing for the opportunity to publish this book.

Utkir A. Rozikov
Tashkent, Uzbekistan
December 2019

Chapter 1

Introduction

In this chapter we give some minimal preliminaries from biology, motivations of studying of population dynamics, and define evolution operators and associated algebras related to free and bisexual populations. Moreover, we formulate the main problems in the consideration of the population dynamics. The next chapters of this book will be devoted to solutions of these problems.

1.1 Background and motivation

A *population* is a set of organisms of the same kind, some long time living in one territory (occupying a particular area) and are completely isolated from other same groups. The term "population" is used in various sections of biology, ecology, demography, medicine and psychometrics.

In the life sciences the *population dynamics* branch studies the size and age composition of populations as dynamical systems. The environmental and biological processes driving populations can be as birth and death rates, and immigration and emigration. Examples are population growth, ageing populations, or population decline.

The population dynamics is a well developed branch of mathematical biology, which has a history of more than two hundred years.

The book [97] is an introduction to population mathematics, which presents deterministic and stochastic models in discrete and continuous time. The flavor of the methods, and the nature of the elementary results are conveyed. It includes inter alia, birth-death processes, branching processes, predator-prey systems and epidemics.

The book [14] contains a short history of applications of mathematics to solving various problems in population dynamics.

The question of how best to model density dependent population growth still has no definitive answer [167]. In [167], [168] this question considered through a number of individual based models of populations expressed using the process algebra Weighted Synchronous Calculus of Communicating Systems. The advantage of these models is that they can be explicitly based on observations of individual interactions. Using some probabilistic models the equations expressing overall population dynamics are derived. These equations are easily compared with the traditionally used deterministic ODE (Ordinary Differential Equation) models and allow evaluation of those ODE models, challenging their assumptions about system dynamics. The approach is applied to epidemiology, combining population growth with disease spread.

Theoretically a population may have unbounded growth, but in the real life, populations cannot grow without bound and that it is competition between individuals for resources which restricts growth. The first principle of population dynamics is a Malthusian growth model [9], [122] based on the idea of the function being proportional to the speed to which the function grows. This model also known as a simple exponential growth model. Another famous examples are the Lotka-Volterra predator-prey equations [131], as well as the alternative Arditi-Ginzburg equations. The computer games (as SimCity, Sim Earth and the MMORPG Ultima Online) also tried to simulate some of these population dynamics.[1]

Last developments of population dynamics has been complemented by evolutionary game theory, developed first by John Maynard-Smith [166], [193]. In these dynamics, evolutionary biology concepts can be given as a deterministic mathematical form. Population dynamics also overlap with following branches in mathematical biology: mathematical epidemiology, the study of infectious disease affecting populations. Several models of viral spread in a population may be applied to health policy decisions.

Population dynamics theory is important to a proper understanding of living populations at all levels. Thus the analysis of gene frequency arrays assuredly has a part to play in anthropological comparisons and observations of animal behavior patterns. Mathematics, as the language of quantitative measurement, is clearly central to these pursuits. The relevant mathematics undoubtedly requires hybridization of nonlinear analysis, nonlinear differential and difference equations, probability theory, compounded stochastic processes modeling, algebra, dynamical systems, innovative

---

[1]https://en.wikipedia.org/wiki/Population_dynamics.

statistical analysis of complex data, and the creative implementation of the gigantic computer methodology and all its ramifications [123].

For detailed background of the population dynamics see [14], where a list of classical models is given including Malthus and Verhulst models (see also [21]), Lotka and Volterra models, May and Yule models, mutation and natural selection models and others. The development of the models is discussed in a chronological order starting with the well known model of rabbits presented by Leonardo of Pisa (defined by Fibonacci's sequence) and the first demographic model and life tables provided by Edmond Halley, through the demographic population models of Euler and Johann Peter Sussmilch and the first epidemic models by Daniel Bernoulli, modern models as May's model and China's one-child policy model by Song Jian.

Now in addition to the traditional movement of mathematics for population dynamics and genetics, inspiration is flowing in the opposite direction, yielding mathematics from these biological systems. The book [155] reflects to some degree both patterns. A pioneer of this synthesis was S. N. Bernstein. He raised and partially solved the problem of characterizing all stationary evolutionary operators, and this work was continued by Yu. I. Lyubich [155]. This problem has not been completely solved, but it appears that only certain operators devoid of any biological significance remain to be addressed.

Moreover, in [155] the population dynamics and their formal analogue are studied. When selection is absent a very effective algebraic approach was introduced by Reiersöl (1961). This approach was extended by Lyubich (1971) to describe explicit solutions of the general evolution equation. Some more abstract algebraic-dynamical theory, discovered by Etherington (1939).

The dynamics of selection require different methods. Here the leading role goes to Fisher's 'fundamental theorem' concerning the increase of mean fitness of a population due to natural selection. An approach based on this theory uses a relatively complicated 'relaxation' technique to establish the global convergence to equilibrium under conditions more general than have been previously achieved. Several topics which could be studied using these methods: polyploidy, overlapping generations, migration etc.

In [254] a new type of evolution algebra is introduced. This algebra also describes some evolution laws of genetics and it is an algebra E over a field K with a countable natural basis $e_1, e_2, \ldots$ and multiplication given by $e_i e_i = \sum_j a_{ij} e_j$, $e_i e_j = 0$ if $i \neq j$. Therefore, $e_i e_i$ is viewed as "self-reproduction". Tian's book motivate the development of the theory of evolution algebras by examples from *biology*: asexual inheritance, the Wright-Fisher model that assumes that the genes from one generation are derived by sampling with replacement from the genes of the previous generation; from *physics*: particles moving in a discrete space, flows in a discrete space (networks), Feynman graphs; from *topology*: motions of particles in a 3-manifold, random walks on braids with negative probabilities; from *probability theory*: stochastic processes.

There exist several classes of non-associative algebras (baric, evolution, Bernstein, train, stochastic, etc.), whose investigation has provided a number of significant contributions to theoretical population genetics. Such classes have been defined different times by several authors, and all algebras belonging to these classes are generally called "genetic." Etherington introduced the formal language of abstract algebra to the study of the genetics (see for example, [49]). In recent years many authors have tried to investigate the difficult problem of classification of these algebras. The most comprehensive references for the mathematical research done in this area are [155, 211, 214, 254, 269].

The paper [123] gives a comprehensive survey of problems in evolutionary theory studied by mathematical methods, both with regard to past developments, the present state of the subject and probable future trends. Many of the models considered are actually within the scope of mathematical population dynamics, although the general ideas discussed have implications for a much wider area.

In recent paper [85] several mathematical problems are formulated which suggested by structural patterns present in biomolecular assemblies. For more background and motivations of the theory of population dynamics see the following recently published books [110], [111], [114], [98], [153], [244], [262]. At the end of each chapter of this book the reader can find a list of literature (related to the chapter) for further reading.

In this book we systematically present recently obtained results for discrete-time dynamical systems and evolution algebras of free populations and sex linked populations. The next section contains very interesting preliminaries from biology.

1.2   Preliminaries from biology

Let us start with some definitions:

A *gene* is the molecular unit of heredity of a living organism.

An *allele* is one of a number of alternative forms of the same gene. For example the gene which determines blood type in humans has three different alleles: A, B and O.

Chromosome: Contains the genetic code of a gene.

Gamete: is a cell that fuses with another cell during fertilization (conception) in organisms that sexually reproduce.

Zygote: is a cell formed by a fertilization event between two gametes.

Remark 1.1. When gametes fuse (or reproduce) to form zygotes a natural "multiplication" operation occurs. This multiplication will allow us to introduce (see the next sections) an algebraic approach in biology.

The action of genes is manifested statistically in sufficiently large communities of matching individuals (belonging to the same species). These communities are called *populations* [155]. The population exists not only in space but also in time, i.e. it has its own life cycle. The basis for this phenomenon is reproduction by mating. Mating in a population can be free or subject to certain restrictions.

A *free population* (or *panmixia*) means random mating in the population. A *panmictic population* is one where all individuals are potential partners. This assumes that there are no mating restrictions, neither genetic or behavioural, upon the population, and that therefore all recombination is possible. The Wahlund effect assumes that the overall population is panmictic [76].

The whole population in space and time comprises discrete generations $F_0, F_1, \ldots$ The generation F_{n+1} is the set of individuals whose parents belong to the F_n generation. A *state* of a population is a distribution of probabilities of the different types of organisms in every generation.

In genetic,[2] *random mating* involves the mating of individuals regardless of any physical, genetic, or social preference. This means that, the mating between two organisms is not influenced by any environmental, hereditary, or social interaction. Hence, potential mates have an equal chance of being selected. Random mating is a factor assumed in the Hardy-Weinberg principle and is distinct from lack of natural selection: in viability selection for instance, selection occurs before mating.

A *sex-determination system*[3] is a biological system that determines the development of sexual characteristics in an organism. Most sexual organisms have two sexes. Occasionally there are hermaphrodites in place of one

---

[2]https://en.wikipedia.org/wiki/Panmixia.

[3]https://en.wikipedia.org/wiki/Sex-determination_system.

or both sexes. There are also some species that are only one sex due to parthenogenesis, the act of a female reproducing without fertilization.

In many cases, sex determination is *genetic*: males and females have different alleles or even different genes that specify their sexual morphology. In animals, this is often accompanied by *chromosomal* differences. Determination genetically is generally through chromosome combinations of XY (for example: humans, mammals), ZW (birds), $X0$ (in this variant of the XY system, females have two copies of the sex chromosome (XX) but males have only one $(X0)$. In this method, the sex is determined by amount of genes expressed across the two chromosomes. This system is observed in a number of insects, including the grasshoppers and crickets of order Orthoptera and in cockroaches. A small number of mammals also lack a Y chromosome), $Z0$ (lepidoptera), WXY (platyfishes). Moreover, some organisms have a multiple sex chromosomes: $X1X2Y$ (for example, Hoplias malabaricus), $X1X2X3X4X5Y$ (for example, Tegenaria ferruginea), $X1X2X3X4X5Y1Y2Y3Y4Y5$ (Ornithorhynchus anatinus) etc. Sexual differentiation is generally started by a main gene, a sex locus, then a multitude of other genes follow in a domino effect. Thus there are the following sex-determination systems

System	Male	Female	Examples
XY	XY	XX	humans, mammals
WZ	ZZ	WZ	birds
X0	X0	XX	insects
Z0	ZZ	Z0	lepidoptera
WXY	XY, YY	XX, WX, WY	platyfishes

The 0 denotes the absence of a second sex chromosome.

See photos of the animals (taken from internet).

In other cases, sex is determined by *environmental* variables such as *temperature*. For example, in some species of reptiles, including alligators, some turtles, the tuatara, and a few birds, sex is determined by the temperature at which the egg is incubated during a temperature sensitive period. For some, this is achieved by hotter temperatures being one sex and cooler temperatures being the other. For others, the extreme temperatures are one sex and the middle temperature is the other. Sex also can be determined by social variables (the *size of an organism* relative to other members of its population).

Environmental sex determination occurred before genetic; it is thought that a temperature-dependent reptile was the common ancestor to sex

Fig. 1.1   Lepidoptera ($Z0$).

Fig. 1.2   Cockroach($X0$).

chromosomes. Some species, such as some snails, practice *sex change*: adults start out male, then become female. In tropical clown fish, the dominant individual in a group becomes female while the other ones are male, and blue head wrasses are the reverse. In the marine worm Bonellia viridis, larvae become males if they make physical contact with a female, and females if they end up on the bare sea floor. This is triggered by the presence of a chemical produced by the females, bonellin.

Some species have no sex-determination system. Hermaphrodites include the common earthworm and certain species of snails. A few species of fish, reptiles, and insects reproduce by parthenogenesis and are female

Fig. 1.3   Platy fish (WXY).

Fig. 1.4   Tegenaria ferruginea $(X1X2X3X4X5Y)$.

altogether. There are some reptiles, such as the boa constrictor and komodo dragon that can reproduce sexually and asexually, depending if a mate is available.

In some arthropods, sex is determined by *infection*, as when bacteria of the genus Wolbachia alter their sexuality; some species consist entirely of ZZ individuals, with sex determined by the presence of Wolbachia. If a

Fig. 1.5   Ornithorhynchus anatinus ($X1X2X3X4X5Y1Y2Y3Y4Y5$).

Fig. 1.6   Alligator and turtle.

male ZZ is infected by the bacteria Wolbachia, $ZZ + w$, then it becomes female. In this case there are three female genotypes: WZ, $WZ + w$ and $ZZ + w$.

Haemophilia is a group of hereditary genetic disorders that impair the body's ability to control blood clotting or coagulation, which is used to stop bleeding when a blood vessel is broken.

Other unusual systems: swordtail fish, the Chironomus midge species, the Platypus has 10 sex chromosomes but lacks the mammalian sex-determining gene SRY, meaning that the process of sex determination

Fig. 1.7   Change from male to female (Clown fish).

Fig. 1.8   Change from female to male (Epinephelus).

in the Platypus remains unknown. Zebrafish go through juvenile herma-phroditism, but what triggers this is unknown. The Platyfish has W, X, and Y chromosomes. This allows WY, WX, or XX females or YY and XY males [242].

Remark 1.2. The details of some sex-determination systems are not yet fully understood. A search in MathSciNet gives about 20 mathematical papers which are related to sex determination models (see for example, [55], [123], [124]). In papers [63], [223] it were attempted to introduce thermodynamic methods in biology. In [263] an algebra associated to a sex change is constructed.

1.3 Free population

In this section we shall give evolution operators and algebras of a free population.

Consider a population consisting of m species. Let $x^0 = (x_1^0, \ldots, x_m^0)$ be the probability distribution (where $x_i^0 = P(i)$ is the probability of i, $i = 1, 2, \ldots, m$) of species in the initial generation, and $P_{ij,k}$ the probability that individuals in the ith and jth species interbreed to produce an individual k, more precisely $P_{ij,k}$ is the conditional probability $P(k|i,j)$ that ith and jth species interbred successfully, then they produce an individual k.

In this section we consider models of free population i.e., there is no difference of sex and in any generation the "parents" ij are independent i.e., $P(i,j) = P(i)P(j) = x_i^0 x_j^0$. Then the probability distribution $x' = (x_1', \ldots, x_m')$ (the state) of the species in the first generation can be found by the total probability

$$x_k' = \sum_{i,j=1}^{m} P(k|i,j)P(i,j) = \sum_{i,j=1}^{m} P_{ij,k} x_i^0 x_j^0, \quad k = 1, \ldots, m. \tag{1.1}$$

This means that the association $x^0 \to x'$ defines a map V called the *evolution operator*.

The population evolves by starting from an arbitrary state x^0, then passing to the state $x' = V(x^0)$ (in the next 'generation'), then to the state $x'' = V(V(x^0))$, and so on. Thus, states of the population described by the following discrete-time dynamical system

$$x^0, \quad x' = V(x^0), \quad x'' = V^2(x^0), \quad x''' = V^3(x^0), \ldots \tag{1.2}$$

where $V^n(x) = \underbrace{V(V(...V(x))...)}_{n}$ denotes the n times iteration of V to x.

Note that V (defined by (1.1)) is a *quadratic stochastic operator* (QSO), and it is higher dimensional if $m \geq 3$. Higher dimensional dynamical systems are important, but there are relatively few dynamical phenomena that are currently understood [40]. For a review on dynamical systems see [213].

The main problem for a given dynamical system is to describe the limit points of $\{x^{(n)}\}_{n=0}^{\infty}$ for arbitrary given $x^{(0)}$.

In [68] we have discussed the recently obtained results on the problem, and also gave several open problems related to the theory of QSOs. See also [155], [183] for more detailed theory of QSOs. Moreover, in [155] an evolution algebra $\mathcal{A}$ associated to the free population is introduced and using this non-associative algebra many results are obtained in explicit form, e.g. the explicit description of stationary quadratic operators, and the explicit solutions of a nonlinear evolutionary equation in the absence of selection, as well as general theorems on convergence to equilibrium in the presence of selection.

In the next chapters of this book we give some recent results related to QSOs.

The algebra $\mathcal{A}$ is defined as follows. Recall that any vector $x \in \mathbb{R}^m$ can be written as $x = \sum_{i=1}^{n} x_i e_i$ with $\{e_1, \ldots, e_m\}$ the canonical basis on $\mathbb{R}^m$ with

$$e_1 = (1, 0, \ldots, 0), \ e_2 = (0, 1, 0, \ldots, 0), \ \ldots, e_m = (0, 0, \ldots, 0, 1) \in \mathbb{R}^m.$$

Now introduce on $\mathbb{R}^m$ a multiplication defined by

$$e_i e_k = \sum_{j=1}^{m} P_{ik,j} e_j. \tag{1.3}$$

Thus we identify the coefficients of inheritance as the structure constants of an algebra $\mathcal{A}$, i.e. a bilinear mapping $\mathbb{R}^m \times \mathbb{R}^m$ to $\mathbb{R}^m$, $x \times y \to xy$. The general formula for multiplication is the extension of (1.3) by bilinearity for multiplication:

$$xy = \sum_{i,k,j=1}^{m} (P_{ik,j} x_i y_k) e_j. \tag{1.4}$$

In particular, the evolution operator defined in coordinate form by (1.1) can be written as $x' = V(x) = x^2$.

See [155] for properties of the algebra $\mathcal{A}$. In the next chapters we shall give some recently obtained properties of the algebra $\mathcal{A}$.

1.4 Bisexual population

Type partition is called differentiation. The simplest example is sex differentiation. In bisexual population (BP) any kind of differentiation must agree with the sex differentiation, i.e. all the organisms of one type must belong to the same sex. Using the above mentioned sex-determination systems, it is possible to speak of male and female types.

Remark 1.3. In this book, following [155], the term *bisexual population* means the population which consists individuals of two type: *female* and *male*, i.e., every individual has only one sex. This should not be confused with other meaning of the term "bisexual" (see for example http://www.lgbtmap.org/bisexual-people).

Evolution operator. In this subsection, following [155], we describe the evolution operator of a BP.

Assuming that the population is bisexual we suppose that the set of females can be partitioned into finitely many different types indexed by $\{1, 2, \ldots, n\}$ and, similarly, that the male types are indexed by $\{1, 2, \ldots, \nu\}$. The number $n+\nu$ is called the dimension of the population. The population is described by its state vector (x, y) in $S^{n-1} \times S^{\nu-1}$, the product of two unit simplexes in $\mathbb{R}^n$ and $\mathbb{R}^\nu$ respectively. Vectors x and y are the probability distributions of the females and males over the possible types:

$$x \in S^{n-1} = \left\{ x \in \mathbb{R}^n : x_i \geq 0, \ \sum_{i=1}^{n} x_i = 1 \right\};$$

$$y \in S^{\nu-1} = \left\{ y \in \mathbb{R}^\nu : y_i \geq 0, \ \sum_{i=1}^{\nu} y_i = 1 \right\}.$$

Denote $S = S^{n-1} \times S^{\nu-1}$. We call the partition into types hereditary if for each possible state $z = (x, y) \in S$ describing the current generation, the state $z' = (x', y') \in S$ is uniquely defined describing the next generation. This means that the association $z \mapsto z'$ defines a map $V : S \to S$ called the evolution operator.

For any point $z^{(0)} \in S$ the sequence $z^{(t)} = V(z^{(t-1)}), t = 1, 2, \ldots$ is called the trajectory of $z^{(0)}$.

Let $P_{ik,j}^{(f)}$ and $P_{ik,l}^{(m)}$ be inheritance coefficients defined as the probability that a female offspring is type j and, respectively, that a male offspring is of type l, when the parental pair is ik ($i, j = 1, \ldots, n$; and $k, l = 1, \ldots, \nu$). We have

$$P_{ik,j}^{(f)} \geq 0, \ \sum_{j=1}^{n} P_{ik,j}^{(f)} = 1; \quad P_{ik,l}^{(m)} \geq 0, \ \sum_{l=1}^{\nu} P_{ik,l}^{(m)} = 1. \tag{1.5}$$

Let $z' = (x', y')$ be the state of the offspring population at the birth stage. This is obtained (similarly to (1.1) from inheritance coefficients as a mapping $W : S \to S$ defined by)

$$W : \quad \begin{aligned} x'_j &= \sum_{i,k=1}^{n,\nu} P_{ik,j}^{(f)} x_i y_k \\ y'_l &= \sum_{i,k=1}^{n,\nu} P_{ik,l}^{(m)} x_i y_k. \end{aligned} \tag{1.6}$$

We see from (1.6) that for a BP the evolution operator W is a quadratic mapping of S into itself.

The main problem: Given an operator W and initial point $z^{(0)} \in S$ what ultimately happens with the trajectory $z^{(n)} = W(z^{(n-1)})$, $n = 1, 2, \ldots$? Does the limit $\lim_{n \to \infty} z^{(n)}$ exist? If not what is the set of limit points of the sequence? Is this set finite or infinite?

In this book we give several recently obtained results related to this main problem.

Evolution algebra of BP. Now following [138] we give an algebra structure on the vector space $\mathbb{R}^{n+\nu}$ which is closely related to the map (1.6).

Consider $\{e_1, \ldots, e_{n+\nu}\}$ the canonical basis on $\mathbb{R}^{n+\nu}$ and divide the basis as $e_i^{(f)} = e_i$, $i = 1, \ldots, n$ and $e_i^{(m)} = e_{n+i}$, $i = 1, \ldots, \nu$.

Now introduce on $\mathbb{R}^{n+\nu}$ a multiplication defined by

$$
\begin{aligned}
e_i^{(f)} e_k^{(m)} = e_k^{(m)} e_i^{(f)} &= \tfrac{1}{2} \left(\sum_{j=1}^{n} P_{ik,j}^{(f)} e_j^{(f)} + \sum_{l=1}^{\nu} P_{ik,l}^{(m)} e_l^{(m)} \right), \\
e_i^{(f)} e_j^{(f)} = 0, \quad i, j = 1, \ldots, n; \quad &e_k^{(m)} e_l^{(m)} = 0, \quad k, l = 1, \ldots, \nu.
\end{aligned}
\tag{1.7}
$$

The general formula for the multiplication is the extension of (1.7) by bilinearity, i.e. for $z, t \in \mathbb{R}^{n+\nu}$,

$$
z = (x, y) = \sum_{i=1}^{n} x_i e_i^{(f)} + \sum_{j=1}^{\nu} y_j e_j^{(m)}, \quad t = (u, v) = \sum_{i=1}^{n} u_i e_i^{(f)} + \sum_{j=1}^{\nu} v_j e_j^{(m)}
$$

using (1.7), we obtain

$$
\begin{aligned}
zt = &\frac{1}{2} \sum_{k=1}^{n} \left(\sum_{i=1}^{n} \sum_{j=1}^{\nu} P_{ij,k}^{(f)}(x_i v_j + u_i y_j) \right) e_k^{(f)} \\
&+ \frac{1}{2} \sum_{l=1}^{\nu} \left(\sum_{i=1}^{n} \sum_{j=1}^{\nu} P_{ij,l}^{(m)}(x_i v_j + u_i y_j) \right) e_l^{(m)}.
\end{aligned}
\tag{1.8}
$$

From (1.8) and using (1.6), in the particular case that $z = t$, i.e. $x = u$ and $y = v$, we obtain

$$
\begin{aligned}
zz = z^2 = &\sum_{k=1}^{n} \left(\sum_{i=1}^{n} \sum_{j=1}^{\nu} P_{ij,k}^{(f)} x_i y_j \right) e_k^{(f)} \\
&+ \sum_{l=1}^{\nu} \left(\sum_{i=1}^{n} \sum_{j=1}^{\nu} P_{ij,l}^{(m)} x_i y_j \right) e_l^{(m)} = V(z)
\end{aligned}
\tag{1.9}
$$

for any $z \in S$.

This algebraic interpretation is very useful. For example, a BP state $z = (x, y)$ is an equilibrium (fixed point, $V(z) = z$) precisely when z is an idempotent element of the set S.

If we write $z^{[t]}$ for the power $(\cdots (z^2)^2 \cdots)$ (t times) with $z^{[0]} \equiv z$ then the trajectory with initial state z is $V^t(z) = z^{[t]}$.

Definition 1.1. The algebra $\mathcal{B} = \mathcal{B}_V$ generated by the evolution operator V (see (1.6)) is called the *evolution algebra of the bisexual population* (EABP).

Remark 1.4. 1. If a population is free then the male and female types are identical and, in particular $n = \nu$, the inheritance coefficients are the same for male and female offsprings, i.e.

$$P_{ik,j} = P_{ik,j}^{(f)} = P_{ik,j}^{(m)} .$$

The evolution algebra $\mathcal{A}$ associated with the free population is commutative when the condition of symmetry $P_{ik,j} = P_{ki,j}$ is satisfied, but it is not in general associative. In [138] it was showed that algebra $\mathcal{B}$ of bisexual population is commutative without any symmetry condition. Hence the algebra $\mathcal{A}$ is a particular case of the algebra $\mathcal{B}$.

2. It is easy to see that the EA introduced in [254] is different from EABP, $\mathcal{B}$.

3. The algebra $\mathcal{B}$ is a natural generalization of a *zygotic algebra* for sex linked inheritance (see [49, 101, 211, 269]).

In the next chapters we give the basic properties of the algebra $\mathcal{B}$.

Gonosomal evolution operator. As it was mentioned above in many cases, the sex determination is genetic, in particular, it is controlled by two chromosomes called gonosomes.

Suppose that the set of female types is $\{1, 2, \ldots, n\}$ and the set of male types is $\{1, 2, \ldots, \nu\}$.

Let $\gamma_{ik,j}^{(f)}$ and $\gamma_{ik,l}^{(m)}$ be some inheritance coefficients (not necessary probabilities) with

$$\sum_{j=1}^{n} \gamma_{ik,j}^{(f)} + \sum_{l=1}^{\nu} \gamma_{ik,l}^{(m)} = 1. \tag{1.10}$$

Note that the condition (1.5) is a particular case of the condition (1.10) which is obtained when $\gamma_{ik,j}^{(\cdot)} = \frac{1}{2} P_{ik,j}^{(\cdot)} \geq 0$.

Consider an evolution operator $W : \mathbb{R}^{n+\nu} \to \mathbb{R}^{n+\nu}$ defined as

$$x_j' = 2 \sum_{i,k=1}^{n,\nu} \gamma_{ik,j}^{(f)} x_i y_k; \quad y_l' = 2 \sum_{i,k=1}^{n,\nu} \gamma_{ik,l}^{(m)} x_i y_k. \tag{1.11}$$

This operator is called gonosomal evolution operator.

Genosomal algebra. An algebra $\mathcal{G}$ is genosomal (see [263]) if there is a basis $B = \{e_i^{(f)}\}_{i=1}^{n} \cup \{e_i^{(m)}\}_{i=1}^{\nu}$ on $\mathcal{G}$ such that for every $1 \leq i, j \leq n$ and

$1 \le p, q \le \nu$, we have

$$
\begin{aligned}
e_i^{(f)} e_p^{(m)} &= e_p^{(m)} e_i^{(f)} = \sum_{j=1}^{n} \gamma_{ip,j}^{(f)} e_j^{(f)} + \sum_{l=1}^{\nu} \gamma_{ip,l}^{(m)} e_l^{(m)}, \\
e_i^{(f)} e_j^{(f)} &= 0, \quad i, j = 1, \ldots, n; \quad e_k^{(m)} e_l^{(m)} = 0, \quad k, l = 1, \ldots, \nu,
\end{aligned}
\tag{1.12}
$$

where $\gamma_{ij,k}^{(\cdot)}$ satisfies (1.10).

The main problem for a given algebra $\mathcal{A}$ related to a population is to study its properties having a biological interpretations and applications.

Remark 1.5. Note that operator (1.11) describes evolution of a haemophilia (a lethal recessive X-linked disorder: a female carrying two alleles for haemophilia die). The dynamical systems generated by gonosomal operator (1.11) have not been studied yet. In the last part of this book we give some recently obtained results concerning to this operator. In [231] a similar operator is considered. Some constructions of the genosomal algebra $\mathcal{G}$ are given in [263].

PART 1

Algebraic approach

Chapter 2

Algebraic preliminaries

In this chapter we give basic definitions related to abstract algebras. Moreover, we study cubic matrices and algebras of such matrices, because the coefficients of *any* evolution operator and matrix of structural constants of *any* evolution algebra studied in this book are defined by cubic matrices. The results of this chapter will be useful for next chapters.

2.1 Basic definitions of abstract algebras

A *field* is a set F on which four binary operations addition, subtraction, multiplication, and division are defined. These operations satisfy the same properties as the corresponding operations on the set of rational numbers $\mathbb{Q}$ and real numbers $\mathbb{R}$ do. Thus examples of fields are $\mathbb{Q}$, $\mathbb{R}$ and the set of complex numbers $\mathbb{C}$.

Define the product $n \cdot a$ of an arbitrary element a of a filed F by a positive integer n to be the n-fold sum $a + a + \cdots + a \in F$. If there is no positive integer such that $n \cdot 1 = 0$, then F is said to have characteristic 0. One can see that the field of rational numbers $\mathbb{Q}$ has characteristic 0 since no positive integer n is zero.

A set V is called a vector space over a field F if upon V we have defined two operations: vector addition (denoted by "+"), and scalar multiplication, which combines a number from F with an element of V and is denoted by juxtaposition, and the following properties hold (for any $x, y, z \subset V$ and any $\alpha, \beta \in F$):

Commutativity: $x + y = y + x$;

Associativity: $x + (y + z) = (x + y) + z$;

Zero vector: There is a vector, 0, called the zero vector, such that $x + 0 = x$;

19

Additive inverses: If for each $x \in V$, there exists a vector $-x \in V$ such that $x + (-x) = 0$;

Scalar multiplication associativity: $\alpha(\beta x) = (\alpha\beta)x$;

Distributivity across vector addition: $\alpha(x + y) = \alpha x + \alpha y$;

Distributivity across scalar addition: $(\alpha + \beta)x = \alpha x + \beta x$;

One: If $x \in V$ then $1x = x$.

The elements in the set V are called vectors, by virtue of being elements of a vector space, no matter what else they might really be.

A set B of vectors (elements) in a vector space V is called a *basis*, if every element of V may be written in a unique way as a linear combination of elements of B. The coefficients of this linear combination are called coordinates on B of the vector. The elements of a basis are called basis vectors.

It is clear that B is a basis if its elements are linearly independent and every element of V is a linear combination of elements of B. A vector space can have several bases, but all the bases have the same number of elements, which is called the *dimension* of the vector space.

Let F be a field, and let $\mathcal{A}$ be a vector space over F equipped with an additional binary operation from $\mathcal{A} \times \mathcal{A}$ to $\mathcal{A}$, denoted by $\cdot$ (we will simply write xy instead $x \cdot y$). Then $\mathcal{A}$ is an *algebra* over F if the following identities hold for every x, y and z in $\mathcal{A}$, and every α and β in F: $(x + y)z = xz + yz$, $x(y + z) = xy + xz$, $(\alpha x)(\beta y) = (\alpha\beta)(xy)$.

In this book we also use the following definitions: If x, y and z denote arbitrary elements of an algebra then

Associative: $(xy)z = x(yz)$.

Commutative: $xy = yx$.

Anticommutative: $xy = -yx$.

Jacobi identity: $(xy)z + (yz)x + (zx)y = 0$.

Jordan identity: $(xy)x^2 = x(yx^2)$.

Power associative: For all x, any three nonnegative powers of x associate. That is if a, b and c are nonnegative powers of x, then $a(bc) = (ab)c$. This is equivalent to saying that $x^m x^n = x^{n+m}$ for all nonnegative integers m and n.

Alternative: $(xx)y = x(xy)$ and $(yx)x = y(xx)$.

Flexible: $x(yx) = (xy)x$.

It is known that these properties are related by

- *associative* implies *alternative* implies *power associative*;
- *associative* implies *Jordan identity* implies *power associative*;

- Each of the associative, commutative, anticommutative properties, Jordan identity, and Jacobi identity individually imply flexible.

For a field with characteristic not two, being both commutative and anti-commutative implies the algebra is just $\{0\}$.

For algebras over a field, the bilinear multiplication from $\mathcal{A} \times \mathcal{A}$ to $\mathcal{A}$, is completely determined by the multiplication of the basis elements of $\mathcal{A}$. Conversely, once a basis for $\mathcal{A}$ has been chosen, the products of the basis elements can be set arbitrarily, and then extended in a unique way to a bilinear operator on $\mathcal{A}$, i.e., the resulting multiplication satisfies the algebra laws.

A *homomorphism* between two algebras, $\mathcal{A}$ and $\mathcal{B}$, over a field F, is a map $f\colon \mathcal{A} \to \mathcal{B}$ such that for all $\lambda \in F$ and $x, y \in \mathcal{A}$,

$$f(\lambda x) = \lambda f(x); \qquad f(x + y) = f(x) + f(y); \qquad f(xy) = f(x)f(y).$$

If f is bijective then f is said to be an isomorphism between $\mathcal{A}$ and $\mathcal{B}$.

Given a field F, any finite-dimensional algebra can be specified up to isomorphism by giving its dimension (say m), and specifying m^3 structure constants c_{ijk}, which are scalars. These structure constants determine the multiplication in $\mathcal{A}$ via the following rule:

$$e_i e_j = \sum_{k=1}^{m} c_{ijk} e_k, \tag{2.1}$$

where $e_1, \ldots, e_m$ form a basis of $\mathcal{A}$.

Definition 2.1. An element x of an algebra $\mathcal{A}$ is called nil if there exists $n(a) \in \mathbb{N}$ such that $(\cdots \underbrace{((x \cdot x) \cdot x) \cdots x}_{n(a)}) = 0$. The algebra $\mathcal{A}$ is called nil if every element of the algebra is nil.

For $k \geq 1$, we introduce the following sequences:

$$\mathcal{A}^{(1)} = \mathcal{A}, \quad \mathcal{A}^{(k+1)} = \mathcal{A}^{(k)}\mathcal{A}^{(k)};$$

$$\mathcal{A}^{<1>} = \mathcal{A}, \quad \mathcal{A}^{<k+1>} = \mathcal{A}^{<k>}\mathcal{A};$$

$$\mathcal{A}^{1} = \mathcal{A}, \quad \mathcal{A}^{k} = \sum_{i=1}^{k-1} \mathcal{A}^i \mathcal{A}^{k-i}.$$

Definition 2.2. An algebra $\mathcal{A}$ is called

(i) solvable if there exists $n \in \mathbb{N}$ such that $\mathcal{A}^{(n)} = 0$ and the minimal such number is called index of solvability;

(ii) right nilpotent if there exists $n \in \mathbb{N}$ such that $\mathcal{A}^{<n>} = 0$ and the minimal such number is called index of right nilpotency;

(iii) nilpotent if there exists $n \in \mathbb{N}$ such that $\mathcal{A}^n = 0$ and the minimal such number is called index of nilpotency.

A non-zero element a in an algebra which satisfies $a^2 = a$ is called an idempotent. Mathematically, the existence of an idempotent in an algebra provides a direct sum decomposition of the algebra. Hence, idempotents play a crucial role in describing the general structure of an algebra.

In addition to their mathematical importance, idempotents also have genetic significance [211]. If a population P satisfies the equation $P^2 = P$, this means that genetic equilibrium has been achieved after one generation of random mating within the population P. That is, the population P^2 has the same genetic pool as the initial population P.

A *character* χ for an algebra $\mathcal{A}$ is a nonzero multiplicative linear form on $\mathcal{A}$, that is, a nonzero algebra homomorphism $\chi \colon \mathcal{A} \to \mathbb{R}$ (see for example, page 73 of [155]).

A pair $(\mathcal{A}, \chi)$ consisting of an algebra $\mathcal{A}$ and a character χ on $\mathcal{A}$ is called a *baric algebra*.

It was showed in [155] that any algebra of free population is baric. But in [211] it is shown that an algebra of bisexual population is not baric. To overcome such complication, Etherington [49] for an algebra of bisexual population (sex linked inheritance) introduced the idea of treating the male and female components of a population separately. In [101] Holgate formalized this concept by introducing (two-dimensional) sex differentiation algebras and a generalization of baric algebras called dibaric algebras:

Definition 2.3. [211, 269] Let $\mathfrak{A} = \langle w, m \rangle_\mathbb{R}$ denote a two dimensional commutative algebra over $\mathbb{R}$ with multiplicative table

$$w^2 = m^2 = 0, \quad wm = \frac{1}{2}(w + m).$$

Then $\mathfrak{A}$ is called the *sex differentiation algebra*.

A *subalgebra* B of an algebra A is a subspace which is closed under multiplication. A subspace B is an *ideal* if it is closed under multiplication by all elements of A. For example, the *square* of the algebra:

$$A^2 = \mathrm{span}\{zt : z, t \in A\}$$

is an ideal.

It is clear that $\mathfrak{A}^2 = \langle w + m \rangle_\mathbb{R}$ is an ideal of $\mathfrak{A}$ which is isomorphic to the field $\mathbb{R}$. Hence the algebra $\mathfrak{A}^2$ is a baric algebra.

Holgate's generalization of a baric algebra is the following:

Definition 2.4. [211] An algebra is called dibaric if it admits a homomorphism onto the sex differentiation algebra $\mathfrak{A}$.

Remark 2.1. *Biological interpretations of algebraic notations*: Let a biological system consists n species then each element of an n-dimensional algebra can be considered as a state of the biological system, which can be a generalized measure (or charge) on the set of species. Because for certain purposes, it is useful to have a 'measure' whose values are not restricted to the non-negative reals or infinity. Therefore each algebraic property can be interpreted as a biological one. For example, the algebraic notions like nilpotency and solvability might be interpreted biologically as a various types of vanishing ('deaths') populations. A subalgebra (in particular an ideal) corresponds to a part of the population which develops independently from the remaining part of the population.

2.2 Cubic matrices

From (2.1) it follows that the multiplication of a finite-dimensional algebra is given by a cubic matrix (c_{ijk}).

In this section following [15, 158, 205] we recall a notion of cubic matrix and different associative multiplication rules of cubic matrices: a cubic matrix $Q = (q_{ijk})_{i,j,k=1}^{m}$ is a m^3-dimensional vector which can be uniquely written as

$$Q = \sum_{i,j,k=1}^{m} q_{ijk} E_{ijk},$$

where E_{ijk} denotes the cubic unit (basis) matrix, i.e. E_{ijk} is a m^3-cubic matrix whose (i, j, k)th entry is equal to 1 and all other entries are equal to 0.

Denote by $\mathfrak{C}$ the set of all cubic matrices over a field F. Then $\mathfrak{C}$ is an m^3-dimensional vector space over F, i.e. for any matrices $A = (a_{ijk}), B = (b_{ijk}) \in \mathfrak{C}, \lambda \in F$, we have

$$A + B := (a_{ijk} + b_{ijk}) \in \mathfrak{C}, \qquad \lambda A := (\lambda a_{ijk}) \in \mathfrak{C}.$$

Denote $I = \{1, 2, \ldots, m\}$.

Following [158] define the following Maksimov's multiplications for basis matrices E_{ijk}:

$$E_{ijk} *_a E_{lnr} = \delta_{kl} E_{ia(j,n)r}, \tag{2.2}$$

where $a\colon I \times I \to I$, $(j,n) \mapsto a(j,n) \in I$, is an arbitrary associative binary operation (see Remark 2.2 below for associative binary operations) and δ_{kl} is the Kronecker symbol.

Denote by $\mathcal{O}_m$ the set of all associative binary operations on I.

Remark 2.2. Note that the number of operations on I with $|I| = m$ is m^{m^2}. Let $\tau(m)$ be the number of distinct associative binary operations on a set of size m, i.e. $\tau(m) = |\mathcal{O}_m|$. It is quickly increasing function of m which can be seen from the following: ([41])[1]

m	1	2	3	4	5	6	7	8
$\tau(m)$	1	8	113	3492	183732	17061118	7743056064	148195347518186

Thus the family of multiplications (2.2) is very rich.

The general formula for the multiplication is the extension of (2.2) by bilinearity, i.e. for any two cubic matrices $A = (a_{ijk}), B = (b_{ijk}) \in \mathfrak{C}$ the matrix $A *_a B = (c_{ijk})$ is defined by

$$c_{ijr} = \sum_{l,n\colon a(l,n)=j} \sum_{k} a_{ilk} b_{knr}. \tag{2.3}$$

Remark 2.3. In [15] the authors provided 15 associative multiplication rules of cubic matrices, some of them coincide with Maksimov's multiplication rules. It was proved that between the 15 algebras corresponding to the multiplications there are 5 non-isomorphic associative algebras. In [205] there are other kinds of multiplications for cubic stochastic matrices.

2.3 Algebras of cubic matrices

In this section following [139] we give some properties of algebras of cubic matrices.

Denote by $\mathfrak{C}_a \equiv \mathfrak{C}_a^m = (\mathfrak{C}, *_a)$, $a \in \mathcal{O}_m$, the algebra of cubic matrices (ACM) given by the multiplication $*_a$ (see (2.2)).

Remark 2.4. The ACM can be considered for more general multiplication: one can fix an $m^3 \times m^3 \times m^3$-cubic matrix $\left(C^{uvw}_{ijk,lnr}\right)$ as a matrix of structural constants and give a multiplication of basis cubic matrices as

$$E_{ijk} E_{lnr} = \sum_{uvw} C^{uvw}_{ijk,lnr} E_{uvw}.$$

---

[1]See also http://math.stackexchange.com/questions/105438/how-many-associative-binary-operations-there-are-on-a-finite-set.

Then the extension of this multiplication by bilinearity to arbitrary cubic matrices gives a general multiplication on the set $\mathfrak{C}$. Under known conditions (see [116]) on structural constants one can make this general ACM as a commutative or/and associative algebra.

Remark 2.5. As it was mentioned above, there is one-to-one correspondence between the set $\mathbf{A}$ of all n-dimensional algebras (with a fixed basis) and the set of all cubic matrices (of size n^3, defined by structural constants). This correspondence can be used to consider $\mathbf{A}$ as an algebra (of algebras), i.e. for any algebra $U \in \mathbf{A}$ with matrix of structural constants $A = (a_{ijk})$, and $V \in \mathbf{A}$ with matrix of structural constants $B = (b_{ijk})$, define $U + V$ as an algebra with matrix of structural constants $A + B$. For $\lambda \in F$ define the algebra λU with matrix of structural constants λA. Fix a multiplication of cubic matrices, say $\star$, and define multiplication $U \star V$ to be an algebra with matrix of structural constants $A \star B$. Thus $(\mathbf{A}, \star)$ is an algebra of algebras. Some properties of the ACM (with multiplication $\star$) can be related to properties of $\mathbf{A}$.

Remark 2.6. Note that baric algebras are useful to describe biological systems of free (one-sex) populations (see [155]). Dibaric algebras are related to systems of bisexual (two-sex) populations (see [138, 211]). As mentioned in the Chapter 1, in biology there are unusual systems having "multiple sexes" instead of having only two. In such systems the accompanying algebra will play the role of the sex differentiation algebra. Then an ACM will play the role of an evolution algebra of such biological systems.

2.3.1 *Isomorphic ACMs*

Let S_m be the group of permutations on I, i.e., S_m is a group whose elements are permutations of the set I and whose group operation is the composition of permutations in S_m (which are thought of as bijective functions from the set I to itself).

Take $a \in \mathcal{O}_m$ and define an action of $\pi \in S_m$ on a (denoted by πa) as

$$\pi a(i, j) = \pi a(\pi^{-1}(i), \pi^{-1}(j)), \qquad \text{for all} \quad i, j \in I.$$

Lemma 2.1. *For any* $a \in \mathcal{O}_m$ *and any* $\pi \in S_m$ *we have* $\pi a \in \mathcal{O}_m$.

Proof. It suffices to show that πa is an associative operation, i.e.

$$\pi a(\pi a(i, j), k) = \pi a(i, \pi a(j, k)), \quad \text{for all} \quad i, j, k \in I. \tag{2.4}$$

For LHS of this equality we have

$$\pi a(\pi a(i,j),k) = \pi a\Big(\pi\big(a(\pi^{-1}(i),\pi^{-1}(j))\big),k\Big)$$
$$= \pi a\Big(a\big(\pi^{-1}(i),\pi^{-1}(j)\big),\pi^{-1}(k)\Big).$$

For RHS of (2.4) we get

$$\pi a(i,\pi a(j,k)) = \pi a\Big(i,\pi\big(a(\pi^{-1}(j),\pi^{-1}(k))\big)\Big)$$
$$= \pi a\Big(\pi^{-1}(i),a\big(\pi^{-1}(j),\pi^{-1}(k)\big)\Big).$$

Since a is associative and π is one-to-one we have

$$\pi a\Big(a\big(\pi^{-1}(i),\pi^{-1}(j)\big),\pi^{-1}(k)\Big) = \pi a\Big(\pi^{-1}(i),a\big(\pi^{-1}(j),\pi^{-1}(k)\big)\Big),$$

i.e. (2.4) holds. $\square$

So, we have an action of the symmetric group, S_m, on $\mathcal{O}_m$, the set of all associative binary operations on I.

The *orbit* of $a \in \mathcal{O}_m$ under the action of S_m is $\mathrm{Orb}(a) = \{\pi a : \pi \in S_m\}$.

It is clear that the set of orbits of (points $a \in \mathcal{O}_m$) under the action of S_m form a partition of $\mathcal{O}_m$. The associated equivalence relation is defined by saying $a \sim b$ if and only if there exists a $\pi \in S_m$ with $\pi a = b$, i.e.,

$$a(j,n) = \pi^{-1}(b(\pi(j),\pi(n))), \qquad \text{for all } \ j,n \in I, \tag{2.5}$$

The orbits are then the equivalence classes under this relation; two operations a and b are equivalent if and only if their orbits are the same.

The following theorem gives a sufficient condition for the isomorphness of ACM.

Theorem 2.1. *If $a \sim b$, then the ACMs $\mathfrak{C}_a$ and $\mathfrak{C}_b$ are isomorphic.*

Proof. By Lemma 2.1 it follows that if $a \in \mathcal{O}_m$ and $b \sim a$ then $b \in \mathcal{O}_m$. Consider $f(E_{ijk}) = E_{\pi(i)\pi(j)\pi(k)}$ and check

$$f(E_{ijk} *_a E_{lnr}) = f(E_{ijk}) *_b f(E_{lnr}).$$

This equality is equivalent to

$$\delta_{kl} E_{\pi(i)\pi(a(j,n))\pi(r)} = \delta_{\pi(k)\pi(l)} E_{\pi(i)b(\pi(j),\pi(n))\pi(r)}. \tag{2.6}$$

Since π is one-to-one map on I, we have $\delta_{kl} = \delta_{\pi(k)\pi(l)}$ for any k,l. Consequently, if $\delta_{kl} = 0$, i.e., $k \neq l$ then (2.6) holds. If $\delta_{kl} = 1$, i.e. $k = l$ then (2.6) is reduced to the form

$$E_{\pi(i)\pi(a(j,n))\pi(r)} = E_{\pi(i)b(\pi(j),\pi(n))\pi(r)}.$$

This is true because of condition (2.5). $\square$

Remark 2.7. In [158] it was proved that the algebra $\mathfrak{C}_a$ is associative, which follows from the associativity of $a \in \mathcal{O}_m$.

Definition 2.5. An operation $a \in \mathcal{O}_m$ is called *symmetric* if its orbit under the action of S_m consists only of a itself, $\mathrm{Orb}(a) = \{a\}$, i.e.

$$a(j,n) = \pi(a(\pi^{-1}(j), \pi^{-1}(n))), \qquad \text{for all } j,n \in I, \quad \text{for all } \pi \in S_m.$$

(Equivalently, a is fixed point for every $\pi \in S_m$.)

Proposition 2.1. *An operation* $a \in \mathcal{O}_m$ *is symmetric if and only if it is one of the following operation:*

$$a(i,j) = j, \qquad \text{for all } i,j \in I, \tag{rs}$$

$$a(i,j) = i, \qquad \text{for all } i,j \in I. \tag{ls}$$

Proof. *Sufficiency.* Assume a is (rs) (the case (ls) is similar). We show that a is symmetric. Take any $\pi \in S_m$. Then for any $i,j \in I$ we have

$$\pi a(i,j) = \pi a(\pi^{-1}(i), \pi^{-1}(j)) = \pi(\pi^{-1}(j)) = j = a(i,j),$$

i.e. a is fixed point for any π.

Necessity. Assume a is different from (rs) and (ls). The assertion is trivial for $|I| = m = 1$. Case $m = 2$ discussed in Example 2.1 below, which shows that proposition is true for $m = 2$. Therefore we consider the case $m \geq 3$. Then the following cases are possible for a which is different from (rs) and (ls):

Case 1: There is $i \in I$ such that $a(i,i) = k$ with $k \neq i$. Consider a permutation π_0 such that $\pi_0(i) = i$, $\pi_0(k) \neq k$. Then

$$\pi_0 a(i,i) = \pi_0 a(\pi_0^{-1}(i), \pi_0^{-1}(i)) = \pi_0(k) \neq k = a(i,i),$$

i.e. a is not fixed point of π_0.

Case 2: There are $i,j \in I$ such that $i \neq j$ and $a(i,j) = k$. For k we have several possibilities:

Subcase 2.1: $k \neq i,j$.

Subsubcase 2.1.1: $m = 3$. Without loss of generality assume $a(1,2) = 3$. Using $a(1,1) = 1$ and associativity of a compute $a(1,3)$. We should have $a(a(1,1),3) = a(1,a(1,3))$. This gives $a(1,3) = a(1,a(1,3))$. Denoting $x = a(1,3)$ we obtain $x = a(1,x)$. From above mentioned equalities for a we see that x can be 1 or 3 but not 2, i.e. $x \neq 2$. Define $\pi_2(1) = 1$, $\pi_2(2) = 3$, $\pi_2(3) = 2$, then

$$\pi_2 a(1,3) = \pi_2 a(\pi_2^{-1}(1), \pi_2^{-1}(3)) = \pi_2(a(1,2)) = \pi_2(3) = 2 \neq x = a(1,3),$$

i.e. a is not fixed point of π_2.

Subsubcase 2.1.2: m > 3. For $k \neq i, j$ we consider a permutation $\tilde{\pi} = \pi_{ij}$ which satisfies $\tilde{\pi}(i) = i, \tilde{\pi}(j) = j$ and $\tilde{\pi}(k) \neq k$ then we have

$$\tilde{\pi}a(i,j) = \tilde{\pi}a(\tilde{\pi}^{-1}(i), \tilde{\pi}^{-1}(j)) = \tilde{\pi}(k) \neq k = a(i,j),$$

i.e. a is not fixed point of $\tilde{\pi}$.

Subcase 2.2: $a(i,j) = k \in \{i,j\}$ for any i, j. After above mentioned cases remain the operations with property $a(i,j) \in \{i,j\}$ for any i,j, because if $a(i,j) \neq i, j$ for some i, j then we come back to Subcase 2.1.

Subsubcase 2.2.1: $a(i,j) = a(j,i) = i$ for some $i \neq j$. (The case $a(i,j) = a(j,i) = j$ is similar.) Consider permutation π_1 with $\pi_1(i) = j$, $\pi_1(j) = i$ then

$$\pi_1 a(i,j) = \pi_1 a(\pi_1^{-1}(i), \pi_1^{-1}(j)) = \pi_1 a(j,i) = \pi_1(i) = j \neq i = a(i,j).$$

Subsubcase 2.2.2: $a(i,j) = i$, $a(j,i) = j$ for any i,j. In this case a is (ls).

Subsubcase 2.2.3: $a(i,j) = j$, $a(j,i) = i$ for any i,j. In this case a is (rs). This completes the proof. $\qquad\square$

Definition 2.6. The symmetric operation (rs) (resp. (ls)) mentioned in Proposition 2.1 is called *right-symmetric* (resp. *left-symmetric*).

Example 2.1.

1. Consider the case $I = \{1,2\}$, then the following 8 operations are associative ([41]):

I	1	2		II	1	2		III	1	2		IV	1	2
1	1	1		1	1	1		1	1	1		1	1	2
2	1	1		2	1	2		2	2	2		2	1	2

V	1	2		VI	1	2		VII	1	2		VIII	1	2
1	1	2		1	1	2		1	2	1		1	2	2
2	2	1		2	2	2		2	1	2		2	2	2

One can check that $I \sim VIII$, $II \sim VI$, $V \sim VII$. Therefore these operations are not symmetric. By Theorem 2.1 the ACMs corresponding to equivalent operations are isomorphic. Operations III and IV are symmetric.

2. In the case $I = \{1,2,3\}$, the set $\mathcal{O}_3$ of associative operations contains 113 elements (see [41]), here some examples

i	1	2	3		ii	1	2	3		iii	1	2	3		iv	1	2	3
1	1	2	3		1	1	2	3		1	1	1	1		1	2	2	1
2	2	2	2		2	2	3	3		2	1	2	3		2	2	2	2
3	3	2	2		3	3	3	3		3	1	3	1		3	1	2	3

v	1 2 3		vi	1 2 3		vii	1 2 3		$viii$	1 2 3
1	1 1 1		1	3 1 3		1	1 2 3		1	1 1 1
2	1 1 2		2	1 2 3		2	1 2 3		2	2 2 2
3	1 2 3		3	3 3 3		3	1 2 3		3	3 3 3

One can check that $\mathrm{Orb}(i) = \{i, ii, iii, iv, v, vi\}$, consequently by Theorem 2.1 these six operations define isomorphic ACMs. By Proposition 2.1 we know that the operations vii and $viii$ are symmetric.

2.3.2 Accompanying algebra of an ACM

Define an analogue of a sex differentiation algebra for an ACM, it will be an m^2-dimensional algebra.

Definition 2.7. Let $\mathfrak{A} = \langle \eta_{ij} : i, j \in I \rangle$ denote an m^2-dimensional algebra with multiplication table

$$\eta_{ij}\eta_{kl} = \delta_{jk}\eta_{il}.$$

Then $\mathfrak{A}$ is called the *accompanying algebra*.

The following proposition is obvious.

Proposition 2.2. *The accompanying algebra $\mathfrak{A}$ is associative. It is commutative iff $m = 1$.*

Now define a generalization of a baric and a dibaric algebra.

Definition 2.8. An algebra is called accompanied if it admits a homomorphism onto the accompanying algebra $\mathfrak{A}$.

Theorem 2.2. *For each $a \in \mathcal{O}_m$ the algebra $\mathfrak{C}_a$ is accompanied.*

Proof. Consider a linear mapping $\varphi \colon \mathfrak{C}_a \to \mathfrak{A}$ defined on basis elements by

$$\varphi(E_{inj}) = \eta_{ij}, \quad i, n, j \in I.$$

On one hand we have

$$\varphi(E_{inj} *_a E_{krl}) = \delta_{jk}\varphi(E_{ia(n,r)l}) = \delta_{jk}\eta_{il}.$$

On the other hand we have

$$\varphi(E_{inj})\varphi(E_{krl}) = \eta_{ij}\eta_{kl} = \delta_{jk}\eta_{il}.$$

Thus

$$\varphi(E_{inj} *_a E_{krl}) = \varphi(E_{inj})\varphi(E_{krl}),$$

i.e. φ generates a homomorphism.

For any $X = (x_{ijk}) \in \mathfrak{C}_a$, i.e.

$$X = \sum_{inj} x_{inj} E_{inj}$$

the homomorphism φ is defined by

$$\varphi(X) = \sum_{inj} x_{inj}\varphi(E_{inj}) = \sum_{ij}\left(\sum_n x_{inj}\right)\eta_{ij}.$$

For arbitrary $u = \sum_{ij} u_{ij}\eta_{ij} \in \mathfrak{A}$ it is easy to see that $\varphi(X) = u$ if $\sum_n x_{inj} = u_{ij}$. Therefore φ is onto. $\qquad\square$

The accompanying algebra will be useful to study subalgebras of ACM, which we give in a next section.

2.3.3   *Commutativity and solvability of equations in ACMs*

Proposition 2.3. *Algebra $\mathfrak{C}_a$ is commutative for each $a \in \mathcal{O}_m$ iff $m = 1$.*

Proof. In case $m = 1$ the algebra is commutative because it is one dimensional, i.e. only with one basis element E_{111}. For $m \geq 2$ taking $k = l$ and $i \neq r$ from (2.2) we get

$$E_{ijk} *_a E_{knr} = E_{ia(j,n)r} \neq 0 = E_{knr} *_a E_{ijk}. \qquad\square$$

An element (cubic matrix) A of $\mathfrak{C}_a$ is called a *left* (resp. *right*) *zero divisor* if there exists a nonzero $X \in \mathfrak{C}_a$ such that $A *_a X = 0$ (resp. $X *_a A = 0$).

For a cubic matrix $A = (a_{ijk}) \in \mathfrak{C}_a$ denote

$$A_{ik} = \sum_{j=1}^{m} a_{ijk}, \quad i,j \in I,$$

and by $\mathbb{B}$ we denote the square matrix $\mathbb{B} = (A_{ik})_{i,k=1}^{m}$, which is called accompanying matrix of the cubic matrix A (see [158]).

Proposition 2.4. *Let $\mathfrak{C}_a$ be ACM over the field of real numbers.*

   (i) If $a \in \mathcal{O}_m$ is a right-symmetric (resp. left-symmetric) operation then a cubic matrix A is a left (resp. right) zero divisor iff $\det \mathbb{B} = 0$.

   (ii) If $a \in \mathcal{O}_m$ is a left-symmetric (resp. right-symmetric) operation then any cubic matrix A is a left (resp. right) zero divisor.

Proof.   (i) For the right-symmetric a we have $a(l,n) = n$ for all l, n. Consequently, the left zero divisibility equation $A *_a X = 0$ can be written as (see (2.3))

$$\sum_{l,n:\, a(l,n)=j} \sum_k a_{ilk} x_{knr} = \sum_{l=1}^{m} \sum_{k=1}^{m} a_{ilk} x_{kjr} = 0, \qquad \text{for all} \ \ i,j,r.$$

Consequently,

$$\sum_{k=1}^{m} A_{ik} x_{kjr} = 0, \qquad \text{for all} \ \ i,j,r.$$

Thus $x_{kjr} \neq 0$ iff $\det \mathbb{B} = 0$. The case of left-symmetric is similar.

   (ii) For the left-symmetric a we have $a(l,n) = l$ for all l, n. Consequently, the left zero divisibility equation $A *_a X = 0$ can be written as (see (2.3))

$$\sum_{l,n:\, a(l,n)=j} \sum_k a_{ilk} x_{knr} = \sum_{n=1}^{m} \sum_{k=1}^{m} a_{ijk} x_{knr} = 0, \qquad \text{for all} \ \ i,j,r.$$

Consequently,

$$\sum_{k=1}^{m} a_{ijk} X_{kr} = 0, \qquad \text{for all} \ \ i,j,r,$$

where

$$X_{kr} = \sum_{n=1}^{m} x_{knr}.$$

Hence independently on values a_{ijk} one can choose x_{knr} (some of them should be non-zero) such that $X_{kr} = 0$ for any k,r. The case of right-symmetric is similar. $\qquad\square$

2.3.4   *Subalgebras of ACM*

In the following theorem we collect several subalgebras of ACM.

Theorem 2.3. *Let $\mathfrak{C}_a$, $a \in \mathcal{O}_m$, be an ACM with basis $\{E_{ijk} : i,j,k \in I\}$.*

1. Let $J_a = \{a(i,j) : i,j \in I\}$ be the image of $a \in \mathcal{O}_m$. If $\tilde{J} \subset J_a$ is an a-invariant (i.e. $a(\tilde{J}, \tilde{J}) \subset \tilde{J}$) then for each fixed i, k, the set $\mathcal{E}_{\tilde{J}}^{ik} = \{E_{ijk} : j \in \tilde{J}\}$ generates a subalgebra, denoted by $\mathfrak{C}_{a,\tilde{J}}^{(ik)}$.

2. If $i \neq \tilde{i}$ or $k \neq \tilde{k}$, then $\mathfrak{C}_{a,\tilde{J}}^{(ik)} \cap \mathfrak{C}_{a,\tilde{J}}^{(\tilde{i}\tilde{k})} = \{0\}$, here 0 is the zero-cubic matrix.

3. If $\tilde{J}$ and $\bar{J}$ are a-invariant sets such that $\tilde{J} \subset \bar{J}$ then $\mathfrak{C}_{a,\tilde{J}}^{(ik)} \subset \mathfrak{C}_{a,\bar{J}}^{(ik)}$.

4. If $\tilde{J}$ and $\bar{J}$ are a-invariant sets such that $\tilde{J} \cap \bar{J} = \emptyset$ then $\mathfrak{C}_{a,\tilde{J}}^{(ik)} \cap \mathfrak{C}_{a,\bar{J}}^{(ik)} = \{0\}$.

5. The set $\{E_{ijk} : i, k \in I, j \in J_a\}$ generates an ideal, denoted by $\mathfrak{I}_a$.

6. The set $\mathfrak{I}_a^0 = \{X = (x_{inj}) \in \mathfrak{C}_a : \sum_n x_{inj} = 0, \quad \text{for all } i,j \in I\}$ is an ideal.

Proof. 1. For any two basis elements $E_{ijk}, E_{ink} \in \mathcal{E}_{\tilde{J}}^{ik}$, we have

$$E_{ijk} *_a E_{ink} = \delta_{ki} E_{ia(j,n)k} \in \mathcal{E}_{\tilde{J}}^{ik},$$

because $a(j,n) \in \tilde{J}$. Thus the set $\mathcal{E}_{\tilde{J}}^{ik}$ generates a subalgebra. Note that the subalgebra $\mathfrak{C}_{a,\tilde{J}}^{(ik)}$ is with zero-multiplication iff $i \neq k$.

Items 2–4 are straightforward.

5. It is clear that $\mathfrak{I}_a$ is a subalgebra. First we prove that it is a right ideal, i.e. for arbitrary $A \in \mathfrak{C}_a$ and $B \in \mathfrak{I}_a$ one has $A *_a B \in \mathfrak{I}_a$. Let $A = \sum_{ijk} a_{ijk} E_{ijk}$ and $B = \sum_{ink:n \in J_a} b_{ink} E_{ink}$. We have

$$(A *_a B)_{ijr} = \sum_{l,n:\, a(l,n)=j} \sum_k a_{ilk} b_{knr},$$

i.e. $(A *_a B)_{ijr} = 0$ if $j \notin J_a$. Consequently, $A *_a B \in \mathfrak{I}_a$. Similarly one can see that $B *_a A \in \mathfrak{I}_a$. Thus $\mathfrak{I}_a$ is an ideal.

6. Note that $\mathfrak{I}_a^0 = \ker \varphi = \{X \in \mathfrak{C}_a : \varphi(X) = 0\}$, where φ is constructed in the proof of Theorem 2.2. Therefore it is an ideal. $\qquad\square$

For $i \in I$ define the sequence

$$i_0 = i, \qquad i_n = a(i_{n-1}, i_{n-1}), \quad n \geq 1. \tag{2.7}$$

Since I is a finite set the sequence i_n may have one of the following forms:

a. *periodic:* there is a number $p \geq 1$ such that $i_p = i$. Denote by $p_a(i) \geq 1$, the minimal period of i and $I_a(i) = \{i_0 = i, i_1, \ldots, i_{p_a(i)-1}\}$.

b. *convergent:* there is a number $q \geq 1$ such that $i_n = i_q$ for any $n \geq q$.

Note that $I_a(i)$ is not a-invariant, in general. This can be seen in the following example.

Example 2.2. Take $a \in \mathcal{O}_3$ as

$$
\begin{array}{c|ccc}
a & 1 & 2 & 3 \\
\hline
1 & 1 & 2 & 3 \\
2 & 2 & 3 & 1 \\
3 & 3 & 1 & 2
\end{array}
$$

Then $I_a(1) = \{1\}$, $I_a(2) = \{2,3\}$ and $I_a(3) = \{3,2\}$. Note that $I_a(1)$ is a-invariant, but $I_a(2)$ is not a-invariant, because $a(2,3) = 1 \notin I_a(2)$.

Define *plenary powers* of an element $A \in \mathfrak{C}_a$ as follows:

$$A^{[1]} = A^{(2)} = A *_a A,$$

$$A^{[2]} = A^{(2^2)} = A^{(2)} *_a A^{(2)},$$

$$\cdots \quad \cdots \quad \cdots$$

$$A^{[n]} = A^{(2^n)} = A^{(2^{n-1})} *_a A^{(2^{n-1})}.$$

For convenience, we denote $A^{[0]} = A$.

Fix a subset $K \subset I$ and construct the following sequens of sets:

$$J_{a,0}(K) = K,$$
$$J_{a,1}(K) = J_{a,0}(K) \cup \{a(s,t) : s,t \in J_{a,0}(K)\},$$

$$\cdots \quad \cdots \quad \cdots$$

$$J_{a,n}(K) = J_{a,n-1}(K) \cup \{a(s,t) : s,t \in J_{a,n-1}(K)\}.$$

Since I is finite there is $n_0 = n_0(K)$ such that

$$J_{a,n}(K) = J_{a,n_0}(K), \text{ for all } n \geq n_0.$$

Proposition 2.5.

1. *For each $i,j \in I$, the sequence $\{E_{jij}^{[n]}\}$ is periodic (resp. convergent) iff i_n constructed for i by (2.7) is periodic (resp. convergent).*
2. *For each $k,n \in I$, $K \subset I$, the set $\{E_{kln} : l \in J_{a,n_0(K)}(K)\}$ generates a subalgebra.*

Proof. 1. One can see that $E_{jij}^{[n]} = E_{ji_nj}$, $i,j \in I$. Thus sequences $E_{jij}^{[n]}$ and i_n have the same behavior.

2. By construction of $J_{a,n_0(K)}(K)$ it is clear that this set is an a-invariant. Therefore the part 1 of Theorem 2.3 completes the proof. $\quad\square$

For $K = \{i\}$ the following example shows that not each a-invariant set of a has the form $J_{a,n_0(i)}(i)$.

Example 2.3. Consider the following operation $a \in \mathcal{O}_3$:

$$
\begin{array}{c|ccc}
a & 1 & 2 & 3 \\
\hline
1 & 1 & 1 & 1 \\
2 & 1 & 2 & 2 \\
3 & 1 & 3 & 3
\end{array}
$$

Then $J_{a,n_0(i)}(i) = \{i\}$, $i \in I = \{1,2,3\}$. But the set $\{i,j\}$, for $i \neq j \in \{1,2,3\}$, and I are also a-invariant. Thus any subset of I is an a-invariant. By above mentioned results the ACM corresponding to this example of operation has at least 7 subalgebras.

2.3.5 *ACM is not baric*

Theorem 2.4. *Each ACM, $\mathfrak{C}_a$, $a \in \mathcal{O}_m$, $m \geq 2$, is not baric.*

Proof. Assume $\chi \colon \mathfrak{C}_a \to \mathbb{R}$ is a character. Then for each cubic matrix $X = (x_{ijk}) \in \mathfrak{C}_a$ it has the form $\chi(X) = \sum_{i,j,k=1}^{m} \alpha_{ijk} x_{ijk}$. We shall prove that $\chi(X) \equiv 0$. Consider it on basis elements:

$$
\chi(E_{ijk}) = \alpha_{ijk} \in \mathbb{R}, \quad i,j,k \in I.
$$

We should check $\chi(E_{ijk} *_a E_{lnr}) = \chi(E_{ijk})\chi(E_{lnr})$. Using (2.2) the last equality can be written as

$$
\alpha_{ijk}\alpha_{lnr} = \delta_{kl}\alpha_{ia(j,n)r}, \qquad \text{for all } i,j,k,l,n,r \in I.
$$

Equivalently,

$$
\alpha_{ijk}\alpha_{lnr} = 0, \qquad \text{for all } i,j,k \neq l,n,r \in I. \tag{2.8}
$$

$$
\alpha_{ijk}\alpha_{knr} = \alpha_{ia(j,n)r}, \qquad \text{for all } i,j,k = l,n,r \in I. \tag{2.9}
$$

In (2.8) take $i \neq k$ and $l = i$, $n = j$, $r = k$, then we obtain

$$
\alpha_{ijk} = 0, \qquad \text{for all } i \neq k, j \in I. \tag{2.10}
$$

Consequently, non-zero coefficient can be only in the form α_{kjk}.

Assume $\alpha_{k_0 j_0 k_0} \neq 0$, for some $k = k_0$ and $j = j_0$. Then from (2.8) we get $\alpha_{lnl} = 0$, for any $l \neq k_0$ and $n \in I$. Using (2.10), from (2.9) for $i = r = k_0$, $k \neq k_0$, we get

$$
0 = \alpha_{k_0 jk}\alpha_{knk_0} = \alpha_{k_0 a(j,n)k_0}, \qquad \text{for all } j,k \neq k_0, n \in I. \tag{2.11}
$$

Write (2.9) for $i = r = k = k_0$:

$$\alpha_{k_0 j k_0} \alpha_{k_0 n k_0} = \alpha_{k_0 a(j,n) k_0}, \qquad \text{for all } j, n \in I. \tag{2.12}$$

By (2.12) for $j = n = j_0$ we get

$$\alpha_{k_0 j_0 k_0}^2 = \alpha_{k_0 a(j_0,j_0) k_0}. \tag{2.13}$$

Using (2.11), for $j = n = j_0$ we obtain $\alpha_{k_0 a(j_0,j_0) k_0} = 0$. Consequently from (2.13) we get $\alpha_{k_0 j_0 k_0} = 0$. Thus $\chi(X) \equiv 0$. $\qquad\square$

Bibliographical notes. This chapter is helpful for reading of next chapters. This chapter is based on [15], [116], [139], [143], [158], [205], [240], [269], and many internet sources. For further reading about abstract algebras and cubic matrices see [83], [125], [145], [151].

Chapter 3

Genetic algebras

In this chapter we systematically present genetic algebras. These algebras are used to model inheritance in genetics. The first section of the chapter contains some variations of these algebras (called baric, train algebras, special train algebras, gametic algebras, Bernstein algebras, copular algebras, zygotic algebras etc.). In other sections we give dibaric and evolution algebras.

3.1 Variations of genetic algebras

The study of genetic algebras was started 80 years ago by Etherington [48]. During these years many variants of the genetic algebras were introduced and studied ([155], [144], [211], [240], [269]). Here we will give some of these algebras.

As it was mentioned in the introduction of this book, in applications to genetics, genetic algebras have a basis corresponding to the genetically different gametes, and the matrix of structural constants of the algebra encode the probabilities of producing offspring of various types. The laws of inheritance are then encoded as algebraic properties of the algebra. Some of such algebras are non associative, this structure naturally occurs as genetic information gets passed down through the generations.

In this section we will give a brief overview of the algebras which arise in genetics and some of their basic properties and relationships.

Let us give simplest example first [211]:

Mendelian Inheritance. Consider Mendelian inheritance for a single gene with two alleles A and a. Then two gametes fusing to form a zygote gives the multiplication (see Remark 1.1) table shown in Table 3.1.

Table 3.1 Alleles passing from gametes to zygotes.

	A	a
A	AA	Aa
a	aA	aa

Since zygotes AA and aa carry two copies of the same allele, they are called homozygous.

In this case, (simple) Mendelian inheritance means that there is no chance involved as to what genetic information will be inherited in the next generation; i.e. AA will pass on the allele A and aa will pass on a. However, the zygotes Aa and aA (which are equivalent) each carry two different alleles. These zygotes are called heterozygous.

The rules of Mendelian inheritance indicate that the next generation will inherit either A or a with equal probability. Thus, when two gametes reproduce, a multiplication is induced which indicates how the hereditary information will be passed down to the next generation. This multiplication is given by the following rules:

$$
\begin{aligned}
A \times A &= A \\
A \times a &= \tfrac{1}{2}(A + a) \\
a \times A &= \tfrac{1}{2}(a + A) \\
a \times a &= a
\end{aligned}
\tag{3.1}
$$

This multiplication table is shown in Table 3.2.

Table 3.2 Multiplication table of the gametic algebra for Mendelian inheritance.

	A	a
A	A	$\tfrac{1}{2}(A + a)$
a	$\tfrac{1}{2}(a + A)$	a

3.1.1 *Baric algebras*

Let us first give a definition of an algebra which has genetic significance [211]:

Definition 3.1. An algebra $\mathcal{A}$ over the real numbers $\mathbb{R}$ which has a basis

$\{e_1, \ldots, e_n\}$ is called an *algebra with genetic realization* if it has a multiplication table

$$e_i e_j = \sum_{k=1}^{n} P_{ij,k} e_k,$$

where $P_{ij,k}$ are the coefficients of heredity and

$$P_{ij,k} \geq 0, \quad \sum_{k=1}^{n} P_{ij,k} = 1, \quad \text{for all} \ \ i, j, k \in \{1, \ldots, n\}.$$

Such a basis is called the natural basis for $\mathcal{A}$.

An element x in a general algebra $\mathcal{A}$ with genetic realization, represents a population, or a gene pool for a population as the following

$$x = \sum_{k=1}^{n} x_k e_k, \quad x_k \geq 0, \quad \sum_{k} x_k = 1.$$

Then x_k is the percentage of the population x which carries the allele e_i.

Definition 3.2. A baric algebra (or weighted algebra) $\mathcal{A}$ over a field F is (a possibly non-associative algebra) together with a homomorphism w, called the weight, from the algebra to F. Denote this baric algebra as $(\mathcal{A}, w)$.

Proposition 3.1. *An n-dimensional algebra of free population $\mathcal{A}$ defined by (1.4) (in particular, the algebra with genetic realization) over the set of real numbers $\mathbb{R}$ is baric.*

Proof. For an $x \in \mathcal{A}$ having representation $x = \sum_{k=1}^{n} x_k e_k$, define the weight $w(x) = \sum_{k=1}^{n} x_k$. Then for $y = \sum_{k=1}^{n} y_k e_k$ we have

$$w(xy) = \sum_{i,j,k} P_{ij,k} x_i y_j = \sum_{i,j} \left(\sum_{k} P_{ij,k} \right) x_i y_j = \sum_{i,j} x_i y_j = w(x) w(y).$$

Thus, w is a homomorphism and $\mathcal{A}$ is a baric algebra. $\qquad\square$

An interesting question to ask about baric algebras is whether or not their weight functions are uniquely determined. The following example shows that in general, they are not.

Example 3.1. [211], [269] Take $\mathcal{A} = \langle e_1, e_2, e_3 \rangle$-three dimensional *commutative* algebra over $\mathbb{R}$ with the multiplication:

$$e_1 e_1 = e_1 + e_2, \quad e_1 e_2 = e_1 e_3 = e_2 e_2 = e_2 e_3 = e_2, \quad e_3 e_3 = e_2 + e_3.$$

Then define $w_1 : \mathcal{A} \to \mathbb{R}$ by $w_1(e_1) = 1$ and $w_1(e_2) = w_1(e_3) = 0$, and define $w_2 : \mathcal{A} \to \mathbb{R}$ by $w_2(e_3) = 1$ and $w_2(e_1) = w_2(e_2) = 0$. Then extending by linearity for arbitrary $x \in \mathcal{A}$ we get

$$w_1(x) = x_1, \quad w_2(x) = x_3.$$

Thus $w_1 \neq w_2$, and they both define homomorphisms.

Recall that a principal power r of an element $x \in \mathcal{A}$ is defined as $x^r = (x^{r-1})x$.

The following proposition gives a sufficient condition for a baric algebra to have a unique weight function.

Proposition 3.2. *Let $(\mathcal{A}, w)$ be a baric algebra over a field F. If $N = \mathrm{Ker}\, w$ is nil,[1] then w is uniquely determined.*

Proof. By the condition of the proposition for any $x \in N$ we have a natural number r such that the principal power $x^r = 0$. Let $\varphi : \mathcal{A} \to F$ be a nontrivial homomorphism. Then

$$\varphi(x^r) = (\varphi(x))^r = \varphi(0) = 0.$$

Consequently, since $\varphi(x) \in F$, we get

$$\varphi(x) = 0, \quad \text{for all } x \in N. \tag{3.2}$$

For any element $y \in \mathcal{A} \setminus N$ (i.e. $w(y) \neq 0$) define the element $\frac{y^2}{w(y)} - y$. Then since w is a homomorphism we get

$$w\left(\frac{y^2}{w(y)} - y\right) = w(y)w\left(\frac{y}{w(y)} - 1\right) = w(y)\left(\frac{w(y)}{w(y)} - 1\right) = 0.$$

Thus $\frac{y^2}{w(y)} - y \in N$ and by (3.2) we have

$$\varphi\left(\frac{y^2}{w(y)} - y\right) = \varphi(y)\left(\frac{\varphi(y)}{w(y)} - 1\right) = 0.$$

Consequently, $\varphi(y) = 0$ or $\varphi(y) = w(y)$. From the first case we get $\varphi(x) \equiv 0$ (a contradiction to assumption that φ is nontrivial). The second case leads to $\varphi = w$. Therefore, the baric function is unique. $\qquad\square$

In general, the existence of an idempotent in $\mathcal{A}$ is not guaranteed. However, if $\mathcal{A}$ does contain an idempotent a, then $w(a) = w(a^2) = w(a)w(a)$, so either $w(a) = 0$ or $w(a) = 1$ [211].

Proposition 3.3. *Let $(\mathcal{A}, w)$ be a baric algebra over a field F. Assume $\mathcal{A}$ contains an idempotent a such that $w(a) = 1$. Then,*

$$\mathcal{A} = Fa \oplus N,$$

where $N = \mathrm{Ker}\, w$.

---

[1] All elements of N are nilpotent.

Proof. By the first homomorphism theorem,[2] $A/N \cong F$ and N is a two sided ideal with codimension one in A. Moreover, by $w(a) = 1$ we have $Fa \cap N = 0$. Note that for any $x \in A$, we have $x - w(x)a \in N$, since $w(a) = 1$. Now the following equality completes the proof:

$$x = w(x)a + (x - w(x)a).$$

$\square$

3.1.2 Train algebras

Let (A, w) be a baric algebra.

Definition 3.3. Let $c_1, \ldots, c_n$ be elements of the field F with

$$1 + c_1 + \cdots + c_n = 0.$$

The formal polynomial

$$x^n + c_1 w(x) x^{n-1} + \cdots + c_n w(x)^n$$

is called a train polynomial.

A train algebra is a special baric algebra defined as.

Definition 3.4. The baric algebra (A, w) is a (principal) train algebra if

$$x^n + c_1 w(x) x^{n-1} + \cdots + c_n w(x)^n = 0, \tag{3.3}$$

for all elements $x \in A$, with x^k defined as principal powers.

Consider a finite dimensional, commutative non associative algebra A over a field F. Let $\{e_1, \ldots, e_n\}$ be a basis of A.

An element $x \in A$ annihilates $y \in A$ if $yx = 0$.

Definition 3.5. A polynomial of minimal degree (in principal powers) that annihilates all elements of A is called the rank polynomial of A.

Denote the rank polynomial by

$$f(x) = x^r + \theta_1 x^{r-1} + \theta_2 x^{r-2} + \cdots + \theta_{r-1} x,$$

where θ_i is a homogeneous polynomial of degree i in the coordinates x_j of the generic element $x = \sum_{k=1}^{n} x_k e_k$.

By the definition the rank polynomial annihilates all elements of the algebra, the equation $f(x) = 0$ is called the rank equation. This equation gives a recurrence relation among the principal powers.

---

[2] https://en.wikipedia.org/wiki/Isomorphism_theorems.

Using condition $1 + c_1 + \cdots + c_n = 0$ of Definition 3.3 we can write the equation (3.3) as the rank equation, i.e. $f(x) = 0$. Thus a (principal) train algebra is baric and has a rank equation in which the coefficients of a general element x depend only on its weight, i.e. $w(x)$.

If the rank polynomial is of degree r, then this recurrence relation indicates how the genetic pool of the r-th generation (in the sequence of principal powers) is related to the genetic pools of previous generations.

If the rank polynomial $f(x)$ has degree r, then corresponding train algebra $\mathcal{A}$ is called a train algebra of rank r.

Proposition 3.4. *Let $(\mathcal{A}, w)$ be a train algebra of rank r. Then every element in $N = \operatorname{Ker} w$ is nilpotent of an index not greater than r.*

Proof. The rank polynomial of $\mathcal{A}$ is

$$f(x) = x^r + c_1 w(x) x^{r-1} + \cdots + c_r w(x)^r.$$

Since $f(x)$ annihilates every element $x \in \mathcal{A}$, for $x \in N$, we have $w(x) = 0$, so the rank equation $f(x) = 0$ becomes $x^r = 0$. Thus, x is nilpotent of index less than or equal to r. $\qquad\qquad\square$

Proposition 3.5. *A train algebra has a uniquely determined weight function.*

Proof. Let $(\mathcal{A}, w)$ be a train algebra (i.e., with weight function w). Then by Proposition 3.4, $N = \operatorname{Ker} w$ is nil. Then Proposition 3.2 implies that w is unique. $\qquad\qquad\square$

Special train algebras. A special train algebra is a baric algebra in which the kernel N of the weight function is nilpotent and the principal powers of N are ideals. This is more precise in the following

Definition 3.6. A baric algebra $(\mathcal{A}, w)$ is called a special train algebra if $N = \operatorname{Ker} w$ is nilpotent and all of the principal power subalgebras N^i of $\mathcal{A}$ (which is defined inductively by $N^1 = N, \quad N^i = N^{i-1}N$ for $i = 2, 3, \ldots$) are ideals of $\mathcal{A}$.

By this definition we get the sequence of descending ideals terminates:

$$A \supset N = N^1 \supset N^2 \supset N^3 \supset N^r \supset N^{r+1} = \langle 0 \rangle.$$

Proposition 3.6. [1] *A commutative train algebra of rank 3 over a field of characteristic zero is necessarily a special train algebra.*

Transformation algebra. Let A be any commutative, non associative algebra over a field F.

For $x \in A$, let $R_x : A \to A$ be right multiplication by x. Since A is commutative, right multiplication by x is equivalent to left multiplication by x.

Definition 3.7. The transformation algebra $T(A)$ of A is the algebra consisting of all polynomials of right multiplication operators on A, as well as the identity operator on A.

Thus any transformation T in the transformation algebra of A can be represented as

$$T = \alpha I + f(R_{x_1}, \ldots, R_{x_s}),$$

where $\alpha \in F$, I is the identity on A, and f is a polynomial.

Definition 3.8. Let $(\mathcal{A}, w)$ be a commutative baric algebra over a field F. Then $\mathcal{A}$ is a genetic algebra if the characteristic function $|\lambda I - T|$ of any transformation $T = \alpha I + f(R_{x_1}, \ldots, R_{x_s})$, in the transformation algebra of $\mathcal{A}$, is a function of $w(x_1), \ldots, w(x_s)$.

This algebra was stated in [241] and it was proved that all of genetic algebras are also train algebras. Moreover, every special train algebra is a genetic algebra and that genetic algebras are closed under commutative duplication. In [102] it was shown that Schafer's definition have relevance genetically. Using tools from Lie algebra theory, Holgate provided some alternative characterizations of Schafer's genetic algebras which illuminated their genetic significance. Therefore the algebras which naturally occur in genetics will satisfy Schafer's condition as long as there is a certain amount of symmetry in the genetic system. As was seen above the most basic genetic systems contain a substantial amount of symmetry.

3.1.3 *Gametic algebra for linked loci*

A gametic algebra is a finite-dimensional real algebra for which all structure constants lie between 0 and 1. Namely, consider a random mating population with n distinct gametes. Denote them $e_1, \ldots, e_n$. Take these n gametes as basis elements of an n-dimensional real vector space. Define multiplication as in Definition 3.1 but with additional condition that $P_{ij,k} = P_{ji,k}$. Thus a gametic algebra is a particular case of the algebra with genetic realization.

Suppose we have an infinite randomly mating population [206], not subject to selection, of diploid individuals which differ in k linked loci. Let $r_m + 1$ be the number of possible alleles in the mth locus. If U' and U'' are complementary subsets of $K = \{1, 2, \ldots, k\}$, then denote by

$$U = (U', U'') = (U'', U')$$

the partition of K determined by U' and U''.

Assume that for a given U, in the zygotes the loci in U' are considered as one block, the loci in U'' as another block, and recombination occurs between the blocks with probability $P(U)$. In [54], [93] and [103] the gametic inheritance for all k loci has been studied by algebraic methods.

Following [206] define to this genetic model the real commutative algebra G called the *gametic algebra for linked loci.*

A canonical basis for G can be represented by the set of monomials

$$X_{i^*} = X_{1i_1} X_{2i_2} \ldots X_{ki_k}, \quad 0 \leq i_m \leq r_m, \quad 1 \leq m \leq k,$$

where i^* is the multi-index $(i_1, \ldots, i_k)$.

Denote $0^* = (0, 0, \ldots, 0)$ and by $s(i^*)$ the set $\{t \in K : i_t \neq 0\}$, and

$$P(i^*, j^*) = \sum P(U),$$

where the sum is taken over the collection of all partitions $U = (U', U'')$ with $s(i^*)$ contained in one of the sets U' or U'' and $s(j^*)$ in the other.

Then the multiplication in G is defined by

$$X_{0^*}^2 = X_{0^*},$$

$$X_{i^*} X_{j^*} = \begin{cases} P(i^*, j^*) X_{i^*+j^*}, & \text{if } s(i^*) \cap s(j^*) = \emptyset \\ 0, & \text{otherwise.} \end{cases}$$

Recall that a derivation d of G is a linear map $d : G \to G$ verifying $d(xy) = d(x)y + xd(y)$. The set $\mathcal{D} = \text{Der}(G)$ of all derivations is closed under the Lie bracket $[d, d'] = dd' - d'd$, and it is a Lie algebra, i.e. $\mathcal{D}$ is a vector space together with the (non-associative operation) Lie bracket, satisfying $[x, y] = -[y, x]$ and the Jacobi identity:

$$[x, [y, z]] + [z, [x, y]] + [y, [z, x]] = 0, \quad \text{for all } x, y, z \in \mathcal{D}.$$

In [104] Holgate gave an interpretation and the genetic meaning of derivations in genetic algebras. Namely, if one has a genetically determined trait with array of values d and d is a derivation, then the above equation $d(xy) = d(x)y + xd(y)$ reflects a kind of symmetry in the time direction.

In [206] Peresi have determined the derivations of G, by using Lie group theory.

3.1.4 *Bernstein's problem and algebras*

A study of genetic crosses depends upon an understanding of Mendel's two laws:[3]

First Law (The principle of segregation): The two members of a gene pair (alleles) segregate (separate) from each other in the formation of gametes. Half the gametes carry one allele, and the other half carry the other allele.

Second Law (The principle of independent assortment): Genes for different traits assort independently of one another in the formation of gametes.

The Bernstein problem (see [155] and [156]) is related to a fundamental statement of population genetics, the so-called stationarity principle. This principle holds provided that the Mendel law is assumed, but it is consistent with other mechanisms of heredity. An adequate mathematical problem is as follows.

A quadratic stochastic operator V defined by (1.1) is called a Bernstein mapping if $V^2 = V$.

This property is just the stationarity principle. This property also is known as Hardy–Weinberg law [99].

The Bernstein's problem is to describe all Bernstein mappings explicitly.

The case $m \leq 2$ (where m is the dimension of simplex) is mathematically trivial and biologically not interesting. In [19] Bernstein solved the above problem for the case $m = 3$ and obtained some results for $m \geq 4$.

In works by Lyubich (see e.g. [155], [156]) the Bernstein problem was solved for all m under the regularity assumption. The regularity means that $V(x)$ depends only on the values $f(x)$, where f runs over all invariant linear forms.

In investigations by Lyubich, the algebra $\mathcal{A} \equiv \mathcal{A}_V$ with the structure constants $P_{ij,k}$ played a very important role (see (1.3) for multiplication table). This algebra is baric with weight $w(x) = x_1 + \cdots + x_m$. Since $V(x) = x^2$, the Bernstein property of V is equivalent to the identity

$$(x^2)^2 = w^2(x)x^2. \tag{3.4}$$

Definition 3.9. A baric algebra $(\mathcal{A}, w)$ is called a Bernstein algebra with respect to the algebra homomorphism $w : \mathcal{A} \to \mathbb{R}$, if for any $x \in \mathcal{A}$ the identity (3.4) is satisfied.

The mapping V is called regular iff the identity

$$x^2 y = w(x)xy$$

holds in the algebra $\mathcal{A}$, by definition, this identity means that $\mathcal{A}$ is regular.

---

[3] http://www.phschool.com/science/biology_place/biocoach/inheritance/laws.html.

Abraham [2] suggests the construction of a generalized Bernstein algebra and the theory of Bernstein algebras has been substantially improved. In this perspective, for an element x of a baric algebra $(\mathcal{A}, w)$, the plenary powers $x^{[m]}$ of x are defined by

$$x^{[1]} = x, \quad x^{[m+1]} = x^{[m]}x^{[m]}, \quad \text{for all} \quad m \geq 1.$$

The plenary powers can be interpreted by saying that they represent random mating between discrete non-overlapping generations. From a geneticist's point of view, the sequence of plenary powers is of greater interest, because the plenary powers more accurately model (compared with principal powers) the way most populations reproduce. This is the viewpoint that motivates the following n-th order Bernstein algebras.

Then a baric algebra $(\mathcal{A}, w)$ is called an n-th order Bernstein algebra if

$$x^{[n+2]} = w(x)^{2^n} x^{[n+1]}, \quad \text{for all} \quad x \in \mathcal{A},$$

where $n \geq 1$ is the smallest such integer.

Note that the second-order Bernstein algebras are Bernstein algebras and first-order Bernstein algebras are also called gametic diploid algebras.

In case when $w(x) = 1$, the interpretation of the equation $x^{[n+2]} = x^{[n+1]}$ is that equilibrium in the population is reached after exactly n generations of intermixing.

Thus, complete knowledge of the structure of the n-th Bernstein (abstract non associative) algebras would provide a great deal of genetically significant information. On the other hand, mathematically speaking, the challenge to classify these algebras is a rich open problem [211].

For genetic properties of Bernstein algebras, see [84], [103] and [170]–[172].

Following [103] and [211] we give some results about the structure of a Bernstein algebra. Let $(\mathcal{B}, w)$ be a Bernstein algebra. Then it must contain at least one idempotent element. Indeed, let x be an element of $\mathcal{B}$ with baric weight 1 (i.e., $w(x) = 1$) and take $e = x^2$. Then the definition of a Bernstein algebra implies that $e^2 = e$. Denote $Z = \text{Ker}\, w$.

Let e be an idempotent chosen once and for all, and $\mathcal{E}$ the subalgebra of its scalar multiples.

If $z \in Z$ then by the definition it satisfies

$$(e + \theta z)^{[3]} - (e + \theta z)^{[2]} = 0. \tag{3.5}$$

If coefficients of powers of θ are equated to zero, then we get the following identities:

$$z^{[3]} = 0, \tag{3.6}$$

$$z^2(ze) = 0, \tag{3.7}$$

$$4(ze)^2 + 2z^2e - z^2 = 0, \tag{3.8}$$

$$2(ze)e - ze = 0. \tag{3.9}$$

By (3.6) it follows that elements of baric value zero also satisfy the equation $x^{[3]} - w^2(x)x^{[2]} = 0$.

Replacing z by $\sum \theta_i z_i$, and equating homogeneous terms to zero, we get the fully linearized forms:

$$(z_1 z_2)(z_3 z_4) + (z_1 z_3)(z_2 z_4) + (z_1 z_4)(z_2 z_3) = 0, \tag{3.10}$$

$$(z_1 z_2)(z_3 e) + (z_1 z_3)(z_2 e) + (z_2 z_3)(z_1 e) = 0, \tag{3.11}$$

$$4(z_1 e)(z_2 e) + 2(z_1 z_2)e - z_1 z_2 = 0. \tag{3.12}$$

Intermediate forms are

$$z_1^2 z_2^2 = -2(z_1 z_2)^2 \tag{3.13}$$

$$z_l^2(z_2 e) = -2(z_1 z_2)(z_1 e). \tag{3.14}$$

From these identities we get

Proposition 3.7. *Z^2 is an ideal in Bernstein algebra $(\mathcal{B}, w)$.*

Proof. Since Z^2 is an ideal in Z, it remains to show that $Z^2 e \subset Z^2$. Rewrite (3.12) as

$$(z_1 z_2)e = \frac{1}{2} z_1 z_2 - 2(z_1 e)(z_2 e). \tag{3.15}$$

Since Z is an ideal in $\mathcal{B}$, the second term on the right as well as the first belongs to Z^2, which completes the proof. $\square$

Recall that a special train algebra is a baric algebra in which the nil ideal is nilpotent, and every power of it is an ideal. From Proposition 3.7 we get

Corollary 3.1. *Let $(\mathcal{B}, w)$ be a Bernstein algebra and $Z = \mathrm{Ker}\, w$. If $Z^3 = 0$, then $\mathcal{B}$ is a special train algebra.*

Let $\mathcal{U}$ be the subspace spanned by the elements $u = ze$, $\mathbf{B}$ the subspace spanned by all products $v = u_i u_j$, and $\mathbf{U}$ the subspace complementary to the union of $\mathcal{E}$, $\mathcal{U}$ and $\mathbf{B}$.

Substituting $u = ze$ in (3.9) we get

$$ue = \frac{1}{2}u. \tag{3.16}$$

The substitution of u_1, u_2 for z_1, z_2 in (3.15) with $v = u_1 u_2$, taken in conjunction with (3.16), leads to

$$ve = \frac{1}{2}u_1 u_2 - 2(u_l e)(u_2 e) = 0. \tag{3.17}$$

By (3.16) and (3.17) we get the following proposition.

Proposition 3.8. *The subspaces $\mathcal{U}$ and $\mathbf{B}$ have no non-zero elements in common.*

The space underlying $\mathbf{B}$, as a vector space, can be decomposed into the direct sum $\mathcal{E} \oplus \mathcal{U} \oplus \mathbf{B} \oplus \mathbf{U}$. Now exploiting equations (3.16) and (3.17) together with identities (3.6)–(3.12) one gets that

$$\mathbf{B}^2 \subset \mathcal{U}, \quad \mathcal{U}\mathbf{B} \subset U, \quad \mathbf{U}^2 \subset \mathcal{U} \oplus \mathbf{B}, \quad \mathbf{U}\mathcal{U} \subset \mathcal{U} \oplus \mathbf{B}, \quad \mathbf{U}\mathbf{B} \subset \mathcal{U}.$$

As a corollary of these relations we get

Proposition 3.9. *We have*

1. *$\mathcal{E} \oplus \mathcal{U} \oplus \mathbf{B}$ is an ideal in $\mathbf{B}$, containing $\mathbf{B}^2$.*
2. *The difference algebra $\mathbf{B} - \mathcal{E} \oplus \mathcal{U} \oplus \mathbf{B}$ is a zero algebra.*

Recall that e is an idempotent of the Bernstein algebra which was fixed. The idempotent elements play an important role, and they are described in the following theorem.

Theorem 3.1. *In a Bernstein algebra, the idempotent elements are precisely those of the form $e + u + u^2$.*

Proof. It is clear that an idempotent can have no component in $\mathbf{U}$. One can see that the elements specified in the theorem are idempotents. If $e + u + v$ is an idempotent, then squaring and equating the components in $\mathbf{B}$ gives $v = u^2$. $\square$

The following proposition gives a conditions under which $\mathbf{B}^2 = 0$.

Proposition 3.10. *If every element in $\mathbf{B}$ is the square of some element in $\mathcal{U}$, then $\mathbf{B}^2 = 0$.*

Proof. If $u = v^2$ then

$$v^2 = u^{[3]} = 0, \quad v_1 v_2 = \frac{1}{2}\left[(v_l - v_2)^2 - v_1^2 - v_2^2\right] = 0.$$

$\square$

Corollary 3.2. *If the algebra* $\mathbf{B}$ *is one-dimensional then* $\mathbf{B}^2 = 0$.

Proof. There must be some element in $\mathcal{U}$ whose square is v. $\square$

The following example shows that the case $\mathbf{B}^2 \neq 0$ can arise.

Example 3.2. Consider the algebra with basis $e, u_1, u_2, u_3, v_1, v_2, v_3$ and multiplication table

$$e^2 = e, \quad eu_i = \tfrac{1}{2}u_i, \quad ev_i = 0, \quad i = 1, 2, 3$$

$$u_1^2 = \theta v_1, \quad u_2^2 = \phi v_2, \quad u_1 u_2 = \psi v_3$$

$$v_1 v_2 = \delta u_3, \quad v_3^2 = \gamma u_3,$$

other products are zero.

An element $x = e + \sum \alpha_i u_i + \sum \beta_i v_i$ with $w(x) = 1$ has a square

$$x^2 = e + \alpha_1 u_1 + \alpha_2 u_2 + (\alpha_3 + \beta_2^2 \gamma + 2\beta_1 \beta_2 \delta)u_3 + \alpha_1^2 \theta v_1 + \alpha_2^2 \phi v_2 + 2\alpha_1 \alpha_2 \psi v_3.$$

The square of this is

$$x^{[3]} = x^2 + 2\alpha_1^2 \alpha_2^2 (\theta \phi \delta + 2\psi^2 \gamma)u_3.$$

Thus if the constants in the multiplication table are chosen so that $\theta \phi \delta + 2\psi^2 \gamma = 0$ the algebra is of Bernstein's type.

For more results on Bernstein algebras see, for example, [92], [103], [171], [172], [211] and the references therein.

Bernstein's classification of populations. Following [103] we give Bernstein's classification of 3- and 4-dimensional populations.

Assume $e_i = (\delta_{1i}, \dots, \delta_{ni})$ (where δ_{ij} is Kronecker delta) be the basis vector with 1 in the i-th position and 0 elsewhere. The i-th type is called pure if $V(e_i) = e_i$ (where V is the evolution operator of the population). In three dimensions the laws of inheritance satisfying the stationarity principle are of five types. The corresponding families of algebras are denoted here by $\mathbf{B}_0$, in which there are no pure types, $\mathbf{B}_2$ and $\mathbf{B}_3$ in which there are two and three pure types, and $\mathbf{B}_{11}$ and $\mathbf{B}_{12}$ the two different laws admitting one pure type.

In all cases except $\mathbf{B}_{12}$ it is possible to find a linear transformation of the algebra, depending on the parameters of the family, which reduces it to a single canonical form. In the case of $\mathbf{B}_{12}$ one of the parameters is accounted for by the transformation, leaving a canonical form involving the other.

The genetic types will be denoted by F, G, H their coefficient in the initial generation by x, y, z, and those in the next by x', y', z'.

Below the transformation and the new multiplication table for case:

Case $\mathbf{B}_0$: The evolution operator is

$$x' = \alpha(x + y + z)^2$$
$$y' = \beta(x + y + z)^2$$
$$z' = \gamma(x + y + z)^2$$

with $\alpha + \beta + \gamma = 1$.

Multiplication table (between genetic types):

$$F^2 = G^2 = H^2 = FG = FH = GH = \alpha F + \beta G + \gamma H.$$

Change of basis:

$$b_0 = \alpha F + \beta G + \gamma H, \quad b_l = F - G, \quad b_2 = F - H.$$

The canonical multiplication table:

$$b_0^2 = b_0, \quad b_i b_j = 0 \quad \text{unless} \quad i = j = 0.$$

For remaining cases we give only the canonical multiplications:

Case $\mathbf{B}_{11}$: The canonical multiplication table:

$$b_0^2 = b_0, \quad b_0 b_1 = \frac{1}{2} b_1, \quad b_1^2 = b_2,$$

other products are zero.

Case $\mathbf{B}_{12}$: The canonical multiplication table:

$$b_0^2 = b_0, \quad b_0 b_1 = \tfrac{1}{2}(1 - \alpha) b_2, \quad b_0 b_2 = \tfrac{1}{2} b_2$$
$$b_1^2 = -\alpha b_2, \quad b_1 b_2 = -\tfrac{1}{2}\alpha b_2, \quad b_1^2 = 0.$$

Case $\mathbf{B}_2$:

$$b_0^2 = b_0, \quad b_0 b_1 = \tfrac{1}{2}(b_1 + b_2), \quad b_0 b_2 = 0$$
$$b_1^2 = b_2, \quad b_1 b_2 = 0, \quad b_1^2 = 0.$$

Case $\mathbf{B}_3$:

$$b_0^2 = b_0, \quad b_0 b_1 = \tfrac{1}{2} b_1, \quad b_0 b_2 = \tfrac{1}{2} b_2$$
$$b_1^2 = b_1 b_2 = b_1^2 = 0.$$

Note that [103] the nil ideal Z is nilpotent of degree 2 in $\mathbf{B}_0$ and $\mathbf{B}_3$, and is nilpotent of degree 3 in $\mathbf{B}_{11}$ and $\mathbf{B}_2$. In the former cases the train equation relating principal powers of elements of weight one is $x^2 - x = 0$, and in the latter it is $x^3 - x^2 = 0$. By the canonical multiplication tables one can seen that in $\mathbf{B}_{11}$, $\mathbf{B}_2$ and $\mathbf{B}_3$, the value $\frac{1}{2}$ is a train root in Gonshor's sense [79] but not a principal train root. The algebra $\mathbf{B}_{12}$ is most interesting in that it is not a special train algebra. Because $Z^k = \{b_2\}$, for $k > 2$. The rank equation satisfied by x, x^2 and x^3 depends on x other than through $w(x)$. The case $\mathbf{B}_3$ is the elementary algebra of [100].

In the four-dimensional case Bernstein's work is a continuation of his concern to find minimal conditions which, when added to his principle of stationarity, ensure that a population is Mendelian [103]. He showed that the only two systems involving four pure types are the Mendelian (whose genetic algebra is the elementary special train algebra of [100]) and the "quadrille law", which has been identified biologically by Lyubich [157].

Denote by g_1, g_2, g_3, g_4 the genetic types. Then the multiplication table of the algebra is

$$g_i^2 = g_1, \quad i = 1, 2, 3, 4$$

$$g_1 g_2 = \tfrac{1}{2}(g_3 + g_4)$$

$$g_3 g_4 = \tfrac{1}{2}(g_1 + g_2)$$

$$g_i g_j = \tfrac{1}{2}(g_i + g_j), \quad i = 1, 2; \quad j = 3, 4.$$

An appropriate canonical basis is

$$e = \tfrac{1}{4}(g_1 + g_2 + g_3 + g_4)$$

$$c_1 = g_1 - g_3$$

$$c_2 = g_1 - g_4$$

$$d = g_1 + g_2 - g_3 - g_4.$$

Then

$$e^2 = e, \quad ec_1 = \frac{1}{2}c_1, \quad ec_2 = \frac{1}{2}c_2, \quad c_1 c_2 = \frac{1}{2}d,$$

other products zero. We have

$$Z = \{c_1, c_2, d\}, \quad Z^2 = \{d\}, \quad Z^3 = 0.$$

Thus the algebra is special train.

Remark 3.1. There are many other variations of genetic algebras (introduced very long time ago): Copular algebra, gametic algebra, zygotic algebra, Multiple allelomorphs (with the gametic and zygotic multiplication), Linked allelomorphs (with gametic multiplication) [48], [49]. For more genetic algebras see [155], [211], [269] and the references therein. In the next sections we give more recently introduced algebras.

3.2 Dibaric algebras

For definition of dibaric algebras see Definition 2.4. In this section we follow [137].

Proposition 3.11. *Let $\mathbf{A}$ be a commutative dibaric algebra over the field $\mathbb{R}$. Then there is an ideal $I \lhd \mathbf{A}$ with $\mathrm{codim}(I) = 2$ such that $\mathbf{A}^2 \nsubseteq I$.*

Proof. Since $\mathbf{A}$ is dibaric algebra there exists a non-zero homomorphism $\varphi \colon \mathbf{A} \to \mathfrak{A}$. Put $I = \ker \varphi$, then since φ is onto one gets $\mathrm{codim}(I) = 2$. Therefore $\mathbf{A}$ can be represented as $\mathbf{A} = \mathbb{R}e_1 + \mathbb{R}e_2 + I$, and we can take $\varphi(e_1) = w$ and $\varphi(e_2) = m$. Then $\varphi(e_1 e_2) = \frac{1}{2}(w + m) \neq 0$, and $e_1 e_2 \in \mathbf{A}^2$, but $e_1 e_2 \notin I$. Thus $\mathbf{A}^2 \nsubseteq I$. $\qquad\square$

Theorem 3.2. *If $\mathbf{A}$ is a commutative algebra over the field $\mathbb{R}$, which satisfies the following conditions:*

(1) There is an ideal $I \lhd \mathbf{A}$ with $\mathrm{codim}(I) = 2$;
(2) There exists $\overline{e} \in \mathbf{A}/I$ such that $\mathbf{A}^2 = \mathbb{R}e + I$ and $e^2 - e \in I$;
(3) There exists $x \in \mathbf{A}$ such that $x^2 + e \in I$.

Then $\mathbf{A}$ is a dibaric algebra.

Proof. The natural homomorphism $\varphi \colon \mathbf{A} \to \mathbf{A}/I$ which is onto. Under conditions (1)–(3) we shall prove that $\varphi(\mathbf{A}) = \mathbf{A}/I$ is isomorphic to $\mathfrak{A}$. Since $\mathfrak{A}$ is isomorphic to an algebra $\langle p, q \rangle$ with multiplication $p^2 = p$, $q^2 = -p$, $pq = 0$, where $p = m + w$, $q = m - w$, from (1) we get $\dim(\mathbf{A}/I) = 2$, i.e. $\mathbf{A}/I = \langle \overline{p}, \overline{q} \rangle$.

We obtain from (2) that $\left(\mathbf{A}/I\right)^2 = \langle \overline{p} \rangle$, where $\overline{p}^2 = \overline{p}$. Therefore, the algebra $\mathbf{A}/I$ has the following table of multiplication:

$$\overline{p}^2 = \overline{p}, \quad \overline{p}\,\overline{q} = \alpha_1 \overline{p}, \quad \overline{q}^2 = \alpha_2 \overline{p}.$$

The change $\overline{q}' = \overline{q} - \alpha_1 \overline{p}$ allows to take $\alpha_1 = 0$. Hence, we consider the table of multiplication:

$$\overline{p}^2 = \overline{p}, \quad \overline{p}\,\overline{q} = 0, \quad \overline{q}^2 = \alpha_2 \overline{p}.$$

The following cases are possible:

Case $\alpha_2 = 0$. In this case, the equation $\overline{x}^2 = -\overline{p}$ is equivalent to $\alpha^2 \overline{p} = -\overline{p}$, which has not solution $\alpha \in \mathbb{R}$, i.e. the condition (3) is not satisfied. Clearly, the corresponding algebra is not isomorphic to $\mathfrak{A}$.

Case $\alpha_2 < 0$. In this case, the change $\overline{q}' = \frac{\overline{q}}{\sqrt{|\alpha_2|}}$ allows to put $\alpha_2 = -1$. Consequently, $\mathbf{A}/I$ is isomorphic to $\mathfrak{A}$.

Case $\alpha_2 > 0$. In this case, the change $\overline{q}' = \frac{\overline{q}}{\sqrt{\alpha_2}}$ allows to put $\alpha_2 = 1$. The equation $\overline{x}^2 = -\overline{p}$ is equivalent to $(\alpha^2 + \beta^2)\overline{p} = -\overline{p}$, which has not solution $\alpha, \beta \in \mathbb{R}$, i.e. the condition (3) is not satisfied. Clearly, the corresponding algebra is not isomorphic to $\mathfrak{A}$.

$\square$

Remark 3.2.

1. From (2) we get $\mathbf{A}^2 \nsubseteq I$, but the converse is not true in general.
2. If $\mathbf{A}/I$ isomorphic to $\langle \overline{p}, \overline{q} \rangle$ with $\overline{p}^2 = 0$, $\overline{p}\,\overline{q} = \overline{p}$, $\overline{q}^2 = \overline{p}$, then the condition (3) of Theorem 3.2 is satisfied, but the condition (2) is not satisfied. Clearly, the corresponding algebra $\mathbf{A}/I$ is not isomorphic to $\mathfrak{A}$.
3. If $\mathbf{A}/I$ isomorphic to $\langle \overline{p}, \overline{q} \rangle$ with $\overline{p}^2 = \overline{p}$, $\overline{p}\,\overline{q} = 0$, $\overline{q}^2 = \overline{p}$, then the condition (2) of Theorem 3.2 is satisfied, but the condition (3) is not satisfied. One can see that the corresponding algebra $\mathbf{A}/I$ is not isomorphic to $\mathfrak{A}$.

Definition 3.10. For a given algebra $\mathbf{A}$, a pair (f, g), of linear forms $f \colon \mathbf{A} \to \mathbb{R}$, $g \colon \mathbf{A} \to \mathbb{R}$ is called bq-homomorphism if

$$f(xy) = g(xy) = \frac{f(x)g(y) + f(y)g(x)}{2} \quad \text{for any} \ \ x, y \in \mathbf{A}. \qquad (3.18)$$

Note that if $f = g$ then the condition (3.18) implies that f is a homomorphism.

A bq-homomorphism (f, g) is called non-zero if $f(z)g(z) \neq 0$, i.e. both f and g are non-zero.

Theorem 3.3. *An algebra $\mathbf{A}$ is dibaric if and only if there is a non-zero bq-homomorphism.*

Proof. Assume $\mathbf{A}$ admits a non-zero bq-homomorphism (f, g). Let $\varphi \colon \mathbf{A} \to \mathfrak{A}$ defined by

$$\varphi(x) = f(x)w + g(x)m.$$

For $x, y \in \mathbf{A}$, we have

$$\varphi(xy) = f(xy)w + g(xy)m = \frac{f(x)g(y) + f(y)g(x)}{2}(w + m).$$

Using $w^2 = m^2 = 0$, $wm = \frac{1}{2}(w + m)$ we get

$$\varphi(x)\varphi(y) = \big(f(x)w + g(x)m\big)\big(f(y)w + g(y)m\big)$$
$$= \big(f(x)g(y) + f(y)g(x)\big)wm = \varphi(xy),$$

i.e. φ is a homomorphism. For arbitrary $u = \alpha w + \beta m \in \mathfrak{A}$ one can see that $\varphi(x) = u$ if $f(x) = \alpha$ and $g(x) = \beta$. Therefore φ is onto. Hence $\mathbf{A}$ is dibaric.

Conversely, if $\mathbf{A}$ is dibaric then there is homomorphism φ from $\mathbf{A}$ onto $\mathfrak{A}$, which has the form $\varphi(x) = W(x)w + M(x)m$. Since φ is onto, $W \neq 0$ and $M \neq 0$. Clearly

$$\varphi(xy) = W(xy)w + M(xy)m = \varphi(x)\varphi(y)$$
$$= \frac{W(x)M(y) + W(y)M(x)}{2}(w + m),$$

which implies that

$$W(xy) = M(xy) = \frac{W(x)M(y) + W(y)M(x)}{2}.$$

Thus for non-zero bq-homomorphism we can take $\big(f(x), g(x)\big) = \big(W(x), M(x)\big)$. $\qquad\square$

A pair $\big(\mathbf{A}, (f, g)\big)$, where $\mathbf{A}$ is an algebra and (f, g) is a non-zero bq-homomorphism, denotes a dibaric algebra.

A linear form, F, on a dibaric algebra, $\big(\mathbf{A}, (f, g)\big)$, is called f-*invariant linear form* (resp. g-*invariant linear form*) if it satisfies the equality

$$F(x^2) = f(x)F(x), \quad \big(\text{resp.} \, F(x^2) = g(x)F(x)\big) \quad x \in \mathbf{A}. \tag{3.19}$$

or equivalently:

$$F(xy) = \tfrac{f(x)F(y)+f(y)F(x)}{2},$$
$$\Big(\text{resp.} \, F(xy) = \tfrac{g(x)F(y)+g(y)F(x)}{2}\Big) \quad x, y \in \mathbf{A}. \tag{3.20}$$

An invariant form defines a conservation law for the dynamical system. The gene conservation laws are examples (see [155] for details).

Denote by J_f (resp. J_g) the set of all f-invariant (resp. g-invariant) linear forms of $\mathbf{A}$. The set J_f and J_g are subspaces of the dual space $\mathbf{A}^*$. Clearly, $F(x) \equiv 0$ is an element of $J_f \cap J_g$. Moreover, $g \in J_f$ and $f \in J_g$. Hence $1 \leq \dim J_f, \dim J_g \leq \dim \mathbf{A}$.

Remark 3.3. A baric algebra $(\mathcal{A}, \sigma)$ is dibaric with $f = g = \sigma$. But a dibaric algebra is not baric in general. For example, an evolution algebra of bisexual population is dibaric but not baric.

Lemma 3.1. *If $\big(\mathbf{A}, (f, g)\big)$ is a dibaric (not baric) algebra then $J_f \cap J_g = \{0\}$.*

Proof. Assume $F \in J_f \cap J_g$ then from (3.20) we get

$$\big(f(x) - g(x)\big)F(y) + \big(f(y) - g(y)\big)F(x) = 0, \tag{3.21}$$

and for $x = y$ we have

$$\big(f(x) - g(x)\big)F(x) = 0. \tag{3.22}$$

Now if $f(x) - g(x) = 0$ then we take y such that $f(y) - g(y) \neq 0$ then from (3.21) it follows that $F(x) = 0$. If $f(x) - g(x) \neq 0$ then from (3.22) it follows that $F(x) = 0$. Thus $F \equiv 0$. $\qquad\square$

Note that in the definition of a bq-homomorphism (f, g), the functions f and g play a "symmetric" role, therefore in the sequel of this section we consider only J_f.

Denote

$$J_f^{\perp} = \{x : F(x) = 0, \quad \text{for all} \quad F \in J_f\},$$
$$\mathrm{ann}\mathbf{A} = \{y \in \mathbf{A} : yx = 0, \quad \text{for all} \quad x \in \mathbf{A}\}.$$

Lemma 3.2. *The following inclusion holds*

$$\mathrm{ann}\mathbf{A} \subseteq J_f^{\perp}.$$

Proof. If $x \in \mathrm{ann}\mathbf{A}$ then using equation (3.20) we obtain $f(x)F(y) + f(y)F(x) = 0$, $y \in \mathbf{A}$. Hence if $f(x) = 0$ then we take y such that $f(y) \neq 0$, consequently $F(x) = 0$. If $f(x) \neq 0$ then from $0 = F(x^2) = f(x)F(x)$ we get $F(x) = 0$ for any $F \in J_f$, i.e. $x \in J_f^{\perp}$. $\qquad\square$

Definition 3.11. The algebra $\mathbf{A}$ is called conservative if

$$J_f^{\perp} = \mathrm{ann}\mathbf{A}. \tag{3.23}$$

Theorem 3.4. *In order to $\mathbf{A}$ being conservative it is necessary and sufficient that the product xy depends only upon the values of the invariant forms on x and y. That is*

$$x \equiv x' \quad \text{and} \quad y \equiv y' (\mathrm{mod}\ J_f^{\perp}) \quad \text{imply} \quad xy = x'y'. \tag{3.24}$$

Proof. Necessariness. Let $\mathbf{A}$ be conservative, then $J_f^{\perp} = \mathrm{ann}\mathbf{A}$. Consider x, y, x', y' with $x - x', y - y' \in \mathrm{ann}\mathbf{A}$. Consequently, $(x - x')z = 0$, $(y - y')z = 0$, $\forall z \in \mathbf{A}$. From $(x - x')(y - y') = 0$ and from these equations for $z = x', y'$ we get $x'y - x'y' = 0$, $xy' - x'y' = 0$, which implies $xy = x'y'$.

Sufficiency. Now assume that (3.24) is satisfied, we shall show that $\mathbf{A}$ is a conservative algebra. Apply (3.24) with $y = y'$, $x' = 0$. We get $x \in J_f^\perp$ which implies $xy = 0$. Hence, $J_f^\perp \subset \mathrm{ann}\mathbf{A}$ which, together with Lemma 3.2, imply (3.23). $\qquad\square$

Corollary 3.3. *Let $\{F_1, \ldots, F_m\}$ be a basis of the subspace J_f of invariant forms for the algebra $\mathbf{A}$. The algebra $\mathbf{A}$ is conservative if and only if there exist vectors $u_{ik} \in \mathbf{A}$ $(1 \leq i, k \leq m)$ with $u_{ik} = u_{ki}$ so that the multiplication is given by the formula*

$$xy = \sum_{i,k=1}^{m} F_i(x)F_k(y)u_{ik}, \quad x, y \in \mathbf{A}.$$

Proposition 3.12. *An algebra $\mathbf{A}$ is conservative if and only if*

$$x^2 y = f(x)xy, \quad x, y \in \mathbf{A}. \tag{3.25}$$

Proof. Using (3.19) one can see that $x^2 - f(x)x$ lies in $J_f^\perp$ for all $x \in \mathbf{A}$, i.e.

$$x^2 = f(x)x \,(\mathrm{mod}\, J_f^\perp).$$

Multiplying both sides by y one gets (3.25) from (3.24) when the algebra is conservative. Conversely, from (3.25) we obtain $x^2 - f(x)x \in \mathrm{ann}\mathbf{A}$. Thus if $F \in (\mathrm{ann}\mathbf{A})^\perp$, then (3.25) implies $F(x^2 - f(x)x) = 0$ for all $x \in \mathbf{A}$, i.e. $F \in J_f$. Moreover, $(\mathrm{ann}\mathbf{A})^\perp \subset J_f$ implies $J_f^\perp \subset \mathrm{ann}\mathbf{A}$. From this inclusion and Lemma 3.2 follows that $\mathbf{A}$ is conservative. $\qquad\square$

Define the following powers of $x \in \mathbf{A}$:

$$x^{[1]} = x^2, \quad x^{[k+1]} = \left(x^{[k]}\right)^2, \, k \geq 1.$$

Proposition 3.13. *In a conservative algebra $\mathbf{A}$ the following identity holds:*

$$x^{[k]} = \left(f(x)\right)^{2^k - 2} x^2, \quad k \geq 1, \,, x \in \mathbf{A}. \tag{3.26}$$

Proof. By (3.24) one obtains

$$(x^2)^2 = \left(f(x)\right)^2 x^2. \tag{3.27}$$

Assume that formula (3.26) is true for $k - 1$ and prove it for k:

$$x^{[k]} = \left(x^{[k-1]}\right)^2 = \left(\left(f(x)\right)^{2^{k-1}-2} x^2\right)^2$$

$$= \left(f(x)\right)^{2^k - 4}(x^2)^2 = \left(f(x)\right)^{2^k - 2} x^2.$$

$\qquad\square$

Recall the algebras $\mathbf{A}$ in which (3.27) holds are called *stationary* or *Bernstein* algebras because of the connection between this class and the problem of Bernstein (see the previous chapter and Section 2.1 and Chapters 4, 5 [155]). For an evolution algebra $\mathcal{B}$, condition (3.27) is equivalent to the stationary principle

$$V^2(z) = V\big(V(z)\big) = V(z),$$

where $V(z) = z^2$ is the evolution operator (1.6) on $S^{n-1} \times S^{\nu-1}$.

Remark 3.4. Notice that each conservative algebra is Bernstein, but in page 99 of [155] an example of a baric algebra $(\mathcal{A}, \sigma)$ was constructed which is Bernstein but is not conservative. If we consider a dibaric algebra $\big(\mathbf{A}, (f, g)\big)$, then the example (for $\sigma = f$) will be an example of a Bernstein algebra which is not conservative.

Definition 3.12. An algebra $\mathbf{A}$ is called induced by a linear operator if there is a linear operator A in the vector space $\mathbf{A}$ such that

$$xy = \frac{f(x)A(y) + f(y)A(x)}{2}, \quad x, y \in \mathbf{A}. \tag{3.28}$$

Proposition 3.14. *The algebra $\mathbf{A}$ induced by a linear operator A is Bernstein if and only if*

$$f(x)A(x) = f\big(A(x)\big)A^2(x). \tag{3.29}$$

Proof. From (3.28) we get $x^2 = f(x)A(x)$; now if $\mathbf{A}$ is Bernstein then from (3.27) we obtain

$$\big(f(x)\big)^2 f\big(A(x)\big)A^2(x) = \big(f(x)\big)^2 x^2. \tag{3.30}$$

Denote

$$W = \{x \in \mathbf{A} : f(x) \neq 0\}.$$

Since the set W is a dense subset of $\mathbf{A}$, from (3.30) we get

$$f\big(A(x)\big)A^2(x) = x^2 = f(x)A(x).$$

Now assume that the condition (3.29) is satisfied. Then

$$(x^2)^2 = \big(f(x)\big)^2 \big(A(x)\big)^2 = \big(f(x)\big)^2 f\big(A(x)\big)A^2(x)$$
$$= \big(f(x)\big)^3 A(x) = \big(f(x)\big)^2 f(x)A(x) = \big(f(x)\big)^2 x^2. \qquad \square$$

For an algebra $\mathbf{A}$ denote by N the subspace of disappearing forms, i.e. the linear forms which vanish on the subalgebra $\mathbf{A}^2$. We have

$$J_f \cap N = \{0\}.$$

Indeed, if $F(x^2) = f(x)F(x)$ ($F \in J_f$) and at the same time $F(x^2) = 0$ ($F \in N$) for all $x \in \mathbf{A}$ then F vanishes on the dense open subset of $\mathbf{A}$ where $f(x) \neq 0$. Consequently, $F = 0$.

It follows that $\dim J_f + \dim N \leq \dim \mathbf{A}$.

Proposition 3.15. *If the algebra is induced by the linear operator A, then $N = (\mathrm{Img}A)^{\perp}$.*

Proof. From (3.28) we get $xy = A\left(\frac{f(x)y + f(y)x}{2}\right)$. Consequently, $\mathbf{A}^2 \subset \mathrm{Img}A$ and $(\mathrm{Img}A)^{\perp} \subset N$. Conversely, for $F \in N$ we have $0 = F(x^2) = F\big(f(x)A(x)\big) = f(x)F\big(A(x)\big)$. Since W is dense in $\mathbf{A}$ we get $F\big(A(x)\big) = 0$ for all x. Hence $F \in (\mathrm{Img}A)^{\perp}$. $\qquad\square$

Given dibaric algebras $\big(\mathbf{A}_1, (f_1, g_1)\big)$ and $\big(\mathbf{A}_2, (f_2, g_2)\big)$, a *dibaric algebra homomorphism* $h\colon \big(\mathbf{A}_1, (f_1, g_1)\big) \to \big(\mathbf{A}_2, (f_2, g_2)\big)$ is an homomorphism $h\colon \mathbf{A}_1 \to \mathbf{A}_2$ such that $h^* f_2 = f_2 h = f_1$. For example, the embedding of a dibaric subalgebra and the quotient map to a dibaric quotient algebra are dibaric homomorphisms. Clearly, the composition of dibaric homomorphisms is dibaric. A bijective dibaric homomorphism is called a *dibaric isomorphism*. The inverse of a dibaric isomorphism is dibaric because $h^* f_2 = f_1$ implies $f_2 = (h^{-1})^* f_1$.

The following proposition says that the definition of dibaric homomorphism does not depend on the choice of f and g.

Proposition 3.16. *For a dibaric algebra homomorphism*

$$h\colon \big(\mathbf{A}_1, (f_1, g_1)\big) \to \big(\mathbf{A}_2, (f_2, g_2)\big)$$

we have $h^* f_2 = f_1$ if and only if $h^* g_2 = g_1$.

Proof. Since f and g play symmetric role it suffices to prove that from $h^* f_2 = f_1$ follows $h^* g_2 = g_1$. So let $h^* f_2 = f_1$. Then we have

$$f_2\big(h(x^2)\big) = f_2\big(h(x)\big)g_2\big(h(x)\big) = f_1(x)g_2\big(h(x)\big)$$
$$f_1(x^2) = f_1(x)g_1(x).$$

Hence

$$f_1(x)\big(g_2(h(x)) - g_1(x)\big) = 0. \tag{3.31}$$

We also have

$$f_1(xy) = f_2(h(xy)) = \frac{f_1(x)g_2(h(y)) + f_1(y)g_2(h(x))}{2},$$

$$f_1(xy) = \frac{f_1(x)g_1(y) + f_1(y)g_1(x)}{2}.$$

Consequently,

$$f_1(x)\Big(g_2(h(y)) - g_1(y)\Big) + f_1(y)\Big(g_2(h(x)) - g_1(x)\Big) = 0. \qquad (3.32)$$

If $f_1(x) \neq 0$ then from (3.31) we get $g_2(h(x)) = g_1(x)$. If $f_1(x) = 0$ we choose y such that $f_1(y) \neq 0$ then from (3.32) we get $g_2(h(x)) = g_1(x)$. $\square$

Theorem 3.5. *Let $\big(\mathbf{A}_1, (f_1, g_1)\big)$ and $\big(\mathbf{A}_2, (f_2, g_2)\big)$ be dibaric algebras with spaces of invariant linear forms $J_1 = J_{f_1}$ and $J_2 = J_{f_2}$ respectively. Assume $h \colon \mathbf{A}_1 \to \mathbf{A}_2$ is a homomorphism.*

(i) h^* *transforms invariant forms to invariant forms, i.e. F invariant on $\mathbf{A}_2$ implies h^*F is invariant on $\mathbf{A}_1$:*

$$h^*(J_2) \subset J_1. \qquad (3.33)$$

(ii) *If $\mathrm{Img}\, h \supset \mathbf{A}_2^2$ then h^* is injective on J_2.*

(iii) *If h is injective, then the subspace $(h^*)^{-1}(J_1)$ of $\mathbf{A}_2^*$ — which contains J_2 by (3.33) — is equal to J_2. Moreover, F is invariant on $\mathbf{A}_2$ if and only if h^*F is invariant on $\mathbf{A}_1$.*

(iv) *If h is surjective and $\ker h \subset J_1^{\perp}$, then h^* maps J_2 bijectively onto J_1.*

Proof. The prove is similar to the proof of Theorem 3.3.12 of [155]. $\square$

Theorem 3.6.

(1) For a dibaric algebra $\big(\mathbf{A}, (f, g)\big)$, let L be a linear subspace of J_f. Then $\mathbf{A}_1 = L^{\perp}$ is a subalgebra.

(2) If $\big(\mathbf{A}, (f, g)\big)$ is a conservative algebra then a linear form F on $\mathbf{A}$ is invariant if and only if its restriction F_1 to $\mathbf{A}_1$ is invariant on $\mathbf{A}_1$.

(3) If J_1 is the space of linear forms on $\mathbf{A}_1$ then

$$\dim J_1 = \dim J_f - \dim L. \qquad (3.34)$$

Proof. (1) If $x, y \in \mathbf{A}_1$ then $F(x) = F(y) = 0$ for any $F \in L$. Hence $F(xy) = 0$ for any $F \in L$, i.e. $xy \in \mathbf{A}_1$.

(2) If $F_1 = F|_{\mathbf{A}_1}$, where $F \in J_f$ then $F(J_f^{\perp}) = 0$, i.e. $F(x^2) = f(x)F(x)$, $\forall x \in \mathbf{A}$. Since $\mathbf{A}_1 \subset \mathbf{A}$ we get $F_1 \in J_1$. Conversely, if F_1 is an invariant on $\mathbf{A}_1$ then $f_1(J_1^{\perp}) = 0$. We have $\mathrm{ann}\mathbf{A} \subseteq \mathrm{ann}\mathbf{A}_1 \subseteq J_1^{\perp}$. Hence

$F_1(\mathrm{ann}\mathbf{A}_1) = 0$ and consequently $F_1(\mathrm{ann}\mathbf{A}) = 0$. Since $\mathbf{A}$ is conservative, we get $F_1(J_f^{\perp}) = 0$, i.e. $F \in J_f$.

(3) Let $i\colon \mathbf{A}_1 \hookrightarrow \mathbf{A}$ be embedding then $i^*\colon \mathbf{A}^* \hookrightarrow \mathbf{A}_1^*$. Hence $i^*\colon J_f \to J_1$, $\ker i^* = L$. By (2) for any $F_1 \in J_1$ there exists $F \in J_f$ such that $F_1 = F|_{\mathbf{A}_1}$, consequently i^* is surjective. Therefore we get (3.34). $\qquad\square$

3.3   Evolution algebras

The evolution algebras are introduced in book of Tian [254], he motivated this algebras by evolution laws of genetics. He considered alleles (or organelles or cells, etc.) as generators of an evolution algebra and define the multiplication of two 'alleles' e_i and e_j by

$$e_i e_j = \begin{cases} 0 & \text{if } i \neq j \\ \sum_k a_{ik} e_k, & \text{if } i = j, \end{cases} \tag{3.35}$$

where the summation is taken over all generators e_j. Thus, a reproduction in genetics is represented by multiplication in algebra.

Definition 3.13. Let $(E, \cdot)$ be an algebra over a field F. If it admits a basis $e_1, e_2, \ldots$, such that (3.35) holds then this algebra is called an *evolution algebra*.

We denote by $A = (a_{ij})$ the matrix of the structural constants of the evolution algebra E.

The basic properties of an evolution algebra are given in [254]: It is proven that an evolution algebra is non associative, but commutative. When the a_{ij}s form Markovian transition probabilities, the properties of algebras are associated with properties of Markov chains. Using Markov chains it is developed an algebraic theory at deeper hierarchical levels than standard algebras. Moreover, new algebraic concepts (algebraic persistency, algebraic transiency, algebraic periodicity, and their relative versions) are given. A hierarchical structures for evolution algebras are described. It is shown that Markov chains can be classified by the skeleton shape classification of their evolution algebras. When applied to non-Mendelian genetics (particularly organelle heredity) evolution algebras can explain establishment of homoplasmy from heteroplasmic cell population and the co-existence of mitochondrial triplasmy, and can also predict all possible mechanisms to establish the homoplasmy of cell population.

In this section we give some recently obtained results related to evolution algebras. These will be based on papers [28], [29], [30], [31], [188], [195].

Remark 3.5. Note that the theory of evolution algebras is currently a very active field of mathematical research. The results which I will mention in this section are by no means complete, because there are many recently published papers (related to evolution algebras) of C. Y. Cabrera, L. M. Camacho, J. M. Casas, A. Elduque, J. R. Gomez, A. Khudayberdiev, A. Labra, M. Ladra, B. A. Omirov, R. M. Turdibaev, J. P. Tian, M. M. Siles, M. V. Velasco and many others, I am not going to collect all results of these papers in this book. Hopefully, someone will write a new book on evolution algebras soon.

3.3.1 *A criterion for an evolution algebra to be baric*

Note that an evolution algebra E introduced is not baric, in general. Following [31], we give the following theorem, which gives a criterion for an evolution algebra E to be baric.

Theorem 3.7. *An n-dimensional evolution algebra E, over field the $\mathbb{R}$, is baric if and only if there is a column $(a_{1i_0}, \ldots, a_{ni_0})^T$ of its structural constants matrix $\mathcal{M} = (a_{ij})_{i,j=1,\ldots,n}$, such that $a_{i_0 i_0} \neq 0$ and $a_{ii_0} = 0$, for all $i \neq i_0$. Moreover, the corresponding weight function is $\sigma(x) = a_{i_0 i_0} x_{i_0}$.*

Proof. Necessity. Take $x, y \in E$ with $x = \sum_{i=1}^{n} x_i e_i$, $y = \sum_{i=1}^{n} y_i e_i$. Assume $\sigma(x) = \sum_{i=1}^{n} \alpha_i x_i$, $x \in E$ is a character. We have

$$\sigma(xy) = \sum_{i=1}^{n} \left(\sum_{j=1}^{n} a_{ij} \alpha_j \right) x_i y_i; \quad \sigma(x)\sigma(y) = \sum_{i=1}^{n} \sum_{j=1}^{n} \alpha_i \alpha_j x_i y_j.$$

From $\sigma(xy) = \sigma(x)\sigma(y)$ we get

$$\alpha_i \alpha_j = 0 \quad \text{for any} \quad i \neq j, \ i, j = 1, \ldots, n; \tag{3.36}$$

$$\sum_{j=1}^{n} a_{ij} \alpha_j = \alpha_i^2 \quad \text{for any} \quad i = 1, \ldots, n. \tag{3.37}$$

One can see that the system (3.36) has a solution $\alpha = (\alpha_1, \ldots, \alpha_n)$ with $\alpha_1^2 + \cdots + \alpha_n^2 > 0$ if and only if exactly one coordinate of α, say α_{i_0}, is not zero, and all others are zeros. Substituting this solution in (3.37) we get

$$a_{ii_0} \alpha_{i_0} = 0, \quad \text{if } i \neq i_0, \ i = 1, \ldots, n;$$

$$a_{i_0 i_0} \alpha_{i_0} = \alpha_{i_0}^2, \text{ if } i = i_0.$$

From the last equations we get $a_{i_0 i_0} \neq 0$, $a_{i i_0} = 0$, for all $i \neq i_0$ and $\alpha_{i_0} = a_{i_0 i_0}$.

Sufficiency. Assume there is a column $(a_{1 i_0}, \ldots, a_{n i_0})^T$, such that $a_{i_0 i_0} \neq 0$ and $a_{i i_0} = 0$, for all $i \neq i_0$. Then one can see that $\sigma(x) = a_{i_0 i_0} x_{i_0}$ is a weight function, therefore E is a baric evolution algebra. $\square$

A baric algebra A may have several weight functions. As a corollary of Theorem 3.7 we have

Corollary 3.4. *If the matrix $\mathcal{M}$, mentioned in Theorem 3.7, has several columns $\left(a_{1 i_j}, \ldots, a_{n i_j}\right)^T$, $j = i_1, \ldots, i_m$, $m \leq n$, which satisfy conditions of Theorem 3.7 then the evolution algebra E has exactly m weight functions $\sigma(x) = a_{i_j i_j} x_{i_j}$, $j = i_1, \ldots, i_m$.*

There are two types of trivial evolution algebras [254]: *zero evolution algebra*, which satisfies $e_i e_j = 0$ for all $i, j = 1, \ldots, n$; *non-zero trivial evolution algebra*, which satisfies $e_i e_j = 0$ for all $i \neq j$ and $e_i^2 = a_{ii} e_i$, where $a_{ii} \in \mathbb{R}$ is non-zero for some $i = 1, \ldots, n$. By Theorem 3.7 we conclude that the zero evolution algebra is not baric, but any non-zero trivial evolution algebra is a baric algebra. Moreover, there are baric evolution algebras which are not trivial.

3.3.2 *Dibaric evolution algebra*

In this section, following [30], we will give some dibaricity properties of evolution algebras.

Theorem 3.8. *Any finite-dimensional nilpotent evolution algebra E is not dibaric.*

Proof. Assume $\varphi = (b_{ij})_{i=1,\ldots,n;\, j=1,2}$ is a homomorphism $\varphi \colon E \to \mathfrak{A}$. We claim the following

$$b_{i1} b_{j2} = b_{i2} b_{j1} = 0, \quad \text{for any} \quad i, j = 1, \ldots, n. \tag{3.38}$$

Indeed, without loss of generality we assume $i \leq j$ and use mathematical induction. Let C_k denote all cases of (3.38) with $2n - (i + j) + 1 = k$. For $k = 1$, i.e. $i = j = n$, from $\varphi(e_n^2) = 0$ we get

$$b_{n1} b_{n2} = 0. \tag{3.39}$$

For $k = 2$ we have $i = n - 1$ and $j = n$. We get

$$0 = \varphi(e_{n-1} e_n) = \frac{1}{2}(b_{n-1,1} b_{n2} + b_{n-1,2} b_{n1})(m + w).$$

This by (3.39) gives

$$b_{n-1,1}b_{n2} = b_{n-1,2}b_{n1} = 0.$$

Assuming that C_k holds, we have to prove C_{k+1}, that is, equation (3.38), for any $i, j = 1, \ldots, n$, $i \le j$, which satisfy $2n - (i+j) + 1 = k + 1$.

Case $i < j$: From $\varphi(e_i e_j) = \frac{1}{2}(b_{i1}b_{j2} + b_{i2}b_{j1})(m + w) = 0$ we get

$$b_{i1}b_{j2} + b_{i2}b_{j1} = 0. \tag{3.40}$$

By assumptions we have $i < j$, $2n - 2j + 1 \le k$ and $b_{j1}b_{j2} = 0$. This by (3.40) gives (3.38).

Case $i = j$: Consider

$$\varphi(e_i^2) = \varphi\left(\sum_{s=i+1}^{n} a_{is}\, e_s\right) = \sum_{s=i+1}^{n} a_{is}\, \varphi(e_s)$$

$$= \left(\sum_{s=i+1}^{n} a_{is}b_{s1}\right) m + \left(\sum_{s=i+1}^{n} a_{is}b_{s2}\right) w$$

$$= \varphi(e_i)^2 = \frac{1}{2}b_{i1}b_{i2}(m + w).$$

Consequently,

$$\begin{cases} 2 \displaystyle\sum_{s=i+1}^{n} a_{is}b_{s1} = b_{i1}b_{i2}, \\[2mm] 2 \displaystyle\sum_{s=i+1}^{n} a_{is}b_{s2} = b_{i1}b_{i2}. \end{cases} \tag{3.41}$$

Since $2n - (i+s) + 1 \le k$ for any $s = i+1, i+2, \ldots, n$, by the assumption of the induction we get

$$b_{s1}b_{i2} = b_{s2}b_{i1} = 0. \tag{3.42}$$

Now, multiplying both sides of the first equation of (3.41) by b_{i2}, then by (3.42) we get $b_{i1}b_{i2} = 0$.

If there exists i_0 such that $b_{i_0 1} \ne 0$ then $b_{j2} = 0$ for all j, i.e., $\varphi(e_i) = b_{i1}m$. Such φ is not onto. $\qquad\square$

Let a two-dimensional real evolution algebra E be given by the matrix of structural constants $A = \begin{pmatrix} a & b \\ c & d \end{pmatrix}$. We shall find a criterion for this algebra to be dibaric.

Proposition 3.17. *The two-dimensional real evolution algebra E is dibaric if and only if one of the following conditions hold*

1. $b = d = 0$ *and* $ac < 0$;

2. $b \neq 0$, $ad = bc$, $D \geq 0$ *and* $B^2 + C^2 \neq 0$, *where* $D = (8a - 1)^2 - 32(bd + a^2)$, $B = 4a^2 + 4bd - a + a\sqrt{D}$ *and* $C = 4a^2 + 4bd - a - a\sqrt{D}$.

Proof.   Let $\varphi = \begin{pmatrix} \alpha & \beta \\ \gamma & \delta \end{pmatrix}$ be a homomorphism. It is onto if $\alpha\delta \neq \gamma\beta$. Moreover, φ must satisfy

$$\begin{cases} 2(a\alpha + b\gamma) = \alpha\beta \\ 2(a\beta + b\delta) = \alpha\beta \\ 2(c\alpha + d\gamma) = \gamma\delta \\ 2(c\beta + d\delta) = \gamma\delta \\ \alpha\delta + \beta\gamma = 0. \end{cases} \tag{3.43}$$

The proof follows from the elementary analysis of the system (3.43). $\square$

3.3.3   *Right nilpotent evolution algebras*

Following [29], we will prove that notions of nil and right nilpotency are equivalent for evolution algebras. Moreover, the matrix A of structural constants for such algebras has upper (or lower, up to permutation of basis of the algebra) triangular form.

Theorem 3.9. *Let E be a nil evolution algebra with basis $\{e_1, \ldots, e_n\}$. Then for the elements of the matrix $A = (a_{ij})$ the following relation holds*

$$a_{i_1 i_2} a_{i_2 i_3} \ldots a_{i_k i_1} = 0, \tag{3.44}$$

for all $i_1, \ldots, i_k \in \{1, \ldots, n\}$ *and* $k \in \{1, \ldots, n\}$, *with* $i_p \neq i_q$ *for* $p \neq q$.

Proof.   Note that $((e_i \cdot e_i) \cdot e_i) = a_{ii} e_i^2$, hence $a_{ii} = 0$ (otherwise the element e_i is not nil). We shall prove the equality (3.44) for right normed terms by induction.

For the element $e_i + e_j$, $1 \leq i, j \leq n$, the following relation can be proved by induction

$$(e_i + e_j)^{2s} = a_{ij}^{s-1} a_{ji}^{s-1} (e_i + e_j)^2.$$

The nil condition for the element $e_i + e_j$ leads to the following equality for the elements of the matrix A:

$$a_{ij} a_{ji} = 0 \text{ or } e_i^2 + e_j^2 = 0.$$

Take in account the fact that $a_{ii} = a_{jj} = 0$ for any i and j and comparing the coefficients at the basic elements, from condition $e_i^2 + e_j^2 = 0$ we obtain

$a_{ij} = a_{ji} = 0$. Hence, the equation $a_{ij}a_{ji} = 0$ for all i, j is obtained and therefore the equality (3.44) is true for $k = 2$.

Let (3.44) be true for $k-1$. We shall prove it for k. For this purpose we consider the element $e_{i_1} + e_{i_2} + \cdots + e_{i_k}$. Without loss of generality instead of this element we can consider the following element $e_1 + e_2 + \cdots + e_k$. Using the hypothesis of the induction it is not difficult to note that

$$\left(\sum_{i=1}^{k} e_i \right)^{s+1} = \sum_{\substack{i_1,\dots,i_s=1 \\ i_p \neq i_q,\ p \neq q}}^{k} a_{i_1 i_2} a_{i_2 i_3} a_{i_3 i_4} \dots a_{i_{s-1} i_s} e_{i_s}^2.$$

Let us take $s = k + 1$ in the above expression, then $i_{s-1} = i_k$. From induction hypothesis the coefficient $a_{i_1 i_2} a_{i_2 i_3} a_{i_3 i_4} \dots a_{i_{s-1} i_s}$ is equal to zero if $i_s \in \{i_2, \dots, i_{s-1}\}$. Therefore we need to consider the case $i_s = i_1$ and the above expression will have the following form

$$\left(\sum_{i=1}^{k} e_i \right)^{k+2} = \sum_{\phi \in S_k} a_{\phi(1)\phi(2)} a_{\phi(2)\phi(3)} \cdots a_{\phi(k)\phi(1)} e_{\phi(1)}^2$$

$$= \sum_{i=1}^{k} \left(\sum_{\phi \in S_k : \phi(1)=i} a_{i\phi(2)} a_{\phi(2)\phi(3)} \cdots a_{\phi(k)i} \right) e_i^2,$$

where S_k denotes the symmetric group of permutations of k elements.

Denote

$$\mathcal{F}_i = \sum_{\phi \in S_k : \phi(1)=i} a_{i\phi(2)} a_{\phi(2)\phi(3)} \cdots a_{\phi(k)i}.$$

We need the following lemmas

Lemma 3.3. *For any $i, j = 1, \dots, k$ we have $\mathcal{F}_i = \mathcal{F}_j$.*

Proof. For $\phi \in S_k$ with $\phi(1) = i$ we construct a unique $\overline{\phi} \in S_k$ such that $\overline{\phi}(1) = j$ and

$$a_{i\phi(2)} a_{\phi(2)\phi(3)} \cdots a_{\phi(k)i} = a_{j\overline{\phi}(2)} a_{\overline{\phi}(2)\overline{\phi}(3)} \cdots a_{\overline{\phi}(k)j} \tag{3.45}$$

as follows. Let s be the number such that $\phi(s) = j$, then the permutation $\overline{\phi}$ is defined as

$$\overline{\phi} = \begin{pmatrix} 1 & 2 & \dots & k-s+1 & k-s+2 & k-s+3 & \dots & k \\ j & \phi(s+1) & \dots & \phi(k) & i & \phi(2) & \dots & \phi(s-1) \end{pmatrix}.$$

Note that for a given ϕ the $\overline{\phi}$ is uniquely defined and satisfies (3.45). Thus we get $\mathcal{F}_i = \mathcal{F}_j$. $\qquad \square$

Put $a = \sum_{i=1}^{k} e_i$.

Lemma 3.4. *If $a^2 = 0$ then $\mathcal{F}_1 = 0$.*

Proof. From $a^2 = 0$ we obtain

$$\sum_{\substack{i=1 \\ i \neq j}}^{k} a_{ij} = 0, \ j = 1, \ldots, n.$$

Using this equality we get

$$\mathcal{F}_1 = \sum_{\phi \in S_k : \phi(1)=1} a_{1\phi(2)} a_{\phi(2)\phi(3)} \cdots a_{\phi(k)1}$$

$$= - \sum_{\phi \in S_k : \phi(1)=1} \sum_{\substack{i=2 \\ i \neq \phi(2)}}^{k} a_{i\phi(2)} a_{\phi(2)\phi(3)} \cdots a_{\phi(k)1} \cdot$$

Since for any $i = 2, \ldots, k$ there exists s_i such that $\phi(s_i) = i$, by the assumption of the induction we get

$$a_{i\phi(2)} a_{\phi(2)\phi(3)} \cdots a_{\phi(k)1}$$

$$= a_{i\phi(2)} a_{\phi(2)\phi(3)} \cdots a_{\phi(s_i-1)i} a_{i\phi(s_i+1)} \cdots a_{\phi(k)1} = 0. \qquad \square$$

Now we continue the proof of theorem. Using Lemma 3.3, we get

$$\left(\sum_{i=1}^{k} e_i \right)^{k+2} = \mathcal{F}_1 a^2 = 0.$$

By Lemma 3.4 we obtain $\mathcal{F}_1 = 0$. Fix an arbitrary $\phi_0 \in S_k$ with $\phi_0(1) = 1$ and multiply both side of $\mathcal{F}_1 = 0$ by $a_{1\phi_0(2)} a_{\phi_0(2)\phi_0(3)} \cdots a_{\phi_0(k)1}$ then (again using the assumption of the induction) we obtain

$$a_{1\phi_0(2)}^2 a_{\phi_0(2)\phi_0(3)}^2 \cdots a_{\phi_0(k)1}^2 = 0,$$

i.e.

$$a_{1\phi_0(2)} a_{\phi_0(2)\phi_0(3)} \cdots a_{\phi_0(k)1} = 0,$$

which completes the induction and the proof of theorem. $\qquad \square$

For an evolution algebra E we introduce the following sequence

$$E^{<1>} = E, \quad E^{<k+1>} = E^{<k>}E, \quad k \geq 1.$$

An evolution algebra is called right nilpotent if there exists some $s \in \mathbb{N}$ such that $E^{<s>} = 0$.

Let evolution algebra E be a right nilpotent algebra, then it is evident that E is a nil algebra. Therefore for the related matrix $A = (a_{ij})_{i,j=1}^{n}$ we have

$$a_{i_1 i_2} a_{i_2 i_3} \ldots a_{i_k i_1} = 0, \tag{3.46}$$

for any $k \in \{1, 2, \ldots, n\}$ and arbitrary $i_1, i_2, \ldots, i_k \in \{1, 2, \ldots, n\}$ with $i_p \neq i_q$ for $p \neq q$.

Lemma 3.5. *Let the matrix A satisfies (3.46). Then for any $j \in \{1, \ldots, n\}$ there is a row π_j of A with j zeros. Moreover, $\pi_{j_1} \neq \pi_{j_2}$ if $j_1 \neq j_2$.*

Proof. First we shall prove that there is a row with $j = n$ zeros, i.e. all entries are zeros. Assume that there is no such a row. Then for any $i \in \{1, \ldots, n\}$ there is a number $\beta(i) \in \{1, \ldots, n\} \setminus \{i\}$ such that $a_{i\beta(i)} \neq 0$. Consider the sequence

$$i_1 = 1, i_2 = \beta(1), i_3 = \beta(\beta(1)), \ldots, i_{n+1} = \underbrace{\beta(\cdots(\beta(1)))}_{n}.$$

Then by the assumption we have $a_{i_m i_{m+1}} \neq 0$, for all $m = 1, \ldots, n$ hence

$$a_{i_1 i_2} a_{i_2 i_3} \ldots a_{i_n i_{n+1}} \neq 0. \tag{3.47}$$

Since $i_j \in \{1, \ldots, n\}, j = 1, \ldots, n+1$ then $i_p = i_q$ for some $p \neq q \in \{1, \ldots, n+1\}$. Thus

$$a_{i_p i_{p+1}} a_{i_{p+1} i_{p+2}} \ldots a_{i_{q-1} i_p} \neq 0. \tag{3.48}$$

So (3.48) is in contradiction with (3.46). Thus there is a row π_n which consists of zeros (n zeros).

Now we shall prove that there is a row $\pi_{n-1} \neq \pi_n$ of A with $n-1$ zeros. Consider A_{π_n}-minor of A which is obtained from A deleting the row π_n and the column π_n. Matrix A_{π_n} is $(n-1) \times (n-1)$ and condition (3.46) implies the condition

$$a_{i_1 i_2} a_{i_2 i_3} \ldots a_{i_k i_1} = 0, \tag{3.49}$$

for any $k \in \{1, \ldots, n-1\}$ and arbitrary $i_1, i_2, \ldots, i_k \in \{1, \ldots, n\} \setminus \{\pi_n\}$ with $i_p \neq i_q$ for all $p \neq q$.

To prove that A has a row with $j = n - 1$ zeros it is sufficient to prove that A_{π_n} has a row with all zeros. But this problem is the same as the case $j = n$, only we must consider condition (3.49) instead of (3.46). Iterating this argument we can show that for any j there exists a row π_j with j zeros. The proof of the lemma is completed. $\qquad\square$

Theorem 3.10. *The following statements are equivalent for an n-dimensional evolution algebra E:*

a) The matrix corresponding to E can be written as

$$\widehat{A} = \begin{pmatrix} 0 & a_{12} & a_{13} & \cdots & a_{1n} \\ 0 & 0 & a_{23} & \cdots & a_{2n} \\ 0 & 0 & 0 & \cdots & a_{3n} \\ \vdots & \vdots & \vdots & \cdots & \vdots \\ 0 & 0 & 0 & \cdots & 0 \end{pmatrix};$$

b) E is a right nilpotent algebra;

c) E is a nil algebra.

Proof.

b) $\Rightarrow$ **a).** Since the equality (3.46) is true for right nilpotent algebra then we are in the conditions of the Lemma 3.5. Consider the permutation of the first indices $\{1, \ldots, n\}$ of the matrix A as

$$\pi = \begin{pmatrix} 1 & 2 & 3 & \cdots & n \\ \pi_1 & \pi_2 & \pi_3 & \cdots & \pi_n \end{pmatrix},$$

where π_j is defined in the proof of the Lemma 3.5. Note that Lemma 3.5 is also true for columns: for any j there is a column τ_j with j zeros. Moreover $\tau_p \neq \tau_q$, $p \neq q$. Now consider permutation of the second indexes $\{1, \ldots, n\}$ as

$$\tau = \begin{pmatrix} 1 & 2 & 3 & \cdots & n \\ \tau_1 & \tau_2 & \tau_3 & \cdots & \tau_n \end{pmatrix}.$$

Then $\tau(\pi(A)) = \widehat{A}$.

The implication **b)** $\Rightarrow$ **c)** is evident since every right nilpotent evolution algebra is a nil algebra.

The implication **a)** $\Rightarrow$ **b, c)** is also true, because the table of the multiplication of the evolution algebra is defined by an upper triangular matrix A will be right nilpotent and nil.

The implication **c)** $\Rightarrow$ **a)** follows from Theorem 3.9. $\qquad\square$

3.3.4   *Maximal nilindex of a nilpotent evolution algebra*

If an algebra is nilpotent then finding the maximal nilindex (nilpotent index) is very important when the algebra associated with dynamics of a population. Following [30] we give the maximal nilindex of a nilpotent evolution algebra.

Lemma 3.6. *Let E be a finite-dimensional evolution algebra and E^j, $j \geq 1$, the evolution subalgebras of E. Then*

$$E^{2^k+i} = E^{2^{k+1}}, \quad i = 1, \ldots, 2^k, \quad k = 0, 1, \ldots. \tag{3.50}$$

Proof. We shall use mathematical induction. We have $E^1 = E$, $E^2 = EE$, and for $k = 1$,

$$E^3 = EE^2 = E^2 E^2, \qquad E^4 = EE^3 + E^2 E^2 = E^2 E^2 = E^3.$$

Assume for k the equalities (3.50) are true. We shall prove for $k+1$. Using $E^{i+j} \subset E^i$, $E^{i+j} E^i = E^{i+j} E^{i+j}$ and assumptions of the induction we get

$$E^{2^{k+1}+i} = \sum_{j=1}^{2^k+\lfloor i/2 \rfloor} E^j E^{2^{k+1}+i-j} = \sum_{j=1}^{2^k+\lfloor i/2 \rfloor} E^j E^{2^k+i}$$

$$= \sum_{j=1}^{2^k+\lfloor i/2 \rfloor} E^j E^{2^{k+1}} = EE^{2^{k+1}} = E^{2^{k+1}} E^{2^{k+1}}.$$

$\square$

Lemma 3.7. *If for an evolution algebra E there exists $s \in \mathbb{N}$ such that $E^{2^s+1} = E^{2^{s+1}+1}$, then $E^k = E^{2^s+1}$ for any $k = 2^s+1, 2^s+2, \ldots, 2^{s+2}+1$.*

Proof. We have

$$E^{2^s+1} \supseteq E^{2^s+2} \supseteq \cdots \supseteq E^{2^{s+1}+1}.$$

Hence by condition of the lemma we get $E^k = E^{2^s+1}$ for any $k = 2^s + 1, 2^s + 2, \ldots, 2^{s+1} + 1$. It remains to prove the equality for $k = 2^{s+1} + i + 1$, $i = 1, \ldots, 2^{s+1}$. We have

$$E^{2^{s+1}+1} = \sum_{j=1}^{2^s} E^j E^{2^{s+1}-j+1} = EE^{2^s+1} = E^{2^s+1}.$$

For $i = 1$ using the above obtained equalities we get

$$E^{2^{s+1}+2} = \sum_{j=1}^{2^s+1} E^j E^{2^{s+1}-j+2} = EE^{2^s+1} = E^{2^s+1}.$$

Now assume the assertion is true for i and we shall show it for $i+1$.

$$E^{2^{s+1}+i+2} = \sum_{j=1}^{2^s+1+\lfloor i/2 \rfloor} E^j E^{2^{s+1}+i-j+2}.$$

Since $2^s + 1 < 2^{s+1} + i - j + 2 \leq 2^{s+1} + i + 1$ for any $j = 1, 2, \ldots 2^s + 1 + \lfloor i/2 \rfloor$, using the assumption of the induction, we get

$$E^{2^{s+1}+i+2} = \sum_{j=1}^{2^s+1+\lfloor i/2 \rfloor} E^j E^{2^s+1} = E E^{2^s+1} = E^{2^s+1}.$$

$\square$

From this lemma we get the following

Corollary 3.5. *If for an evolution algebra E there exists $s \in \mathbb{N}$ such that $E^{2^s+1} = E^{2^{s+1}+1}$, then $E^k = E^{2^s+1}$ for any $k \geq 2^s + 1$.*

Proof. If the condition of Lemma 3.7 is satisfied for s, then it is satisfied for $s + 1$. So, iterating the lemma we get $E^k = E^{2^s+1}$ for any $k \geq 2^s + 1$.

$\square$

From this corollary it follows that an evolution algebra E satisfying the condition of Lemma 3.6 is not nilpotent.

Theorem 3.11. *An n-dimensional nilpotent evolution algebra E has maximal nilpotent index, $2^{n-1} + 1$, if and only if*

$$a_{12}a_{23} \ldots a_{n-1,n} \neq 0.$$

Proof. *Necessity.* Assume $a_{12}a_{23} \ldots a_{n-1,n} = 0$ then $\dim E^2 \leq n - 2$. Since E is nilpotent, by Lemma 3.6, for any k we have $E^{2^k+1} \supsetneq E^{2^{k+1}+1}$. Consequently, $\dim E^{2^k+1} \leq n - 2 - k$. Hence $E^{2^{n-2}+1} = 0$, i.e., E has not maximal nilpotent index.

Sufficiency was proved in [28].

$\square$

Remark 3.6. Note that the algebraic notions like nilpotency, right nilpotency and solvability might be interpreted in a biological way as a various types of vanishing (deaths) populations. The maximal nilpotent index gives the maximal time of living duration of the population.

3.3.5   *Classification of complex 2-dimensional evolution algebras*

In this section we give the classification of 2-dimensional complex evolution algebras.

Let E and E' be evolution algebras and $\{e_i\}$ a natural basis of E. A linear map $\varphi \colon E \to E'$ is called an *homomorphism* of evolution algebras if it is an algebraic map and if the set $\{\varphi(e_i)\}$ can be complemented to a natural basis of E'. Moreover, if φ is bijective, then it is called an *isomorphism*.

Let E be a 2-dimensional complex evolution algebra and $\{e_1, e_2\}$ be a basis of the algebra E.

It is evident that if $\dim E^2 = 0$ then E is an abelian algebra, i.e. an algebra with all products equal to zero.

Theorem 3.12. *Any 2-dimensional complex evolution algebra E is isomorphic to one of the following pairwise non isomorphic algebras:*

(1) $\dim E^2 = 1$

- $E_1: \quad e_1 e_1 = e_1, \ e_2 e_2 = 0,$
- $E_2: \quad e_1 e_1 = e_1, \quad e_2 e_2 = e_1,$
- $E_3: \quad e_1 e_1 = e_1 + e_2, \quad e_2 e_2 = -e_1 - e_2,$
- $E_4: \quad e_1 e_1 = e_2, \ e_2 e_2 = 0.$

(2) $\dim E^2 = 2$

- $E_5(a_2, a_3): \quad e_1 e_1 = e_1 + a_2 e_2, \quad e_2 e_2 = a_3 e_1 + e_2, \quad 1 - a_2 a_3 \neq 0.$
  Moreover, $E_5(a_2, a_3) \cong E_5'(a_3, a_2),$
- $E_6(a_4): \quad e_1 e_1 = e_2, \quad e_2 e_2 = e_1 + a_4 e_2.$ *Moreover, for* $a_4 \neq 0,$
  $E_6(a_4) \cong E_6(a_4') \ \Leftrightarrow \ \frac{a_4'}{a_4} = \cos \frac{2\pi k}{3} + i \sin \frac{2\pi k}{3}$ *for some* $k = 0, 1, 2.$

Proof. For an evolution algebra E we have

$$e_1 e_1 = a_1 e_1 + a_2 e_2, \quad e_2 e_2 = a_3 e_1 + a_4 e_2, \quad e_1 e_2 = e_2 e_1 = 0.$$

(1) Since $\dim E^2 = 1$, then

$$e_1 e_1 = c_1(a_1 e_1 + a_2 e_2), \quad e_2 e_2 = c_2(a_1 e_1 + a_2 e_2), \quad e_1 e_2 = e_2 e_1 = 0.$$

Evidently $(c_1, c_2) \neq (0, 0)$, because otherwise the algebra will be abelian.

Since e_1 and e_2 are symmetric we can suppose that $c_1 \neq 0$, then by simple change of basis we can suppose that $c_1 = 1$.

Case 1. $a_1 \neq 0$. Then we take an appropriate change of basis in the following form:

$$e_1' = a_1 e_1 + a_2 e_2, \quad e_2' = A e_1 + B e_2,$$

where $a_1 B - a_2 A \neq 0$.

Consider the product

$$0 = e_1' e_2' = (a_1 e_1 + a_2 e_2)(A e_1 + B e_2) = a_1 A(a_1 e_1 + a_2 e_2)$$
$$+ a_2 B c_2(a_1 e_1 + a_2 e_2) = (a_1 A + a_2 B c_2)(a_1 e_1 + a_2 e_2).$$

Therefore, $a_1 A + a_2 Bc = 0$, i.e. $A = -\frac{a_2 B c_2}{a_1}$ and $a_1 B - a_2 A = a_1 B + \frac{a_2^2 B c_2}{a_1} \neq 0$.

It means that in the case when $a_1^2 + a_2^2 c_2 \neq 0$ we can take the above change.

Consider the products

$$e_1' e_1' = (a_1 e_1 + a_2 e_2)(a_1 e_1 + a_2 e_2) = a_1^2(a_1 e_1 + a_2 e_2) + a_2^2 c_2(a_1 e_1 + a_2 e_2)$$
$$= (a_1^2 + a_2^2 c_2)(a_1 e_1 + a_2 e_2) = (a_1^2 + a_2^2 c_2)e_1',$$
$$e_2' e_2' = (Ae_1 + Be_2)(Ae_1 + Be_2) = A^2(a_1 e_1 + a_2 e_2) + B^2 c_2(a_1 e_1 + a_2 e_2)$$
$$= (A^2 + B^2 c_2)(a_1 e_1 + a_2 e_2) = \left(\frac{a_2^2 B^2 c_2^2}{a_1^2} + B^2 c_2 \right) e_1'$$
$$= \frac{B^2 c_2(a_1^2 + a_2^2 c_2)}{a_1^2} e_1'.$$

Case 1.1. $c_2 = 0$. Then $e_1 e_1 = a_1^2 e_1$, $e_2 e_2 = e_1 e_2 = e_2 e_1 = 0$. Taking $e_1' = \frac{e_1}{a_1^2}$ we obtain the algebra E_1.

Case 1.2. $c_2 \neq 0$. Then taking $B = \sqrt{\frac{a_1^2}{c_2}}$ we obtain

$$e_1 e_1 = (a_1^2 + a_2^2 c_2)e_1, \quad e_2 e_2 = (a_1^2 + a_2^2 c_2)e_1.$$

If $a_1^2 + a_2^2 c_2 \neq 0$, the following change of basis

$$e_1' = \frac{e_1}{a_1^2 + a_2^2 c_2}, \quad e_2' = \frac{e_2}{a_1^2 + a_2^2 c_2}$$

brings us to the algebra with multiplication:

$$e_1 e_1 = e_1, \quad e_2 e_2 = e_1.$$

If $a_1^2 + a_2^2 c_2 = 0$, then $c_2 = -\frac{a_1^2}{a_2^2}$ and we have $e_1 e_1 = a_1 e_1 + a_2 e_2$, $e_2 e_2 = -\frac{a_1^3}{a_2^2} e_1 - \frac{a_1^2}{a_2} e_2$.

The change of basis $e_1' = \frac{e_1}{a_1}$, $e_2' = \frac{a_2}{a_1^2} e_2$ gives the algebra E_3.

Case 2. $a_1 = 0$. Then we have the products $e_1 e_1 = a_2 e_2$, $e_2 e_2 = c_2 a_2 e_2$, where $a_2 \neq 0$.

If $c_2 = 0$, then by the change $e_1' = \frac{e_1}{\sqrt{a_2}}$ we get again the algebra E_4.

If $c_2 \neq 0$, then by $e_1' = \frac{e_1}{\sqrt{c_2 a_2^2}}$, $e_2' = \frac{e_2}{c_2 a_2}$ we get the algebra $e_1 e_1 = e_2$, $e_2 e_2 = e_2$ which is isomorphic to the algebra E_2.

(2) Now we consider algebras with $\dim E^2 = 2$. Let us write the table of multiplication:

$$e_1 e_1 = a_1 e_1 + a_2 e_2, \quad e_2 e_2 = a_3 e_1 + a_4 e_2,$$

where $a_1 a_4 - a_2 a_3 \neq 0$.

Case 1. $a_1 \neq 0$ and $a_4 \neq 0$. Then we may suppose $a_1 = a_4 = 1$. Therefore we have the two-parametric family $E_5(a_2, a_3)$:

$e_1 e_1 = e_1 + a_2 e_2, \quad e_2 e_2 = a_3 e_1 + e_2, \quad 1 - a_2 a_3 \neq 0.$

Let us take the general change of basis of the form

$$e_1' = A_1 e_1 + A_2 e_2, \quad e_2' = B_1 e_1 + B_2 e_2,$$

where $A_1 B_2 - A_2 B_1 \neq 0$.

Consider the product

$0 = e_1' e_2' = (A_1 e_1 + A_2 e_2)(B_1 e_1 + B_2 e_2) = A_1 B_1(e_1 + a_2 e_2) + A_2 B_2(a_3 e_1 + e_2) = (A_1 B_1 + A_2 B_2 a_3)e_1 + (A_1 B_1 a_2 + A_2 B_2)e_2.$

Since in this new basis the algebra should be also an evolution algebra we have

$$A_1 B_1 + A_2 B_2 a_3 = 0, \quad A_1 B_1 a_2 + A_2 B_2 = 0.$$

From which we have $A_2 B_2(1 - a_2 a_3) = 0, \quad A_1 B_1(1 - a_2 a_3) = 0$. Since $1 - a_2 a_3 \neq 0$, then we have $A_1 B_1 = A_2 B_2 = 0$.

Case 1.1. $A_2 = 0$. Then $B_1 = 0$.

Consider the products

$e_1' e_1' = A_1^2(e_1 + a_2 e_2) = e_1' + a_2' e_2' = A_1 e_1 + a_2' B_2 e_2 \Rightarrow A_1^2 = A_1, A_1^2 a_2 = a_2' B_2 \Rightarrow A_1 = 1,$
$e_2' e_2' = B_2^2(a_3 e_1 + e_2) = a_3' e_1' + e_2' = a_3' A_1 e_1 + B_2 e_2 \Rightarrow B_2^2 a_3 = a_3' A_1, B_2^2 = B_2 \Rightarrow B_2 = 1.$

Case 1.2. $A_1 = 0$. Then $B_2 = 0$ and from the family of algebras $E_5(a_2, a_3)$ we get the family $E_5(a_3, a_2)$.

Case 2. $a_1 = 0$ or $a_4 = 0$. Since e_1 and e_2 are symmetric then without loss of generality we can suppose that $a_1 = 0$.

$$e_1 e_1 = a_2 e_2, \quad e_2 e_2 = a_3 e_1 + a_4 e_2,$$

where $a_2 a_3 \neq 0$.

Taking the change of basis $e_1' = \sqrt[3]{\dfrac{1}{a_2^2 a_3}}\, e_1, \quad e_2' = \sqrt[3]{\dfrac{1}{a_2 a_3^2}}\, e_2$ we obtain the one-parametric family of algebras $E_6(a_4)$:

$$e_1 e_1 = e_2, \quad e_2 e_2 = e_1 + a_4 e_2.$$

Let us take the general change of basis

$$e_1' = A_1 e_1 + A_2 e_2, \quad e_2' = B_1 e_1 + B_2 e_2,$$

where $A_1 B_2 - A_2 B_1 \neq 0$.

Consider the product

$$0 = e_1'e_2' = (A_1e_1 + A_2e_2)(B_1e_1 + B_2e_2) = A_1B_1e_2 + A_2B_2(e_1 + a_4e_2).$$

Therefore $A_1B_1 + A_2B_2a_4 = 0$, $A_2B_2 = 0 \Rightarrow A_1B_1 = 0$, $A_2B_2 = 0$.

Without loss of generality we can assume that $A_2 = 0$. Then $B_1 = 0$.

Consider the product

$$e_1'e_1' = A_1^2e_2 = e_2' = B_2e_2 \Rightarrow A_1^2 = B_2,$$
$$e_2'e_2' = B_2^2(e_1 + a_4e_2) = e_1' + a_4'e_2' = A_1e_1 + a_4'B_2e_2 \Rightarrow B_2^2 = A_1, \ B_2^2a_4 = B_2a_4'.$$

From these equalities we have $B_2^3 = 1$, $B_2a_4 = a_4'$.

If $\frac{a_4'}{a_4} = \cos\frac{2\pi k}{3} + i\sin\frac{2\pi k}{3}$ for some $k = 0, 1, 2$, then putting $B_2 = \cos\frac{2\pi k}{3} + i\sin\frac{2\pi k}{3}$ we obtain the isomorphism between the algebras $E_6(a_4)$ and $E_6(a_4')$.

The obtained algebras are pairwise non-isomorphic this may be checked by comparison of the algebraic properties listed in the following table.

	$\dim E^2$	Right Nilpotency	$\dim(\text{Annihilator})$	Nil Elements
E_1	1	No	1	Yes
E_2	1	No	0	Yes
E_3	1	No	0	No
E_4	1	Yes	1	Yes
E_5	2	No	0	No
E_6	2	No	0	Yes

$\square$

3.3.6 *Classification of real 2-dimensional evolution algebras*

The next theorem gives the classification of the two-dimensional real evolution algebras.

Theorem 3.13. [188] *Any two-dimensional real evolution algebra E is isomorphic to one of the following pairwise non-isomorphic algebras:*

(i) $\dim E^2 = 1$:

$E_1 : e_1e_1 = e_1, \quad e_2e_2 = 0;$

$E_2 : e_1e_1 = e_1, \quad e_2e_2 = e_1;$

$E_3 : e_1e_1 = e_1 + e_2, \quad e_2e_2 = -e_1 - e_2;$

$E_4 : e_1e_1 = e_2, \quad e_2e_2 = 0;$

$E_5 : e_1e_1 = e_2, \quad e_2e_2 = -e_2;$

(ii) $\dim E^2 = 2$:

$E_6 : e_1 e_1 = e_1 + a_2 e_2, \quad e_2 e_2 = a_3 e_1 + e_2, \quad 1 - a_2 a_3 \neq 0, a_2, a_3 \in \mathbb{R}$, *where* $E_6(a_2, a_3) \cong E_6'(a_3, a_2)$;

$E_7 : e_1 e_1 = e_2, \quad e_2 e_2 = e_1 + a_4 e_2, \quad$ *where* $a_4 \in \mathbb{R}$.

Remark 3.7.

- The classification of the two-dimensional complex evolution algebras consists of a complex variant of the algebras E_i, $i = 1, 2, 3, 4, 6, 7$ (see Theorem 3.12). But E_5 is present only in the real case.
- For the classification of the three-dimensional complex evolution algebras see [26].
- For the classification of the finite-dimensional complex evolution algebras with maximal nilpotent index see [30].

Bibliographical notes. This chapter mainly deals with genetic algebras, the study of these algebras was started by Etherington (1939). Thus they have 80 years history. In this chapter I have very briefly presented these algebras based on [1], [2], [19]–[38], [48]–[54], [79]–[82], [92], [93], [100]–[104], [137], [155]–[157], [170]–[172], [188]–[211], [223], [241], [269] and many internet sources. For further reading see [8], [24], [25], [27], [39], [44], [45], [50], [51], [65], [67], [91], [107], [112], [130], [147], [169], [175], [191], [192], [196], [209], [236], [254]–[258].

Chapter 4

Algebras of bisexual population

The genetic algebras discussed in the previous chapter concerned with inheritance which is symmetric with respect to sex, in that properties are determined by genes located at autosomal loci, and it is assumed that the segregation pattern is the same in males and females. In this chapter we consider asymmetric situations, and give algebras of bisexual populations. These algebras are complicated by their higher dimensions and by fact that the passage from the gametic to the zygotic algebra no longer quite corresponds to the process of duplication, as it does in the symmetric case. We give some basic properties of the algebras of bisexual populations.

4.1 Evolution algebra of bisexual population

The genetic algebras related with sex linkage were studied by many authors. In [49] Etherington considered the gametic and zygotic algebras of a single sex linked diallelic locus, and its properties were discussed by Gonshor [79]. In [80] Gonshor studied sex linkage in the case of multiple alleles, choosing a canonical basis and ideal structure of the algebra. In [101] Holgate for a single multiallelic locus, linked to the sex determining locus, obtained canonical forms for the gametic and zygotic algebras.

In this section following [138] we give properties of EABP (see Definition 1.1).

4.1.1 *Basic properties of the EABP*

The following theorem gives basic properties of the EABP.

Theorem 4.1. *We have*

77

1) Algebra $\mathcal{B}$ is not associative, in general.

2) Algebra $\mathcal{B}$ is commutative, flexible.

3) $\mathcal{B}$ is not power-associative, in general.

Proof.   1) Take $e_i^{(f)}$, $e_j^{(m)}$ such that $P_{ij,s}^{(f)} \neq 0$ for some s and take $e_k^{(m)}$ such that $P_{sk,r}^{(f)} \neq 0$ for some r then

$$(e_i^{(f)} e_j^{(m)}) e_k^{(m)} = \frac{1}{2} \sum_{q=1}^{n} P_{ij,q}^{(f)} e_q^{(f)} e_k^{(m)}$$

$$= \frac{1}{4} \left(P_{ij,s}^{(f)} e_s^{(f)} e_k^{(m)} + \sum_{\substack{q=1 \\ q \neq s}}^{n} P_{ij,q}^{(f)} e_q^{(f)} e_k^{(m)} \right)$$

$$= \frac{1}{4} P_{ij,s}^{(f)} P_{sk,r}^{(f)} e_r^{(f)} + \text{non-negative terms} \neq 0 .$$

But

$$e_i^{(f)} (e_j^{(m)} e_k^{(m)}) = 0, \quad \text{i.e.} \quad (e_i^{(f)} e_j^{(m)}) e_k^{(m)} \neq e_i^{(f)} (e_j^{(m)} e_k^{(m)}) .$$

2) Commutativity of $\mathcal{B}$ follows from formula (1.8). An algebra is called *flexible* if it satisfies $z(tz) = (zt)z$ for any z, t. One can see that a commutative algebra is flexible.

3) To show that $\mathcal{B}$ is not a power-associative, in general, we shall construct an example of z such that $(zz)(zz) \neq ((zz)z)z$. Consider $n = 1$, $\nu = 2$. In this case

$$P_{11,1}^{(f)} = P_{12,1}^{(f)} = 1, \quad P_{11,1}^{(m)} + P_{11,2}^{(m)} = 1, \quad P_{12,1}^{(m)} + P_{12,2}^{(m)} = 1 .$$

Denote $a = P_{11,1}^{(m)}$, $b = P_{12,1}^{(m)}$. Take $z = e_1^{(f)} + e_1^{(m)}$. Then we have

$$z^2 = e_1^{(f)} + a e_1^{(m)} + (1-a) e_2^{(m)} .$$

$$z^2 z^2 = e_1^{(f)} + \left(a^2 + (1-a)b \right) e_1^{(m)} + (1-a)(a-b+1) e_2^{(m)}. \tag{4.1}$$

$$z^2 z = e_1^{(f)} + \frac{1}{2} \left(a^2 + (1-a)b + a \right) e_1^{(m)} + \frac{1}{2}(1-a)(a-b+2) e_2^{(m)} .$$

$$(z^2 z)z = e_1^{(f)} + \frac{1}{4} \left(a(a-b)^2 + (a+b)(a-b+2) \right) e_1^{(m)} \tag{4.2}$$

$$+ \frac{1}{4}(1-a) \left(3 + (a-b+1)^2 \right) e_2^{(m)} .$$

Assume $a = P_{11,1}^{(m)} \neq 1$ and $a \neq b = P_{12,1}^{(m)}$. Then from (4.1) and (4.2) we get $(zz)(zz) \neq ((zz)z)z$. $\square$

4.1.2 $\mathcal{B}$ *is not a baric algebra*

As mentioned in the previous chapter (see also [155]) an algebra of free population has a character $\sigma(x) = \sum_i x_i$, therefore this algebra is baric. But the following theorem says that the EABP, i.e. $\mathcal{B}$ is not baric.

Theorem 4.2. *The EABP, $\mathcal{B}$, has no a nonzero character.*

Proof. Assume $\sigma(z) = \sum_{i=1}^{n} a_i x_i + \sum_{j=1}^{\nu} b_j y_j$, $z = (x, y) \in \mathcal{B}$ is a character. We shall prove that $\sigma(z) \equiv 0$. For $z = (x, y)$, $t = (u, v)$ we have

$$\sigma(zt) = \frac{1}{2} \sum_{i=1}^{n} \sum_{j=1}^{\nu} \left(\sum_{p=1}^{n} a_p P_{ij,p}^{(f)} + \sum_{q=1}^{\nu} b_q P_{ij,q}^{(m)} \right) (x_i v_j + u_i y_j);$$

$$\sigma(z)\sigma(t) = \sum_{i=1}^{n} \sum_{j=1}^{n} a_i a_j x_i u_j + \sum_{i=1}^{n} \sum_{j=1}^{\nu} a_i b_j (x_i v_j + u_i y_j) + \sum_{i=1}^{\nu} \sum_{j=1}^{\nu} b_i b_j y_j v_j.$$

From $\sigma(zt) = \sigma(z)\sigma(t)$ we get

$$\begin{cases} a_i a_j = 0 & \text{for any} \quad i, j = 1, \ldots, n; \\ b_i b_j = 0 & \text{for any} \quad i, j = 1, \ldots, \nu. \end{cases}$$

This system of equations has only the solution $a_1 = \cdots = a_n = b_1 = \cdots = b_\nu = 0$. $\qquad\square$

Remark 4.1. In [155] the baric EA of a free population is classified as algebra induced by a linear operator; unit algebra; constant algebra; Bernstein (stationary) algebra; genetic algebra; train algebra, etc. But since the EABP is not baric there are not similar algebras for bisexual population.

4.1.3 $\mathcal{B}$ *is a dibaric algebra*

Theorem 4.3. *The algebra $\mathcal{B}$ is dibaric.*

Proof. Consider mapping $\varphi \colon \mathcal{B} \to \mathfrak{A}$ defined by

$$\varphi(e_i^{(f)}) = w, \quad i = 1, \ldots, n; \quad \varphi(e_k^{(m)}) = m, \quad i = 1, \ldots, \nu. \tag{4.3}$$

For $z, t \in \mathbb{R}^{n+\nu}$,

$$z = (x, y) = \sum_{i=1}^{n} x_i e_i^{(f)} + \sum_{j=1}^{\nu} y_j e_j^{(m)}, \quad t = (u, v) = \sum_{i=1}^{n} u_i e_i^{(f)} + \sum_{j=1}^{\nu} v_j e_j^{(m)},$$

using linearity of φ, (4.3) and (1.5) we get

$$\varphi(zt) = \sum_{i=1}^{n} \sum_{j=1}^{\nu} (x_i v_j + u_i y_j) \left(\frac{1}{2}(w + m) \right).$$

We also have

$$\varphi(z)\varphi(t) = \left(\sum_{i=1}^{n} x_i w + \sum_{j=1}^{\nu} y_j m\right)\left(\sum_{i=1}^{n} u_i w + \sum_{j=1}^{\nu} v_j m\right)$$

$$= \sum_{i=1}^{n}\sum_{j=1}^{\nu} (x_i v_j + u_i y_j)\left(\frac{1}{2}(w+m)\right) = \varphi(zt),$$

i.e. φ is a homomorphism. For arbitrary $u = \alpha w + \beta m \in \mathfrak{A}$ it is easy to see that $\varphi(z) = u$ if $z = \sum_{i=1}^{n} x_i e_i^{(f)} + \sum_{j=1}^{\nu} y_j e_j^{(m)} \in \mathcal{B}$ with $\sum_{i=1}^{n} x_i = \alpha$ and $\sum_{j=1}^{\nu} y_j = \beta$. Therefore φ is onto. $\qquad\square$

It is known that if an algebra A is dibaric, then A^2 is baric. Therefore, as a corollary of this Proposition and Theorem 4.3 we have

Corollary 4.1. *The subalgebra $\mathcal{B}^2$ is a baric algebra.*

Since $\mathcal{B}^2$ is a baric algebra all types of particular algebras mentioned in Remark 4.1 can be defined and studied for the algebra $\mathcal{B}^2$.

Let $\mathcal{B}$ be an algebra of bisexual population. Denote by $\mathcal{B}^*$ the dual space of $\mathcal{B}$, i.e. consisting of all linear functionals on $\mathcal{B}$, together with the vector space structure of pointwise addition and scalar multiplication by constants.

For $z = (x, y) = \sum_{i=1}^{n} x_i e_i^{(f)} + \sum_{j=1}^{\nu} y_j e_j^{(m)} \in \mathcal{B}$ denote

$$\tilde{\mathcal{B}}^* = \left\{(f,g) \in \mathcal{B}^* \times \mathcal{B}^* : f(z^2) = g(z^2) = f(z)g(z)\right\},$$

$$\tilde{\mathcal{B}}_{01}^* = \left\{(0,g) \in \mathcal{B}^* \times \mathcal{B}^* : g(z) = \sum_{i=1}^{n} \gamma_i x_i + \sum_{j=1}^{\nu} \delta_j y_j,\ \text{with}\right.$$

$$\left.\sum_{k=1}^{n} P_{ij,k}^{(f)}\gamma_k + \sum_{l=1}^{\nu} P_{ij,l}^{(m)}\delta_l = 0, \forall i,j\right\},$$

$$\tilde{\mathcal{B}}_{10}^* = \left\{(f,0) \in \mathcal{B}^* \times \mathcal{B}^* : f(z) = \sum_{i=1}^{n} \alpha_i x_i + \sum_{j=1}^{\nu} \beta_j y_j,\ \text{with}\right.$$

$$\left.\sum_{k=1}^{n} P_{ij,k}^{(f)}\alpha_k + \sum_{l=1}^{\nu} P_{ij,l}^{(m)}\beta_l = 0, \forall i,j\right\},$$

$$\tilde{\mathcal{B}}_{12}^* = \left\{(f,g) \in \mathcal{B}^* \times \mathcal{B}^* : f(z) = \sum_{i=1}^{n} \alpha_i x_i,\ g(z) = \sum_{j=1}^{\nu} \delta_j y_j,\ \text{with} \right.$$

$$\left. \sum_{k=1}^{n} P_{ij,k}^{(f)} \alpha_k = \sum_{l=1}^{\nu} P_{ij,l}^{(m)} \delta_l = \alpha_i \delta_j,\ \forall i,j \right\},$$

$$\tilde{\mathcal{B}}_{21}^* = \left\{(f,g) \in \mathcal{B}^* \times \mathcal{B}^* : f(z) = \sum_{i=1}^{\nu} \beta_i y_i,\ g(z) = \sum_{j=1}^{n} \gamma_j x_j,\ \text{with} \right.$$

$$\left. \sum_{k=1}^{n} P_{ij,k}^{(f)} \gamma_k = \sum_{l=1}^{\nu} P_{ij,l}^{(m)} \beta_l = \gamma_i \beta_j,\ \forall i,j \right\}.$$

The following theorem describes the set $\tilde{\mathcal{B}}^*$, [137]:

Theorem 4.4. *The set $\tilde{\mathcal{B}}^*$ has the form*

$$\tilde{\mathcal{B}}^* = \big\{(0,0)\big\} \cup \tilde{\mathcal{B}}_{01}^* \cup \tilde{\mathcal{B}}_{10}^* \cup \tilde{\mathcal{B}}_{12}^* \cup \tilde{\mathcal{B}}_{21}^*.$$

Proof. Let

$$f(z) = \sum_{i=1}^{n} \alpha_i x_i + \sum_{j=1}^{\nu} \beta_j y_j,\ g(z) = \sum_{i=1}^{n} \gamma_i x_i + \sum_{j=1}^{\nu} \delta_j y_j.$$

From $f(z^2) = g(z^2)$ we obtain

$$\sum_{k=1}^{n} P_{ij,k}^{(f)} (\alpha_k - \gamma_k) + \sum_{l=1}^{\nu} P_{ij,l}^{(m)} (\beta_l - \delta_l) = 0,\ \forall i,j.$$

From $f(z^2) = f(z)g(z)$ we get

$$\alpha_i \gamma_j = 0,\ \forall i,j = 1,\ldots,n;\quad \beta_i \delta_j = 0,\ \forall i,j = 1,\ldots,\nu; \qquad (4.4)$$

$$\sum_{k=1}^{n} P_{ij,k}^{(f)} \alpha_k + \sum_{l=1}^{\nu} P_{ij,l}^{(m)} \beta_l = \alpha_i \delta_j + \gamma_i \beta_j,\ \text{for all } i = 1,\ldots,n;\ j = 1,\ldots,\nu.$$

By (4.4) one can see that if there exists i_0 such that $\alpha_{i_0} \neq 0$ then $\gamma_j = 0$ for any $j = 1,\ldots,n$. Similarly, if there exists i_1 such that $\delta_{i_1} \neq 0$ then $\beta_j = 0$ for all $j = 1,\ldots,\nu$. Therefore if f depends only on x (resp. y) then g depends only on y (resp. x). Moreover if f (resp. g) depends on both x and y then $g = 0$ (resp. $f = 0$). $\qquad \square$

For $z = (x,y) \in \mathcal{B}$ denote $X(z) = \sum_{i=1}^{n} x_i$ and $Y(z) = \sum_{i=1}^{\nu} y_j$. One can see that $\big(X(z), Y(z)\big) \in \tilde{\mathcal{B}}_{12}^*$, hence $\big(X(z), Y(z)\big)$ is a bq-homomorphism.

In this section we consider only X-invariant linear forms, the Y-linear forms can be obtained from X-linear forms by replacing n with ν and x with y.

Denote by $J = J_X$ the set of all X-invariant linear forms of $\mathcal{B}$. The set J is a subspace of the dual space $\mathcal{B}^*$.

Denote $\mathbb{P}_i^{(m)} = \left(P_{ij,k}^{(m)} \right)_{j,k=1,\dots,\nu}$, $i = 1,\dots,n$.

Proposition 4.1.

(1) $\det \left(\mathbb{P}_i^{(m)} - I \right) = 0$ *for any* $i = 1,\dots,n$, *where* I *is the* $(\nu \times \nu)$-*identity matrix.*

(2) For any $d \in \{1,\dots,\nu\}$ *there is* $\mathbb{P}_i^{(m)}$ *such that* $\dim J = d$.

Proof. (1) From (1.5) it follows that sum of all columns of $\mathbb{P}_i^{(m)} - I$ is equal to zero for any i. Consequently, the columns are linearly dependent.

(2) Take $F(z) = \sum_{i=1}^{n} \beta_i x_i + \sum_{j=1}^{\nu} \alpha_j y_j$. Then we get $\beta_1 = \cdots = \beta_n = 0$ and

$$\sum_{l=1}^{\nu} P_{ij,l}^{(m)} \alpha_l = \alpha_j, \quad \text{for all} \ \ i = 1,\dots,n; \ j = 1,\dots,\nu. \tag{4.5}$$

Using (1) we get that the system (4.5) has a set $W_i \neq \{0\}$ of solutions for any $i = 1,\dots,n$. Moreover one can see that $\alpha_1 = \cdots = \alpha_\nu = $ constant is a solution of the system for any i. Therefore $W = \cap_{i=1}^{n} W_i \neq 0$. Consequently,

$$J = \left\{ f(z) = \sum_{j=1}^{\nu} \alpha_j y_j : \alpha = (\alpha_1,\dots,\alpha_\nu) \in W \right\}, \quad \dim J = \dim W.$$

For a given $d = 1,\dots,\nu$ one can choose $\mathbb{P}_i^{(m)}$ such that $\dim J = d$. $\qquad\square$

From this proposition we get

Corollary 4.2. *Any X-invariant (resp. Y-invariant) linear form $F(z) = F(x,y)$ depends only on y (resp. on x).*

Let us consider an example.

Example 4.1. Consider the case $n = 1$, $\nu = 2$. In this case $P_{11,1}^{(f)} = P_{12,1}^{(f)} = 1$. Denote $P_{11,1}^{(m)} = a$, $P_{11,2}^{(m)} = 1 - a$, $P_{12,1}^{(m)} = b$, $P_{12,2}^{(m)} = 1 - b$. In this case the system (4.5) gets the following form

$$(a - 1)(\alpha_1 - \alpha_2) = 0, \quad b(\alpha_1 - \alpha_2) = 0.$$

If $a = 1$, $b = 0$ then any (α_1, α_2), $\alpha_1, \alpha_2 \in \mathbb{R}$, is a solution to the system; if $1 - a + b \neq 0$ then (α_1, α_1), $\alpha_1 \in \mathbb{R}$, is the only solution. Therefore

$$\dim J = \begin{cases} 2, & \text{if } a = 1, b = 0; \\ 1, & \text{otherwise.} \end{cases}$$

The following example shows that the invariant linear forms on $S^{n-1} \times S^{\nu-1}$ may depend on both x and y.

Example 4.2. Consider the case $n = 3$, $\nu = 2$. Consider the evolution operator $V \colon S^3 \times S^2 \to S^3 \times S^2$, $z = (x_1, x_2, x_3, y_1, y_2) \mapsto z' = (x_1', x_2', x_3', y_1', y_2')$ given by:

$$\begin{aligned} x_1' &= x_1 y_1 + \frac{1}{2} x_3 y_1 \\ x_2' &= x_2 y_2 + \frac{1}{2} x_3 y_2 \\ x_3' &= x_1 y_2 + x_2 y_1 + \frac{1}{2} x_3 y_1 + \frac{1}{2} x_3 y_2 \\ y_1' &= x_1 y_1 + x_1 y_2 + \frac{1}{2} x_3 y_1 + \frac{1}{2} x_3 y_2 \\ y_2' &= x_2 y_1 + x_2 y_2 + \frac{1}{2} x_3 y_1 + \frac{1}{2} x_3 y_2 \,. \end{aligned} \tag{4.6}$$

For the algebra $\mathcal{B}$ corresponding to the operator (4.6) we have the following invariant linear forms:

$$f_1(z) = \frac{1}{3}(2x_1 + x_3 + y_1), \quad f_2(z) = \frac{1}{3}(2x_2 + x_3 + y_2).$$

Lemma 4.1.

(1) $z = (x, y) \in \mathrm{ann}\mathcal{B}$ *if and only if*

$$\sum_{i=1}^{n} P_{ij,k}^{(f)} x_i = 0, \ k = 1, \ldots, n; \quad \sum_{i=1}^{n} P_{ij,l}^{(m)} x_i = 0, \ l = 1, \ldots, \nu, \ \text{ for all } \ j;$$

$$\sum_{j=1}^{\nu} P_{ij,k}^{(f)} y_j = 0, \ k = 1, \ldots, n; \quad \sum_{j=1}^{\nu} P_{ij,l}^{(m)} y_j = 0, \ l = 1, \ldots, \nu \ \text{ for all } \ i.$$

$$\tag{4.7}$$

(2) If $z \in \mathrm{ann}\mathcal{B}$ *then* $X(z) = Y(z) = 0$.

Proof. (1) Let $z = (x, y), t = (u, v) \in \mathbb{R}^{n+\nu}$ with $z \in \text{ann}\mathcal{B}$. Using (1.8) from $zt = 0$ we get

$$
\begin{aligned}
\sum_{i=1}^{n}\sum_{j=1}^{\nu} P_{ij,k}^{(f)}(x_i v_j + u_i y_j) &= 0, \quad k = 1, \ldots, n, \\
\sum_{i=1}^{n}\sum_{j=1}^{\nu} P_{ij,l}^{(m)}(x_i v_j + u_i y_j) &= 0, \quad l = 1, \ldots, n.
\end{aligned}
\tag{4.8}
$$

Since (4.8) must be true for any t, we take $t = (u, v)$ with $u_1 = \cdots = u_n = 0$, $v_{j_0} = 1$, for some j_0 and $v_j = 0, j \neq j_0$. Then we get

$$
\sum_{i=1}^{n} P_{ij_0,k}^{(f)} x_i = 0, \qquad \sum_{i=1}^{n} P_{ij_0,l}^{(m)} x_i = 0.
$$

By arbitrariness of j_0 we get the first line of the condition (4.7). The second line can be obtained similarly.

Now assume that z satisfies the condition (4.7). Then from (1.8) it easily follows that $zt = 0$ for any $t \in \mathcal{B}$, i.e. $z \in \text{ann}\mathcal{B}$.

(2) From equations (4.7) we get

$$
\sum_{k=1}^{n}\left(\sum_{i=1}^{n} P_{ij,k}^{(f)} x_i \right) = 0,
$$

i.e.

$$
\sum_{i=1}^{n}\left(\sum_{k=1}^{n} P_{ij,k}^{(f)} \right) x_i = \sum_{i=1}^{n} x_i = X(z) = 0.
$$

Equality $Y(z) = 0$ can be obtained by a similar way. $\qquad \square$

Proposition 4.2. *An EABP $\mathcal{B}$ is an algebra induced by a linear operator if and only if*

$$
P_{ij,k}^{(f)} = P_{1j,k}^{(f)}, \quad P_{ij,l}^{(m)} = P_{1j,l}^{(m)} \ \text{ for all } \ i, k = 1, \ldots, n; \quad j, l = 1, \ldots, \nu. \tag{4.9}
$$

Moreover the linear operator A has the form $A(z) = 2ze_1^{(f)}$.

Proof. From equation (3.28) for $z = t = e_i^{(f)}$, we get $A(e_i^{(f)}) = 0$, $i = 1, \ldots, n$. For $z = (x, y), t = (u, v) \in \mathcal{B}$ from (3.28) we get

$$2 \left(\sum_{i=1}^{n} x_i e_i^{(f)} + \sum_{j=1}^{\nu} y_j e_j^{(m)} \right) \left(\sum_{i=1}^{n} u_i e_i^{(f)} + \sum_{j=1}^{\nu} v_j e_j^{(m)} \right)$$

$$= 2 \sum_{i=1}^{n} \sum_{j=1}^{\nu} (x_i v_j + u_i y_j) e_i^{(f)} e_j^{(m)}$$

$$= \sum_{i=1}^{n} u_i \left(\sum_{i=1}^{n} x_i A(e_i^{(f)}) + \sum_{j=1}^{\nu} y_j A\left(e_j^{(m)}\right) \right)$$

$$+ \sum_{i=1}^{n} x_i \left(\sum_{i=1}^{n} u_i A(e_i^{(f)}) + \sum_{j=1}^{\nu} v_j A\left(e_j^{(m)}\right) \right).$$

Consequently,

$$2 \sum_{i=1}^{n} \sum_{j=1}^{\nu} (x_i v_j + u_i y_j) e_i^{(f)} e_j^{(m)} = \sum_{i=1}^{n} u_i \sum_{j=1}^{\nu} y_j A(e_j^{(m)}) + \sum_{i=1}^{n} x_i \sum_{j=1}^{\nu} v_j A(e_j^{(m)})$$

$$= \sum_{i=1}^{n} \sum_{j=1}^{\nu} (x_i v_j + u_i y_j) A(e_j^{(m)}).$$

Therefore, the last equality is true iff $2 e_i^{(f)} e_j^{(m)} = A(e_j^{(m)})$ for any $i = 1, \ldots, n$; $j = 1, \ldots, \nu$. This equality is satisfied iff $e_i^{(f)} e_j^{(m)} = e_1^{(f)} e_j^{(m)}$ for any $i = 1, \ldots, n$; $j = 1, \ldots, \nu$ which is equivalent to the condition (4.9). Using $A(e_i^{(f)}) = 0$ and $A(e_j^{(m)}) = 2 e_1^{(f)} e_j^{(m)}$ one gets that $A(z) = 2 z e_1^{(f)}$. $\square$

Remark 4.2. In the class of baric algebras the algebra induced by a linear operator A is Bernstein if and only if $A^2 = A$, i.e. A is a projection, and in this case the algebra is necessarily conservative. But in the non-baric case this property is not true. Indeed, for $z = (x, y) \in \mathcal{B}$ we get $X(z) = \sum_{i=1}^{n} x_i$, $X(A(z)) = X(2 z e_1^{(f)}) = \sum_{j=1}^{\nu} y_j$, i.e. $X(z) \neq X(A(z))$ in general.

Given two EABP algebras $\mathcal{B}_1$, $\mathcal{B}_2$ a homomorphism $h \colon \mathcal{B}_1 \to \mathcal{B}_2$ is a linear mapping with $h(zt) = h(z)h(t)$ and $X(h(z)) = X(z)$.

Theorem 4.5. *Let $\mathcal{B}_1$ and $\mathcal{B}_2$ be EABPs. If $h \colon \mathcal{B}_1 \to \mathcal{B}_2$ is a homomorphism then*

$$X(h(e_i^{(f)})) = 1, \qquad X(h(e_j^{(m)})) = 0,$$

$$Y(h(e_i^{(f)})) = 0, \qquad Y(h(e_j^{(m)})) = 1, \qquad i = 1, \ldots, n; \quad j = 1, \ldots, \nu.$$

Proof. The first two equalities easily follow from $X(z) = X\big(h(z)\big)$. Now we shall prove the third and fourth equalities. Assume h on basis elements is given as follows

$$h\big(e_i^{(f)}\big) = \sum_{j=1}^{n} \alpha_{ij} e_j^{(f)} + \sum_{k=1}^{\nu} \beta_{ik} e_k^{(m)}, \quad i = 1, \ldots, n,$$

$$h\big(e_j^{(m)}\big) = \sum_{i=1}^{n} \lambda_{ji} e_i^{(f)} + \sum_{l=1}^{\nu} \mu_{jl} e_l^{(m)}, \quad j = 1, \ldots, \nu.$$

We have $X\big(h(e_i^{(f)})\big) = \sum_{j=1}^{n} \alpha_{ij} = 1$, $X\big(h(e_j^{(m)})\big) = \sum_{i=1}^{n} \lambda_{ji} = 0$. From $h\big(e_i^{(f)} e_i^{(f)}\big) = 0$ we get

$$\sum_{j=1}^{n} \sum_{k=1}^{\nu} \alpha_{ij} \beta_{ik} P_{jk,l}^{(f)} = 0, \, l = 1, \ldots, n, \quad \sum_{j=1}^{n} \sum_{k=1}^{\nu} \alpha_{ij} \beta_{ik} P_{jk,q}^{(m)} = 0, \, q = 1, \ldots, \nu.$$

$$(4.10)$$

From the first equality of (4.10) we get

$$\sum_{l=1}^{n} \left(\sum_{j=1}^{n} \sum_{k=1}^{\nu} \alpha_{ij} \beta_{ik} P_{jk,l}^{(f)} \right) = \sum_{j=1}^{n} \alpha_{ij} \sum_{k=1}^{\nu} \beta_{ik} = \sum_{k=1}^{\nu} \beta_{ik} = Y\big(e_i^{(f)}\big) = 0.$$

One can check that the condition $h\big(e_i^{(f)} e_s^{(m)}\big) = h\big(e_i^{(f)}\big) h\big(e_s^{(m)}\big)$ is equivalent to the following equations

$$\sum_{j=1}^{n} \sum_{l=1}^{\nu} \alpha_{ij} \mu_{sl} P_{jl,q}^{(f)} + \sum_{t=1}^{n} \sum_{k=1}^{\nu} \lambda_{st} \beta_{ik} P_{tk,q}^{(f)} = \sum_{j=1}^{n} \alpha_{jq} P_{is,j}^{(f)} + \sum_{l=1}^{\nu} \lambda_{lq} P_{is,l}^{(m)},$$

$$(4.11)$$

$$\sum_{j=1}^{n} \sum_{l=1}^{\nu} \alpha_{ij} \mu_{sl} P_{jl,\eta}^{(m)} + \sum_{t=1}^{n} \sum_{k=1}^{\nu} \lambda_{st} \beta_{ik} P_{tk,\eta}^{(m)} = \sum_{j=1}^{n} \beta_{j\eta} P_{is,j}^{(f)} + \sum_{l=1}^{\nu} \mu_{l\eta} P_{is,l}^{(m)}.$$

Summing the equation (4.11) by $q = 1, \ldots, n$ and using above obtained relations we get $Y\big(e_s^{(m)}\big) = \sum_{l=1}^{\nu} \mu_{sl} = 1$, $s = 1, \ldots, \nu$. $\square$

Let $\mathcal{B}$ be an EABP algebra with basis set

$$e_i^{(f)}, \, i = 1, \ldots, n; \; e_j^{(m)}, \, j = 1, \ldots, \nu.$$

One says $e_i^{(f)}$ (resp. $e_j^{(m)}$) occurs in $z \in \mathcal{B}$, if the coefficient α_i (resp. β_j) is nonzero in $z = \sum_i \alpha_i e_i^{(f)} + \sum_j \beta_j e_j^{(m)}$.

The following example shows that in $h(e_i^{(f)})$ may occur $e_j^{(f)}$ for some j and $e_k^{(m)}$ for some k.

Example 4.3. Consider the EABP of Example 4.1, i.e. case $n = 1$, $\nu = 2$. In this case

$$e_1^{(f)} e_1^{(m)} = e_1^{(f)} + a e_1^{(m)} + (1 - a) e_2^{(m)}, \quad e_1^{(f)} e_2^{(m)} = e_1^{(f)} + b e_1^{(m)} + (1 - b) e_2^{(m)}.$$

One can see that $h(e_1^{(f)}) = e_1^{(f)} + \alpha e_1^{(m)} - \alpha e_2^{(m)}$, and $\alpha \neq 0$ if $a = b$.

4.1.4 *The derivations of $\mathcal{B}$*

There are many papers devoted on the subject of derivations of genetic algebras (see e.g. [38, 82, 104, 254]). As we mention in the previous chapter, in [104] an explanation of the genetic meaning of a derivation of a genetic algebra is given. For any algebra, it is known that the space of its derivations is a Lie algebra. The Lie algebra of derivations of a given algebra is an important tool for studying its structure, particularly in the non-associative case; so this is a natural development.

In this section we describe the set of all derivations of $\mathcal{B}$. Let $D \in \mathrm{Der}(\mathcal{B})$ and suppose

$$D(e_k^{(f)}) = \sum_{i=1}^{n} d_{ki}^{ff} e_i^{(f)} + \sum_{l=1}^{\nu} d_{kl}^{fm} e_l^{(m)}, \quad k = 1, \dots, n.$$

$$D(e_k^{(m)}) = \sum_{i=1}^{n} d_{ki}^{mf} e_i^{(f)} + \sum_{l=1}^{\nu} d_{kl}^{mm} e_l^{(m)}, \quad k = 1, \dots, \nu.$$

By the definition of derivation $D(zt) - D(z)t + zD(t)$, we have

$$D(e_k^{(f)} e_j^{(f)}) = D(e_k^{(f)}) e_j^{(f)} + e_k^{(f)} D(e_j^{(f)})$$

$$= \left(\sum_{i=1}^{n} d_{ki}^{ff} e_i^{(f)} + \sum_{l=1}^{\nu} d_{kl}^{fm} e_l^{(m)} \right) e_j^{(f)} + e_k^{(f)} \left(\sum_{i=1}^{n} d_{ji}^{ff} e_i^{(f)} + \sum_{l=1}^{\nu} d_{jl}^{fm} e_l^{(m)} \right)$$

$$= \frac{1}{2} \sum_{i=1}^{n} \left(\sum_{l=1}^{\nu} \left(d_{kl}^{fm} P_{jl,i}^{(f)} + d_{jl}^{fm} P_{kl,i}^{(f)} \right) \right) e_i^{(f)}$$

$$+ \frac{1}{2} \sum_{q=1}^{\nu} \left(\sum_{l=1}^{\nu} \left(d_{kl}^{fm} P_{jl,q}^{(m)} + d_{jl}^{fm} P_{kl,q}^{(m)} \right) \right) e_q^{(m)} = 0.$$

Consequently,

$$\begin{cases} \sum_{l=1}^{\nu} \left(d_{kl}^{fm} P_{jl,i}^{(f)} + d_{jl}^{fm} P_{kl,i}^{(f)} \right) = 0, \, i, j, k = 1, \dots, n; \\ \sum_{l=1}^{\nu} \left(d_{kl}^{fm} P_{jl,q}^{(m)} + d_{jl}^{fm} P_{kl,q}^{(m)} \right) = 0, \, j, k = 1, \dots, n; \, q = 1, \dots, \nu. \end{cases} \tag{4.12}$$

Similarly, from $D(e_k^{(m)} e_j^{(m)}) = D(e_k^{(m)})e_j^{(m)} + e_k^{(m)} D(e_j^{(m)}) = 0$ we obtain

$$\begin{cases} \sum_{i=1}^{n} \left(d_{ki}^{mf} P_{ij,s}^{(f)} + d_{ji}^{mf} P_{ik,s}^{(f)} \right) = 0, \ s = 1, \ldots, n; \ j, k = 1, \ldots, \nu; \\[2mm] \sum_{i=1}^{n} \left(d_{ki}^{mf} P_{ij,q}^{(m)} + d_{ji}^{mf} P_{ik,q}^{(m)} \right) = 0, \ j, k, q = 1, \ldots, \nu. \end{cases} \tag{4.13}$$

The equality $D(e_i^{(f)} e_j^{(m)}) = D(e_i^{(f)})e_j^{(m)} + e_i^{(f)} D(e_j^{(m)})$ gives the following conditions

$$\begin{cases} \sum_{p=1}^{n} \left(d_{ip}^{ff} P_{pj,s}^{(f)} - d_{ps}^{ff} P_{ij,p}^{(f)} \right) \\[1mm] \quad + \sum_{q=1}^{\nu} \left(d_{jq}^{mm} P_{iq,s}^{(f)} - d_{qs}^{mf} P_{ij,q}^{(m)} \right) = 0, \ i, s = 1, \ldots, n; \ j = 1, \ldots, \nu; \\[3mm] \sum_{p=1}^{n} \left(d_{ip}^{ff} P_{pj,t}^{(m)} - d_{pt}^{fm} P_{ij,p}^{(f)} \right) \\[1mm] \quad + \sum_{q=1}^{\nu} \left(d_{jq}^{mm} P_{iq,t}^{(m)} - d_{qt}^{mm} P_{ij,q}^{(m)} \right) = 0, \ i = 1, \ldots, n; \ j, t = 1, \ldots, \nu. \end{cases} \tag{4.14}$$

Since for any $z = (x, y) \in \mathcal{B}$ we have $D(z) = \sum_{i=1}^{n} x_i D(e_i^{(f)}) + \sum_{j=1}^{\nu} y_j D(e_j^{(m)})$, D is uniquely defined by the matrix $\mathcal{D} = \mathcal{D}(D) = \left(d_{ij}^{ff}, d_{ip}^{fm}, d_{pi}^{mf}, d_{pq}^{mm} \right)_{\substack{i,j=1,\ldots,n \\ p,q=1,\ldots,\nu}}$.

     Hence,

$$\mathrm{Der}(\mathcal{B}) = \{ D : \mathcal{D}(D) \ \text{satisfies} \ (4.12), (4.13), (4.14) \}.$$

Example 4.4. Consider the case $n = 1$, $\nu = 2$. In this case $P_{11,1}^{(f)} = P_{12,1}^{(f)} = 1$. Denote $P_{11,1}^{(m)} = a$, $P_{11,2}^{(m)} = 1 - a$, $P_{12,1}^{(m)} = b$, $P_{12,2}^{(m)} = 1 - b$. Assume $a = 1 - b$, $a \neq \frac{1}{2}$. Then using (4.12), (4.13), (4.14) we get

$$D(e_1^{(f)}) = 0; \quad D(e_1^{(m)}) = \alpha \cdot (e_1^{(m)} - e_2^{(m)}); \quad D(e_2^{(m)}) = \alpha \cdot (-e_1^{(m)} + e_2^{(m)}),$$

where $\alpha \in \mathbb{R}$. Consequently, for arbitrary element $z = (x, y_1, y_2) \in \mathbb{R}^{1+2}$ we have

$$D(z) = \alpha \cdot (y_1 - y_2)(e_1^{(m)} - e_2^{(m)}).$$

Thus if $n = 1$, $\nu = 2$ and $a = 1 - b$, $a \neq \frac{1}{2}$ then

$$\mathrm{Der}(\mathcal{B}) = \left\{ D : D(z) = \alpha(y_1 - y_2)(e_1^{(m)} - e_2^{(m)}), z = (x, y_1, y_2) \in \mathcal{B}, \alpha \in \mathbb{R} \right\}.$$

4.1.5   *Dynamics of the operator (1.6)*

Extend the operator (1.6) on $\mathbb{R}^{n+\nu}$, i.e. consider the operator $V : \mathbb{R}^{n+\nu} \to \mathbb{R}^{n+\nu}$, $z = (x, y) \mapsto V(z) = z' = (x', y')$ defined as

$$x_j' = \sum_{i,k=1}^{n,\nu} P_{ik,j}^{(f)} x_i y_k; \quad y_l' = \sum_{i,k=1}^{n,\nu} P_{ik,l}^{(m)} x_i y_k. \tag{4.15}$$

Consider the following linear form $X\colon \mathbb{R}^n \to \mathbb{R}$ $(Y\colon \mathbb{R}^\nu \to \mathbb{R})$ defined by

$$X(x) = \sum_{i=1}^{n} x_i, \quad \left(Y(y) = \sum_{k=1}^{\nu} y_k \right). \qquad (4.16)$$

Denote

$$H_i = \{ z = (x,y) : X(x) = Y(y) = i \}, \quad i = 0,1, \qquad (4.17)$$

the product of the i-hyperplanes in $\mathbb{R}^n$ and $\mathbb{R}^\nu$, respectively.

A point $z = (x,y) \in \mathbb{R}^{n+\nu}$ is called a *fixed point* (resp. *zero point*) of V if $V(z) = z$ (resp. $V(z) = 0$).

Proposition 4.3.

(1) If z is a fixed point then $z \in H_0 \cup H_1$.
(2) If z is a zero point then $z \in \{ z = (x,y) : X(x)Y(y) = 0 \}$.

Proof. From (4.15) and using (1.5) we get

$$\begin{aligned}
X(x') &= \textstyle\sum_{j=1}^{n} x'_j = \sum_{i,k=1}^{n,\nu} \left(\sum_{j=1}^{n} P_{ik,j}^{(f)} \right) x_i y_k = X(x)Y(y)\,; \\
Y(y') &= \textstyle\sum_{l=1}^{\nu} y'_l = \sum_{i,k=1}^{n,\nu} \left(\sum_{l=1}^{\nu} P_{ik,l}^{(m)} \right) x_i y_k = X(x)Y(y)\,.
\end{aligned} \qquad (4.18)$$

(1) If z is a fixed point then $X(x') = X(x)$ and $Y(y') = Y(y)$, this by (4.18) gives that $X(x) = Y(y)$ and $X(x) = (X(x))^2$. Hence $X(x) = Y(y) = 0$ or 1.

(2) If z is a zero point then $X(x') = Y(y') = 0$, and (4.18) gives $X(x)Y(y) = 0$.

$\square$

Let $z^{(0)} = (x^{(0)}, y^{(0)}) \in \mathbb{R}^{n+\nu}$ be an initial point. Its *trajectory* is defined by $z^{(t)} = (x^{(t)}, y^{(t)}) = V(z^{(t-1)})$, $t = 1, 2, \ldots$. Denote by $\omega(z^{(0)})$ the set of limit points of the trajectory $\{z^{(t)}\}_{t=0}^{\infty}$. If $\omega(z^{(0)})$ consists of a single point, then the trajectory converges, and $\omega(z^{(0)})$ is a fixed point of the operator V.

The following theorem gives an upper estimate of the set $\omega(z^{(0)})$.

Theorem 4.6. *We have*

$$\omega(z^{(0)}) \subset \begin{cases} H_0 & \text{if } \ \left| X(x^{(0)})Y(y^{(0)}) \right| < 1 \\ H_1 & \text{if } \ \left| X(x^{(0)})Y(y^{(0)}) \right| = 1 \\ H_\infty & \text{if } \ \left| X(x^{(0)})Y(y^{(0)}) \right| > 1, \end{cases}$$

where $H_\infty = \{ z = (x,y) : X(x) = Y(y) = +\infty \}$.

Proof.   Using (4.18) we get $X(x^{(t)}) = Y(y^{(t)}) = X(x^{(t-1)})Y(y^{(t-1)})$ for any $t = 1, 2, \ldots$. Iterating this recurrent equation we obtain

$$X(x^{(t)}) = Y(y^{(t)}) = \left(X(x^{(0)})Y(y^{(0)}) \right)^{2^{t-1}}. \tag{4.19}$$

Consequently,

$$\lim_{t \to \infty} X(x^{(t)}) = \lim_{t \to \infty} Y(y^{(t)}) = \begin{cases} 0 & \text{if } \left| X(x^{(0)})Y(y^{(0)}) \right| < 1 \\ 1 & \text{if } \left| X(x^{(0)})Y(y^{(0)}) \right| = 1 \\ +\infty & \text{if } \left| X(x^{(0)})Y(y^{(0)}) \right| > 1. \end{cases}$$

$\square$

4.1.6   *A special case of an EABP*

In this section we consider a special case of an EABP giving an additional condition on heredity coefficients (1.5), i.e. consider the coefficients as follows

$$P_{ik,j}^{(f)} = \begin{cases} a_{ij} & \text{if } k = 1 \\ 1 & \text{if } k \neq 1, j = 1 \\ 0 & \text{if } k \neq 1, j \neq 1 \end{cases}, \quad P_{ik,l}^{(m)} = \begin{cases} b_{kl} & \text{if } i = 1 \\ 1 & \text{if } i \neq 1, l = 1 \\ 0 & \text{if } i \neq 1, l \neq 1. \end{cases} \tag{4.20}$$

The matrices $A = (a_{ij})$ and $B = (b_{kl})$ by (1.5) satisfy the following conditions

$$a_{ij} \geq 0, \quad \sum_{j=1}^{n} a_{ij} = 1, \quad i = 1, \ldots, n; \quad b_{kl} \geq 0, \quad \sum_{l=1}^{\nu} b_{kl} = 1, \quad k = 1, \ldots, \nu, \tag{4.21}$$

i.e. both matrices are stochastic.

Remark 4.3. The condition (4.20) is taken to simplify computations; in this way we reduced both cubic matrices $\left(P_{ik,j}^{(f)} \right)$, $\left(P_{ik,l}^{(m)} \right)$ to quadratic matrices. But that condition has very clear biological treatment: the type "1" of females and the type "1" of males have preference, i.e. any type of female (male) can be born if its father (mother) has type "1". If the father (mother) has type $\neq 1$ then only type "1" female (male) can be born.

Under condition (4.20) the multiplication (1.7) became as

$$
e_i^{(f)} e_k^{(m)} = e_k^{(m)} e_i^{(f)} = \frac{1}{2}
\begin{cases}
\sum_{j=1}^{n} a_{1j} e_j^{(f)} + \sum_{l=1}^{\nu} b_{1l} e_l^{(m)} & \text{if } i = 1, k = 1 \\
\sum_{j=1}^{n} a_{ij} e_j^{(f)} + e_1^{(m)} & \text{if } i \neq 1, k = 1 \\
e_1^{(f)} + \sum_{l=1}^{\nu} b_{kl} e_l^{(m)} & \text{if } i = 1, k \neq 1 \\
e_1^{(f)} + e_1^{(m)} & \text{if } i \neq 1, k \neq 1,
\end{cases}
$$

$$
e_i^{(f)} e_j^{(f)} = 0, \quad i, j = 1, \dots, n; \qquad e_k^{(m)} e_l^{(m)} = 0, \quad k, l = 1, \dots, \nu .
$$

$$(4.22)$$

Operator (1.6) has the following form

$$
x_1' = \sum_{i=1}^{n} \left(a_{i1} y_1 + \sum_{k=2}^{\nu} y_k \right) x_i
$$

$$
x_j' = y_1 \sum_{i=1}^{n} a_{ij} x_i, \quad j \neq 1
$$

$$(4.23)$$

$$
y_1' = \sum_{k=1}^{\nu} \left(b_{k1} x_1 + \sum_{i=2}^{n} x_i \right) y_k
$$

$$
y_l' = x_1 \sum_{k=1}^{\nu} b_{kl} y_k, \quad l \neq 1 .
$$

Denote by $\mathcal{B}_1$ the EABP defined by the multiplication table (4.22).

4.1.6.1 *Idempotent elements of* $\mathcal{B}_1$

Recall that an element $z \in \mathcal{B}$ is called *idempotent* if $z^2 = z$; such points of an EABP are especially important, because they are the fixed points of the evolution map V, i.e. $V(z) = z$. We denote by $\mathcal{I}d(\mathcal{B})$ the idempotent elements of an algebra $\mathcal{B}$. Clearly, $0 \in \mathcal{I}d(\mathcal{B})$ and this set is an algebraic variety. By Proposition 4.3 we have $\mathcal{I}d(\mathcal{B}) \subset H_0 \cup H_1$. In this subsection we shall describe idempotent elements of $\mathcal{B}_1$. First we describe the set of idempotent elements which belong in $H_0 \cap \mathcal{I}d(\mathcal{B}_1)$ and after that we shall describe elements of $H_1 \cap \mathcal{I}d(\mathcal{B}_1)$.

Idempotents in H_0.

Using (4.23) and the condition that $z \in H_0$, from the equation $V(z) = z^2 = z$ we obtain

$$
x_j = y_1 \sum_{i=2}^{n} (a_{ij} - a_{1j}) x_i, \quad j = 2, \dots, n;
$$

$$
y_l = x_1 \sum_{k=2}^{\nu} (b_{kl} - b_{1l}) y_k, \quad l = 2, \dots, \nu .
$$

$$(4.24)$$

Case $x_1 y_1 = 0$: If $x_1 = 0$ then $y_2 = \cdots = y_\nu = 0$, consequently, $y_1 = 0$. Similarly, if $y_1 = 0$ we get $x_1 = \cdots = x_n = 0$. Hence in the case $x_1 y_1 = 0$ we have a unique idempotent $z = 0$.

Case $x_1 y_1 \neq 0$: Consider matrices $C_y = (c_{ij})_{i,j=2,\dots,n}$ and $D_x = (d_{kl})_{k,l=2,\dots,\nu}$ such that

$$
c_{ij} =
\begin{cases}
(a_{ij} - a_{1j}) y & \text{if } i \neq j \\
(a_{ij} - a_{1j}) y - 1 & \text{if } i = j
\end{cases}, \quad
d_{kl} =
\begin{cases}
(b_{kl} - b_{1l}) x & \text{if } k \neq l \\
(b_{kl} - b_{1l}) x - 1 & \text{if } k = l .
\end{cases}
$$

Then equation (4.24) can be written as

$$C_{y_1} x = 0, \quad D_{x_1} y = 0, \tag{4.25}$$

where $x = (x_2, \ldots, x_n)$, $y = (y_2, \ldots, y_\nu)$.

Consider now $x_1 \in \mathbb{R} \setminus \{0\}$ as a parameter, then equation $D_{x_1} y = 0$ has a unique solution $y = 0$ if $\det(D_{x_1}) \neq 0$, which gives $y_1 = 0$, i.e. this is a contradiction to the assumption that $y_1 \neq 0$.

If $\det(D_{x_1}) = 0$ then we fix a solution $x_1 = x_1^* \neq 0$ of the equation $\det(D_{x_1}) = 0$. In this case there are infinitely many solutions $y^* = (y_2^*, \ldots, y_\nu^*)$ of $D_{x_1} y = 0$. Substituting the solution $y_1^* = -\sum_{k=2}^{\nu} y_k^*$ in $C_{y_1} x = 0$, we get

$$x_j = y_1^* \sum_{i=2}^{n} (a_{ij} - a_{1j}) x_i = 0, j = 2, \ldots, n. \tag{4.26}$$

This system has a unique solution $x_2 = \cdots = x_n = 0$ if $\det(C_{y_1^*}) \neq 0$ but in this case we get $x_1 = 0$ which is a contradiction to the assumption that $x_1 \neq 0$. If $\det(C_{y_1^*}) = 0$ then we have infinitely many solutions $x^* = (x_2^*, \ldots, x_n^*)$.

Hence we have proved the following

Proposition 4.4. *We have*

$$H_0 \cap \mathcal{I}d(\mathcal{B}_1) = \{0\} \cup \{((x_1^*, \ldots, x_n^*), (y_1^*, \ldots, y_\nu^*)) :$$
$$C_{y_1^*} x^* = 0, \quad D_{x_1^*} y^* = 0, \det(D_{x_1^*}) = \det(C_{y_1^*}) = 0\}.$$

Idempotents in H_1. Using the condition that $z \in H_1$, the equation $V(z) = z^2 = z$ can be written as

$$x_1 = \sum_{i=1}^{n} \left((a_{i1} - 1)y_1 + 1\right) x_i;$$
$$x_j = y_1 \sum_{i=1}^{n} a_{ij} x_i, \quad j \neq 1. \tag{4.27}$$

$$y_1 = \sum_{k=1}^{\nu} \left((b_{k1} - 1)x_1 + 1\right) y_k;$$
$$y_l = x_1 \sum_{k=1}^{\nu} b_{kl} y_k, \quad l \neq 1. \tag{4.28}$$

Consider several cases.

Case $x_1 = y_1 = 0$. In this case we get $z = 0$ which is not in H_1.

Case $x_1 \neq 0$, $y_1 = 0$. In this case $x_1 = 1$, $x_j = 0$, $j = 2, \ldots, n$. Substituting $x_1 = 1$ in the system of equations (4.28) we get

$$0 = \sum_{k=2}^{\nu} b_{k1} y_k$$
$$y_l = \sum_{k=2}^{\nu} b_{kl} y_k, \quad l = 2, \ldots, \nu. \tag{4.29}$$

Denote $B_2 = (b'_{kl})_{k,l=2,\ldots,\nu}$, with

$$b'_{kl} = \begin{cases} b_{kl} - 1 & \text{if } k = l \\ b_{kl} & \text{if } k \neq l. \end{cases}$$

If $\det(B_2) \neq 0$ then $B_2 y = 0$ gives $y_1 = \cdots = y_\nu = 0$ but this does not satisfy $\sum_{k=2}^{\nu} y_k = 1$. If $\det(B_2) = 0$ then $B_2 y = 0$ has infinitely many solutions $y = (y_2, \ldots, y_\nu)$. We must take these solutions which satisfy $\sum_{k=2}^{\nu} b_{k1} y_k = 0$. So in this case we have the following idempotent elements (which belong to H_1),

$$I_0 = \begin{cases} ((1,0,\ldots,0),(0,y_2,\ldots,y_\nu)) : B_2 y = 0, \sum_{k=2}^{\nu} b_{k1} y_k = 0\} \\ \qquad \text{if } \det(B_2) = 0, \\ \emptyset \qquad \text{if } \det(B_2) \neq 0. \end{cases}$$

Case $x_1 = 0$, $y_1 \neq 0$. This case is similar to the previous case. Denote $A_2 = \left(a'_{ij}\right)_{i,j=2,\ldots,n}$, with

$$a'_{ij} = \begin{cases} a_{ij} - 1 & \text{if } i = j \\ a_{ij} & \text{if } i \neq j. \end{cases}$$

Then we have the following idempotent elements (which belong to H_1),

$$I_1 = \begin{cases} ((0,x_1,\ldots,x_n),(1,0,\ldots,0)) : A_2 x = 0, \sum_{i=2}^{n} a_{i1} x_i = 0\} \\ \qquad \text{if } \det(A_2) = 0, \\ \emptyset \qquad \text{if } \det(A_2) \neq 0. \end{cases}$$

Case $x_1 \neq 0$, $y_1 \neq 0$. Here, to avoid many special cases we assume that $a_{11} \neq 1$. Take y_1 as a parameter and solve the system (4.27) which is equivalent to the following

$$\sum_{i=2}^{n} \left(\frac{a_{1j}((a_{i1} - 1)y_1 + 1)}{1 - a_{11}} + a_{ij} y_1 \right) x_i = x_j, \quad j = 2, \ldots, n. \qquad (4.30)$$

Denote $U_y = (u_{ij})_{i,j=2,\ldots,n}$ with

$$u_{ij} = \begin{cases} \frac{a_{1j}((a_{i1}-1)y+1)}{1-a_{11}} + a_{ij} y & \text{if } i \neq j \\ \frac{a_{1j}((a_{i1}-1)y+1)}{1-a_{11}} + a_{ij} y - 1 & \text{if } i = j. \end{cases}$$

Subcase $\det(U_y) \neq 0$. In this case we have $x_2 = \cdots = x_n = 0$, consequently, $x_1 = 1$. For $x_1 = 1$ from the equation (4.28) we get $B_1 y = 0$ where $B_1 = (b^*_{kl})_{k,l=1,\ldots,\nu}$, with

$$b^*_{kl} = \begin{cases} b_{kl} - 1 & \text{if } k = l \\ b_{kl} & \text{if } k \neq l. \end{cases}$$

So in this case we have the following set of idempotent elements of H_1.

$$I_2 = \begin{cases} ((1,0,\ldots,0),(y_1,\ldots,y_\nu)) : B_1 y = 0, \det(U_y) \neq 0\} \\ \qquad\qquad\qquad\qquad \text{if } \det(B_1) = 0 \\ \emptyset \qquad\qquad\qquad\qquad\qquad \text{if } \det(B_1) \neq 0. \end{cases}$$

Subcase $\det(U_y) = 0$. In this case we fix a solution $y_1 = y_1^*$ of $\det(U_y) = 0$. We have infinitely many solutions $(x_2^*,\ldots,x_n^*)$. Denote $C_x = (c_{kl}^*)_{k,l=1,\ldots,\nu}$, with

$$c_{kl}^* = \begin{cases} (b_{11}-1)x & \text{if } k = l = 1; \\ (b_{k1}-1)x + 1 & \text{if } k \neq 1, l = 1; \\ b_{ll}x - 1 & \text{if } k = l \neq 1; \\ b_{k1}x & \text{if } k \neq l, l = 2,\ldots,\nu. \end{cases}$$

Now one has the following set of idempotent elements

$$I_3 = \{((x_1^*,\ldots,x_n^*),(y_1^*,\ldots,y_\nu^*)) : C_{x_1^*} y^* = 0,$$
$$U_{y_1^*} x^* = 0, \det(U_{y_1^*}) = \det(C_{x_1^*}) = 0\}.$$

Thus we have the following

Theorem 4.7. *If $a_{11} \neq 1$ then the full set of the idempotent elements which belong to H_1 is*

$$\mathcal{I}d(\mathcal{B}_1) \cap H_1 = I_0 \cup I_1 \cup I_2 \cup I_3.$$

Remark 4.4. By Proposition 4.3 we can conclude that Proposition 4.4 and Theorem 4.7 give the full set $\mathcal{I}d(\mathcal{B}_1)$. But conditions described in the Proposition and Theorem are complicated in general. In the sequel of this subsection we shall consider additional conditions on (a_{ij}), (b_{ij}) and under these conditions we explicitly describe the set $\mathcal{I}d(\mathcal{B}_1)$.

Now we consider a particular case and describe the full set of idempotent elements of $\mathcal{B}_1$.

If we consider the case

$$a_{ij} = \begin{cases} 1 & \text{if } i = j \\ 0 & \text{if } i \neq j \end{cases}, \quad b_{kl} = \begin{cases} 1 & \text{if } k = l \\ 0 & \text{if } k \neq l, \end{cases} \tag{4.31}$$

then the following is true

Proposition 4.5. *If the condition (4.31) is satisfied then*

$$
\mathcal{I}d(\mathcal{B}_1) = \{0\} \cup \left\{ ((1, x_2, \ldots, x_n), (1, y_2, \ldots, y_\nu)) : \sum_{i=2}^{n} x_i = \sum_{k=2}^{\nu} y_k = 0 \right\}
$$

$$
\cup \left\{ ((1, x_2, \ldots, x_n), (1, y_2, \ldots, y_\nu)) : \sum_{i=2}^{n} x_i = \sum_{k=2}^{\nu} y_k = -1 \right\}
$$

$$
\cup \left\{ ((1, 0, \ldots, 0), (y_1, \ldots, y_\nu)) : \sum_{k=1}^{\nu} y_k = 1, y_1 \neq 1 \right\}
$$

$$
\cup \left\{ ((x_1, \ldots, x_n), (1, 0, \ldots, 0)) : \sum_{i=1}^{n} x_i = 1, x_1 \neq 1 \right\} .
$$

Proof. The equation $z^2 = z$, for $z = (x, y) \in \mathbb{R}^{n+\nu}$, using condition (4.31) can be written as

$$
\begin{aligned}
(1 - y_1)x_1 &= \sum_{k=2}^{\nu} y_k \sum_{i=1}^{n} x_i, \\
(1 - y_1)x_j &= 0, \quad j = 2, \ldots, n, \\
(1 - x_1)y_1 &= \sum_{i=2}^{n} x_i \sum_{k=1}^{\nu} y_k, \\
(1 - x_1)y_l &= 0, \quad l = 2, \ldots, \nu.
\end{aligned}
\tag{4.32}
$$

The simple analysis of the system (4.32) gives the set of all idempotents.

$\square$

4.1.6.2 *Absolute nilpotent elements of $\mathcal{B}_1$*

The element z is called an *absolute nilpotent* if $z^2 = 0$, i.e. z is zero-point of the evolution operator V. For $\mathcal{B}_1$ the equation $z^2 = 0$ is equivalent to the following system of quadratic equations

$$
\begin{aligned}
\sum_{i=1}^{n} \left(a_{i1} y_1 + \sum_{k=2}^{\nu} y_k \right) x_i &= 0; \\
y_1 \sum_{i=1}^{n} a_{ij} x_i &= 0, \quad j = 2, \ldots, n; \\
\sum_{k=1}^{\nu} \left(b_{k1} x_1 + \sum_{i=2}^{n} x_i \right) y_k &= 0; \\
x_1 \sum_{k=1}^{\nu} b_{kl} y_k &= 0, \quad l = 2, \ldots, \nu.
\end{aligned}
\tag{4.33}
$$

By Proposition 4.3 we have $\sum_{i=1}^{n} x_i \sum_{k=1}^{\nu} y_k = 0$. There are the following three cases.

- *Case* $\sum_{i=1}^{n} x_i = \sum_{k=1}^{\nu} y_k = 0$. In this case, from the system of quadratic equations (4.33) we get

$$y_1 \sum_{i=1}^{n} a_{ij} x_i = 0, \quad j = 1, \ldots, n;$$

$$x_1 \sum_{k=1}^{\nu} b_{kl} y_k = 0, \quad l = 1, \ldots, \nu.$$

(4.34)

Subcase $y_1 = x_1 = 0$. In this case one easily gets the following set of solutions

$$\mathcal{N}_{00}^0 = \left\{ ((0, x_2, \ldots, x_n), (0, y_2, \ldots, y_\nu)) : \sum_{i=2}^{n} x_i = 0, \sum_{k=2}^{\nu} y_k = 0 \right\}.$$

Subcase $x_1 = 0, y_1 \neq 0$. Denote $A_0 = (a_{ij})_{i,j=2,\ldots,n}$. One can see that the set of solutions is as follows

$$\mathcal{N}_{01}^0 = \begin{cases} \mathcal{N}^1 & \text{if } \det(A_0) \neq 0; \\ \mathcal{N}^0 & \text{if } \det(A_0) = 0, \end{cases}$$

where

$$\mathcal{N}^1 = \left\{ ((0, \ldots, 0), (y_1, \ldots, y_\nu)) : y_1 \neq 0, \sum_{k=1}^{\nu} y_k = 0 \right\},$$

$$\mathcal{N}^0 = \left\{ ((0, x_2, \ldots, x_n), (y_1, \ldots, y_\nu)) : \sum_{i=2}^{n} x_i = \sum_{i=2}^{n} a_{i1} x_i = 0, A_0 x = 0, \right.$$

$$\left. y_1 \neq 0, \sum_{k=1}^{\nu} y_k = 0 \right\}.$$

Subcase $x_1 \neq 0, y_1 = 0$. Denote $B_0 = (b_{kl})_{k,l=2,\ldots,\nu}$. In this case the set of solutions is

$$\mathcal{N}_{10}^0 = \begin{cases} \mathcal{N}^2 & \text{if } \det(B_0) \neq 0; \\ \mathcal{N}^3 & \text{if } \det(B_0) = 0, \end{cases}$$

where

$$\mathcal{N}^2 = \left\{ ((x_1, \ldots, x_n), (0, \ldots, 0)) : x_1 \neq 0, \sum_{i=1}^{n} x_i = 0 \right\},$$

$$\mathcal{N}^3 = \left\{ ((x_1, \ldots, x_n), (0, y_2, \ldots, y_\nu)) : x_1 \neq 0, \sum_{i=1}^{n} x_i = 0, \right.$$

$$\left. B_0 y = 0, \sum_{k=2}^{\nu} y_k = \sum_{k=2}^{\nu} b_{k1} y_k = 0 \right\}.$$

Subcase $x_1 \neq 0, y_1 \neq 0$. In this case it is easy to get the following

$$\mathcal{N}_{11}^0 = \begin{cases} \emptyset & \text{if } \det(A) \neq 0 \text{ or } \det(B) \neq 0; \\ \mathcal{N}^4 & \text{if } \det(A) = \det(B) = 0, \end{cases}$$

where

$$\mathcal{N}^4 = \left\{ ((x_1, \ldots, x_n), (y_1, \ldots, y_\nu)) : Ax = 0, By = 0 \right\}.$$

- *Case* $\sum_{i=1}^{n} x_i = 0, \sum_{k=1}^{\nu} y_k \neq 0$. In this case the system of quadratic equations (4.33) can be written as

$$\sum_{i=1}^{n} \left(a_{i1} y_1 + \sum_{k=2}^{\nu} y_k \right) x_i = 0\,;$$

$$y_1 \sum_{i=1}^{n} a_{ij} x_i = 0\,, \quad j = 2, \ldots, n\,;$$

$$x_1 \sum_{k=1}^{\nu} (b_{k1} - 1)\, y_k = 0\,;$$

$$x_1 \sum_{k=1}^{\nu} b_{kl} y_k = 0\,, \quad l = 2, \ldots, \nu\,.$$

(4.35)

Subcase $y_1 = x_1 = 0$. In this case we have the following set of absolute nilpotents

$$\mathcal{N}_{00}^1 = \left\{ ((0, x_2, \ldots, x_n), (0, y_2, \ldots, y_\nu)) : \sum_{i=2}^{n} x_i = 0, \sum_{k=2}^{\nu} y_k \neq 0 \right\}.$$

Subcase $x_1 = 0, y_1 \neq 0$. From (4.35) we get

$$\sum_{i=2}^{n} \left(a_{i1} y_1 + \sum_{k=2}^{\nu} y_k \right) x_i = 0\,;$$

$$\sum_{i=1}^{n} a_{ij} x_i = 0\,, \quad j = 2, \ldots, n\,.$$

(4.36)

This has the following set of solutions

$$\mathcal{N}_{01}^1 = \begin{cases} \mathcal{N}^5 & \text{if} \quad \det(A_0) \neq 0\,; \\ \mathcal{N}^6 & \text{if} \quad \det(A_0) = 0\,, \end{cases}$$

where

$$\mathcal{N}^5 = \left\{ ((0, \ldots, 0), (y_1, \ldots, y_\nu)) : y_1 \neq 0, \sum_{k=1}^{\nu} y_k \neq 0 \right\},$$

$$\mathcal{N}^6 = \left\{ ((0, x_2, \ldots, x_n), (y_1, \ldots, y_\nu)) : \sum_{i=2}^{n} x_i \right.$$

$$\left. = \sum_{i=2}^{n} \left(a_{i1} y_1 + \sum_{k=2}^{\nu} y_k \right) x_i = 0, A_0 x = 0, y_1 \neq 0, \sum_{k=1}^{\nu} y_k \neq 0 \right\}.$$

Subcase $x_1 \neq 0, y_1 = 0$. In this case from (4.35) we get

$$\sum_{k=2}^{\nu} (b_{k1} - 1)\, y_k = 0\,;$$

$$\sum_{k=2}^{\nu} b_{kl} y_k = 0\,, \quad l = 2, \ldots, \nu\,.$$

(4.37)

If $\det(B_0) \neq 0$ we get $y_2 = \cdots = y_\nu = 0$, this is impossible since we have condition $\sum_{k=1}^{\nu} y_k \neq 0$. For $\det(B_0) = 0$ the set of absolute nilpotents will be

$$\mathcal{N}_{10}^1 = \left\{ ((x_1, \ldots, x_n), (0, y_2, \ldots, y_\nu)) : x_1 \neq 0, \sum_{i=1}^{n} x_i = 0, \right.$$

$$\left. B_0 y = 0, \sum_{k=2}^{\nu} y_k \neq 0, \sum_{k=2}^{\nu} (b_{k1} - 1) y_k = 0 \right\}.$$

Subcase $x_1 \neq 0, y_1 \neq 0$. Denote $\mathbf{A} = (\mathbf{a}_{ij})_{i,j=1,\ldots,n}$, with

$$\mathbf{a}_{ij} = \begin{cases} 1 & \text{if } j = 1; \\ a_{ij} & \text{if } j \neq 1, \end{cases}$$

and $\mathbf{B} = (\mathbf{b}_{kl})_{k,l=1,\ldots,\nu}$, with

$$\mathbf{b}_{kl} = \begin{cases} b_{k1} - 1 & \text{if } l = 1; \\ b_{kl} & \text{if } l \neq 1. \end{cases}$$

In this case we have

$$\mathcal{N}_{11}^1 = \begin{cases} \emptyset & \text{if } \det(\mathbf{A}) \neq 0 \text{ or } \det(\mathbf{B}) \neq 0; \\ \mathcal{N}^9 & \text{if } \det(\mathbf{A}) = \det(\mathbf{B}) = 0, \end{cases}$$

where

$$\mathcal{N}^9 = \left\{ ((x_1, \ldots, x_n), (y_1, \ldots, y_\nu)) : \mathbf{A}x = 0, \mathbf{B}y = 0, \right.$$

$$\left. \sum_{i=1}^{n} \left(a_{i1} y_1 + \sum_{k=2}^{\nu} y_k \right) x_i = 0, \sum_{k=1}^{\nu} y_k \neq 0 \right\}.$$

- *Case* $\sum_{i=1}^{n} x_i \neq 0, \sum_{k=1}^{\nu} y_k = 0$. This case is similar to the previous case, to get the set of nilpotents, one has to change sets $\mathcal{N}_{ij}^1$, $i, j = 0, 1$ as follows. Replace x and y, also rearrange parameters a_{ij} with b_{kl}. Denote resulting sets by $\mathcal{N}_{ij}^2$, $i, j = 0, 1$, respectively.

Thus we have proved the following theorem

Theorem 4.8. *The full set $\mathcal{N}$ of absolute nilpotent elements of the algebra $\mathcal{B}_1$ is*

$$\mathcal{N} = \bigcup_{i=0}^{2} \left(\mathcal{N}_{00}^i \cup \mathcal{N}_{01}^i \cup \mathcal{N}_{10}^i \cup \mathcal{N}_{11}^i \right).$$

4.2 Algebra of "chicken" population

4.2.1 *Definition and basic properties of the EACP*

We consider a set $\{h_i, i = 1, \ldots, n\}$ (the set of "hen"s) and r (a "rooster").

Definition 4.1. Let $(\mathcal{C}, \cdot)$ be an algebra over a field K (with characteristic $\neq 2$). If it admits a basis $\{h_1, \ldots, h_n, r\}$, such that

$$h_i r = r h_i = \frac{1}{2}\left(\sum_{j=1}^{n} a_{ij} h_j + b_i r\right),$$

$$h_i h_j = 0, \qquad i, j = 1, \ldots, n; \quad rr = 0,$$
(4.38)

then this algebra is called an evolution algebra of a "chicken" population (EACP). We call the basis $\{h_1, \ldots, h_n, r\}$ a natural basis.

Remark 4.5. If

$$\sum_{j=1}^{n} a_{ij} = 1; \quad b_i = 1; \quad \text{for all} \quad i = 1, 2, \ldots, n$$
(4.39)

then the corresponding $\mathcal{C}$ is a particular case of an evolution algebra of a bisexual population, $\mathcal{B}$. The study of the algebra $\mathcal{B}$ is difficult, since it is determined by two cubic matrices. While the algebra $\mathcal{C}$ is simpler, since it is defined by a rectangular $n \times (n+1)$-matrix

$$M = \begin{pmatrix} a_{11} & a_{12} & \ldots & a_{1n} & b_1 \\ a_{21} & a_{22} & \ldots & a_{2n} & b_2 \\ \vdots & \vdots & \vdots & \vdots & \vdots \\ a_{n1} & a_{n2} & \ldots & a_{nn} & b_n \end{pmatrix},$$

which is called the matrix of structural constants of the algebra $\mathcal{C}$. This simplicity allows to obtain deeper results on $\mathcal{C}$ than on $\mathcal{B}$. Moreover, here we do not require the condition (4.39).

The general formula for the multiplication is the extension of (4.38) by bilinearity, i.e. for $x, y \in \mathbb{C}$,

$$x = \sum_{i=1}^{n} x_i h_i + ur, \quad y = \sum_{i=1}^{n} y_i h_i + vr$$

we obtain

$$xy = \frac{1}{2}\sum_{j=1}^{n}\left(\sum_{i=1}^{n}(vx_i + uy_i)a_{ij}\right)h_j + \frac{1}{2}\left(\sum_{i=1}^{n}(vx_i + uy_i)b_i\right)r$$
(4.40)

and

$$x^2 = xx = \sum_{j=1}^{n} \left(\sum_{i=1}^{n} (ux_i)a_{ij} \right) h_j + \left(\sum_{i=1}^{n} (ux_i)b_i \right) r. \qquad (4.41)$$

Now we shall give conditions on the matrix M under which $\mathbb{C}$ will be associative.

Theorem 4.9. *The algebra $\mathbb{C}$ is associative iff the elements of the corresponding matrix M satisfy the following*

$$\sum_{j=1}^{n} a_{ij}a_{jk} = 0, \quad b_i = 0, \quad \text{for any} \ \ i, k = 1, 2, \dots, n. \qquad (4.42)$$

Proof. *Necessity.* Assume the algebra $\mathbb{C}$ is associative. First we consider the equality $(xy)z = x(yz)$ for basis elements. If $x, y, z \in \{h_1, \dots, h_n\}$ or $x = y = z = r$ then the equality is obvious.

From $(h_i r)h_j = h_i(rh_j)$ we get

$$a_{jk}b_i = a_{ik}b_j, \quad \text{for all} \ \ i, j, k = 1, \dots, n.$$

The equality $(h_i h_j)r = h_i(h_j r)$ gives

$$a_{ik}b_j = 0, \quad b_i b_j = 0 \quad \text{for all} \ \ i, j, k = 1, \dots, n.$$

From $(rh_i)r = r(h_i r) = (h_i r)r = h_i(rr) = 0$ we obtain

$$\sum_{j=1}^{n} a_{ij}a_{jk} = 0, \quad \text{for all} \ \ i, k = 1, \dots, n.$$

These conditions imply the condition (4.42).

Sufficiency. Assume the condition (4.42) is satisfied, then the following lemma shows that the algebra $\mathbb{C}$ is associative. $\qquad \square$

Lemma 4.2. *If the condition* (4.42) *is satisfied then*

$$xyz = 0, \quad \text{for all} \ \ x, y, z \in \mathbb{C}.$$

Proof. Under condition (4.42) from (4.40) we get

$$xy = \frac{1}{2} \sum_{j=1}^{n} \left(\sum_{i=1}^{n} (vx_i + uy_i)a_{ij} \right) h_j.$$

Consequently, for $z = \sum_{m=1}^{n} z_m h_m + wr$, using condition (4.42) we get

$$(xy)z = \frac{1}{2} \sum_{j=1}^{n} w \left(\sum_{i=1}^{n} (vx_i + uy_i)a_{ij} \right) (h_j r)$$

$$= \frac{1}{4} \sum_{m=1}^{n} \left(w \sum_{j=1}^{n} \left(\sum_{i=1}^{n} (vx_i + uy_i)a_{ij} \right) a_{jm} \right) h_m$$

$$= \frac{1}{4} \sum_{m=1}^{n} \left(\sum_{i=1}^{n} \left[w(vx_i + uy_i) \left(\sum_{j=1}^{n} a_{ij} a_{jm} \right) \right] \right) h_m = 0.$$

By this lemma and above mentioned properties we get

Corollary 4.3. *If the condition* (4.42) *is satisfied then algebra* $\mathbb{C}$ *is alternative, power associative, satisfies Jacobi and Jordan identities.*

Note that the conditions (4.39) and (4.42) cannot be satisfied simultaneously, so the corresponding algebra $\mathcal{B}$ of a bisexual population is not associative.

Example 4.5. The following matrix M (for $n = 2$) satisfies the condition (4.42):

$$M = \begin{pmatrix} a & b & 0 \\ c & -a & 0 \end{pmatrix},$$

for any a, b, c with $a^2 = -bc$.

Note that for an EACP notions as nil, nilpotent and right nilpotent algebras are equivalent. However, the indexes of nility, right nilpotency and nilpotency do not coincide in general.

The following is also a corollary of Lemma 4.2.

Corollary 4.4. *If the condition* (4.42) *is satisfied then algebra* $\mathbb{C}$ *is nilpotent with nilpotency index equal 3.*

Recall that an algebra is unital or unitary if it has an element e with $ex = x = xe$ for all x in the algebra.

Proposition 4.6. *The algebra* $\mathbb{C}$ *is not unital.*

Proof. Assume $e = \sum_{i=1}^{n} a_i h_i + br$ is a unity element. We then have $eh_i = h_i$ which gives

$$ba_{jj} = 1; \quad ba_{jm} = 0, \, m \neq j; \quad bb_j = 0, \quad \text{for any } j = 1, \ldots, n. \quad (4.43)$$

From $er = r$ we get

$$\sum_{i=1}^{n} a_i a_{ij} = 0, \quad \text{for any} \quad j = 1, \ldots, n; \quad \sum_{i=1}^{n} a_i b_i = 1. \qquad (4.44)$$

From system (4.43) we get $b \neq 0$ and $b_i = 0$ for all i. But for this b_i the second equation of the system (4.44) is not satisfied. This completes the proof. $\qquad \square$

An algebra $\mathcal{A}$ is a division algebra if for every $a, b \in \mathcal{A}$ with $a \neq 0$ the equations $ax = b$ and $xa = b$ are solvable in $\mathcal{A}$.

Proposition 4.7. *The algebra $\mathbb{C}$ is not a division algebra.*

Proof. Since $\mathbb{C}$ is a commutative algebra we shall check only $ax = b$. For coordinates of any $a = \sum_{i=1}^{n} \alpha_i h_i + \alpha r$, $b = \sum_{i=1}^{n} \beta_i h_i + \beta r$, $x = \sum_{i=1}^{n} x_i h_i + ur$ the equation $ax = b$ has the following form

$$\left(\sum_{i=1}^{n} a_{ij} \alpha_i \right) u + \alpha \sum_{i=1}^{n} a_{ij} x_i = 2\beta_j,$$

$$\left(\sum_{i=1}^{n} b_i \alpha_i \right) u + \alpha \sum_{i=1}^{n} b_i x_i = 2\beta, \qquad j = 1, \ldots, n.$$

So this is a linear system with $n + 1$ unknowns $x_1, \ldots, x_n, u$. This system can be written as $\mathbf{M}y = B$ where $y^T = (x_1, \ldots, x_n, u)$, $B = 2(\beta_1, \ldots, \beta_n, \beta)$ and

$$\mathbf{M} = \alpha^n \cdot \begin{pmatrix} a_{11} & a_{21} & \cdots & a_{n1} & \sum_{i=1}^{n} a_{i1}\alpha_i \\ a_{12} & a_{22} & \cdots & a_{n2} & \sum_{i=1}^{n} a_{i2}\alpha_i \\ \vdots & \vdots & \vdots & \vdots & \vdots \\ a_{1n} & a_{2n} & \cdots & a_{nn} & \sum_{i=1}^{n} a_{in}\alpha_i \\ b_1 & b_2 & \cdots & b_n & \sum_{i=1}^{n} b_i\alpha_i \end{pmatrix}.$$

By the very known Kronecker–Capelli theorem the system of linear equations $\mathbf{M}y = B$ has a solution if and only if the rank of the matrix $\mathbf{M}$ is equal to the rank of its augmented matrix $(\mathbf{M}|B)$. Since the last column of the matrix $\mathbf{M}$ is a linear combination of the other columns of the matrix, we have $\det(\mathbf{M}) = 0$. Consequently $\operatorname{rank}\mathbf{M} \leq n$. Moreover, since the dimension of the algebra $\mathbb{C}$ is $n + 1$ one can choose b, i.e. the vector B such that $\operatorname{rank}(\mathbf{M}|B) = 1 + \operatorname{rank}\mathbf{M}$. Then for such b the equation $ax = b$ is not solvable. This completes the proof. $\qquad \square$

4.2.2 *Evolution subalgebras and operator of* $\mathbb{C}$

By analogues of Definition 4, page 23 of [254] we give the following

Definition 4.2.

(1) Let $\mathbb{C}$ be an EACP, and $\mathbb{C}_1$ be a subspace of $\mathbb{C}$. If $\mathbb{C}_1$ has a natural basis, $\{h'_1, h'_2, \ldots, h'_m, r'\}$, with multiplication table like (1.7), we call $\mathbb{C}_1$ an evolution subalgebra of a CP.
(2) Let $I \subset \mathbb{C}$ be an evolution subalgebra of a CP. If $\mathbb{C}I \subseteq I$, we call I an evolution ideal of a CP.
(3) Let $\mathbb{C}$ and $\mathcal{D}$ be EACPs, a linear homomorphism f from $\mathbb{C}$ to $\mathcal{D}$ is called an evolution homomorphism if f is an algebraic map and for a natural basis $\{h_1, \ldots, h_n, r\}$ of $\mathbb{C}$, $\{f(r), f(h_i), i = 1, \ldots, n\}$ spans an evolution subalgebra of a CP in $\mathcal{D}$. Furthermore, if an evolution homomorphism is one to one and onto, it is an evolution isomorphism.
(4) An EACP $\mathbb{C}$ is simple if it has no proper evolution ideals.
(5) $\mathbb{C}$ is irreducible if it has no proper subalgebras.

The following proposition gives some evolution subalgebras of a CP.

Proposition 4.8. *Let $\mathbb{C}$ be an EACP with the natural basis $\{h_1, \ldots, h_n, r\}$ and matrix*

$$
M = \begin{pmatrix}
a_{11} & 0 & 0 & \ldots & 0 & b_1 \\
a_{21} & a_{22} & 0 & \ldots & 0 & b_2 \\
\vdots & \vdots & \vdots & \vdots & \vdots & \vdots \\
a_{n1} & a_{n2} & a_{n3} & \ldots & a_{nn} & b_n
\end{pmatrix}.
$$

Then for each m, $1 \leq m \leq n$, the algebra $\mathbb{C}_m = \langle h_1, \ldots, h_m, r \rangle \subset \mathbb{C}$ is an evolution subalgebra of a CP.

Proof. For given M one can see that $\mathbb{C}_m$ is closed under multiplication. The chosen subset of the natural basis of $\mathbb{C}$ satisfies (4.38). $\qquad\square$

The following is an example of a subalgebra of $\mathbb{C}$, which is not an evolution subalgebra of a CP.

Example 4.6. Let $\mathbb{C}$ be EACP with basis $\{h_1, h_2, h_3, r\}$ and multiplication defined by $h_i r = h_i + r$, $i = 1, 2, 3$. Take $u_1 = h_1 + r$, $u_2 = h_2 + r$. Then

$$
(au_1 + bu_2)(cu_1 + du_2) = acu_1^2 + (ad + bc)u_1u_2 + bdu_2^2
$$
$$
= (2ac + ad + bc)u_1 + (2bd + ad + bc)u_2.
$$

Hence, $F = Ku_1 + Ku_2$ is a subalgebra of $\mathbb{C}$, but it is not an evolution subalgebra of a CP. Indeed, assume v_1, v_2 be a basis of F. Then $v_1 = au_1 + bu_2$ and $v_2 = cu_1 + du_2$ for some $a, b, c, d \in K$ such that $D = ad - bc \neq 0$. We have $v_1^2 = (2a^2 + 2ab)u_1 + (2b^2 + 2ab)u_2$ and $v_2^2 = (2c^2 + 2cd)u_1 + (2d^2 + 2cd)u_2$. We must have $v_1^2 = v_2^2 = 0$, i.e.

$$a^2 + ab = 0, \quad b^2 + ab = 0, \quad c^2 + cd = 0, \quad d^2 + cd = 0.$$

From this we get $a = -b$ and $c = -d$. Then $D = 0$, a contradiction. If $a = 0$ then $b = 0$ (resp. $c = 0$ then $d = 0$), we reach the same contradiction. Hence $v_1^2 \neq 0$ and $v_2^2 \neq 0$, and consequently F is not an evolution subalgebra of a CP.

Let $\mathbb{C}$ be an EACP on the field $K = \mathbb{R}$, with a basis set $\{h_1, \ldots, h_n, r\}$ and $x = \sum_{i=1}^{n} x_i h_i + ur \in \mathbb{C}$. Formula (4.41) can be written as

$$x^2 = V(x) = \sum_{j=1}^{n} x_j' h_j + u'r,$$

where the evolution operator $V : x \in \mathbb{C} \to x' = V(x) \in \mathbb{C}$ is defined as the following

$$V : \begin{cases} x_j' = u \sum_{i=1}^{n} a_{ij}x_i, \quad j = 1, \ldots, n, \\ u' = u \sum_{i=1}^{n} b_i x_i. \end{cases}$$

If we write $x^{[k]}$ for the power $(\cdots (x^2)^2 \cdots)$ (k times) with $x^{[0]} = x$ then the trajectory with initial x is given by k times iteration of the operator V, i.e. $V^k(x) = x^{[k]}$. This algebraic interpretation of the trajectory is useful to connect powers of an element of the algebra and with the dynamical system generated by the evolution operator V. For example, zeros of V, i.e. $V(x) = 0$ correspond to absolute nilpotent elements of $\mathbb{C}$ and fixed points of V, i.e. $V(x) = x$ correspond to idempotent elements of $\mathbb{C}$.

For $x = \sum_{i=1}^{n} x_i h_i + ur$ define a functional $\mathbf{b}$ as

$$\mathbf{b}(x) = \sum_{i=1}^{n} b_i x_i.$$

The following proposition fully describes the set $\mathcal{N}$ of absolute nilpotent elements of $\mathbb{C}$ with the ground field $K = \mathbb{R}$.

Proposition 4.9. *We have*
$$\mathcal{N} = \{(x, u) \in \mathbb{C} : u = 0\}$$

$$\cup \begin{cases} \{(0, \ldots, 0, u) \in C : u \neq 0\}, & if \quad \det(\mathbf{A}) \neq 0, \\ \{(x, u) \in \mathbb{C} : u \neq 0, \quad \mathbf{A}x = 0, \quad \mathbf{b}(x) = 0\} & if \quad \det(\mathbf{A}) = 0, \end{cases}$$

where $x = (x_1, \ldots, x_n)$, $\mathbf{A} = (a_{ij})$.

Proof. An absolute nilpotent element $(x_1, \ldots, x_n, u)$ satisfies

$$\begin{cases} u \sum_{i=1}^n a_{ij} x_i = 0, & j = 1, \ldots, n, \\ u \sum_{i=1}^n b_i x_i = 0. \end{cases}$$

The proof follows from a simple analysis of this system. $\qquad\square$

Now we shall describe idempotent elements of $\mathbb{C}$, these are solutions to $x^2 = x$. Such an element $x = (x_1, \ldots, x_n, u)$ satisfies the following

$$\begin{cases} u \sum_{i=1}^n a_{ij} x_i = x_i, & j = 1, \ldots, n, \\ u \sum_{i=1}^n b_i x_i = u. \end{cases} \tag{4.45}$$

Case $u = 0$. If $u = 0$ then from (4.45) we get $x_i = 0$ for all $i = 1, \ldots, n$. Hence $x = 0$ is a unique idempotent element.

Case $u \neq 0$. Consider a matrix $T_u = (t_{ij})_{i,j=1,\ldots,n}$ such that

$$t_{ij} = \begin{cases} u a_{ji} & \text{if } i \neq j, \\ u a_{ii} - 1 & \text{if } i = j. \end{cases}$$

Then first n equations of the system (4.45) can be written as

$$T_u x = 0, \tag{4.46}$$

where $x = (x_1, \ldots, x_n)$.

Consider now $u \in \mathbb{R} \setminus \{0\}$ as a parameter, then equation (4.46) has a unique solution $x = 0$ if $\det(T_u) \neq 0$, which gives $u = 0$, i.e. this is a contradiction to the assumption that $u \neq 0$.

If $\det(T_u) = 0$ then we fix a solution $u = u_* \neq 0$ of the equation $\det(T_u) = 0$. In this case there are infinitely many solutions $x^* = (x_1^*, \ldots, x_n^*)$ of $T_{u_*} x = 0$. Substituting a solution x^* in the last equation of the system (4.45), we get

$$\mathbf{b}(x^*) = \sum_{i=1}^n b_i x_i^* = 1.$$

Denote by $\mathcal{I}d(\mathbb{C})$ the set of idempotent elements of $\mathbb{C}$. Hence we have proved the following

Proposition 4.10. *We have*

$$\mathcal{I}d(\mathbb{C}) = \{0\} \cup \{(x_1^*, \ldots, x_n^*, u_*) : u_* \neq 0, T_{u_*} x^* = 0, \mathbf{b}(x^*) = 1, \det(T_{u_*}) = 0\}.$$

4.2.3   *The enveloping algebra of an EACP*

For a given algebra $\mathcal{A}$ with ground field K, we recall that multiplication by elements of $\mathcal{A}$ on the left or on the right give rise to left and right K-linear transformations of $\mathcal{A}$ given by $L_a(x) = ax$ and $R_a(x) = xa$. The *enveloping algebra*, denoted by $\mathcal{E}(\mathcal{A})$, of a non-associative algebra $\mathcal{A}$ is the subalgebra of the full algebra of K-endomorphisms of $\mathcal{A}$ which is generated by the left and right multiplication maps of $\mathcal{A}$. This enveloping algebra is necessarily associative, even though $\mathcal{A}$ may be non-associative. In a sense this makes the enveloping algebra "the smallest associative algebra containing $\mathcal{A}$".

Since an EACP, $\mathbb{C}$, is a commutative algebra the right and left operators coincide, so we use only L_a.

Theorem 4.10. *Let $\mathbb{C}$ be an EACP with a natural basis $\{h_1, \ldots, h_n, r\}$ and matrix of structural constants $M = \mathbf{A} \oplus \mathbf{b}$. If $\det(\mathbf{A}) \neq 0$ then $\{L_1, \ldots, L_n, L_r\}$ (where $L_i = L_{h_i}$) spans a linear space, denoted by $\mathrm{span}(L, \mathbb{C})$, which is the set of all operators of left multiplication. The vector space $\mathrm{span}(L, \mathbb{C})$ and $\mathbb{C}$ have the same dimension.*

Proof. For $x = \sum_{i=1}^{n} x_i h_i + ur \in \mathbb{C}$ by linearity of multiplication in $\mathbb{C}$ we can write L_x as the following

$$L_x = \sum_{i=1}^{n} x_i L_i + u L_r.$$

If $L_x = L_y$, for $y = \sum_{i=1}^{n} y_i h_i + vr \in \mathbb{C}$, then

$$\left(\sum_{i=1}^{n} x_i h_i + ur \right) h_j = \left(\sum_{i=1}^{n} y_i h_i + vr \right) h_j$$

implies $urh_j = vrh_j$, i.e. $u = v$. From

$$\left(\sum_{i=1}^{n} x_i h_i + ur \right) r = \left(\sum_{i=1}^{n} y_i h_i + vr \right) r$$

we get

$$\sum_{j=1}^{n} \left(\sum_{i=1}^{n} (x_i - y_i) a_{ij} \right) h_j + \left(\sum_{i=1}^{n} (x_i - y_i) b_i \right) r = 0.$$

Hence

$$\sum_{i=1}^{n} (x_i - y_i) a_{ij} = 0, \quad \sum_{i=1}^{n} (x_i - y_i) b_i = 0.$$

By assumption $\det(\mathbf{A}) \neq 0$ from the last system we get $x_i = y_i$ for all $i = 1, \ldots, n$. Thus $x = y$. This means that L_x is an injection. So the linear space that is spanned by all operators of left multiplication can be spanned by the set $\{L_i, i = 1, \ldots, n, r\}$. This set is a basis for $\mathrm{span}(L, \mathbb{C})$. $\square$

Proposition 4.11. *For any $x \in \mathbb{C}$ and any $i, i_1, i_2, \ldots, i_m \in \{1, 2, \ldots, n\}$ the following hold*

$$L_{i_m} \circ L_{i_{m-1}} \circ \cdots \circ L_{i_1}(x) = \left(\frac{1}{2^{m-1}} \prod_{j=1}^{m-1} b_{i_j} \right) L_{i_m}(x), \qquad (4.47)$$

$$L_r \circ L_i(x) = \frac{1}{2} \sum_{j=1}^{n} a_{ij} L_j(x), \qquad (4.48)$$

$$L_i \circ L_r(x) = \frac{\mathbf{b}(x)}{2} L_i(r). \qquad (4.49)$$

Proof. For $x = \sum_{i=1}^{n} x_i h_i + ur$ note that $L_j(x) = u h_j r$, $j = 1, \ldots, n$.

(1) To prove (4.47) we use mathematical induction over m. For $m = 2$ we have

$$(L_{i_2} \circ L_{i_1})(x) = h_{i_2}(h_{i_1} x) = h_{i_2}(u h_{i_1} r)$$

$$= h_{i_2} \frac{u}{2} \left(\sum_{i=1}^{n} a_{i_1 j} h_j + b_{i_1} r \right) = \frac{b_{i_1}}{2} (u h_{i_2} r) = \frac{b_{i_1}}{2} L_{i_2}(x).$$

Assume now that the formula (4.47) is true for m, we shall prove it for $m + 1$:

$$L_{i_{m+1}} \circ L_{i_m} \circ \cdots \circ L_{i_1}(x) = L_{i_{m+1}} \circ \left(\frac{1}{2^{m-1}} \prod_{j=1}^{m-1} b_{i_j} L_{i_m}(x) \right)$$

$$= \left(\frac{1}{2^{m-1}} \prod_{j=1}^{m-1} b_{i_j} \right) L_{i_{m+1}} \circ L_{i_m}(x) = \left(\frac{1}{2^{m}} \prod_{j=1}^{m} b_{i_j} \right) L_{i_{m+1}}(x).$$

(2) Proof of (4.48):

$$L_r \circ L_i(x) = r(h_i x) = r(u h_i r) = \frac{1}{2} \sum_{j=1}^{n} a_{ij}(u r h_j) = \frac{1}{2} \sum_{j=1}^{n} a_{ij} L_j(x).$$

(3) Proof of (4.49):

$$L_i \circ L_r(x) = h_i(rx) = h_i \left(\sum_{j=1}^{n} x_j(r h_j) \right)$$

$$= \frac{h_i}{2} \left(\sum_{m=1}^{n} \left[\sum_{j=1}^{n} a_{im} x_j \right] h_m + \left(\sum_{j=1}^{n} x_j b_j \right) r \right) = \frac{\mathbf{b}(x)}{2} L_i(r). \qquad \square$$

4.2.4 *The centroid of an EACP*

We recall (see [254]) that the centroid $\Gamma(\mathcal{A})$ of an algebra $\mathcal{A}$ is the set of all linear transformations $T \in \mathrm{Hom}(\mathcal{A}, \mathcal{A})$ that commute with all left and right multiplication operators

$$TL_x = L_x T, \qquad TR_y = R_y T, \quad \text{for all} \quad x, y \in \mathcal{A}.$$

An algebra $\mathcal{A}$ over a field K is centroidal if $\Gamma(\mathcal{A}) \cong K$.

Theorem 4.11. *Let $\mathbb{C}$ be an EACP with a natural basis $\{h_1, \ldots, h_n, r\}$ and matrix of structural constants $M = \mathbf{A} \oplus \mathbf{b}$. If $\det(\mathbf{A}) \neq 0$ then $\mathbb{C}$ is centroidal.*

Proof.   Let $T \in \Gamma(\mathbb{C})$. Assume

$$T(h_i) = \sum_{j=1}^{n} t_{ij} h_j + t_i r, \quad T(r) = \sum_{k=1}^{n} \tau_k h_k + \tau r.$$

We have

$$TL_j(h_i) = T(h_j h_i) = 0 = L_j T(h_i) = h_j \left(\sum_{k=1}^{n} t_{ik} h_k + t_i r \right) = t_i h_j r.$$

This gives

$$t_i = 0, \quad \text{for all} \quad i = 1, \ldots, n. \tag{4.50}$$

Now consider

$$TL_j(r) = T(h_j r) = \frac{1}{2} T \left(\sum_{m=1}^{n} a_{jm} h_m + b_j r \right) = \frac{1}{2} \sum_{m=1}^{n} a_{jm} T(h_m) + \frac{b_j}{2} T(r)$$

$$= \frac{1}{2} \sum_{k=1}^{n} \left(\sum_{m=1}^{n} a_{jm} t_{mk} + b_j \tau_k \right) h_k + \frac{1}{2} \left(\sum_{m=1}^{n} a_{jm} t_m + b_j \tau \right) r.$$

In another way we have

$$L_j T(r) = h_j \left(\sum_{k=1}^{n} \tau_k h_k + \tau r \right) = \tau h_j r = \frac{\tau}{2} \left(\sum_{k=1}^{n} a_{jk} h_k + b_j r \right).$$

According to (4.50) we should have

$$\sum_{m=1}^{n} a_{jm} t_{mk} + b_j \tau_k = \tau a_{jk}. \tag{4.51}$$

Furthermore,

$$TL_r(h_j) = \frac{1}{2} \sum_{k=1}^{n} \left[\sum_{m=1}^{n} a_{jm} t_{mk} + b_j \tau_k \right] h_k + \frac{1}{2} \left[\sum_{m=1}^{n} a_{jm} t_m + b_j \tau \right] r$$

and

$$L_r T(h_j) = \frac{1}{2} \sum_{k=1}^{n} \left[\sum_{m=1}^{n} a_{mk} t_{jm} \right] h_k + \frac{1}{2} \left[\sum_{m=1}^{n} t_{jm} b_m \right] r.$$

These equalities imply

$$\sum_{m=1}^{n} a_{jm} t_{mk} + b_j \tau_k = \sum_{m=1}^{n} a_{mk} t_{jm}$$

$$\sum_{m=1}^{n} a_{jm} t_m + b_j \tau = \sum_{m=1}^{n} t_{jm} b_m. \tag{4.52}$$

Finally,

$$TL_r(r) = T(rr) = 0 = L_r T(r) = \frac{1}{2} \sum_{j=1}^{n} \left(\sum_{k=1}^{n} a_{kj} \tau_k \right) h_j + \frac{1}{2} \left(\sum_{k=1}^{n} \tau_k b_k \right) r.$$

Consequently,

$$\sum_{k=1}^{n} a_{kj} \tau_k = 0, \quad j = 1, \ldots, n; \qquad \sum_{k=1}^{n} b_k \tau_k = 0. \tag{4.53}$$

Since $\det(\mathbf{A}) \neq 0$ from (4.53) we get

$$\tau_i = 0, \quad \text{for all} \quad i = 1, \ldots, n. \tag{4.54}$$

Using (4.54), from (4.51) we obtain

$$\sum_{m=1}^{n} a_{jm} t_{mk} = \tau a_{jk}, \quad j, k = 1, \ldots, n. \tag{4.55}$$

Again using $\det(\mathbf{A}) \neq 0$, by the Cramer's rule, from (4.55) we get the following solution

$$t_{mk} = \begin{cases} 0, & \text{if } m \neq k \\ \tau, & \text{if } m = k. \end{cases} \tag{4.56}$$

Note that solutions (4.50), (4.54) and (4.56) satisfy system (4.52). Hence we obtain

$$T(h_i) = \tau h_i, \quad T(r) = \tau r,$$

where τ is a scalar in the ground field K. That is T is a scalar multiplication. Consequently, $\Gamma(\mathbb{C}) \cong K$ and $\mathbb{C}$ is centroidal. $\qquad \square$

4.2.5   *The structure of EACP*

Let $\mathcal{C}$ and $\mathcal{D}$ be EACPs, following [42], a linear homomorphism f from $\mathcal{C}$ to $\mathcal{D}$ is an evolution *homomorphism* if f is an algebraic map and for a natural basis $\{h_1, \ldots, h_n, r\}$ of $\mathcal{C}$, $\{f(r), f(h_i), i = 1, \ldots, n\}$ spans an evolution subalgebra in $\mathcal{D}$. If an evolution homomorphism is one to one and onto, it is an evolution *isomorphism*.

Theorem 4.12. *If $b_i = 0$ for any $i = 1, \ldots n$, then $\mathcal{C}$ is a solvable EACP and it is isomorphic to one of the following pairwise non isomorphic algebras:*

$$h_{\sum_{i=1}^{s-1} n_i + 1} r = \frac{1}{2}\left(a_s h_{\sum_{i=1}^{s-1} n_i + 1} + h_{\sum_{i=1}^{s-1} n_i + 2}\right),$$

$$h_{\sum_{i=1}^{s-1} n_i + 2} r = \frac{1}{2}\left(a_s h_{\sum_{i=1}^{s-1} n_i + 2} + h_{\sum_{i=1}^{s-1} n_i + 3}\right),$$

$$\cdots\cdots\cdots\cdots\cdots\cdots\cdots$$

$$h_{\sum_{i=1}^{s-1} n_i + n_s - 1} r = \frac{1}{2}\left(a_s h_{\sum_{i=1}^{s-1} n_i + n_s - 1} + h_{\sum_{i=1}^{s-1} n_i + n_s}\right),$$

$$h_{\sum_{i=1}^{s-1} n_i + n_s} r = \frac{1}{2}\left(a_s h_{\sum_{i=1}^{s-1} n_i + n_s}\right),$$

where $n = n_1 + \cdots + n_k$, $s = 1, \ldots, k$ *and* $(a_1, a_2, \ldots, a_k) = (1, a_2, \ldots, a_k)$ *or* $(a_1, a_2, \ldots, a_k) = (0, 0, \ldots, 0)$.

Proof.   If $b_i = 0$ for any $i = 1, \ldots n$, then we have the multiplication

$$h_i r = \frac{1}{2}\left(\sum_{j=1}^{n} a_{i,j} h_j\right), \quad h_i h_j = rr = 0.$$

It follows that $\mathcal{C}^{(2)} \subseteq \langle h_1, h_2, \ldots, h_n\rangle$, which implies $\mathcal{C}^{(3)} = 0$. Hence, $\mathcal{C}$ is solvable.

Since $\mathcal{C}$ is a complex EACP, then by Jordan decomposition[1] we conclude that there exists a basis of $\mathcal{C}$ such that the matrix of the operator R_r has Jordan form, i.e. diagonal block consist of Jordan blocks. We denote $R_r = J_1 \oplus J_2 \oplus \ldots J_k$, where a_i are diagonal elements of J_i and n_i are sizes of blocks J_i, $1 \leq i \leq k$. Obviously, $n_1 + n_2 + \cdots + n_k = n$.

Then the table of multiplication of the algebra is follows:

$$h_{\sum_{i=1}^{s-1} n_i + 1} r = \frac{1}{2}\left(a_s h_{\sum_{i=1}^{s-1} n_i + 1} + h_{\sum_{i=1}^{s-1} n_i + 2}\right),$$

---

[1] See https://en.wikipedia.org/wiki/Jordan_normal_form.

$$h_{\sum_{i=1}^{s-1} n_i+2}r = \frac{1}{2}\left(a_s h_{\sum_{i=1}^{s-1} n_i+2} + h_{\sum_{i=1}^{s-1} n_i+3}\right),$$

$$\cdots\cdots\cdots\cdots\cdots\cdots$$

$$h_{\sum_{i=1}^{s-1} n_i+n_s-1}r = \frac{1}{2}\left(a_s h_{\sum_{i=1}^{s-1} n_i+n_s-1} + h_{\sum_{i=1}^{s-1} n_i+n_s}\right),$$

$$h_{\sum_{i=1}^{s-1} n_i+n_s}r = \frac{1}{2}\left(a_s h_{\sum_{i=1}^{s-1} n_i+n_s}\right),$$

where $s = 1, \ldots, k$.

If $a_{s_0} \neq 0$, for some $1 \leq s_0 \leq k$, then without loss of generality we can assume that $a_1 \neq 0$.

In this case putting: $a_s' = \frac{a_s}{a_1}$, $2 \leq s \leq k$ and

$$h'_{\sum_{i=1}^{s-1} n_i+j} = \frac{h_{\sum_{i=1}^{s-1} n_i+j}}{a_1^j}, \quad 1 \leq j \leq n_s,\ 1 \leq s \leq k,$$

one can assume that $a_1 = 1$. $\qquad\square$

Remark 4.6. From the above description we get that for any s, $1 \leq s \leq k$

$$< h_{\sum_{i=1}^{s-1} n_i+1}, h_{\sum_{i=1}^{s-1} n_i+2}, \ldots, h_{\sum_{i=1}^{s-1} n_i+n_s-1}, h_{\sum_{i=1}^{s-1} n_i+n_s} >$$

is an ideal of the EACP.

Moreover, if $(a_1, a_2, \ldots, a_k) = (0, 0, \ldots, 0)$, then EACP is nilpotent.

Suppose that there exists $b_{i_0} \neq 0$, $i_0 \in \{1, \ldots n\}$, then by shifting of the basis elements h_i, $1 \leq i \leq n$ we can assume that $b_1 \neq 0$. Moreover, by the scaling $h_1' = \frac{1}{b_1}h_1$ we can assume that $b_1 = 1$. So, we have

$$h_1 r = \frac{1}{2}\left(\sum_{j=1}^{n} a_{1,j}h_j + r\right), \qquad h_i r = \frac{1}{2}\left(\sum_{j=1}^{n} a_{i,j}h_j + b_i r\right), \quad 2 \leq i \leq n.$$

In obtained table of multiplication we take the following basis transformation

$$h_i' = h_i - b_i h_1,\ 2 \leq i \leq n.$$

Therefore, we obtain the table of multiplication

$$h_1 r = \frac{1}{2}\left(\sum_{j=1}^{n} a_{1,j}h_j + r\right), \qquad h_i r = \frac{1}{2}\left(\sum_{j=1}^{n} a_{i,j}h_j\right), \quad 2 \leq i \leq n.$$

Thus, we obtain the following result

Proposition 4.12. *Let $\mathcal{C}$ be an EACP, then there exists a basis $\{h_1, h_2, \ldots, h_n, r\}$ such that $\mathcal{C}$ on this basis is represented by the table of multiplication as follows*

$$h_1 r = \frac{1}{2}\left(\sum_{j=1}^{n} a_{1,j}h_j + \delta r\right), \delta \in \{0, 1\},\ h_i r = \frac{1}{2}\left(\sum_{j=1}^{n} a_{i,j}h_j\right), 2 \leq i \leq n.$$

4.2.6 *Classification of 2 and 3-dimensional EACP*

Let $\mathbb{C}$ be a 2-dimensional EACP and $\{h, r\}$ be a basis of this algebra.

It is evident that if $\dim \mathbb{C}^2 = 0$ then $\mathbb{C}$ is an abelian algebra, i.e. an algebra with all products equal to zero.

Proposition 4.13. *Any 2-dimensional, non-trivial EACP $\mathbb{C}$ is isomorphic to one of the following pairwise non isomorphic algebras:*

$\mathbb{C}_1$: $rh = hr = h$, $h^2 = r^2 = 0$,
$\mathbb{C}_2$: $rh = hr = \frac{1}{2}(h + r)$, $h^2 = r^2 = 0$.

Proof. For an EACP $\mathbb{C}$ we have
$$hr = \frac{1}{2}(ah + br), \qquad h^2 = r^2 = 0.$$

Case: $a \neq 0$, $b = 0$. By change of basis $h' = h$ and $r' = \frac{2}{a}r$ we get the algebra $\mathbb{C}_1$.

Case: $a = 0$, $b \neq 0$. Take $h' = r$ and $r' = \frac{2}{b}h$ then we get the algebra $\mathbb{C}_1$.

Case: $a \neq 0$, $b \neq 0$. The change $h' = \frac{1}{a}r$, and $r' = \frac{1}{b}h$ implies the algebra $\mathbb{C}_2$.

Since $\mathbb{C}_1^2\mathbb{C}_1^2 = 0$ and $\mathbb{C}_2^2\mathbb{C}_2^2 \neq 0$, the algebras $\mathbb{C}_1$ and $\mathbb{C}_2$ are not isomorphic. $\qquad\qquad\square$

Note that the algebra $\mathbb{C}_2$ is known as the sex differentiation algebra [211].

In the following theorem we present the classification of three-dimensional EACP.

Theorem 4.13. *An arbitrary three dimensional complex EACP $\mathcal{C}$ is isomorphic to one of the following pairwise non-isomorphic algebras*

1) If $\dim \mathcal{C}^2 = 1$ then
$$\mathcal{C}_1 : h_1 r = \tfrac{1}{2}r$$
$$\mathcal{C}_2 : h_1 r = \tfrac{1}{2}h_2;$$
$$\mathcal{C}_3 : h_1 r = \tfrac{1}{2}h_1 + \tfrac{1}{2}r.$$

2) If $\dim \mathcal{C}^2 = 2$ then

$$\mathcal{C}_4 : \qquad h_1 r = \tfrac{1}{2}(h_1 + h_2), \qquad\qquad h_2 r = \tfrac{1}{2}h_2;$$
$$\mathcal{C}_5(\beta) : \quad h_1 r = \tfrac{1}{2}h_1, \qquad\qquad\quad h_2 r = \tfrac{\beta}{2}h_2, \ \beta \neq 0;$$
$$\mathcal{C}_6(\alpha, \beta) : h_1 r = \tfrac{1}{2}(\alpha h_1 + \beta h_2 + r), \ h_2 r = \tfrac{1}{2}h_1;$$
$$\mathcal{C}_7(\alpha) : \quad h_1 r = \tfrac{1}{2}(\alpha h_1 + r), \qquad\quad h_2 r = \tfrac{1}{2}h_2;$$
$$\mathcal{C}_8 : \qquad h_1 r = \tfrac{1}{2}(h_1 + h_2 + r), \quad h_2 r = \tfrac{1}{2}h_2.$$

where one of non-zero parameter α, β in the algebra $C_6(\alpha, \beta)$ can be assumed to be equal to 1.

Proof. See [42], [136], [142]. $\qquad\square$

Using the explicit form of algebras C_i, $i = 1, \ldots, 8$ mentioned in Theorem 4.13 one can prove the following

Proposition 4.14.

- *The algebras C_i, $i = 1, 4, 5$ are solvable.*
- *The algebra C_2 is nilpotent.*
- *The algebras C_i, $i = 3, 6, 7, 8$ are non-solvable.*

4.2.7 *The description of complex EACP*

In this section we give the description of some $n + 1$-dimensional EACP.

According to Proposition 4.12, we obtain that any $n + 1$-dimensional EACP admits a basis $\{h_1, h_2, \ldots, h_n, r\}$ such that the table of multiplication of this basis has the form

$$h_1 r = \frac{1}{2}\left(\sum_{j=1}^{n} a_{1,j} h_j + \delta r\right), \delta \in \{0, 1\}, \quad h_i r = \frac{1}{2}\left(\sum_{j=1}^{n} a_{i,j} h_j\right), 2 \leq i \leq n.$$

Since the result of Theorem 4.12 gives a description in the case of $\delta = 0$, we shall consider only the case $\delta = 1$.

Let us introduce denotations

$$V_2 =< h_2, \ldots, h_n >, \quad R_{\bar{r}} = R_r|_{V_2}.$$

Then we have

$$h_i \bar{r} = \frac{1}{2}\left(\sum_{j=1}^{n} a_{i,j} h_j\right), \quad 2 \leq i \leq n. \tag{4.57}$$

Considering the equality (4.57) by modulo space $< h_1 >$, we obtain

$$h_i \bar{r} = \frac{1}{2}\left(\sum_{j=2}^{n} a_{i,j} h_j\right), \quad 2 \leq i \leq n \; mod(< h_1 >). \tag{4.58}$$

Similarly to the proof of Theorem 4.12 we obtain the following products

$$h_{\sum_{i=1}^{s-1} n_i + 2}\, r = \frac{1}{2}\left(a_s h_{\sum_{i=1}^{s-1} n_i + 2} + h_{\sum_{i=1}^{s-1} n_i + 3}\right),$$

$$h_{\sum_{i=1}^{s-1} n_i + 3}\, r = \frac{1}{2}\left(a_s h_{\sum_{i=1}^{s-1} n_i + 3} + h_{\sum_{i=1}^{s-1} n_i + 4}\right),$$

$$\cdots\cdots\cdots\cdots\cdots\cdots\cdots\cdots\cdots\cdots\cdots$$

$$h_{\sum_{i=1}^{s-1} n_i + n_s - 1}\, r = \frac{1}{2}\left(a_s h_{\sum_{i=1}^{s-1} n_i + n_s - 1} + h_{\sum_{i=1}^{s-1} n_i + n_s}\right),$$

$$h_{\sum_{i=1}^{s-1} n_i + n_s}\, r = \frac{1}{2}\left(a_s h_{\sum_{i=1}^{s-1} n_i + n_s}\right),$$

where $s = 1,\ldots k$, $n_1 + n_2 + \cdots + n_k = n - 1$ and $(a_1, a_2, \ldots, a_k) = (1, a_2, \ldots, a_k)$ or $(a_1, a_2, \ldots, a_k) = (0, 0, \ldots, 0)$.

Now go up to the space V_1, and obtain the following result:

Theorem 4.14. *There exist a basis $\{h_1, h_2, \ldots, h_n, r\}$ of EACP, such that the table of multiplication of this basis has the following form*

$$h_1 r = \frac{1}{2}\left(\sum_{j=1}^{n} a_{1,j} h_j + r\right),$$

$$h_{\sum_{i=1}^{s-1} n_i + 2}\, r = \frac{1}{2}\left(a_{\sum_{i=1}^{s-1} n_i + 2, 1} h_1 + a_s h_{\sum_{i=1}^{s-1} n_i + 2} + h_{\sum_{i=1}^{s-1} n_i + 3}\right),$$

$$h_{\sum_{i=1}^{s-1} n_i + 3}\, r = \frac{1}{2}\left(a_{\sum_{i=1}^{s-1} n_i + 3, 1} h_1 + a_s h_{\sum_{i=1}^{s-1} n_i + 3} + h_{\sum_{i=1}^{s-1} n_i + 4}\right),$$

$$\cdots\cdots\cdots\cdots\cdots\cdots\cdots\cdots\cdots\cdots\cdots$$

$$h_{\sum_{i=1}^{s-1} n_i + n_s - 1}\, r = \frac{1}{2}\left(a_{\sum_{i=1}^{s-1} n_i + n_s - 1, 1} h_1 + a_s h_{\sum_{i=1}^{s-1} n_i + n_s - 1} + h_{\sum_{i=1}^{s-1} n_i + n_s}\right),$$

$$h_{\sum_{i=1}^{s-1} n_i + n_s}\, r = \frac{1}{2}\left(a_{\sum_{i=1}^{s-1} n_i + n_s, 1} h_1 + a_s h_{\sum_{i=1}^{s-1} n_i + n_s}\right),$$

where $s = 1, \ldots, k$ and $a_1 = 1$.

The case of $k = 1$. Now we shall investigate the case of $k = 1$. By Theorem 4.14 we obtain the multiplication:

$$\begin{cases} h_1 r = \frac{1}{2}\left(\sum_{j=1}^{n} a_{1,j} h_j + r\right), \\[2mm] h_i r = \frac{1}{2}(a_{i,1} h_1 + \alpha h_i + h_{i+1}), & 2 \le i \le n - 1, \\[2mm] h_n r = \frac{1}{2}(a_{n,1} h + \alpha h_n), \end{cases}$$

where $\alpha \in \{0; 1\}$.

Assume there exists some m, $2 \leq m \leq n$ such that $a_{1,m} \neq 0$ and $a_{1,m}$ is the first non zero parameter. Putting

$$h'_i = \sum_{j=m}^{n} a_{1,j} h_{j-m+i}, \quad 2 \leq i \leq n,$$

we can assume that the table of multiplication has the form

$$\mathcal{C}^\alpha(a_i, m_1): \begin{cases} h_1 r = a_1 h_1 + \alpha h_{m_1} + r, & 2 \leq m_1 \leq n, \\ h_i r = a_i h_1 + \alpha h_i + h_{i+1}, & 2 \leq i \leq n-1, \\ h_n r = a_n h_1 + h_n. \end{cases}$$

where $\alpha \in \{0; 1\}$.

Consider two algebras $\mathcal{C}^1(a_i, m_1)$ and $\mathcal{C}^1(b_i, m_2)$, with basis $\{h_1, h_2, \ldots, h_n, r\}$ and $\{h'_1, h'_2, \ldots, h'_n, r'\}$, respectively. If $m_1 \neq m_2$, then without loss of generality we can suppose $m_1 < m_2$.

Theorem 4.15. *Two algebras $\mathcal{C}^1(a_i, m_1)$ and $\mathcal{C}^1(b_i, m_2)$ with $m_1 < m_2$ are isomorphic if and only if*

1) $a_i = b_i = 0$, $m_1 \leq i \leq n$;
2) $a_1 = b_1$, $b_1 \neq 1$, and $b_1 \neq a_{m_1-1} + 1$ in the case of $b_{m_1-1} \neq 0$; $b_1 \neq 1$, or $a_{m_1-2} \neq -1$ in the case of $b_{m_1-1} = 0$, $b_{m_1-2} \neq 0$;
3) there exist $B_2, B_3, \ldots, B_{m_1-1} \in \mathbb{C}$, with $B_2 \neq 0$, such that

$$b_i = \sum_{k=i}^{m_1-1} B_{k-i+2} u_k, \quad 2 \leq i \leq m_1 - 1.$$

Proof. *Necessity.* Let f be an isomorphism $f : \mathcal{C}^1(b_i, m_2) \to \mathcal{C}^1(a_i, m_1)$. Then

$$f(h'_i) = \sum_{j=1}^{n} A_{i,j} h_j + C_i r, \ 1 \leq i \leq n, \quad f(r') = \sum_{j=1}^{n} D_j h_j + C_{n+1} r.$$

Note that, without loss of generality we can assume $C_{n+1} \neq 0$.
From the following chain of equalities

$$0 = f(h'_i) f(h'_i)$$

$$= \left(\sum_{j=1}^{n} A_{i,j} h_j + C_i r \right) \left(\sum_{j=1}^{n} A_{i,j} h_j + C_i r \right) = 2C_i \sum_{j=1}^{n} A_{i,j} h_j r$$

$$= C_i \left[A_{i,2}h_2 + A_{i,1}h_{m_1} + A_{i,1}r + \sum_{j=1}^{n} A_{i,j}a_j h_1 + \sum_{j=3}^{n} (A_{i,j-1} + A_{i,j})h_j \right],$$

$$0 = f(r')f(r')$$

$$= \left(\sum_{j=1}^{n} D_j h_j + D_{n+1}r \right) \left(\sum_{j=1}^{n} D_j h_j + C_{n+1}r \right) = 2C_{n+1} \sum_{j=1}^{n} D_j h_j r$$

$$= C_{n+1} \left[D_2 h_2 + D_1 h_{m_1} + D_1 r + \sum_{j=1}^{n} D_j a_j h_1 + \sum_{j=3}^{n} (D_{j-1} + D_j)h_j \right],$$

we obtain the restrictions:

$$C_i \sum_{j=1}^{n} A_{i,j}a_j = 0, \ C_i A_{i,k} = 0, \quad 1 \le i \le n, \ 1 \le k \le n,$$

$$C_{n+1} \sum_{j=1}^{n} D_j a_j = 0, \ C_{n+1}D_k = 0, \quad 1 \le k \le n.$$

Since $C_{n+1} \ne 0$, we obtain $D_i = 0$ for $1 \le i \le n$. Moreover, it is not difficult to obtain that $C_i = 0$ for $1 \le i \le n$. Indeed, if there exist $C_{i_0} \ne 0$, for $2 \le i_0 \le n$, then it implies $A_{i_0,j} = 0$ for $1 \le j \le n$. It is a contradiction with the condition of the matrix of the general change is not singular, since $D_i = 0$ for $1 \le i \le n$.

Thus,

$$f(h'_i) = \sum_{j=1}^{n} A_{i,j}h_j, \ \ 1 \le i \le n, \quad f(r') = r.$$

Consider the multiplication

$$f(h'_n)f(r') = \sum_{j=1}^{n} A_{n,j}h_j r$$

$$= A_{n,2}h_2 + A_{n,1}h_{m_1} + A_{n,1}r + \sum_{j=1}^{n} A_{n,j}a_j h_1 + \sum_{j=3}^{n} (A_{n,j-2} + A_{n,j})h_j.$$

On the other hand,

$$f(h'_n)f(r') = b_n f(h'_1) + f(h'_n) = \sum_{j=1}^{n} (b_n A_{1,j} + A_{n,j})h_j.$$

Comparing the coefficients at the basis elements we obtain

$$\begin{cases} A_{n,1} = 0, \qquad \sum_{j=2}^{n} A_{n,j} a_j = b_n A_{1,1}, \\ 0 = b_n A_{1,2}, \\ A_{n,k} = b_{n,1} A_{1,k+1}, \ 2 \le k \le n-1. \end{cases} \tag{4.59}$$

Analogously, considering products

$$f(h'_i) f(r') = \sum_{j=1}^{n} A_{i,j} h_j r$$

$$= A_{i,2} h_2 + A_{i,1} h_{m_1} + A_{i,1} r + \sum_{j=1}^{n} A_{i,j} a_j h_1 + \sum_{j=3}^{n} (A_{i,j-1} + A_{i,j}) h_j$$

$$= b_i f(h'_1) + f(h'_i) + f(h'_{i+1}) = \sum_{j=1}^{n} (b_i A_{1,j} + A_{i,j} + A_{i+1,j}) h_j$$

for $2 \le i \le n-1$ we get

$$\begin{cases} A_{i,1} = 0, \\ \sum_{j=2}^{n} A_{i,j} a_j = b_i A_{1,1}, \\ b_i A_{1,2} + A_{i+1,2} = 0, \\ A_{i,k} = b_i A_{1,k+1} + A_{i+1,k+1}, \ 2 \le k \le n-1. \end{cases} \tag{4.60}$$

Consider the product

$$f(h'_1) f(r') = \sum_{j=1}^{n} A_{1,j} h_j r$$

$$= A_{1,2} h_2 + A_{1,1} h_{m_1} + A_{1,1} r + \sum_{j=1}^{n} A_{1,j} a_j h_1 + \sum_{j=3}^{n} (A_{1,j-1} + A_{1,j}) h_j$$

$$= b_1 f(h'_1) + f(h'_{m_2}) + f(r') = r + \sum_{j=1}^{n} (b_1 A_{1,j} + A_{m_2,j}) h_j.$$

From this we derive

$$\begin{cases} A_{1,1} = 1, \qquad \sum_{j=1}^{n} A_{1,j} a_j = b_1, \\ A_{1,2} = b_1 A_{1,2} + A_{m_2,2}, \\ A_{1,k} + A_{1,k+1} = b_1 A_{1,k+1} + A_{m_2,k+1}, \qquad 2 \le k \le n-1, \ k \ne m_1 - 1, \\ 1 + A_{1,m_1-1} + A_{1,m_1} = b_1 A_{1,m_1} + A_{m_2,m_1}. \end{cases}$$

$$\tag{4.61}$$

From $b_n A_{1,2} = 0$, we obtain $b_n = 0$. Indeed, if $b_n \neq 0$ then $A_{1,2} = 0$ and according to (4.60) we get $A_{i,2} = 0$ for $3 \leq i \leq n$. From $A_{n,2} = b_n A_{1,3}$ and by (4.60) we get

$$A_{1,3} = 0, \quad A_{2,2} = A_{3,3}, \quad A_{i,3} = 0, \quad 4 \leq i \leq n.$$

Recurrently, we obtain

$$A_{1,k} = 0, \quad A_{k,k} = A_{2,2}, \quad A_{i,k} = 0, \quad 2 \leq k \leq m_1, \ k+1 \leq i \leq n.$$

But since $m_1 < m_2$, from $1 + A_{1,m_1-1} + A_{1,m_1} = b_1 A_{1,m_1} + A_{m_2,m_1}$ we get incorrect equality $1 = 0$. It is a contradiction with assumption $b_n \neq 0$. Thus, $b_n = 0$. From this we have $A_{n,i} = 0$, $2 \leq i \leq n-1$ and $a_n = 0$. Continuing this process we obtain the condition 1), i.e.,

$$b_i = a_i = 0, \quad m_1 \leq i \leq n.$$

- If $b_{m_1-1,1} \neq 0$, then it is not difficult to obtain that

$$\begin{aligned}
A_{1,j} &= 0, \ 2 \leq j \leq m_1 - 1, \\
A_{i,j} &= 0, \ 3 \leq i \leq m_1 - 1, \ 2 \leq j \leq i - 1.
\end{aligned} \tag{4.62}$$

Indeed, from $b_{m_1-1,1} A_{1,2} = 0$ we have $A_{1,2} = 0$, which implies $A_{i,2} = 0$ for $3 \leq i \leq m_1 - 1$. So, the equality (4.62) is true for $j = 2$. Recurrently we obtain the equality (4.62) for any j ($2 \leq j \leq m_1 - 1$).

Thus, we obtain $a_1 = b_1$, $\displaystyle\sum_{k=i}^{m_1-1} A_{i,k} a_k = b_i$, $2 \leq i \leq m_1 - 1$, and

$$\begin{cases}
1 + A_{1,m_1} = b_1 A_{1,m_1}, \\[4pt]
A_{1,k} + A_{1,k+1} = b_1 A_{1,k+1}, \quad m_1 \leq k \leq m_2 - 2, \\[4pt]
\begin{aligned} A_{1,k} + A_{1,k+1} \\ = b_1 A_{1,k+1} + A_{m_1,m_1-m_2+k+1}, \end{aligned} \ m_2 - 1 \leq k \leq n - 1, \\[4pt]
A_{i,k} = A_{i+1,k+1}, \quad 2 \leq i \leq m_1 - 2, \ i \leq k \leq m_1 - 2, \\[4pt]
A_{i,k} = b_i A_{1,k+1} + A_{i+1,k+1}, \quad 2 \leq i \leq m_1 - 1, \ m_1 - 1 \leq k \leq n - 1, \\[4pt]
A_{i,k} = A_{i+1,k+1}, \quad m_1 \leq i \leq n - 1, \ i \leq k \leq n - 1.
\end{cases} \tag{4.63}$$

Taking into account the equality $1 + A_{1,m_1} = b_1 A_{1,m_1}$, we have $b_1 \neq 1$, and $A_{1,m_1} = \frac{1}{b_1-1}$. From the fifth equality of (4.63) for $k = m_1 - 1$, we have

$$A_{m_1,m_1} = A_{m_1-1,m_1-1} - b_{m_1-1} A_{1,m_1} = b_{m_1-1}\left(\frac{1}{a_{m_1-1}} - \frac{1}{b_1 - 1}\right).$$

Since, $A_{m_1,m_1} \neq 0$ we obtain $b_1 \neq a_{m_1-1} + 1$, i.e., the condition 2) is satisfied.

Taking into account, the fourth equation of the system (4.63), putting $B_k = A_{2,k}$, $2 \leq k \leq m_1 - 1$ we obtain the condition 3).

- In the case of $b_{m_1-1} = 0$, using the similar argument for the first non zero element b_t from the set $\{b_{m_1-2}, \ldots, b_2\}$ we obtain the equality $a_1 = b_1$, $\sum_{k=i}^{t} A_{i,k} a_k = b_i$, $2 \leq i \leq t$, and

$$
\begin{cases}
A_{1,t+1} = b_1 A_{1,t+1}, \\[4pt]
A_{1,k} + A_{1,k+1} = b_1 A_{1,k+1}, & t+1 \leq k \leq m_1 - 2, \\[4pt]
1 + A_{1,m_1-1} + A_{1,m_1} = b_1 A_{1,m_1}, \\[4pt]
A_{1,k} + A_{1,k+1} = b_1 A_{1,k+1}, & m_1 \leq k \leq m_2 - 2, \\[4pt]
A_{1,k} + A_{1,k+1} \\
\quad = b_1 A_{1,k+1} + A_{t+1,t-m_2+k+2}, & m_2 - 1 \leq k \leq n - 1, \\[4pt]
A_{i,k} = A_{i+1,k+1}, & 2 \leq i \leq t - 1, \; i \leq k \leq t - 1, \\[4pt]
A_{i,k} = b_i A_{1,k+1} + A_{i+1,k+1}, & 2 \leq i \leq t, \; t \leq k \leq n - 1, \\[4pt]
A_{i,k} = A_{i+1,k+1}, & t+1 \leq i \leq n - 1, \; i \leq k \leq n - 1.
\end{cases}
$$
$$(4.64)$$

If $t = m_1 - 2$, then in the case of $b_1 = 1$ and $a_{m_1-2} = -1$, we obtain $A_{m_1-1,m_1-1} = b_{m_1-2}(\frac{1}{a_{m_1-2}} + 1) = 0$, which is a contradiction with the existence of isomorphism f. Therefore, if $t = m_1 - 2$, then $b_1 \neq 1$ or $a_{m_1-2} \neq -1$., i.e., the condition 2) is satisfied.

Putting $B_k = A_{2,k}$, $2 \leq k \leq m_1 - 1$ we obtain the condition 3).

Sufficiency. Let the conditions 1), 2) and 3) are satisfied. From the previous proof it follows that the existence of an isomorphism $f : \mathcal{C}^1(b_i, m_2) \to \mathcal{C}^1(a_i, m_1)$ is equivalent to the solvability of system of equation (4.63) ((4.64) if $b_{m_1-1} = 0$).

In the case of $b_{m_1-1} \neq 0$, and $b_1 \neq 1$, $b_1 \neq a_{m_1-1} + 1$ we find a solution of (4.63) as follows:

$$A_{i,k} = B_{k-i+2}, \quad 2 \leq i \leq m_1 - 1, \; i \leq k \leq m_1 - 1,$$

$$A_{1,k} = \frac{1}{(b_1 - 1)^{k-m_1+1}}, \quad m_1 \leq k \leq m_2 - 1,$$

$$A_{i,i} = b_{m_1-1}\left(\frac{1}{a_{m_1-1}} - \frac{1}{b_1 - 1}\right), \quad m_1 \leq i \leq n.$$

Then from the system of equations (4.63) we obtain other parameters $A_{i,k}$.

The case $b_{m_1-1} = 0$ is similar to the case $b_{m_1-1} \neq 0$. $\square$

Corollary 4.5. $\mathcal{C}^1(0, m) \cong \mathcal{C}^1(0, 2)$ *for any* m $(3 \leq m \leq n)$. *And as an isomorphism we can take*

$$f(h_1') = h_1 + \sum_{k=1}^{m-1} (-1)^k h_k + \sum_{k=m}^{m} ((-1)^{k-1} + (-1)^{k-m}) h_k,$$

$$f(h_i') = h_i, \quad 2 \leq i \leq n, \qquad f(r') = r.$$

Analogously, we obtain the following theorem for the class of algebras $\mathcal{C}^0(a_i, m)$

Theorem 4.16. *Two algebras* $\mathcal{C}^0(a_i, m_1)$ *and* $\mathcal{C}^0(b_i, m_2)$ *with* $m_1 < m_2$ *are isomorphic if and only if*

1) $a_i = b_i = 0$, $m_1 \leq i \leq n$;

2) $a_1 = b_1$ *and* $b_1 \neq 0$, *and* $b_1 \neq a_{m_1-1}$ *in the case of* $b_{m_1-1} \neq 0$; $b_1 \neq 0$, *or* $a_{m_1-2} \neq -1$ *in the case of* $b_{m_1-1} = 0$, $b_{m_1-2} \neq 0$;

3) there exist $B_2, B_3, \ldots, B_{m_1-1} \in \mathbb{C}$, *with* $B_2 \neq 0$, *such that*

$$b_i = \sum_{k=i}^{m_1-1} B_{k-i+2} a_k, \quad 2 \leq i \leq m_1 - 1.$$

4.3   Generalization of the EACP

In this section following [136] define a generalized EACP, which can be an infinite dimensional.

Consider the set H (the set of "hen") and r (a "rooster").

Definition 4.3. Let $(\mathcal{G}, \cdot)$ be an algebra over a field K of characteristic $\neq 2$. If $\mathcal{G} = H \oplus Kr$ admits a multiplication given by

$$\begin{aligned} xr = rx &= \varphi(x) + \mu(x)r, & \forall\, x \in H, \\ xy = 0, \quad rr &= 0, & \forall\, x, y \in H, \end{aligned} \tag{4.65}$$

where $\varphi \in \mathrm{End}_K(H)$ and $\mu \in H^*$, is a linear map then this algebra is called a generalized evolution algebra of a "chicken" population (GEACP).

Note that if H is finite-dimensional and $\{h_1, \ldots, h_n\}$ is a basis of H and $\{h_1, \ldots, h_n, r\}$ the basis of $\mathcal{G}$, then taking $\varphi_1 = \frac{1}{2}\varphi, \mu_1 = \frac{1}{2}\mu$, we get EACP considered in the previous section.

Recall the following definitions:

Power-associative: For each element a the subalgebra generated by a is associative, that is $a^n a^m = a^{m+n}$, for all nonnegative integers n, m.

Fourth power-associative: Each element a satisfies the identity $a^2 a^2 = a^4$. It is known that if the characteristic of K satisfies $\mathrm{Char} K \neq 2, 3, 5$ and the algebra is flexible (particularly commutative), then being fourth power-associative implies being power-associative (see [6]).

Now we shall give conditions under which $\mathcal{G}$ will be associative and fourth power-associative.

Theorem 4.17. $\mathcal{G}$ *alternative implies* $\mathcal{G}^3 = \{0\}$.

Proof. $\mathcal{G}$ alternative implies that for all $x \in H$, we have $(xr)r = xr^2 = 0$ and $x(xr) = x^2 r = 0$. Thus,

$$0 = (\varphi(x) + \mu(x)r)r = \varphi(x)r = \varphi^2(x) + \mu(\varphi(x))r,$$

and

$$0 = x(\varphi(x) + \mu(x)r) = \mu(x)xr = \mu(x)(\varphi(x) + \mu(x)r) = \mu(x)\varphi(x) + \mu(x)^2 r.$$

Therefore, $\varphi^2(x) = 0$ and $\mu(x) = 0, \forall\, x \in H$.

Let $a = x + \alpha r, b = y + \beta r, c = z + \gamma r$ elements in $\mathcal{G}$. Since $\mu = 0$ we have that $ab = (x + \alpha r)(y + \beta r) = \beta xr + \alpha yr = \beta \varphi(x) + \alpha \varphi(y)$.

Therefore, using that $\mu = 0$ and $\varphi^2 = 0$, we get

$$(ab)c = \big(\beta \varphi(x) + \alpha \varphi(y)\big)(z + \gamma r) = \beta\gamma\varphi(x)r + \alpha\gamma\varphi(y)r$$

$$= \beta\gamma\varphi^2(x) + \alpha\gamma\varphi^2(y) = 0,$$

and so $\mathcal{G}^3 = \{0\}$. $\qquad\square$

Corollary 4.6. $\mathcal{G}$ *alternative if and only if* $\mathcal{G}$ *is associative if and only if* $\mu = 0$ *and* $\varphi^2 = 0$. *Moreover,* $\mathcal{G}$ *satisfies Jacobi and Jordan identities.*

Theorem 4.18. $\mathcal{G}$ *is fourth power-associative if and only if* $\mu = 0$ *and* $\varphi^3 = 0$, *that is,* φ *is a nilpotent operator.*

Proof. Let $a = x + \alpha r$ be an element in $\mathcal{G}$. Then

$$a^2 = 2\alpha xr = 2\alpha(\varphi(x) + \mu(x)r),$$

$$a^2 a^2 = \big(2\alpha(\varphi(x) + \mu(x)r)\big)^2 = 4\alpha^2(2\mu(x))\varphi(x)r = 8\alpha^2\mu(x)(\varphi^2(x) + \mu(\varphi(x))r).$$

On the other hand,

$$\begin{aligned}
a^3 = a^2 a &= 2\alpha(\varphi(x) + \mu(x)r)(x + \alpha r) = 2\alpha(\alpha\varphi(x)r + \mu(x)xr) \\
&= 2\alpha\left[\alpha\varphi^2(x) + \alpha\mu(\varphi(x))r + \mu(x)\varphi(x) + \mu(x)^2 r\right] \\
&= 2\alpha\left[\alpha\varphi^2(x) + \mu(x)\varphi(x) + \left(\alpha\mu(\varphi(x)) + \mu(x)^2\right)r\right],
\end{aligned}$$

and

$$\begin{aligned}
a^4 = a^3 a &= 2\alpha\left[\alpha\varphi^2(x) + \mu(x)\varphi(x) + \left(\alpha\mu(\varphi(x)) + \mu(x)^2\right)r\right](x + \alpha r) \\
&= 2\alpha\left[\alpha^2\varphi^2(x)r + \alpha\mu(x)\varphi(x)r + \left(\alpha\mu(\varphi(x)) + \mu(x)^2\right)xr\right] \\
&= 2\alpha\left[\alpha^2\left(\varphi^3(x) + \mu(\varphi^2(x))r\right) + \alpha\mu(x)\left(\varphi^2(x) + \mu(\varphi(x))r\right)\right. \\
&\quad \left. + \alpha\mu(\varphi(x))\left(\varphi(x) + \mu(x)r\right) + \mu(x)^2\left(\varphi(x) + \mu(x)r\right)\right] \\
&= 2\alpha\left[\alpha^2\varphi^3(x) + \alpha\mu(x)\varphi^2(x) + \alpha\mu(\varphi(x))\varphi(x) + \mu(x)^2\varphi(x)\right. \\
&\quad \left. + \left(\alpha^2\mu(\varphi^2(x)) + 2\alpha\mu(x)\mu(\varphi(x)) + \mu(x)^3\right)r\right].
\end{aligned}$$

Therefore, since $\mathrm{Char}K \neq 2$, we get $a^2 a^2 = a^4$ for all $a \in \mathcal{G}$ if and only if

$$\begin{aligned}
4\alpha\mu(x)\left(\varphi^2(x) + \mu(\varphi(x))r\right) &= \alpha^2\varphi^3(x) + \alpha\mu(x)\varphi^2(x) + \alpha\mu(\varphi(x))\varphi(x) \\
&\quad + \mu(x)^2\varphi(x) + \left(\alpha^2\mu(\varphi^2(x)) + 2\alpha\mu(x)\mu(\varphi(x)) + \mu(x)^3\right)r,
\end{aligned}$$

for all $x \in H, \alpha \in K$. Thus, $\mu(x)^3 = 0$ and $\mu(x) = 0$. So, $\varphi^3(x) = 0$, that is $\varphi^3 = 0$ and Theorem 4.18 follows. $\qquad\square$

Example 4.7. The GEACP of dimension 3 given by the matrix

$$\begin{pmatrix} 0 & 0 & 1 & 0 \\ 0 & 0 & 0 & 0 \\ 0 & 1 & 0 & 0 \end{pmatrix},$$

is fourth power-associative ($\mu = 0$ and $\varphi^3 = 0$) but not associative, since it does not satisfy $\varphi^2 = 0$. Moreover, if $\mathrm{Char}K \neq 2, 3, 5$, then the algebra is power-associative.

From Theorem 4.17 it follows

Corollary 4.7. *$\mathcal{G}$ alternative implies that $\mathcal{G}$ is nilpotent with nilindex 3.*

Note that if $\mu = 0$ then $\mathcal{G}$ is solvable. Indeed, if $\mu = 0$ and $a = x + \alpha r \in \mathcal{G}$ then $a^2 = 2\alpha\varphi(x)$. Therefore $\mathcal{G}^{(2)} = \varphi(H) \subset H$ and $\mathcal{G}^{(2)}\mathcal{G}^{(2)} \subset H^2 = 0$. So, $\mathcal{G}$ is solvable.

Proposition 4.15. *Let $\mathcal{G}$ be a non alternative algebra. If $\mathcal{G}$ is fourth power-associative then $\mathcal{G}$ is nilpotent with nilindex 4.*

Proof. It suffices to prove that $\mathcal{G}$ is right nilpotent. From Theorem 4.18, $\mathcal{G}$ is fourth power-associative if and only if $\mu = 0$ and $\varphi^3 = 0$. Since $\mu = 0$ we have

$$\mathcal{G}^{<2>} = \varphi(H),$$
$$\mathcal{G}^{<3>} = \mathcal{G}^{<2>}\mathcal{G} = \varphi(H)\mathcal{G} = \varphi(H)r = \varphi^2(H),$$
$$\mathcal{G}^{<4>} = \mathcal{G}^{<3>}\mathcal{G} = \varphi^2(H)\mathcal{G} = \varphi^2(H)r = \varphi^3(H) = 0,$$

since $\varphi^3 = 0$. $\qquad\square$

Corollary 4.8. *Let $\mathcal{G}$ be a non alternative algebra. Then $\mathcal{G}$ is a Jordan algebra if and only if $\mathcal{G}$ power-associative algebra.*

Proof. It is known that a Jordan algebra is power-associative. Conversely, let $\mathcal{G}$ power-associative then $\mathcal{G}$ is fourth power-associative and Proposition 4.15 implies $\mathcal{G}^4 = 0$ and $\mathcal{G}$ satisfies the Jordan identity. $\qquad\square$

Recall that an algebra is *unital* or *unitary* if it has an element a with $ab = b = ba$ for all b in the algebra.

Proposition 4.16. *The algebra $\mathcal{G}$ is not unital.*

Proof. Assume $a = x + \alpha r$ is a unity element. We then have $ar = r$ which gives $\varphi(x) + \mu(x)r = r$, so $\varphi(x) = 0$. From $ax = x$ we get $\alpha x r = x$. Since $\varphi(x) = 0$, we obtain $\alpha\mu(x)r = x$, which is a contradiction. This completes the proof. $\qquad\square$

The following proposition describes the nonzero absolute nilpotent elements $(a^2 = 0)$ of $\mathcal{G}$.

Proposition 4.17. *$a = x + \alpha r$ is a nonzero absolute nilpotent element of $\mathcal{G}$ if and only if $\alpha = 0$ or $(\varphi(x) = 0$ and $\mu(x) = 0)$.*

Proof. Let $a = x + \alpha r$ be a nonzero absolute nilpotent element of $\mathcal{G}$. Then $\alpha \neq 0$ or $x \neq 0$. Therefore, $a^2 = (x + \alpha r)^2 = 2\alpha x r = 2\alpha(\varphi(x) + \mu(x)r)$. Then $a^2 = 0 \iff 2\alpha(\varphi(x) + \mu(x)r) = 0 \iff \alpha = 0$ or $(\varphi(x) = 0$ and $\mu(x) = 0)$. $\qquad\square$

Now we shall describe idempotent elements of $\mathcal{G}$, these are solutions to $a^2 = a$.

Proposition 4.18. *$e = x + \alpha r$ is a nonzero idempotent of $\mathcal{G}$ if and only if $\alpha \neq 0, \mu(x) = \frac{1}{2}$ and x is an eigenvector of φ with eigenvalue $\frac{1}{2\alpha}$.*

Proof. Let $e = x + \alpha r$ be a nonzero idempotent of $\mathcal{G}$. Then $\alpha \neq 0$ or $x \neq 0$. Therefore, $e^2 = e \iff 2\alpha\varphi(x) + 2\alpha\mu(x)r = x + \alpha r \iff 2\alpha\varphi(x) = x$ and $2\alpha\mu(x) = \alpha \iff \varphi(x) = \frac{1}{2\alpha}x$ and $\mu(x) = \frac{1}{2}$ and proposition follows. $\square$

Let x be an element of H such that $\varphi(x) = \lambda x$ (for some $\lambda \neq 0$) and $\mu(x) \neq 0$, then $e = \frac{1}{2\mu(x)}x + \frac{1}{2\lambda}r$ is a nonzero idempotent of $\mathcal{G}$.

Bibliographical notes. In this chapter we considered recently introduced algebras. The chapter based on [6], [30], [42], [136], [137], [138], [142], [155], [211], [214], [269] and many internet sources. For further reading see [20], [93]–[105], [255], [257], [263].

Chapter 5

Flows of algebras

This chapter is devoted to flows (chains) of algebras. A flow of algebras (FA) is a particular case of a continuous-time dynamical system whose states are algebras, the matrix (in general may be cubic matrix) of structural constants of which (depending on time) satisfy an analogue of the Kolmogorov–Chapman equation (KCE). Since there are several kinds of multiplications between cubic matrices we fix a multiplication and then consider the KCE with respect to the fixed multiplication. The existence of a solution to the KCE provides the existence of a FA. The aim is to construct FAs with respect to several specially chosen multiplications of matrices (known as Maksimov's multiplications). Moreover, some time-dependent behavior properties of such FAs are given.

5.1 Chains of evolution algebras

5.1.1 *Definitions*

Following [31] we give a chain of evolution algebras (CEA). Let

$$\left\{ E^{[s,t]} : \ s,t \in \mathbb{R}, \ 0 \le s \le t \right\}$$

be a family of n-dimensional evolution algebras over the field $\mathbb{R}$, with basis $e_1, \ldots, e_n$ and multiplication table

$$e_i e_i = M_i^{[s,t]} = \sum_{j=1}^{n} a_{ij}^{[s,t]} e_j, \quad i = 1, \ldots, n; \quad e_i e_j = 0, \quad i \ne j. \qquad (5.1)$$

Here parameters s, t are considered as time.

Denote by $\mathcal{M}^{[s,t]} = \left(a_{ij}^{[s,t]} \right)_{i,j=1,\ldots,n}$ the matrix of structural constants.

125

Definition 5.1. A family $\left\{ E^{[s,t]} : s,t \in \mathbb{R},\, 0 \leq s \leq t \right\}$ of n-dimensional evolution algebras over the field $\mathbb{R}$ is called a chain of evolution algebras (CEA) if the matrix $\mathcal{M}^{[s,t]}$ of structural constants satisfies the Chapman–Kolmogorov equation

$$\mathcal{M}^{[s,t]} = \mathcal{M}^{[s,\tau]}\mathcal{M}^{[\tau,t]}, \quad \text{for any } s < \tau < t. \tag{5.2}$$

If ρ_i is a projection map of $E^{[s,t]}$, which maps every element of $E^{[s,t]}$ to its e_i component, then equation (5.2) can be written as

$$M_i^{[s,t]} = \sum_{j=1}^{n} \rho_j(M_i^{[s,\tau]})M_j^{[\tau,t]}, \quad \text{for any } s < \tau < t. \tag{5.3}$$

Definition 5.2. A CEA is called a time-homogeneous CEA if the matrix $\mathcal{M}^{[s,t]}$ depends only on $t - s$. In this case we write $\mathcal{M}^{[t-s]}$.

Definition 5.3. A CEA is called periodic if its matrix $\mathcal{M}^{[s,t]}$ is periodic with respect to at least one of the variables s, t, i.e. (periodicity with respect to t) $\mathcal{M}^{[s,t+P]} = \mathcal{M}^{[s,t]}$ for all values of t. The constant P is called the period, and is required to be nonzero.

Remark 5.1. In general, an algebra $\mathcal{A}^{[s,t]}$ can be given by a cubic matrix $\mathcal{M}^{[s,t]} = \left(a_{ijk}^{[s,t]} \right)_{i,j,k=1,\ldots,n}$ of structural constants. The Definition 5.1 can be extended to $\mathcal{A}^{[s,t]}$ using analogues of the Chapman–Kolmogorov equations for quadratic operators (see [60], [61], [183], [237]). Since in the general case there are two types of the Chapman–Kolmogorov equations: type A and type B [183], one also can define two types of chain of (general) algebras using the Chapman–Kolmogorov equations of type A and type B, respectively. In this section we shall only consider CEA, which is more simple than general case, because it is defined by quadratic matrices. In the following sections we define chains (flows) for general algebras, i.e., the algebras defined by cubic matrices.

Note that any Markov process (chain) defines a CEA. Indeed, let $\left\{ \mathcal{M}^{[s,t]},\ 0 \leq s \leq t \right\}$ be a family of stochastic matrices which satisfies the equation (5.2), then it gives the fundamental relationship between the probability transitions (kernels) (see e.g. [251]) and if each element of a family of matrices satisfying the Chapman–Kolmogorov equation is stochastic, then it generates a Markov process. Thus we have

Theorem 5.1. *For each Markov process, there is a CEA whose structural constants are transition probabilities of the process, and whose generator set (basis) is the state space of the Markov process.*

The following is an example of CEA corresponding to a Markov process.

Example 5.1. To show a time dependent CEA we use the following example of time homogeneous Markov process (see [132]): for $n = 3$ consider

$$a_{ii}^{[t]} = \frac{2}{3}e^{-\frac{3}{2}At}\cos(\alpha t) + \frac{1}{3}, \quad i = 1, 2, 3;$$

$$a_{12}^{[t]} = a_{23}^{[t]} = a_{31}^{[t]} = e^{-\frac{3}{2}At}\left(\frac{1}{\sqrt{3}}\sin(\alpha t) - \frac{1}{3}\cos(\alpha t)\right) + \frac{1}{3};$$

$$a_{21}^{[t]} = a_{32}^{[t]} = a_{13}^{[t]} = -e^{-\frac{3}{2}At}\left(\frac{1}{\sqrt{3}}\sin(\alpha t) + \frac{1}{3}\cos(\alpha t)\right) + \frac{1}{3},$$

where $A > 0$, $\alpha = \frac{\sqrt{3}}{2}A$.

Let $E^{[t]}, t \geq 0$ be the corresponding CEA. One can see that $E^{[t]}$ has an oscillation behavior depending on time t. Moreover $\lim_{t\to+\infty} E^{[t]} = E$, where E is an evolution algebra with the multiplication table

$$e_1^2 = e_2^2 = e_3^2 = \frac{1}{3}(e_1 + e_2 + e_3), \quad e_i e_j = 0, i \neq j.$$

If $\mathcal{M}^{[s,t]}$ does not depend on time (i.e. $= \mathcal{M}$) then the CEA contains only one evolution algebra E. Note that for a Markov chain defined by $\mathcal{M}$ the corresponding E has been studied in [254].

5.1.2 *Two-dimensional CEAs*

Now following [31] and [218] we give several examples of CEAs, many of them correspond to a family of matrices which do not define a Markov process.

To construct a chain of two-dimensional evolution algebra one has to solve equation (5.2) for 2×2 matrix $\mathcal{M}^{[s,t]}$. This equation gives the following system of functional equations (with four unknown functions):

$$\begin{aligned}
a_{11}^{[s,t]} &= a_{11}^{[s,\tau]}a_{11}^{[\tau,t]} + a_{12}^{[s,\tau]}a_{21}^{[\tau,t]}, \\
a_{12}^{[s,t]} &= a_{11}^{[s,\tau]}a_{12}^{[\tau,t]} + a_{12}^{[s,\tau]}a_{22}^{[\tau,t]}, \\
a_{21}^{[s,t]} &= a_{21}^{[s,\tau]}a_{11}^{[\tau,t]} + a_{22}^{[s,\tau]}a_{21}^{[\tau,t]}, \\
a_{22}^{[s,t]} &= a_{21}^{[s,\tau]}a_{12}^{[\tau,t]} + a_{22}^{[s,\tau]}a_{22}^{[\tau,t]}.
\end{aligned} \tag{5.4}$$

In general, the analysis of the system (5.4) is difficult. Therefore, we shall consider several cases where the system is solvable:

Case 1. $a_{11}^{[s,t]} = a_{22}^{[s,t]} = \alpha(s,t)$, $a_{12}^{[s,t]} = a_{21}^{[s,t]} = \beta(s,t)$. Then equation (5.4) is reduced to

$$\alpha(s,t) = \alpha(s,\tau)\alpha(\tau,t) + \beta(s,\tau)\beta(\tau,t);$$

$$\beta(s,t) = \alpha(s,\tau)\beta(\tau,t) + \beta(s,\tau)\alpha(\tau,t).$$

Denote

$$f(s,t) = \alpha(s,t) + \beta(s,t), \quad \varphi(s,t) = \alpha(s,t) - \beta(s,t), \tag{5.5}$$

then the last system of functional equations can be written as

$$f(s,t) = f(s,\tau)f(\tau,t), \quad \varphi(s,t) = \varphi(s,\tau)\varphi(\tau,t), \ s \leq \tau \leq t.$$

Both these equations are known as Cantor's second equation[1] which has very rich family of solutions:

a) $f(s,t) \equiv 0$;
b) $f(s,t) = \frac{\Phi(t)}{\Phi(s)}$, where Φ is an arbitrary function with $\Phi(s) \neq 0$;
c)

$$f(s,t) = \begin{cases} 1, \text{ if } \ s \leq t < a, \\ 0, \text{ if } \ t \geq a, \end{cases} \quad \text{where } \ a > 0.$$

Similarly,

a') $\varphi(s,t) \equiv 0$;
b') $\varphi(s,t) = \frac{\Psi(t)}{\Psi(s)}$, where Ψ is an arbitrary function with $\Psi(s) \neq 0$;
c')

$$\varphi(s,t) = \begin{cases} 1, \text{ if } \ s \leq t < b, \\ 0, \text{ if } \ t \geq b, \end{cases} \quad \text{where } \ b > 0.$$

Substituting these solutions into (5.5) and finding $\alpha(s,t)$ and $\beta(s,t)$ we get the following matrices

$$\mathcal{M}_0^{[s,t]} = \begin{pmatrix} 0 & 0 \\ 0 & 0 \end{pmatrix}; \quad \mathcal{M}_1^{[s,t]} = \frac{1}{2}\begin{pmatrix} \frac{\Psi(t)}{\Psi(s)} & -\frac{\Psi(t)}{\Psi(s)} \\ -\frac{\Psi(t)}{\Psi(s)} & \frac{\Psi(t)}{\Psi(s)} \end{pmatrix};$$

$$\mathcal{M}_2^{[s,t]} = \frac{1}{2}\begin{cases} \begin{pmatrix} 1 & -1 \\ -1 & 1 \end{pmatrix}, \text{ if } \ s \leq t < b; \\ \begin{pmatrix} 0 & 0 \\ 0 & 0 \end{pmatrix}, \quad \text{ if } \ t \geq b \end{cases};$$

---

[1] For Cantor's first and second equations, see
http://eqworld.ipmnet.ru/en/solutions/eqindex/eqindex-fe.htm.

$$\mathcal{M}_3^{[s,t]} = \frac{1}{2}\begin{pmatrix} \frac{\Phi(t)}{\Phi(s)} & \frac{\Phi(t)}{\Phi(s)} \\[4pt] \frac{\Phi(t)}{\Phi(s)} & \frac{\Phi(t)}{\Phi(s)} \end{pmatrix}.$$

$$\mathcal{M}_4^{[s,t]} = \frac{1}{2}\begin{pmatrix} \frac{\Phi(t)}{\Phi(s)} + \frac{\Psi(t)}{\Psi(s)} & \frac{\Phi(t)}{\Phi(s)} - \frac{\Psi(t)}{\Psi(s)} \\[4pt] \frac{\Phi(t)}{\Phi(s)} - \frac{\Psi(t)}{\Psi(s)} & \frac{\Phi(t)}{\Phi(s)} + \frac{\Psi(t)}{\Psi(s)} \end{pmatrix}.$$

Remark 5.2. In the next subsections we will study some properties of the CEA corresponding to

$$\mathcal{M}_4'^{[s,t]} = \frac{1}{2}\begin{pmatrix} \lambda^t + \mu^t & \lambda^t - \mu^t \\ \lambda^t - \mu^t & \lambda^t + \mu^t \end{pmatrix}.$$

Denote by $E_{4'}^{[t]}, t \geq 0$ the corresponding CEA. Depending on parameters λ and μ we get distinct behavior of $E_{4'}^{[t]}$ for $t \to +\infty$, i.e.

$$\lim_{t \to +\infty} E_{4'}^{[t]} = \begin{cases} E_0 & \text{if } 0 < \lambda, \mu < 1, \\ E_1 & \text{if } \lambda = \mu = 1, \\ E_{1/2} & \text{if } \lambda = 1, 0 \leq \mu < 1, \\ E_{-1/2} & \text{if } \mu = 1, 0 \leq \lambda < 1, \\ E_\infty & \text{otherwise}, \end{cases}$$

where E_0 is an evolution algebra with zero multiplication; E_1 is an evolution algebra with multiplication table

$$e_1^2 = e_1, e_2^2 = e_2, \quad e_1 e_2 = 0;$$

$E_{1/2}$ is an evolution algebra with multiplication table

$$e_1^2 = e_2^2 = \frac{1}{2}(e_1 + e_2), \quad e_1 e_2 = 0;$$

$E_{-1/2}$ is an evolution algebra with multiplication table

$$e_1^2 = \frac{1}{2}(e_1 - e_2), \quad e_2^2 = -\frac{1}{2}(e_1 - e_2), \quad e_1 e_2 = 0;$$

and E_∞ is a vector space which has "infinity multiplication", or we can say that in E_∞ an algebra structure is not defined. This example shows that a limit of a CEA can be non evolution algebra.

$$
\mathcal{M}_5^{[s,t]} = \frac{1}{2}
\begin{cases}
\begin{pmatrix} \frac{\Phi(t)}{\Phi(s)}+1 & \frac{\Phi(t)}{\Phi(s)}-1 \\ \frac{\Phi(t)}{\Phi(s)}-1 & \frac{\Phi(t)}{\Phi(s)}+1 \end{pmatrix}, & \text{if } s \le t < b; \\[2em]
\begin{pmatrix} \frac{\Phi(t)}{\Phi(s)} & \frac{\Phi(t)}{\Phi(s)} \\ \frac{\Phi(t)}{\Phi(s)} & \frac{\Phi(t)}{\Phi(s)} \end{pmatrix}, & \text{if } t \ge b
\end{cases}
\;;
$$

$$
\mathcal{M}_6^{[s,t]} = \frac{1}{2}
\begin{cases}
\begin{pmatrix} 1 & 1 \\ 1 & 1 \end{pmatrix}, & \text{if } s \le t < a; \\[1.5em]
\begin{pmatrix} 0 & 0 \\ 0 & 0 \end{pmatrix}, & \text{if } t \ge a
\end{cases}
\;;
$$

$$
\mathcal{M}_7^{[s,t]} = \frac{1}{2}
\begin{cases}
\begin{pmatrix} 1+\frac{\Psi(t)}{\Psi(s)} & 1-\frac{\Psi(t)}{\Psi(s)} \\ 1-\frac{\Psi(t)}{\Psi(s)} & 1+\frac{\Psi(t)}{\Psi(s)} \end{pmatrix}, & \text{if } s \le t < a; \\[2em]
\begin{pmatrix} \frac{\Psi(t)}{\Psi(s)} & -\frac{\Psi(t)}{\Psi(s)} \\ -\frac{\Psi(t)}{\Psi(s)} & \frac{\Psi(t)}{\Psi(s)} \end{pmatrix}, & \text{if } t \ge a
\end{cases}
\;;
$$

$$
\mathcal{M}_8^{[s,t]} = \frac{1}{2}
\begin{cases}
\begin{pmatrix} 2 & 0 \\ 0 & 2 \end{pmatrix}, & \text{if } s \le t < \min\{a,b\}; \\[1.5em]
\begin{pmatrix} 1 & -1 \\ -1 & 1 \end{pmatrix}, & \text{if } a \le t < b,\ a < b; \\[1.5em]
\begin{pmatrix} 1 & 1 \\ 1 & 1 \end{pmatrix}, & \text{if } b \le t < a,\ a > b; \\[1.5em]
\begin{pmatrix} 0 & 0 \\ 0 & 0 \end{pmatrix}, & \text{if } t \ge \max\{a,b\}
\end{cases}
\;.
$$

Thus in this case we have nine CEAs: $E_i^{[s,t]}$, $0 \le s \le t$ which correspond to $\mathcal{M}_i^{[s,t]}$, $i = 0, 1, \ldots, 8$ listed above.

Case 2. $a_{11}^{[s,t]} = a_{22}^{[s,t]} = \alpha(s,t)$, $a_{12}^{[s,t]} = -a_{21}^{[s,t]} = \beta(s,t)$. Then equation (5.4) is reduced to

$$
\begin{aligned}
\alpha(s,t) &= \alpha(s,\tau)\alpha(\tau,t) - \beta(s,\tau)\beta(\tau,t); \\
\beta(s,t) &= \alpha(s,\tau)\beta(\tau,t) + \beta(s,\tau)\alpha(\tau,t).
\end{aligned}
\tag{5.6}
$$

One can check that this system of functional equations has a solution

$$
\alpha(s,t) = \cos(t-s), \quad \beta(s,t) = \sin(t-s).
$$

This gives the following matrix

$$\mathcal{M}_9^{[s,t]} = \begin{pmatrix} \cos(t-s) & \sin(t-s) \\ -\sin(t-s) & \cos(t-s) \end{pmatrix}. \tag{5.7}$$

Note that this matrix defines a periodic CEA. We do not know another solution (with $\alpha\beta \neq 0$) of the system (5.6).

We denote by $E_9^{[s,t]} = E_9^{[t-s]}$ the CEA which corresponds to $\mathcal{M}_9^{[s,t]}$.

Since this CEA is periodic, for any fixed s, the limit $\lim_{t\to+\infty} E_9^{[s,t]}$ does not exist in general, moreover its set of limit points (evolution algebras) can be a continuum set. We shall make this point clear as follows. Denoting $t - s$ by t we rewrite the matrix as

$$\mathcal{M}_9^{[t]} = \begin{pmatrix} \cos t & \sin t \\ -\sin t & \cos t \end{pmatrix}. \tag{5.8}$$

Since that matrix is periodic with period $P = 2\pi$, the corresponding CEA $E_9^{[t]}$ is also periodic. Moreover this CEA is very interesting: for arbitrary 2-dimensional evolution algebra E_a^+, or E_a^-, $a \in [-1,1]$ with structural constants matrix

$$\mathcal{M}_a^{\pm} = \begin{pmatrix} a & \pm\sqrt{1-a^2} \\ \mp\sqrt{1-a^2} & a \end{pmatrix}$$

respectively, there is a sequence $t_n = t_n(a)$ of times such that $\lim_{n\to\infty} E^{[t_n]} = E_a^+$ or E_a^-. We have $E_a^{\pm} \neq E_b^{\pm}$ if $a \neq b$. Moreover the following is true

Proposition 5.1.

1) *For any* $a,b \in [-1,1]$, $a \neq \pm b$, *the algebras* E_a^+ *and* E_b^+ *are not isomorphic. The algebras* E_a^+ *and* E_{-a}^+ *are isomorphic.*

2) *For any* $a,b \in [-1,1]$, $a \neq \pm b$, *the algebras* E_a^- *and* E_b^- *are not isomorphic. The algebras* E_a^- *and* E_{-a}^- *are isomorphic.*

Proof. 1) Let $\varphi = \begin{pmatrix} \alpha & \beta \\ \gamma & \delta \end{pmatrix}$ be an isomorphism of the evolution algebra E_a^+ to the evolution algebra E_b^+. Here $\det(\varphi) \neq 0$. By the multiplication table of the evolution algebras, we get the following relation between matrices $\mathcal{M}_a^+$ and $\mathcal{M}_b^+$:

$$\mathcal{M}_b^+ = \frac{1}{\det(\varphi)} \times$$

$$\begin{pmatrix} (a\delta - T\gamma)\alpha^2 - (a\gamma + T\delta)\beta^2 & (a\alpha + T\beta)\beta^2 - (a\beta - T\alpha)\alpha^2 \\ (a\delta - T\gamma)\gamma^2 - (a\gamma + T\delta)\delta^2 & (a\alpha + T\beta)\delta^2 - (a\beta - T\alpha)\gamma^2 \end{pmatrix},$$

where $T = \sqrt{1-a^2}$.

Since $\det(\mathcal{M}_a^+) = 1$, it is easy to see that there are two classes of isomorphisms:

$$\mathcal{C}_1 = \left\{ \begin{pmatrix} \alpha & 0 \\ 0 & \delta \end{pmatrix} : \alpha\delta \neq 0 \right\}, \quad \mathcal{C}_2 = \left\{ \begin{pmatrix} 0 & \beta \\ \gamma & 0 \end{pmatrix} : \beta\gamma \neq 0 \right\}.$$

For the class $\mathcal{C}_1$ the matrix $\mathcal{M}_b^+$ must satisfy the following

$$\mathcal{M}_b^+ = \begin{pmatrix} b & \sqrt{1-b^2} \\ -\sqrt{1-b^2} & b \end{pmatrix} = \begin{pmatrix} a\alpha & \sqrt{1-a^2}\frac{\alpha^2}{\delta} \\ -\sqrt{1-a^2}\frac{\delta^2}{\alpha} & a\delta \end{pmatrix}.$$

From this equality we get $\alpha = \delta = \sqrt{\frac{1-b^2}{1-a^2}} = \frac{b}{a}$ if $a \neq 0, \pm 1$ which is satisfied iff $a = b$. Hence the isomorphisms from the class $\mathcal{C}_1$ cannot give an isomorphism from $\mathcal{M}_a^+$ to $\mathcal{M}_b^+$. For $a = 0$ we get $b = 0$. One can take $\alpha = \delta = \mp 1$ if $a = \pm 1$ and $b = \mp 1$. Hence $E_{\pm 1}^+$ is isomorph to $E_{\mp 1}^+$.

For the class $\mathcal{C}_2$ the matrix $\mathcal{M}_b^+$ must satisfy the following

$$\mathcal{M}_b^+ = \begin{pmatrix} b & \sqrt{1-b^2} \\ -\sqrt{1-b^2} & b \end{pmatrix} = \begin{pmatrix} a\beta & -\sqrt{1-a^2}\frac{\beta^2}{\gamma} \\ \sqrt{1-a^2}\frac{\gamma^2}{\beta} & a\gamma \end{pmatrix}.$$

From this equality we get $\beta = \gamma = -\sqrt{\frac{1-b^2}{1-a^2}} = \frac{b}{a}$ if $a \neq 0, \pm 1$ which is satisfied iff $a = -b$. Hence the isomorphisms from the class $\mathcal{C}_2$ can only give an isomorphism from $\mathcal{M}_a^+$ to $\mathcal{M}_{-a}^+$.

2) The proof of 2) is similar to the proof of 1). $\qquad\square$

Consider now discrete time n, $n \in \mathbb{N}$ and the CEA $\{E^{[n]}, n \in \mathbb{N}\}$ given by matrix (5.7).

Proposition 5.2. *The discrete time CEA $E^{[n]}$, $n \in \mathbb{N}$, is dense in the set $\{E_a^{\pm}, a \in [-1,1]\}$ of evolution algebras, i.e. for an arbitrary evolution algebra $E_a^{\pm}$ there exists a sequence $\{n_k\}_{k=1,2,\dots}$ of natural numbers such that $\lim_{k\to\infty} E^{[n_k]} = E_a^+$ or E_a^-.*

Proof. It is known that the sequences $\{\sin n\}$ and $\{\cos n\}$, $n \in \mathbb{N}$, are dense in $[-1,1]$ (see e.g. [57], [213]). Hence for any $a \in [-1,1]$ there is a sequence $\{n_k\}_{k=1,2,\dots}$ of natural numbers such that $\lim_{k\to\infty} \cos(n_k) = a$. The same sequence can be used to get $\lim_{k\to\infty} E^{[n_k]} = E_a^+$ or E_a^-. $\qquad\square$

Case 3. $a_{11}^{[s,t]} = a_{21}^{[s,t]} = \alpha(s,t)$, $a_{12}^{[s,t]} = a_{22}^{[s,t]} = \beta(s,t)$. In this case the equation (5.4) is reduced to

$$\alpha(s,t) = \alpha(\tau,t)(\alpha(s,\tau) + \beta(s,\tau));$$

$$\beta(s,t) = \beta(\tau,t)(\alpha(s,\tau) + \beta(s,\tau)).$$

Denote

$$\gamma(s,t) = \alpha(s,t) + \beta(s,t), \quad \delta(s,t) = \alpha(s,t) - \beta(s,t), \tag{5.9}$$

then the last system of functional equations can be written as

$$\gamma(s,t) = \gamma(s,\tau)\gamma(\tau,t), \quad \delta(s,t) = \gamma(s,\tau)\delta(\tau,t), \ \ s \le \tau \le t.$$

The first equation is Cantor's second equation which has solutions:

a) $\gamma(s,t) \equiv 0$;
b) $\gamma(s,t) = \frac{h(t)}{h(s)}$, where h is an arbitrary function with $h(s) \ne 0$;
c)

$$\gamma(s,t) = \begin{cases} 1, & \text{if } \ s \le t < a, \\ 0, & \text{if } \ t \ge a, \end{cases} \qquad \text{where } \ a > 0.$$

Using these solution from the second equation we find δ:

a') $\delta(s,t) \equiv 0$;
b') $\delta(s,t) = \frac{g(t)}{h(s)}$, where g is an arbitrary function;
c')

$$\delta(s,t) = \begin{cases} \psi(t), & \text{if } \ s \le t < a, \\ 0, & \text{if } \ t \ge a, \end{cases} \qquad \text{where } \ a > 0, \ \psi(t) \text{ is an arbitrary function.}$$

Substituting these solutions into (5.9) we get the following (non-zero) matrices

$$\mathcal{M}_{10}^{[s,t]} = \frac{1}{2} \begin{pmatrix} \frac{h(t)+g(t)}{h(s)} & \frac{h(t)-g(t)}{h(s)} \\ \frac{h(t)+g(t)}{h(s)} & \frac{h(t)-g(t)}{h(s)} \end{pmatrix}.$$

$$\mathcal{M}_{11}^{[s,t]} = \frac{1}{2} \begin{cases} \begin{pmatrix} 1+\psi(t) & 1-\psi(t) \\ 1+\psi(t) & 1-\psi(t) \end{pmatrix}, & \text{if } \ s \le t < a; \\ \begin{pmatrix} 0 & 0 \\ 0 & 0 \end{pmatrix}, & \text{if } \ t \ge a \end{cases}.$$

Thus in this case we have two new CEAs: $E_i^{[s,t]}$, $0 \le s \le t$ which correspond to $\mathcal{M}_i^{[s,t]}$, $i = 10, 11$ listed above.

Case 4. $a_{11}^{[s,t]} = a_{12}^{[s,t]} = \alpha(s,t)$, $a_{21}^{[s,t]} = a_{22}^{[s,t]} = \beta(s,t)$. In this case the equation (5.4) is reduced to

$$\alpha(s,t) = \alpha(s,\tau)(\alpha(\tau,t) + \beta(\tau,t));$$

$$\beta(s,t) = \beta(s,\tau)(\alpha(\tau,t) + \beta(\tau,t)).$$

Denote

$$\gamma(s,t) = \alpha(s,t) + \beta(s,t), \quad \delta(s,t) = \alpha(s,t) - \beta(s,t), \tag{5.10}$$

then the last system of functional equations can be written as

$$\gamma(s,t) = \gamma(s,\tau)\gamma(\tau,t), \quad \delta(s,t) = \delta(s,\tau)\gamma(\tau,t), \quad s \le \tau \le t.$$

The first equation has solutions:

a) $\gamma(s,t) \equiv 0$;

b) $\gamma(s,t) = \frac{h(t)}{h(s)}$, where h is an arbitrary function with $h(s) \ne 0$;

c)

$$\gamma(s,t) = \begin{cases} 1, & \text{if } s \le t < a, \\ 0, & \text{if } t \ge a, \end{cases} \qquad \text{where } a > 0.$$

Using these solution from the second equation we find δ:

a') $\delta(s,t) \equiv 0$;

b') $\delta(s,t) = g(s)h(t)$, where g is an arbitrary function;

c')

$$\delta(s,t) = \begin{cases} \psi(s), & \text{if } s \le t < a, \\ 0, & \text{if } t \ge a, \end{cases} \qquad \text{where } a > 0,\ \psi(t) \text{ is an arbitrary function.}$$

Consequently, we get the following (non-zero) new matrices

$$\mathcal{M}_{12}^{[s,t]} = \frac{1}{2}\begin{pmatrix} h(t)\left(\frac{1}{h(s)} + g(s)\right) & h(t)\left(\frac{1}{h(s)} + g(s)\right) \\ h(t)\left(\frac{1}{h(s)} - g(s)\right) & h(t)\left(\frac{1}{h(s)} - g(s)\right) \end{pmatrix};$$

$$\mathcal{M}_{13}^{[s,t]} = \frac{1}{2}\begin{cases} \begin{pmatrix} 1 + \psi(s) & 1 + \psi(s) \\ 1 - \psi(s) & 1 - \psi(s) \end{pmatrix}, & \text{if } s \le t < a; \\ \begin{pmatrix} 0 & 0 \\ 0 & 0 \end{pmatrix}, & \text{if } t \ge a \end{cases}.$$

Hence in this case we have two new CEAs: $E_i^{[s,t]}$, $0 \le s \le t$ which correspond to $\mathcal{M}_i^{[s,t]}$, $i = 12, 13$ listed above.

Case 5. $a_{12}^{[s,t]} \equiv 0$, (the case $a_{21}^{[s,t]} \equiv 0$ is similar) then the system (5.4) reduced to following

$$a_{11}^{[s,t]} = a_{11}^{[s,\tau]} a_{11}^{[\tau,t]},$$

$$a_{21}^{[s,t]} = a_{21}^{[s,\tau]} a_{11}^{[\tau,t]} + a_{22}^{[s,\tau]} a_{21}^{[\tau,t]}, \tag{5.11}$$

$$a_{22}^{[s,t]} = a_{22}^{[s,\tau]} a_{22}^{[\tau,t]}.$$

The first and the third equations of the system (5.11) are Cantor's second equations. Substituting solutions of these equations into the second equation of the system (5.11) we find function $a_{21}^{[s,t]}$. Note that in many cases the second equation of the system (5.11) will be reduced to Cantor's first equation:

$$\gamma(s, t) = \gamma(s, \tau) + \gamma(\tau, t),$$

which also has very rich family of solutions: $\gamma(s, t) = \Psi(t) - \Psi(s)$, where Ψ is an arbitrary function. Thus solving the system (5.11) we obtain the following new matrices:

$$\mathcal{M}_{14}^{[s,t]} = \begin{pmatrix} \frac{\Phi(t)}{\Phi(s)} & 0 \\ \Phi(t)\psi(s) & 0 \end{pmatrix};$$

$$\mathcal{M}_{15}^{[s,t]} = \begin{cases} \begin{pmatrix} 1 & 0 \\ \psi(s) & 0 \end{pmatrix}, & \text{if } s \le t < a; \\ \begin{pmatrix} 0 & 0 \\ 0 & 0 \end{pmatrix}, & \text{if } t \ge a \end{cases};$$

$$\mathcal{M}_{16}^{[s,t]} = \begin{pmatrix} 0 & 0 \\ \frac{g(t)}{\psi(s)} & \frac{\psi(t)}{\psi(s)} \end{pmatrix}.$$

$$\mathcal{M}_{17}^{[s,t]} = \begin{pmatrix} \frac{\Phi(t)}{\Phi(s)} & 0 \\ \frac{\Phi(t)}{\psi(s)}(g(t) - g(s)) & \frac{\psi(t)}{\psi(s)} \end{pmatrix}.$$

$$\mathcal{M}_{18}^{[s,t]} = \begin{cases} \begin{pmatrix} 1 & 0 \\ \frac{h(t)-h(s)}{\psi(s)} & \frac{\psi(t)}{\psi(s)} \end{pmatrix}, & \text{if } s \le t < a; \\ \begin{pmatrix} 0 & 0 \\ \frac{h(t)}{\psi(s)} & \frac{\psi(t)}{\psi(s)} \end{pmatrix}, & \text{if } t \ge a \end{cases};$$

$$
\mathcal{M}_{19}^{[s,t]} = \begin{cases} \begin{pmatrix} 0 & 0 \\ h(t) & 1 \end{pmatrix}, & \text{if } s \leq t < b; \\ \begin{pmatrix} 0 & 0 \\ 0 & 0 \end{pmatrix}, & \text{if } t \geq b \end{cases} ;
$$

$$
\mathcal{M}_{20}^{[s,t]} = \begin{cases} \begin{pmatrix} \frac{\Phi(t)}{\Phi(s)} & 0 \\ \Phi(t)(v(t) - v(s)) & 1 \end{pmatrix}, & \text{if } s \leq t < b; \\ \begin{pmatrix} \frac{\Phi(t)}{\Phi(s)} & 0 \\ \Phi(t)w(s) & 0 \end{pmatrix}, & \text{if } t \geq b \end{cases} ;
$$

$$
\mathcal{M}_{21}^{[s,t]} = \begin{cases} \begin{pmatrix} 1 & 0 \\ v(t) - v(s) & 1 \end{pmatrix}, & \text{if } s \leq t < \min\{a, b\}; \\ \begin{pmatrix} 1 & 0 \\ v(s) & 0 \end{pmatrix}, & \text{if } b \leq t < a, \, a > b; \\ \begin{pmatrix} 0 & 0 \\ v(t) & 1 \end{pmatrix}, & \text{if } a \leq t < b, \, a < b; \\ \begin{pmatrix} 0 & 0 \\ 0 & 0 \end{pmatrix}, & \text{if } t \geq \max\{a, b\} \end{cases} .
$$

Hence, in this case we have eight new CEAs: $E_i^{[s,t]}$, $0 \leq s \leq t$ which correspond to $\mathcal{M}_i^{[s,t]}$, $i = 14, \ldots, 21$ listed above.

Case 6. $a_{22}^{[s,t]} \equiv 0$, (the case $a_{11}^{[s,t]} \equiv 0$ is similar) then the system (5.4) reduced to following

$$
\begin{aligned}
a_{11}^{[s,t]} &= a_{11}^{[s,\tau]} a_{11}^{[\tau,t]} + a_{12}^{[s,\tau]} a_{21}^{[\tau,t]}, \\
a_{12}^{[s,t]} &= a_{11}^{[s,\tau]} a_{12}^{[\tau,t]}, \\
a_{21}^{[s,t]} &= a_{21}^{[s,\tau]} a_{11}^{[\tau,t]}, \\
0 &= a_{21}^{[s,\tau]} a_{12}^{[\tau,t]}.
\end{aligned}
\tag{5.12}
$$

Multiplying the second and the third equations and using the fourth one we get $a_{12}^{[s,t]} a_{21}^{[s,t]} = 0$. Thus this case reduced to the Case 5, hence does not give any new CEA.

Case 7. Assume $a_{11}^{[s,t]} + a_{12}^{[s,t]} = a_{21}^{[s,t]} + a_{22}^{[s,t]} = 1$. Denote $\alpha(s,t) = a_{11}^{[s,t]}$, $\beta(s,t) = a_{21}^{[s,t]}$ then from (5.4) we get

$$\alpha(s,t) = \alpha(s,\tau)\alpha(\tau,t) + (1 - \alpha(s,t))\beta(\tau,t);$$

$$\beta(s,t) = \beta(s,\tau)\alpha(\tau,t) + (1 - \beta(s,\tau))\beta(\tau,t).$$

Putting

$$\gamma(s,t) = \alpha(s,t) + \beta(s,t), \quad \delta(s,t) = \alpha(s,t) - \beta(s,t), \tag{5.13}$$

we obtain

$$\gamma(s,t) = \gamma(s,\tau)\delta(\tau,t) + \gamma(\tau,t) - \delta(\tau,t);$$
$$\delta(s,t) = \delta(s,\tau)\delta(\tau,t). \tag{5.14}$$

The second equation of (5.14) has the following solutions:
1) $\delta(s,t) \equiv 0$;
2) $\delta(s,t) = \frac{\theta(t)}{\theta(s)}$, with $\theta(t) \neq 0$;
3)

$$\delta(s,t) = \begin{cases} 1, & \text{if } s \leq t < a, \\ 0, & \text{if } t \geq a, \end{cases} \quad \text{where } a > 0.$$

Substituting these solutions into the first equation of (5.14) we get the following
1') $\gamma(s,t) = f(t)$, where f is an arbitrary function;
2') For $\delta(s,t) = \frac{\theta(t)}{\theta(s)}$ we get

$$\tilde{\gamma}(s,t) = \tilde{\gamma}(s,\tau) + \tilde{\gamma}(\tau,t) - \frac{1}{\theta(\tau)}, \tag{5.15}$$

where $\tilde{\gamma}(s,t) = \frac{\gamma(s,t)}{\theta(t)}$. We shall find a solution of the equation (5.15) which has the form

$$\tilde{\gamma}(s,t) = \lambda \cdot u(t) - \mu \cdot u(s), \quad \lambda, \mu \in \mathbb{R}, \lambda \neq \mu.$$

We do not know another kind of solutions of the equation (5.15).

Substituting this function into equation (5.15) we get $u(t) = \frac{1}{(\lambda-\mu)\theta(t)}$. Consequently, we obtain

$$\gamma(s,t) = \frac{\lambda}{\lambda - \mu} - \frac{\mu\theta(t)}{(\lambda - \mu)\theta(s)}.$$

3') In the case of 3) we get the following equation

$$\gamma(s,t) = \begin{cases} \gamma(s,\tau) + \gamma(\tau,t) - 1, & \text{if } s \leq t < a, \\ \gamma(\tau,t), & \text{if } t \geq a. \end{cases}$$

This equation has the following solution

$$\gamma(s,t) = \begin{cases} 1, & \text{if } s \le t < a, \\ g(t), & \text{if } t \ge a, \end{cases}$$

where $g(t)$ is an arbitrary function.

Using the obtained solutions γ and δ by (5.13) we get the following matrices

$$\mathcal{M}_{22}^{[s,t]} = \begin{pmatrix} f(t) & 1 - f(t) \\ f(t) & 1 - f(t) \end{pmatrix};$$

$$\mathcal{M}_{23}^{[s,t]}(\lambda,\mu) = \begin{pmatrix} 1 - \frac{\lambda-2\mu}{2(\lambda-\mu)}\left(1 - \frac{\theta(t)}{\theta(s)}\right) & \frac{\lambda-2\mu}{2(\lambda-\mu)}\left(1 - \frac{\theta(t)}{\theta(s)}\right) \\ \frac{\lambda}{2(\lambda-\mu)}\left(1 - \frac{\theta(t)}{\theta(s)}\right) & 1 - \frac{\lambda}{2(\lambda-\mu)}\left(1 - \frac{\theta(t)}{\theta(s)}\right) \end{pmatrix};$$

$$\mathcal{M}_{24}^{[s,t]} = \begin{cases} \begin{pmatrix} 1 & 0 \\ 0 & 1 \end{pmatrix}, & \text{if } s \le t < a; \\[2mm] \begin{pmatrix} g(t) & 1 - g(t) \\ g(t) & 1 - g(t) \end{pmatrix}, & \text{if } t \ge a \end{cases}.$$

A two-state evolution. Another class of solutions under condition of Case 7 can be given as follows. Consider matrix $\mathcal{M}^{[s,t]} = \left(a_{ij}^{[s,t]}\right)_{i,j=1,2}$ with

$$a_{11}^{[s,t]} = \tfrac{1}{2}\left(1 + \alpha(s,t) + \beta(s,t)\right), \quad a_{12}^{[s,t]} = \tfrac{1}{2}\left(1 - \alpha(s,t) - \beta(s,t)\right),$$

$$a_{21}^{[s,t]} = \tfrac{1}{2}\left(1 + \alpha(s,t) - \beta(s,t)\right), \quad a_{22}^{[s,t]} = \tfrac{1}{2}\left(1 - \alpha(s,t) + \beta(s,t)\right). \tag{5.16}$$

In this case the equation (5.2) is equivalent to (see [88])

$$\alpha(s,t) = \alpha(\tau,t) + \alpha(s,\tau)\beta(\tau,t),$$
$$\beta(s,t) = \beta(s,\tau)\beta(\tau,t), \quad s < \tau < t. \tag{5.17}$$

The second equation of the system (5.17) has solutions: $\beta(s,t) = \frac{\Phi(t)}{\Phi(s)}$, where Φ is an arbitrary function with $\Phi(s) \ne 0$. Using this function β for the function α we obtain

$$\frac{\alpha(s,t)}{\Phi(t)} = \frac{\alpha(\tau,t)}{\Phi(t)} + \frac{\alpha(s,\tau)}{\Phi(\tau)}.$$

Now denote $\gamma(s,t) = \frac{\alpha(s,t)}{\Phi(t)}$ then the last equation gets the following form

$$\gamma(s,t) = \gamma(s,\tau) + \gamma(\tau,t).$$

This is Cantor's first equation which has solutions: $\gamma(s,t) = \Psi(t) - \Psi(s)$, where Ψ is an arbitrary function. Hence a solution, denoted by $\mathcal{M}_{25}^{[s,t]} = \left(a_{ij}^{[s,t]} \right)_{i,j=1,2}$ to the equation (5.2) is given by

$$a_{11}^{[s,t]} = \frac{1}{2} \left(1 + \Phi(t)(\Psi(t) - \Psi(s)) + \frac{\Phi(t)}{\Phi(s)} \right),$$

$$a_{12}^{[s,t]} = \frac{1}{2} \left(1 - \Phi(t)(\Psi(t) - \Psi(s)) - \frac{\Phi(t)}{\Phi(s)} \right),$$

$$a_{21}^{[s,t]} = \frac{1}{2} \left(1 + \Phi(t)(\Psi(t) - \Psi(s)) - \frac{\Phi(t)}{\Phi(s)} \right),$$

$$a_{22}^{[s,t]} = \frac{1}{2} \left(1 - \Phi(t)(\Psi(t) - \Psi(s)) + \frac{\Phi(t)}{\Phi(s)} \right).$$

Thus we constructed four new CEAs: $E_i^{[s,t]}$, $0 \le s \le t$ which correspond to $\mathcal{M}_i^{[s,t]}$, $i = 22, 23, 24, 25$ listed above.

Note that each above mentioned CEAs varies by parameters, for example, if in $E_{25}^{[s,t]}$, $0 \le s \le t$ we assume $t = s$ then we get $E_{25}^{[t,t]} = E$ with multiplication table $e_1^2 = e_1, e_2^2 = e_2, e_1 e_2 = 0$. Moreover, choosing functions Φ and Ψ one can variate the limit behavior of $E_{25}^{[t,t]}$. For example, if Φ and Ψ such that $\lim_{t \to +\infty} \Phi(t)\Psi(t) = \lim_{t \to +\infty} \Phi(t) = 0$, then for a fixed s we have $\lim_{t \to +\infty} E^{[s,t]} = E_{1/2}$, where $E_{1/2}$ is an evolution algebra with multiplication table

$$e_1^2 = e_2^2 = \frac{1}{2}(e_1 + e_2), \quad e_1 e_2 = 0.$$

5.1.3 *A construction of chains of three-dimensional EAs*

Following [197] we construct several chains of three-dimensional evolution algebras. Consider a matrix of structural constants having the form

$$\mathcal{M}^{[s,t]} = \begin{pmatrix} a_1^{[s,t]} & a_2^{[s,t]} & a_3^{[s,t]} \\ 0 & a_4^{[s,t]} & a_5^{[s,t]} \\ 0 & 0 & a_6^{[s,t]} \end{pmatrix}. \tag{5.18}$$

Then the equation (5.2) can be written as

$$a_1^{[s,\tau]}a_1^{[\tau,t]} = a_1^{[s,t]}$$
$$a_1^{[s,\tau]}a_2^{[\tau,t]} + a_2^{[s,\tau]}a_4^{[\tau,t]} = a_2^{[s,t]}$$
$$a_1^{[s,\tau]}a_3^{[\tau,t]} + a_2^{[s,\tau]}a_5^{[\tau,t]} + a_3^{[s,\tau]}a_6^{[\tau,t]} = a_3^{[s,t]}$$
$$a_4^{[s,\tau]}a_4^{[\tau,t]} = a_4^{[s,t]} \qquad\qquad (5.19)$$
$$a_4^{[s,\tau]}a_5^{[\tau,t]} + a_5^{[s,\tau]}a_6^{[\tau,t]} = a_5^{[s,t]}$$
$$a_6^{[s,\tau]}a_6^{[\tau,t]} = a_6^{[s,t]}.$$

Note that the first, the fourth and the last equations of the system (5.19) are Cantor's second equations and they have the following solutions ($i = 1, 4, 6$):

(1_i)

$$a_i^{[s,t]} = \frac{\Phi_i(t)}{\Phi_i(s)}, \qquad\qquad (5.20)$$

where Φ_i is an arbitrary function with $\Phi_i(s) \neq 0$, $i = 1, 4, 6$;

(2_i)

$$a_i^{[s,t]} = \begin{cases} 1, & \text{if } s \leq t < \alpha_i, \\ 0, & \text{if } t \geq \alpha_i, \end{cases} \qquad \text{where } \alpha_i > 0, \, i = 1, 4, 6. \qquad (5.21)$$

Thus we have eight possibilities for $(a_1^{[s,t]}, a_4^{[s,t]}, a_6^{[s,t]})$. Consider these possibilities:

Case (1_1), (1_4), (1_6): The second equation of (5.19) becomes

$$\frac{\Phi_1(\tau)}{\Phi_1(s)}a_2^{[\tau,t]} + \frac{\Phi_4(t)}{\Phi_4(\tau)}a_2^{[s,\tau]} = a_2^{[s,t]}.$$

From this equation denoting $\gamma(s,t) = \frac{\Phi_1(s)}{\Phi_4(t)}a_2^{[s,t]}$ we get

$$\gamma(s,t) = \gamma(s,\tau) + \gamma(\tau,t).$$

This equation is Cantor's first equation which has solutions

$$\gamma(s,t) = \xi(t) - \xi(s), \quad \text{where } \xi \text{ is an arbitrary function.}$$

Hence

$$a_2^{[s,t]} = \frac{\Phi_4(t)}{\Phi_1(s)}(\xi(t) - \xi(s)). \qquad\qquad (5.22)$$

Similarly, one gets that

$$a_5^{[s,t]} = \frac{\Phi_6(t)}{\Phi_4(s)}(f(t) - f(s)), \qquad\qquad (5.23)$$

where f is an arbitrary function. Using these functions from the third equation of (5.19) we obtain

$$\varphi^{[s,\tau]} + \varphi^{[\tau,t]} = \varphi^{[s,t]} + (f(\tau) - f(t))(\xi(\tau) - \xi(s)), \qquad (5.24)$$

where $\varphi^{[s,t]} = \frac{\Phi_1(s)}{\Phi_6(t)} a_3^{[s,t]}$. We shall find solutions to (5.24) which have the form

$$\varphi^{[s,t]} = af(s)\xi(s) + bf(s)\xi(t) + cf(t)\xi(s) + df(t)\xi(t).$$

Substituting this function in (5.24) we obtain $b = 0$, $c = -1$ and $a + d = 1$. Hence the following function is a solution to (5.24):

$$\varphi_0^{[s,t]} = af(s)\xi(s) - f(t)\xi(s) + (1-a)f(t)\xi(t),$$

where $a \in \mathbb{R}$. Moreover, one can see that $\varphi^{[s,t]} = \varphi_0^{[s,t]} + \gamma(t) - \gamma(s)$ is also solution to (5.24), where γ is an arbitrary function.

Consequently,

$$a_3^{[s,t]} = \frac{\Phi_6(t)}{\Phi_1(s)} \left[af(s)\xi(s) - f(t)\xi(s) + (1-a)f(t)\xi(t) + \gamma(t) - \gamma(s) \right].$$

$$(5.25)$$

Thus we have proved the following

Proposition 5.3. *The matrix (5.18) with elements given by (5.20), (5.22), (5.23) and (5.25) generates a CEA.*

Case $(1_1), (1_4), (2_6)$: In this case the functions $a_i^{[s,t]}$, $i = 1, 2, 4$ coincide with functions given in the previous case. The function $a_6^{[s,t]}$ is given by (5.21). Here we have to find $a_3^{[s,t]}$ and $a_5^{[s,t]}$. From (5.19) we have

$$a_5^{[s,t]} = \begin{cases} a_5^{[s,\tau]} + \frac{\Phi_4(\tau)}{\Phi_4(s)} a_5^{[\tau,t]}, & \text{if } s \le t < \alpha_6 \\[2mm] \frac{\Phi_4(\tau)}{\Phi_4(s)} a_5^{[\tau,t]}, & \text{if } t \ge \alpha_6. \end{cases} \qquad (5.26)$$

Denoting $\psi^{[s,t]} = \Phi_4(s)a_5^{[s,t]}$ from (5.26) we get

$$\psi^{[s,t]} = \begin{cases} \psi^{[s,\tau]} + \psi^{[\tau,t]}, & \text{if } s \le t < \alpha_6 \\[2mm] \psi^{[\tau,t]}, & \text{if } t \ge \alpha_6. \end{cases}$$

This equation has the following solution

$$\psi^{[s,t]} = \begin{cases} \beta(t) - \beta(s), & \text{if } s \le t < \alpha_6 \\[2mm] \gamma(t), & \text{if } t \ge \alpha_6, \end{cases}$$

where β and γ are arbitrary functions. Therefore, we have

$$a_5^{[s,t]} = \frac{1}{\Phi_4(s)} \begin{cases} \beta(t) - \beta(s), & \text{if } s \le t < \alpha_6 \\ \gamma(t), & \text{if } t \ge \alpha_6, \end{cases} \tag{5.27}$$

Now for $a_3^{[s,t]}$ we have

$$a_3^{[s,t]} = \frac{1}{\Phi_1(s)} \begin{cases} \Phi_1(s)a_3^{[s,\tau]} + \Phi_1(\tau)a_3^{[\tau,t]} \\ +(\xi(\tau) - \xi(s))(\beta(t) - \beta(\tau)), & \text{if } s \le t < \alpha_6 \\ \Phi_1(\tau)a_3^{[\tau,t]} + (\xi(\tau) - \xi(s))\gamma(t), & \text{if } t \ge \alpha_6. \end{cases} \tag{5.28}$$

Denoting $\eta(s,t) = \Phi_1(s)a_3^{[s,t]}$ and using arguments of the previous case we get

$$\eta(s,t) = \begin{cases} a\beta(s)\xi(s) - \beta(t)\xi(s) \\ +(1-a)\beta(t)\xi(t) + \theta(t) - \theta(s), & \text{if } s \le t < \alpha_6 \\ \gamma(t)(v(t) + b\xi(t) - \xi(s)), & \text{if } t \ge \alpha_6, \end{cases}$$

where $b \in \mathbb{R}$ and θ, v are arbitrary functions. Consequently,

$$a_3^{[s,t]} = \frac{1}{\Phi_1(s)} \begin{cases} a\beta(s)\xi(s) - \beta(t)\xi(s) \\ +(1-a)\beta(t)\xi(t) + \theta(t) - \theta(s), & \text{if } s \le t < \alpha_6 \\ \gamma(t)(v(t) + b\xi(t) - \xi(s)), & \text{if } t \ge \alpha_6. \end{cases} \tag{5.29}$$

Thus we have proved the following

Proposition 5.4. *The matrix (5.18) with elements given by (5.20) for $i = 1, 4$; (5.21) for $i = 6$; (5.22), (5.27) and (5.29) generates a CEA.*

Case $(1_1), (2_4), (2_6)$: From the second equation of (5.19) we get

$$a_2^{[s,t]} = \frac{1}{\Phi_1(s)} \begin{cases} \omega(t) - \omega(s), & \text{if } s \le t < \alpha_4 \\ \delta(t), & \text{if } t \ge \alpha_4, \end{cases} \tag{5.30}$$

where ω and δ are arbitrary functions.

Now we shall find $a_5^{[s,t]}$. Without loss of generality we assume $\alpha_4 < \alpha_6$. Then we obtain the following equation

$$a_5^{[s,t]} = \begin{cases} a_5^{[s,\tau]} + a_5^{[\tau,t]}, & \text{if } s \le \tau < \min\{\alpha_4, t\} \\ a_5^{[s,\tau]}, & \text{if } \alpha_4 \le \tau < t < \alpha_6 \\ 0, & \text{if } \alpha_4 \le \tau, \ \alpha_6 \le t. \end{cases}$$

This equation has the following solutions

$$
a_5^{[s,t]} = \begin{cases} \zeta(t) - \zeta(s), & \text{if } s \le \tau < \min\{\alpha_4, t\} \\[2mm] p(s), & \text{if } \alpha_4 \le \tau < t < \alpha_6 \\[2mm] 0, & \text{if } \alpha_4 \le \tau, \ \alpha_6 \le t, \end{cases} \tag{5.31}
$$

where ζ and p are arbitrary functions.

In this case $a_3^{[s,t]}$ satisfies the following equation

$$
a_3^{[s,t]} = \begin{cases} a_3^{[s,\tau]} + \dfrac{\Phi_1(\tau)}{\Phi_1(s)} a_3^{[\tau,t]} \\[1mm] \quad + \dfrac{1}{\Phi_1(s)} (\omega(\tau) - \omega(s))(\zeta(t) - \zeta(\tau)), & \text{if } s \le \tau < \min\{\alpha_4, t\} \\[3mm] a_3^{[s,\tau]} + \dfrac{\Phi_1(\tau)}{\Phi_1(s)} a_3^{[\tau,t]} + \dfrac{1}{\Phi_1(s)} \delta(\tau)p(\tau), & \text{if } \alpha_4 \le \tau < t < \alpha_6 \\[3mm] \dfrac{\Phi_1(\tau)}{\Phi_1(s)} a_3^{[\tau,t]}, & \text{if } \alpha_4 \le \tau, \ \alpha_6 \le t. \end{cases}
$$

Similarly as above one can see that this equation has the following solution

$$
a_3^{[s,t]} = \frac{1}{\Phi_1(s)} \begin{cases} a\zeta(s)\omega(s) - \zeta(t)\omega(s) \\[1mm] \quad + (1-a)\zeta(t)\omega(t) + d(t) - d(s), & \text{if } s \le \tau < \min\{\alpha_4, t\} \\[3mm] a\delta(s)p(s) - (1+a)\delta(t)p(t) + e(t) - e(s), & \text{if } \alpha_4 \le \tau < t < \alpha_6, \\[3mm] m(t), & \text{if } \alpha_4 \le \tau, \ \alpha_6 \le t, \end{cases}
$$
$$\tag{5.32}$$

where d, e and m are arbitrary functions, $a \in \mathbb{R}$.

Thus we have proved the following

Proposition 5.5. *The matrix (5.18) with elements given by (5.20) for $i = 1$; (5.21) for $i = 4, 6$; (5.30), (5.31) and (5.32) generates a CEA.*

Case $(2_1), (2_4), (2_6)$: Assume $\alpha_1 < \alpha_4 < \alpha_6$. In this case for $a_2^{[s,t]}$ we have

$$
a_2^{[s,t]} = \begin{cases} a_2^{[s,\tau]} + a_2^{[\tau,t]}, & \text{if } s \le \tau < \min\{\alpha_1, t\} \\[2mm] a_2^{[s,\tau]}, & \text{if } \alpha_1 \le \tau < t < \alpha_4 \\[2mm] 0, & \text{if } \alpha_1 \le \tau, \ \alpha_4 \le t. \end{cases}
$$

This has the following solution

$$
a_2^{[s,t]} = \begin{cases} l(t) - l(s), & \text{if } s \le \tau < \min\{\alpha_1, t\} \\ w(s), & \text{if } \alpha_1 \le \tau < t < \alpha_4 \\ 0, & \text{if } \alpha_1 \le \tau, \ \alpha_4 \le t, \end{cases} \tag{5.33}
$$

where l and w are arbitrary functions.

Similarly

$$
a_5^{[s,t]} = \begin{cases} A(t) - A(s), & \text{if } s \le \tau < \min\{\alpha_4, t\} \\ B(s), & \text{if } \alpha_4 \le \tau < t < \alpha_6 \\ 0, & \text{if } \alpha_4 \le \tau, \ \alpha_6 \le t, \end{cases} \tag{5.34}
$$

where A and B are arbitrary functions. For $a_3^{[s,t]}$ we have the following equation

$$
a_3^{[s,t]} = \begin{cases} a_3^{[s,\tau]} + a_3^{[\tau,t]} + (l(\tau) - l(s))(A(t) - A(\tau)), & \text{if } s \le \tau < \min\{\alpha_1, t\} \\ a_3^{[s,\tau]} + w(s)(A(t) - A(\tau)), & \text{if } \alpha_1 \le \tau < t < \alpha_4 \\ a_3^{[\tau,t]}, & \text{if } \alpha_4 \le t < \alpha_6 \\ 0, & \text{if } t \ge \alpha_6. \end{cases}
$$

This equation has the following solution

$$
a_3^{[s,t]} = \begin{cases} aA(s)l(s) - A(t)l(s) \\ \quad +(1-a)A(t)l(t) + c(t) - c(s), & \text{if } s \le \tau < \min\{\alpha_1, t\} \\ (A(t) + bA(s) + g(s))w(s), & \text{if } \alpha_1 \le \tau < t < \alpha_4, \\ q(s), & \text{if } \alpha_4 \le t < \alpha_6 \\ 0, & \text{if } t \ge \alpha_6, \end{cases} \tag{5.35}
$$

where c, g and q are arbitrary functions, $a, b \in \mathbb{R}$.

Thus we have proved the following

Proposition 5.6. *The matrix (5.18) with elements given by (5.21) for $i = 1, 4, 6$; (5.33), (5.34) and (5.35) generates a CEA.*

Remark 5.3. Note that the cases $\{(1_1), (2_4), (1_6)\}$, $\{(2_1), (1_4), (1_6)\}$ are similar to the case $\{(1_1), (1_4), (2_6)\}$ and $\{(2_1), (2_4), (1_6)\}$, $\{(2_1), (1_4), (2_6)\}$ are similar to the case $\{(1_1), (2_4), (2_6)\}$. Hence these cases do not give any new CEA.

5.1.4 *High dimensional CEAs*

5.1.4.1 *A n-dimensional time non-homogeneous CEA*

For arbitrary n, following [31], we shall give an example of time non-homogeneous CEA. Let $\{A^{[t]}, t \geq 0\}$ be a family of invertible (for all t), $n \times n$ matrices. Define the following matrix

$$\mathcal{M}_{26}^{[s,t]} = A^{[s]}(A^{[t]})^{-1}, \qquad (5.36)$$

where $(A^{[t]})^{-1}$ is the inverse of $A^{[t]}$.

This matrix satisfies the equation (5.2). Indeed, using associativity of the multiplication of matrices we get

$$\mathcal{M}_{26}^{[s,\tau]}\mathcal{M}_{26}^{[\tau,t]} = A^{[s]}\left((A^{[\tau]})^{-1}A^{[\tau]}\right)(A^{[t]})^{-1} = A^{[s]}(A^{[t]})^{-1} = \mathcal{M}_{26}^{[s,t]}.$$

Thus each family (with one parameter) of invertible $n \times n$ matrices defines a CEA $E_{26}^{[s,t]}$ which is time non-homogeneous, in general. But will be a time homogeneous CEA, for example, if $A^{[t]}$ is equal to tth power of an invertible matrix A.

Construction of a family of invertible $n \times n$ matrices $A^{[t]}$ is not difficult, for example, one can take $A^{[t]}$ as a triangular $n \times n$ matrix of the form

$$A^{[t]} = \begin{pmatrix} a_{11}^{[t]} & 0 & 0 & \dots & 0 \\ a_{21}^{[t]} & a_{22}^{[t]} & 0 & \dots & 0 \\ a_{31}^{[t]} & a_{32}^{[t]} & \ddots & \dots & 0 \\ \vdots & \vdots & \ddots & \ddots & 0 \\ a_{n1}^{[t]} & a_{n2}^{[t]} & \dots & a_{nn-1}^{[t]} & a_{nn}^{[t]} \end{pmatrix}$$

which is called lower triangular matrix or one can take an upper triangular matrix. Then the matrices are invertible iff $a_{ii}^{[t]} \neq 0$, for all $i = 1, \dots, n$ and t. So this example also gives a very rich class of CEAs.

5.1.4.2 *The CEA corresponding to a permutation*

Let S_n be the group of permutations of the set $\{1, 2, \dots, n\}$. For a quadratic matrix $A = \{a_{ij}\}_{i,j=1}^n$, following [197], the matrix corresponds to a permutation π if

$$a_{ij} = \begin{cases} 0, & \text{if } j \neq \pi(i) \\ \neq 0, & \text{if } j = \pi(i). \end{cases} \qquad (5.37)$$

Let A_π, B_π be two matrices corresponding to a permutation π. One can see that

$$A_\pi B_\pi = C_{\pi^2}. \tag{5.38}$$

For a fixed $\pi \in S_n$, following [197], we shall construct CEAs corresponding to $\mathcal{M}_\pi^{[s,t]}$.

Proposition 5.7. *Let $\pi \in S_n$ and $\left\{\mathcal{M}_\pi^{[s,t]}, \ 0 \le s \le t\right\}$ be a family of matrices which satisfies the equation (5.2). Then*

(i) $\mathcal{M}_\pi^{[s,t]} \equiv 0$ if $\{i : \pi(i) = i\} = \emptyset$.
(ii) $\mathcal{M}_\pi^{[s,t]} \ne 0$ if $\{i : \pi(i) = i\} \ne \emptyset$.

Proof.   (i) Let $\mathcal{M}_\pi^{[s,\tau]} \mathcal{M}_\pi^{[\tau,t]} = (c_{ij})_{i,j=1}^n$. We have by (9.2) that

$$c_{ij} = \begin{cases} 0, & \text{if } \ j \ne \pi^2(i) \\ \ne 0, & \text{if } \ j = \pi^2(i). \end{cases} \tag{5.39}$$

From (5.2) we get

$$\begin{cases} 0 \ne c_{i\pi^2(i)} = a_{i\pi(i)}^{[s,\tau]} a_{\pi(i)\pi(\pi(i))}^{[\tau,t]} = a_{i\pi(i)}^{[s,t]}, \\ a_{ij}^{[s,t]} = 0, & \text{if } \ j \ne \pi(i), \ \ \forall s,t. \end{cases} \tag{5.40}$$

Hence $a_{i\pi(i)}^{[s,t]} = 0$ iff $\pi(i) \ne \pi^2(i)$, i.e., $i \ne \pi(i)$.

(ii) Now assume $\{i : \pi(i) = i\} \ne \emptyset$. Without loss of generality we can take $\{i : \pi(i) = i\} = \{1, 2, \ldots, k\}$. Then from (5.40) we obtain

$$a_{jj}^{[s,\tau]} a_{jj}^{[\tau,t]} = a_{jj}^{[s,t]}, \ \ j = 1, \ldots, k.$$

Thus we again get Cantor's second equation which has very rich family of solutions:

a) $a_{jj}^{[s,t]} = \frac{\Phi_j(t)}{\Phi_j(s)}$, where Φ_j is an arbitrary function with $\Phi_j(s) \ne 0$;

b)

$$a_{jj}^{[s,t]} = \begin{cases} 1, & \text{if } \ s \le t < a_j, \\ 0, & \text{if } \ t \ge a_j, \end{cases} \quad \text{where } \ a_j > 0.$$

The CEAs $E_\pi^{[s,t]}$ corresponding to solutions a)–b) have matrix of structural constants in the following form:

$$\mathcal{M}_\pi^{[s,t]} = \begin{pmatrix} a_{11}^{[s,t]} & 0 & 0 & \ldots & \ldots & 0 \\ 0 & a_{22}^{[s,t]} & 0 & \ldots & \ldots & 0 \\ \vdots & & \ldots & \ldots & \ldots & \vdots \\ 0 & 0 & \ldots & a_{kk}^{[s,t]} & \ldots & 0 \\ 0 & 0 & \ldots & 0 & \ldots & 0 \\ \vdots & & \ldots & \ldots & \ldots & \vdots \\ 0 & 0 & \ldots & 0 & \ldots & 0 \end{pmatrix}.$$

This completes the proof. $\qquad\square$

5.1.4.3 *CEAs corresponding to symmetric matrices*

Now following [197] consider $\mathcal{M}^{[s,t]} = \left(a_{ij}^{[s,t]}\right)_{i,j=1,\ldots,n}$ with $a_{ij}^{[s,t]} = a_{ji}^{[s,t]}$ and solve the equation (5.2) for such matrices. The equation (5.2) has the following form

$$\sum_{k=1}^{n} a_{ik}^{[s,\tau]} a_{kj}^{[\tau,t]} = a_{ij}^{[s,t]}, \quad i,j = 1,\ldots,n. \tag{5.41}$$

We introduce the following functions

$$f_i(s,t) = \sum_{j=1}^{n} a_{ij}^{[s,t]}, \quad i = 1,\ldots,n. \tag{5.42}$$

$$g_{ik}(s,t) = \sum_{j=1}^{n} a_{ij}^{[s,t]} - a_{ik}^{[s,t]}, \quad i,k = 1,\ldots,n. \tag{5.43}$$

Using (5.41) from (5.42) we get

$$f_i(s,t) = \sum_{j=1}^{n} \sum_{k=1}^{n} a_{ik}^{[s,\tau]} a_{kj}^{[\tau,t]}$$

$$= \sum_{k=1}^{n} \left(a_{ik}^{[s,\tau]} \sum_{j=1}^{n} a_{kj}^{[\tau,t]} \right) = \sum_{k=1}^{n} a_{ik}^{[s,\tau]} f_k(\tau,t), \quad i = 1,\ldots,n.$$

Consequently, using symmetry of the matrix we get

$$\sum_{i=1}^{n} f_i(s,t) = \sum_{k=1}^{n} \left(\sum_{i=1}^{n} a_{ik}^{[s,\tau]} \right) f_k(\tau,t) = \sum_{k=1}^{n} f_k(s,\tau) f_k(\tau,t).$$

Hence

$$\sum_{i=1}^{n} (f_i(s,t) - f_i(s,\tau) f_i(\tau,t)) = 0. \tag{5.44}$$

Now we shall derive an equation for functions g_{ik}. Using (5.41) from (5.43) we get

$$g_{ik}(s,t) = \sum_{j=1}^{n} \sum_{p=1}^{n} a_{ip}^{[s,\tau]} a_{pj}^{[\tau,t]} - \sum_{p=1}^{n} a_{ip}^{[s,\tau]} a_{pk}^{[\tau,t]}$$

$$= \sum_{p=1}^{n} \left(a_{ip}^{[s,\tau]} \left(\sum_{j=1}^{n} a_{pj}^{[\tau,t]} - a_{pk}^{[\tau,t]} \right) \right).$$

Hence

$$g_{ik}(s,t) = \sum_{p=1}^{n} a_{ip}^{[s,\tau]} g_{pk}(\tau,t), \quad i,k = 1,\ldots,n.$$

Using the symmetry of the matrix from the last equality we get

$$\sum_{i=1}^{n} g_{ik}(s,t) = \sum_{p=1}^{n} \left(\sum_{i=1}^{n} a_{ip}^{[s,\tau]} \right) g_{pk}(\tau,t)$$

$$= \sum_{p=1}^{n} f_p(s,\tau) g_{pk}(\tau,t), \quad k = 1,\ldots,n.$$

Thus we obtain

$$\sum_{i=1}^{n} (g_{ik}(s,t) - f_i(s,\tau) g_{ik}(\tau,t)) = 0, \quad k = 1,\ldots,n. \tag{5.45}$$

Note that (5.44) and (5.45) are general equations for functions f_i and g_{ik}. These equations are satisfied, in particular, if

$$f_i(s,t) = f_i(s,\tau) f_i(\tau,t), \quad \text{for any } i = 1,\ldots,n; \tag{5.46}$$

$$g_{ik}(s,t) = f_i(s,\tau) g_{ik}(\tau,t), \quad \text{for any } i,k = 1,\ldots,n. \tag{5.47}$$

Equations (5.46) have the following solutions

$$f_i(s,t) = \frac{\Phi_i(t)}{\Phi_i(s)}, \quad i = 1,\ldots,n,$$

where Φ_i are arbitrary functions with $\Phi_i(s) \neq 0$. Substituting this solution in (5.47) we obtain

$$g_{ik}(s,t) = \frac{\gamma_{ik}(t)}{\Phi_i(s)}, \quad i,k = 1,\ldots,n,$$

where γ_{ik} are arbitrary functions. By (5.42) and (5.43) we have

$$a_{ik}^{[s,t]} = a_{ki}^{[s,t]} = f_i(s,t) - g_{ik}(s,t) = \frac{\Phi_i(t) - \gamma_{ik}(t)}{\Phi_i(s)}, \quad i,k = 1,\ldots,n. \tag{5.48}$$

Thus we have proved the following

Theorem 5.2. *A matrix* $\mathcal{M}^{[s,t]} = \left(a_{ik}^{[s,t]}\right)_{i,k=1,\ldots,n}$ *given by (5.48) generates an n-dimensional CEA.*

5.1.4.4 *CEAs corresponding to block diagonal matrices*

Recall that a block diagonal matrix is a block matrix which is a square matrix, and having main diagonal blocks square matrices, such that the off-diagonal blocks are zero matrices. A block diagonal matrix M has the form

$$M = \begin{pmatrix} M_1 & 0 & \cdots & 0 \\ 0 & M_2 & \cdots & 0 \\ \vdots & \vdots & \ddots & \vdots \\ 0 & 0 & \cdots & M_m \end{pmatrix},$$

where M_k is a square matrix. It can also be indicated as $M_1 \oplus M_2 \oplus \cdots \oplus M_m$.

Lemma 5.1. *Let* $\mathcal{M}^{[s,t]} = \mathcal{M}_1^{[s,t]} \oplus \mathcal{M}_2^{[s,t]} \oplus \cdots \oplus \mathcal{M}_m^{[s,t]}$ *be a block diagonal matrix. This satisfies (5.2) if and only if each $\mathcal{M}_i^{[s,t]}$, $i = 1,2,\ldots,m$ satisfies the equation (5.2).*

Proof. It follows from the fact that the product of diagonal matrices amounts to simply multiplying corresponding diagonal elements together. $\qquad\square$

Using above constructed CEAs, by Lemma 5.1, one can construct new chains of arbitrary dimensional evolution algebras, i.e. the following is true

Theorem 5.3. *If $\mathcal{M}^{[s,t]} = \mathcal{M}_1^{[s,t]} \oplus \mathcal{M}_2^{[s,t]} \oplus \cdots \oplus \mathcal{M}_m^{[s,t]}$ is a block diagonal matrix, where $\mathcal{M}_i^{[s,t]}$ correspond to a known CEA. Then $\mathcal{M}^{[s,t]}$ generates a CEA.*

Note that the set of CEAs is very rich: taking matrices $\mathcal{M}^{[s,t]}$ constructed in Theorem 5.3 as blocks of a new matrix $\tilde{\mathcal{M}}^{[s,t]}$ one can construct new CEAs. Repeating the argument one can construct very rich class of CEAs.

5.1.4.5 *Nilpotent CEAs*

This subsection is devoted to the following

Proposition 5.8. *A chain of evolution algebras $E^{[s,t]}$ is nilpotent for any (s,t) if and only if $\mathcal{M}^{[s,t]} = 0$ for any (s,t).*

Proof. Necessity. By Theorem 3.10 we have that the matrix should have the following form

$$\mathcal{M}^{[s,t]} = \begin{pmatrix} 0 & a_{12}^{[s,t]} & a_{13}^{[s,t]} & \ldots & a_{1n}^{[s,t]} \\ 0 & 0 & a_{23}^{[s,t]} & \ldots & a_{2n}^{[s,t]} \\ 0 & 0 & 0 & \ldots & a_{3n}^{[s,t]} \\ \vdots & \vdots & \vdots & \cdots & \vdots \\ 0 & 0 & 0 & \cdots & 0 \end{pmatrix}. \tag{5.49}$$

From equation (5.2) we get

$$\mathcal{M}^{[s,t]} = \begin{pmatrix} 0 & 0 & a_{13}^{[s,t]} & a_{14}^{[s,t]} & \ldots & a_{1n}^{[s,t]} \\ 0 & 0 & 0 & a_{24}^{[s,t]} & \ldots & a_{2n}^{[s,t]} \\ 0 & 0 & 0 & 0 & \ldots & a_{3n}^{[s,t]} \\ \vdots & \vdots & \vdots & & \cdots & \vdots \\ 0 & 0 & 0 & 0 & \cdots & 0 \end{pmatrix}. \tag{5.50}$$

Iterating we get $\mathcal{M}^{[s,t]} = 0$.

Sufficiency. Straightforward. $\qquad\qquad\square$

5.2 Property transitions of CEA

It is known that, the behavior of phases (states) of a system in physics, depends on temperature $T > 0$, if for some values of T there is a unique phase and for other values there are several phases, then the physical system has a phase transition [77]. Similar transitions of a property can be seen for systems of biology, chemistry, etc. Here following [31] we shall define a notion of property transition (dynamics) for CEA.

Definition 5.4. Assume a CEA, $E^{[s,t]}$, has a property, denoted by P, at pair of times (s_0, t_0); then one says that the CEA has P property transition if there is a pair $(s, t) \neq (s_0, t_0)$ at which the CEA has no the property P.

Denote

$$\mathcal{T} = \{(s,t) : 0 \leq s \leq t\};$$

$$\mathcal{T}_P = \{(s,t) \in \mathcal{T} : E^{[s,t]} \text{ has property } P\};$$

$$\mathcal{T}_P^0 = \mathcal{T} \setminus \mathcal{T}_P = \{(s,t) \in \mathcal{T} : E^{[s,t]} \text{ has no property } P\}.$$

Definition 5.5. We call the set
$\mathcal{T}_P$-the duration of the property P;
$\mathcal{T}_P^0$-the lost duration of the property P;
The partition $\{\mathcal{T}_P, \mathcal{T}_P^0\}$ of the set $\mathcal{T}$ is called P property diagram.

For example, if $P =$ commutativity then since any evolution algebra is commutative, we conclude that any CEA has not commutativity property transition.

5.2.1 *Baric property transition*

Since a CEA is not a baric algebra, in general, we give baric property diagram for the above given examples of CEAs.

For the case of Example 5.1, by Theorem 3.7 we have that $E^{[t]}$ is baric iff

$$a_{ii}^{[t]} = \frac{2}{3} e^{-\frac{3}{2} At} \cos(\alpha t) + \frac{1}{3} = 1.$$

This has unique solution $t = 0$. Consequently, $\mathcal{T}_{\mathrm{baric}} = \{0\}$, $\mathcal{T}_{\mathrm{baric}}^0 = \{t : t > 0\}$. Thus the CEA $E^{[t]}$ is baric (even non zero trivial) evolution algebra only at initial time, and it loses baricity as soon as the time turned on.

Denote by $\mathcal{T}_b^{(i)}$ the baric property duration of the CEA $E_i^{[s,t]}$, $i = 0, \ldots, 25$.

Theorem 5.4.

(i) (There is no non-baric property transition.) *The algebras* $E_i^{[s,t]}$, $i = 0, 1, 2, 3, 6, 10, 11, 14, 22$ *are not baric for any time* $(s,t) \in \mathcal{T}$;

(ii) (There is no baric property transition.) *The algebras* $E_i^{[s,t]}$, $i = 16, 17, 18$ *and* $E_{23}^{[s,t]}(0,\mu)$, $E_{23}^{[s,t]}(2\mu,\mu)$, $\mu \neq 0$ *are baric for any time* $(s,t) \in \mathcal{T}$;

(iii) (There is baric property transition.) *The CEAs* $E_i^{[s,t]}$, $i = 4, 5, 7, 8, 9, 12, 13, 15, 19, 20, 21, 24$ *and* $E_{23}^{[s,t]}(\lambda,\mu)$, *with* $\lambda \notin \{0, \mu, 2\mu\}$ *have baric property transition with baric property duration sets as the following*

$$\mathcal{T}_b^{(4)} = \left\{ (s,t) \in \mathcal{T} : \frac{\Phi(s)}{\Psi(s)} = \frac{\Phi(t)}{\Psi(t)} \right\};$$

$$\mathcal{T}_b^{(5)} = \{(s,t) \in \mathcal{T} : s \leq t < b,\ \Phi(s) = \Phi(t)\};$$

$$\mathcal{T}_b^{(7)} = \{(s,t) \in \mathcal{T} : s \leq t < a,\ \Psi(s) = \Psi(t)\};$$

$$\mathcal{T}_b^{(8)} = \{(s,t) \in \mathcal{T} : s \leq t < \min\{a, b\}\};$$

$$\mathcal{T}_b^{(9)} = \{(s,t) \in \mathcal{T} : t = s + \pi k, k \in \mathbb{Z}\};$$

$$\mathcal{T}_b^{(12)} = \left\{ (s,t) \in \mathcal{T} : g(s) = \pm\frac{1}{h(s)} \right\};$$

$$\mathcal{T}_b^{(13)} = \{(s,t) \in \mathcal{T} : s \leq t < a,\ \psi(s) = \pm 1\};$$

$$\mathcal{T}_b^{(15)} = \{(s,t) \in \mathcal{T} : s \leq t < a,\ \psi(s) = 0\};$$

$$\mathcal{T}_b^{(19)} = \{(s,t) \in \mathcal{T} : s \leq t < a\};$$

$$\mathcal{T}_b^{(20)} = \{(s,t) \in \mathcal{T} : s \leq t < b\} \cup \{(s,t) \in \mathcal{T} : t \geq b, w(s) = 0\};$$

$$\mathcal{T}_b^{(21)} = \{(s,t) \in \mathcal{T} : s \leq t < \max\{a, b\}\};$$

$$\mathcal{T}_b^{(23)}(\lambda,\mu) = \{(s,t) \in \mathcal{T} : \theta(t) = \theta(s)\}, \lambda \neq 0, \mu, 2\mu.$$

$$\mathcal{T}_b^{(24)} = \{(s,t) \in \mathcal{T} : s \leq t < a\}.$$

Proof. By Theorem 3.7 a two-dimensional evolution algebra $E^{[s,t]}$ is baric if and only if $a_{11}^{[s,t]} \neq 0$, $a_{21}^{[s,t]} = 0$ or $a_{22}^{[s,t]} \neq 0$, $a_{12}^{[s,t]} = 0$. The assertions of the theorem are results of the detailed checking of these conditions. $\square$

Note that sets $\mathcal{T}_b^{(i)}$, $i = 8, 9, 19, 21, 24$ do not depend on any parameter function. But $\mathcal{T}_b^{(i)}$, $i = 4, 5, 7, 12, 13, 15, 20, 23$ depend on some parameter functions and can be controlled by choosing the corresponding parameter functions Φ, Ψ, g, h, ψ, w. These functions are called *baric property controllers* of the CEAs. Because, they really control the baric duration set, for example, if some of them is a strong monotone function then the duration is "minimal", i.e. the line $s = t$, but if a function is a constant function then the baric duration set is "maximal", i.e. it is $\mathcal{T}$. Since these functions are arbitrary functions, we have a rich class of controller functions, therefore we have a "powerful" control on the property to be baric.

Now we shall compute the Lebesgue measure ν of the sets $\mathcal{T}_b^{(i)}$. One can see that

$$\nu\left(\mathcal{T}_b^{(8)}\right) = \frac{1}{2}(\min\{a,b\})^2; \quad \nu\left(\mathcal{T}_b^{(9)}\right) = 0; \quad \nu\left(\mathcal{T}_b^{(19)}\right) = \frac{1}{2}a^2;$$

$$\nu\left(\mathcal{T}_b^{(20)}\right) \geq \frac{1}{2}b^2; \quad \nu\left(\mathcal{T}_b^{(21)}\right) = \frac{1}{2}(\max\{a,b\})^2; \quad \nu\left(\mathcal{T}_b^{(24)}\right) = \frac{1}{2}a^2.$$

The Lebesgue measure of sets $\mathcal{T}_b^{(i)}$, $i = 4, 5, 7, 12, 13, 15, 20, 23$ depend on the corresponding controller functions.

For a given functions g, h, ψ one can easily compute $\nu\left(\mathcal{T}_b^{(i)}\right)$, $i = 12, 13, 15$. For example, if g, h, ψ are elementary functions which do not have "constant parts" in their graphs then $\nu\left(\mathcal{T}_b^{(i)}\right) = 0$, $i = 12, 13, 15$.

Baric property transition for a two-state evolution.

Now we shall consider baricity of the CEA $E_{25}^{[s,t]}$, in more detail. In this case, using Theorem 3.7, we obtain that $\mathcal{T}_b^{(25)}$ is the set of (s,t) such that

$$1 + \Phi(t)(\Psi(t) - \Psi(s)) - \frac{\Phi(t)}{\Phi(s)} = 0 \quad \text{or} \quad 1 - \Phi(t)(\Psi(t) - \Psi(s)) - \frac{\Phi(t)}{\Phi(s)} = 0.$$

These equations can be rewritten as

$$\theta(t) = \theta(s), \quad \theta^-(t) = \theta^-(s),$$

where

$$\theta(t) = \frac{1}{\Phi(t)} + \Psi(t), \ \theta^-(t) = \frac{1}{\Phi(t)} - \Psi(t). \tag{5.51}$$

Thus

$$\mathcal{T}_b^{(25)} = \mathcal{T}_b^{(25)}(\theta) \cup \mathcal{T}_b^{(25)}(\theta^-),$$

here $\mathcal{T}_b^{(25)}(\theta) = \{(s,t) \in \mathcal{T} : \theta(t) = \theta(s)\}$.

Remark 5.4. To describe the set $\mathcal{T}_b^{(25)}$ one has to describe the sets $\mathcal{T}_b^{(25)}(\theta)$ and $\mathcal{T}_b^{(25)}(\theta^-)$, both of which are defined by the parameter functions Φ and Ψ. Note that if we replace Φ with $-\Phi$ or Ψ with $-\Psi$ then these sets transfer to each other. Since Φ and Ψ are arbitrary functions, it will be enough to describe only $\mathcal{T}_b^{(25)}(\theta)$ for arbitrary θ. Thus in the sequel of this subsection we shall deal with description of $\mathcal{T}_b^{(25)}(\theta)$.

The function $\theta(t)$ is the baric property controller of the CEA $E_{25}^{[s,t]}$. For a special choose of θ we have

Proposition 5.9. *If* $\Phi(t) = \lambda^t$, $\lambda > 0$ *and* $\Psi(t) = ct$, $c \in \mathbb{R}$. *Then*

$$\mathcal{T}_b^{(25)}(\theta) = \mathcal{T}_b^{(25)}(\lambda, c) = \{(s,t) : s = t\} \cup$$

$$\begin{cases} \emptyset & \text{if } 0 < \lambda \leq 1, c \geq \ln\lambda; \ \text{ or} \\ & \lambda > 1, c \in (-\infty, 0] \cup [\ln\lambda, +\infty), \\ \{(s,t) : 0 \leq s \leq t_c, t_c \leq t \leq t'_c, \theta(s) = \theta(t)\} & \text{if } 0 < \lambda \leq 1, c < \ln\lambda; \ \text{ or} \\ & \lambda > 1, c \in (0, \ln\lambda), \end{cases}$$

where t_c *and* t'_c *serve as critical times, which defined by* $t_c = \frac{1}{\ln\lambda}\ln\left(\frac{\ln\lambda}{c}\right)$ *and* $t'_c > 0$ *is a unique solution to* $\theta(t'_c) = 1$.

Proof. Under the conditions of the proposition we have $\theta(t) = \lambda^{-t} + ct$, and the simple analysis of the equation $\theta(s) = \theta(t)$ for this θ gives the full set $\mathcal{T}_b^{(25)}(\lambda, c)$. $\qquad\square$

In Fig. 5.1, the baric property diagram is given.

As a corollary of Proposition 5.9 we have

Corollary 5.1.

1) *For any fixed* s, *with* $0 \leq s < t_c$ *(resp.* $t_c \leq s \leq t$*), the time* t *has two (resp. one) critical values:* $t_c^{(1)} = s$ *(resp. s) and* $t_c^{(2)}$ *which is a unique solution of* $\theta(t_c^{(2)}) = \theta(s)$.

2) *For any fixed* t, *with* $0 \leq s \leq t \leq t_c$ *or* $t'_c < t$ *(resp.* $t_c < t \leq t'_c$*), the time* s *has one (resp. two) critical values:* $s_c^{(1)} = t$ *(resp.* $s_c^{(1)} = t$ *and* $s_c^{(2)}$ *which is a unique solution of* $\theta(t_c^{(2)}) = \theta(s)$*).*

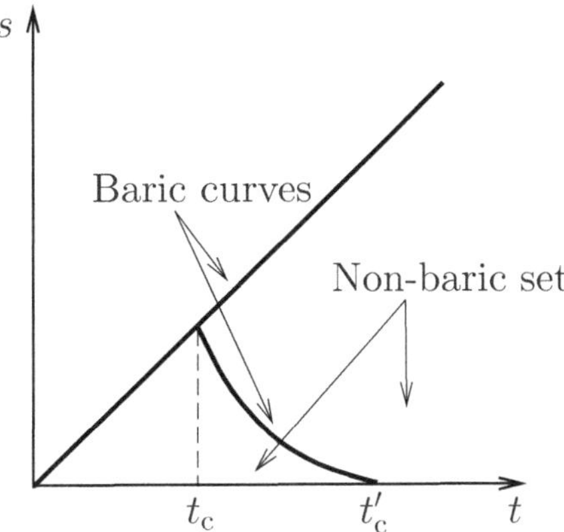

Fig. 5.1   The baric property diagram for $\theta(t) = \lambda^{-t} + ct$.

Let us discuss some more examples of the controller θ. If $\theta(t) = \tan(t)$ then $\tan(s) = \tan(t)$ has solution $t = s + \pi k$, $k \in \mathbb{Z}$. The intersection of this family of lines with $\mathcal{T}$ gives the family of half lines, i.e.

$$\mathcal{T}_b^{(25)}(\tan(t)) = \bigcup_{k=0,1,2,\ldots} \{(s,t) \in \mathcal{T} : s = t - \pi k\}.$$

If $\theta(t) = \sin(t)$ then $\sin(s) = \sin(t)$ has two family of solutions: $s = t + 2\pi k$, $k \in \mathbb{Z}$ and $s = -t + (2k+1)\pi$, $k \in \mathbb{Z}$. The intersection of these families of lines with $\mathcal{T}$ is

$$\mathcal{T}_b^{(25)}(\sin(t)) = \bigcup_{k=0,1,2,\ldots} \{(s,t) \in \mathcal{T} : t = s + 2\pi k \ \ \text{or} \ \ t = -s + (2k+1)\pi\}.$$

In all above considered examples we obtained a set $\mathcal{T}_b^{(25)}(\theta)$ which has zero Lebesgue measure. But there is controllers for which this set has non-zero Lebesgue measure, for example, if $\theta(t)$ is a controller function with the graph as shown in Fig. 5.2, then the corresponding baric property diagram is as shown in Fig. 5.3. Thus any "constant part" of the graph of the controller gives a full triangle in the diagram, moreover, any "non-constant part" gives several curves. In this case the set $\mathcal{T}_b^{(25)}(\theta)$ has a non-zero Lebesgue measure.

Let $\theta(t) = D(t)$ be the Dirichlet function defined by

$$D(t) = \begin{cases} 1 & \text{if } t \text{ rational}; \\ 0 & \text{if } t \text{ irrational}. \end{cases}$$

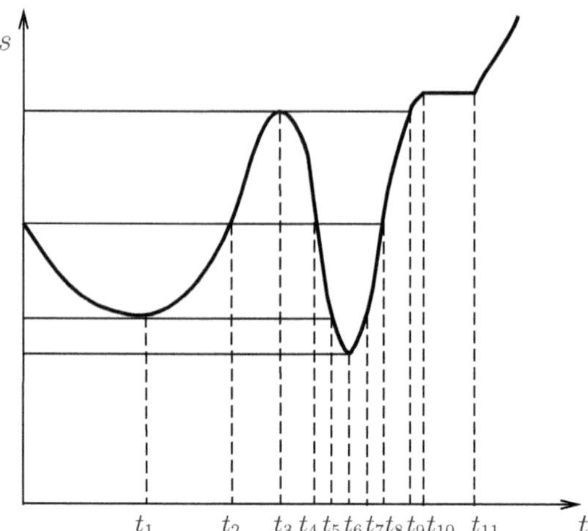

Fig. 5.2   An example of controller θ.

In this case we have a rich set of baric property duration, i.e.

$$\mathcal{T}_b^{(25)}(D(t)) = \{(s,t) \in \mathcal{T} : t \text{ and } s \text{ rational}\} \cup \{(s,t) \in \mathcal{T} : t \text{ and } s \text{ irrational}\}.$$

Definition 5.6. A function θ defined on $\mathbb{R}$ is called a function of *countable variation* if it has the following properties:

1. it is continuous except at most on a countable set, (which is denoted by $X_c = \{x_1, x_2, \dots\}$), it has only jump-type discontinuities (denote the one-sided limit from the negative direction by $\theta(x_i^-)$ and from the positive direction by $\theta(x_i^+)$, $i = 1, 2, \dots$);

2. it has at most a countable set of singular (extremum) points (which is denoted by $X_e = \{y_1, y_2, \dots\}$).

Note that any function of countable variation has not "constant parts" in its graph.

The following theorem gives a characteristics of the baric property duration set.

Theorem 5.5. *If the controller θ (see (5.51)) is a function of countable variation, then the baric duration set $\mathcal{T}_b^{(25)}(\theta)$ has zero Lebesgue measure, that is the corresponding CEA is not baric almost surely.*

Proof.   Using the (finite or infinite) sequences X_c and X_e we construct the sequences $\{t_{i,k}^-\}_{k=1,2,\dots}$, $i = 1, 2, \dots$ with $\theta(t_{i,k}^-) = \theta(x_i^-)$ for all k;

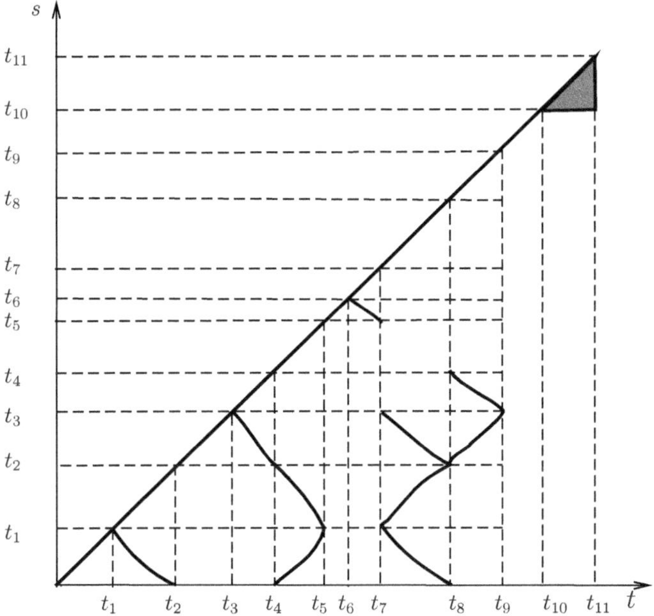

Fig. 5.3 The baric property diagram for the controller θ with graph as in Fig. 5.2.

$\{t_{i,q}^{+}\}_{q=1,2,\ldots}$, $i = 1,2,\ldots$ with $\theta(t_{i,q}^{+}) = \theta(x_i^{+})$ and $\{t_{j,l}^{e}\}_{l=1,2,\ldots}$, $j = 1,2,\ldots$, where $\theta(t_{j,l}^{e}) = \theta(y_j)$ for all l. Now define a sequence $\{t_i\}_{i=1,2,\ldots}$, with $t_1 < t_2 < t_3 < \ldots$ as follows

$$\{t_i\}_{i=1,2,\ldots} = X_c \cup X_e \bigcup_i \left(\{t_{i,k}^{-}\}_{k=1,2,\ldots} \cup \{t_{i,q}^{+}\}_{q=1,2,\ldots} \right) \bigcup_j \{t_{j,l}^{e}\}_{l=1,2,\ldots}.$$

Since θ is a function of countable variation, the sequence $\{t_i\}_{i=1,2,\ldots}$ is at most countable. In a case, if it is a bounded sequence (in particular, a finite sequence), then we add the last term to be $+\infty$. Consider rectangles

$$\mathcal{T}_{ij} = \left\{ (s,t) \in \mathbb{R}^2 : t_i \leq s \leq t_{i+1},\, t_j \leq t \leq t_{j+1} \right\}.$$

Denote $G(\theta) = \{(t,y) : y = \theta(t)\}$.

By the construction, the rectangles have the following properties:

– The set of all rectangles is at most a countable set;

– The intersection $G(\theta) \cap \mathcal{T}_{ij}$ is empty or contains a monotone part of the graph $G(\theta)$.

If $G(\theta) \cap \mathcal{T}_{ij}$ is empty then $\mathcal{T}_{ij}$ is called empty too.

Now we shall construct the set $\mathcal{T}_b^{(25)}(\theta)$. Fix i, j such that the rectangle $\mathcal{T}_{ij}$ is not-empty (an empty rectangle does not give any contribution to the set $\mathcal{T}_b^{(25)}(\theta)$), then we have

$$\mathcal{T}_b^{(25)}(\theta) \cap \mathcal{T}_{kj} = \text{a curve giving an one-to-one corespondence between}$$

$$[t_j, t_{j+1}] \quad \text{and} \quad [t_k, t_{k+1}] \quad \text{if} \quad \mathcal{T}_{ik} \neq \emptyset, \ k = 1, \ldots, j-1.$$

Thus we have

$$\mathcal{T}_b^{(25)}(\theta) = \bigcup_{kj} \left(\mathcal{T}_b^{(25)}(\theta) \cap \mathcal{T}_{kj} \right).$$

Since there are a countable set of rectangles and in each rectangle we may have at most a curve which has Lebesgue measure zero (because, these curves give one-to-one correspondences), we conclude that the set $\mathcal{T}_b^{(25)}(\theta)$ also has zero Lebesgue measure. $\qquad \square$

The following theorem gives a characteristics of the baric property duration set for other CEAs.

Theorem 5.6. *If the controller function of $\mathcal{T}_b^{(i)}$, $i = 4, 5, 7, 23$ (i.e. $\frac{\Phi(t)}{\Psi(t)}$ for $i = 4$; $\Phi(t)$ for $i = 5$; $\Psi(t)$ for $i = 7$; $\theta(t)$ for $i = 23$) is a function of countable variation, then the baric duration set $\mathcal{T}_b^{(i)}$ has zero Lebesgue measure, that is the corresponding CEA is not baric almost surely.*

Proof. Is similar to the proof of Theorem 5.5. $\qquad \square$

Consider now the CEA $E_{26}^{[s,t]}$ given by a family of invertible lower (or upper) triangular matrices $A^{[t]}$, $t \geq 0$, see (5.36).

Theorem 5.7. *For any pair of time (s, t) the n-dimensional evolution algebra $E_{26}^{[s,t]}$, constructed by a family of (lower or upper) triangular invertible matrices is baric. Moreover, $E_{26}^{[s,t]}$ has a weight function $\sigma(x) = \mathcal{M}_{26,nn}^{[s,t]} x_n$, where $\mathcal{M}_{26,ii}^{[s,t]}$, $i = 1, \ldots, n$ are diagonal entries of $\mathcal{M}_{26}^{[s,t]} = A^{[s]}(A^{[t]})^{-1}$.*

Proof. It is known that the standard operations on triangular matrices conveniently preserve the triangular form: the sum and product of two lower triangular matrices is again lower triangular. The inverse of a lower triangular matrix is also lower triangular, and of course we can multiply a lower triangular matrix by a constant and it will still be lower triangular. This means that the lower triangular matrices form a subalgebra of the ring of square matrices for any given size. The analogous result holds for upper triangular matrices. Using these properties we get that $\mathcal{M}_{26}^{[s,t]}$ is also a triangular matrix. Moreover, since $A^{[t]}$ is invertible, its determinant is non-zero for all t. Thus

$$\det(\mathcal{M}_{26}^{[s,t]}) = \prod_{i=1}^{n} \mathcal{M}_{26,ii}^{[s,t]} = \det(A^{[s]})\det((A^{[t]})^{-1}) \neq 0.$$

Consequently, all diagonal entries of the matrix are non-zero. In particular, $\mathcal{M}_{26,nn}^{[s,t]} \neq 0$, and Theorem 3.7 completes the proof. $\qquad\square$

Corollary 5.2. *The CEA* $E_{26}^{[s,t]}$ *constructed by triangular invertible matrices has not baric property transition.*

5.2.2 *Absolute nilpotent elements transition*

Recall that an element x of an algebra A is called an *absolute nilpotent* if $x^2 = 0$.

Let $E = \mathbb{R}^n$ be an evolution algebra over the field $\mathbb{R}$ with structural constant coefficients matrix $\mathcal{M} = (a_{ij})$, then for arbitrary $x = \sum_i x_i e_i$ and $y = \sum_i y_i e_i \in \mathbb{R}^n$ we have

$$xy = \sum_j \left(\sum_i a_{ij} x_i y_i \right) e_j, \quad x^2 = \sum_j \left(\sum_i a_{ij} x_i^2 \right) e_j.$$

For a n-dimensional evolution algebra $\mathbb{R}^n$ consider operator $V \colon \mathbb{R}^n \to \mathbb{R}^n$, $x \mapsto V(x) = x'$ defined as

$$x_j' = \sum_{i=1}^{n} a_{ij} x_i^2, \quad j = 1, \ldots, n. \tag{5.52}$$

We have $V(x) = x^2$, hence the equation $V(x) = x^2 = 0$ is given by the following system

$$\sum_i a_{ij} x_i^2 = 0, \quad j = 1, \ldots, n. \tag{5.53}$$

If $\det(\mathcal{M}) \neq 0$ then the system (5.53) has unique solution $(0, \ldots, 0)$. If $\det(\mathcal{M}) = 0$ and $\operatorname{rank}(\mathcal{M}) = r$ then we can assume that the first r rows of $\mathcal{M}$ are linearly independent, consequently, the system of equations (5.53) can be written as

$$x_i^2 = -\sum_{j=r+1}^{n} d_{ij} x_j^2, \quad i = 1, \ldots, r, \tag{5.54}$$

where $d_{ij} = \frac{\det(\mathcal{M}_{ij})}{\det(\mathcal{M}_r)}$ with $\mathcal{M}_r = (a_{ij})_{i,j=1,\dots,r}$,

$$
\mathcal{M}_{ij} = \begin{pmatrix}
a_{11} & \dots & a_{i-1,1} & a_{j1} & a_{i+1,1} & \dots & a_{r1} \\
a_{12} & \dots & a_{i-1,2} & a_{j2} & a_{i+1,2} & \dots & a_{r2} \\
& \dots & & \dots & & \dots & \\
a_{1r} & \dots & a_{i-1,r} & a_{jr} & a_{i+1,r} & \dots & a_{rr}
\end{pmatrix}.
$$

An interesting problem is to find a necessary and sufficient condition on matrix $D = (d_{ij})_{\substack{i=1,\dots,r \\ j=r+1,\dots,n}}$ under which the system (5.54) has unique solution. The difficulty of the problem depends on rank r, here we shall consider the case $r = n - 1$.

Proposition 5.10. *1) If* $\det(\mathcal{M}) \neq 0$ *then the finite dimensional evolution algebra* $\mathbb{R}^n$ *has unique absolute nilpotent* $(0, \dots, 0)$.

2) If $\det(\mathcal{M}) = 0$ *and* $\mathrm{rank}(\mathcal{M}) = n - 1$ *then the evolution algebra* R^n *has unique absolute nilpotent* $(0, \dots, 0)$ *if and only if*

$$
\det(\mathcal{M}_{i_0 n}) \cdot \det(\mathcal{M}_{n-1}) > 0, \tag{5.55}
$$

for some $i_0 \in \{1, \dots, n-1\}$.

Proof. 1) Straightforward.

2) If $\mathrm{rank}(\mathcal{M}) = n - 1$ then from (5.54) we get

$$
x_i^2 = -\frac{\det(\mathcal{M}_{in})}{\det(\mathcal{M}_{n-1})} x_n^2, \quad i = 1, \dots, n - 1. \tag{5.56}
$$

From (5.56) it follows that the condition (5.55) is necessary and sufficient to have unique solution $(0, \dots, 0)$. $\qquad\square$

For a CEA $E^{[s,t]}$ with matrix $\mathcal{M}^{[s,t]}$ denote

$$
\mathcal{T}_{nil} = \{(s,t) \in \mathcal{T} : E^{[s,t]} \text{ has unique absolute nilpotent}\}, \quad \mathcal{T}_{nil}^0 = \mathcal{T} \setminus \mathcal{T}_{nil}.
$$

The following theorem gives an answer on problem of existence of "uniqueness of absolute nilpotent element" property transition.

Theorem 5.8. *1) There are CEAs which have not "uniqueness of absolute nilpotent element" property transition.*

2) There is CEA which has "uniqueness of absolute nilpotent element" property transition.

Proof. Denote $d(s,t) = \det(\mathcal{M}^{[s,t]})$. We have

$$d(s,t) = d(s,\tau)d(\tau,t), \quad \text{for all} \quad \tau, \; s < \tau < t. \tag{5.57}$$

As it was mentioned above, the equation (5.57) is known as Cantor's second equation.

1) The equation (5.57) has solutions $d(s,t) = \frac{\Phi(s)}{\Phi(t)}$, where $\Phi(t) \neq 0$ is an arbitrary function. Thus for such solutions we conclude that if $d(s_0, t_0) \neq 0$ for some (s_0, t_0) then $d(s,t) \neq 0$ for any (s,t). Consequently, corresponding CEAs have not "uniqueness of absolute nilpotent element" property transition.

2) Note that the equation (5.57) has solution $d(s,t) = f(t)$, where $f(t) = 1$ for $t < 1$ and $f(t) = 0$ otherwise. For this solution we have $d(s,t) = 1, s < t < 1$ and $d(s,t) = 0, t \geq 1$. For some $t \geq 1$ one can construct a matrix $\mathcal{M}^{[s,t]}$ which does not satisfy uniqueness condition mentioned in part 2) of Proposition 5.10. Indeed let us consider the matrix $\mathcal{M}^{[s,t]} = \left(a_{ij}^{[s,t]} \right)_{i,j=1,2}$ with entries as in (5.16). The second equation of the system (5.17) has a solution:

$$\beta(s,t) = \begin{cases} 1, & \text{if } s < t < 1 \\ 0, & \text{if } t \geq 1. \end{cases}$$

Substituting this solution in the first equation of (5.17) we obtain

$$\alpha(s,t) = \begin{cases} \psi(t) - \psi(s), & \text{if } s < t < 1 \\ g(t), & \text{if } t \geq 1, \end{cases}$$

where ψ and g are arbitrary functions. The corresponding matrix has the following form

$$\mathcal{M}^{[s,t]} = \frac{1}{2} \begin{pmatrix} 2 + \psi(t) - \psi(s) & -\psi(t) + \psi(s) \\ \psi(t) - \psi(s) & 2 - \psi(t) + \psi(s) \end{pmatrix}, \quad \text{if } s < t < 1,$$

and

$$\mathcal{M}^{[s,t]} = \frac{1}{2} \begin{pmatrix} 1 + g(t) & 1 - g(t) \\ 1 + g(t) & 1 - g(t) \end{pmatrix}, \quad \text{if } t \geq 1. \tag{5.58}$$

We have

$$d(s,t) = \det(\mathcal{M}^{[s,t]}) = \begin{cases} 1, & \text{if } s < t < 1 \\ 0, & \text{if } t \geq 1. \end{cases}$$

Assume $g(t) \neq -1$ then for (5.58) the equation (5.56) has the form

$$x_1^2 = -\frac{1-g(t)}{1+g(t)}x_2^2.$$

This equation has infinitely many solutions if $|g(t)| > 1$ for some $t \geq 1$. Thus corresponding CEA has "uniqueness of absolute nilpotent element" property transition. $\hfill\square$

Now we shall study consider $E_i^{[s,t]}$, $i = 0, \ldots, 25$.

For a CEA $E_i^{[s,t]}$ with matrix $\mathcal{M}_i^{[s,t]}$ denote

$$\mathcal{T}_{nil}^{(i)} = \{(s,t) \in \mathcal{T} : E_i^{[s,t]} \text{ has unique absolute nilpotent}\}, \quad \mathcal{T}_{nil}^0 = \mathcal{T} \setminus \mathcal{T}_{nil}.$$

The following theorem gives an answer on problem of existence of "uniqueness of absolute nilpotent element" property transition.

Theorem 5.9.

(1) *The CEAs* $E_i^{[s,t]}$, $i = 3, 4, 5, 9, 10, 17, 22, 23, 24, 25, 26$ *have unique absolute nilpotent element* $(0,0)$ *for any time* $(s,t) \in \mathcal{T}$.

(2) *There CEAs* $E_i^{[s,t]}$, $i = 0, 1, 2, 16, 19$ *have infinitely many of absolute nilpotent elements for any time* $(s,t) \in \mathcal{T}$.

(3) *The CEAs* $E_i^{[s,t]}$, $i = 6, 7, 8, 11, 12, 13, 14, 15, 18, 20, 21$ *have "uniqueness of absolute nilpotent element" property transition with the property duration sets as the following*

$$\mathcal{T}_{nil}^{(i)} = \{(s,t) \in \mathcal{T} : t < a\}, \quad a > 0, \quad i = 6, 7, 8, 11, 18.$$
$$\mathcal{T}_{nil}^{(12)} = \left\{(s,t) \in \mathcal{T} : g^2(t) \leq \tfrac{1}{h^2(s)}\right\};$$
$$\mathcal{T}_{nil}^{(13)} = \left\{(s,t) \in \mathcal{T} : s \leq t < a, \, \psi^2(s) \leq 1\right\};$$
$$\mathcal{T}_{nil}^{(14)} = \{(s,t) \in \mathcal{T} : \Phi(s)\psi(s) > 0\};$$
$$\mathcal{T}_{nil}^{(15)} = \{(s,t) \in \mathcal{T} : s \leq t < a, \psi(s) > 0\};$$
$$\mathcal{T}_{nil}^{(20)} = \left\{(s,t) \in \mathcal{T} : s \leq t < b\right\} \cup \left\{(s,t) \in \mathcal{T} : t \geq b, \, \tfrac{w(s)}{\Phi(s)} > 0\right\};$$
$$\mathcal{T}_{nil}^{(21)} = \{(s,t) \in \mathcal{T} : s \leq t < \min\{a,b\}\}$$
$$\cup\{(s,t) \in \mathcal{T} : b \leq t < a, \, b < a, \, v(s) > 0\}.$$

Proof. The proof consists the simple analysis of the solutions of the system (5.53) for each $E_i^{[s,t]}$, $i = 0, \ldots, 26$. $\hfill\square$

Remark 5.5. The above mentioned examples of "uniqueness of nilpotent element" property transition of CEAs with time-parameter are similar to the "uniqueness of Gibbs phase" property transition, i.e. phase transition of physical systems with respect to temperature-parameter, $T > 0$. Usually there is a phase transition if the temperature is very low ($T \sim 0$) or if it is very high ($T \sim +\infty$) (see [77]).

5.2.3 *Idempotent elements transition*

A element x of an algebra $\mathcal{A}$ is called *idempotent* if $x^2 = x$; such points of an evolution algebra are especially important, because they are the fixed points (i.e. $V(x) = x$) of the evolution operator V, (5.52).

By (5.52) the equation $x^2 = x$ can be written as

$$x_j = \sum_{i=1}^{n} a_{ij} x_i^2, \quad j = 1, \dots, n. \tag{5.59}$$

The general analysis of the solutions of the system (5.59) is very difficult. Let us give solutions of this problem for the CEAs $E_i^{[s,t]}$, $i = 0, \dots, 24$.

Denote by $\mathcal{I}d(E)$ the idempotent elements of an algebra E.

The following theorem gives the time-dynamics of the idempotent elements for algebras $E_i^{[s,t]}$, $i = 0, \dots, 24$.

Theorem 5.10. *The following assertions hold*

(1) *The algebras $E_i^{[s,t]}$, $i = 0, 1, 2$, have unique idempotent $(0,0)$ any time $(s,t) \in \mathcal{T}$.*

(2) *The algebras $E_i^{[s,t]}$, $i = 3, 10, 12, 14, 16, 22$ have two idempotents $(0,0)$, $(x_i(s,t), y_i(s,t))$ any time $(s,t) \in \mathcal{T}$. Moreover explicit formula of each $x_i(s,t)$ and $y_i(s,t)$ can be given.*

(3) *We have* $\mathcal{I}d\left(E_4^{[s,t]}\right) =$

$$\begin{cases} \{0, z_1, z_2, z_3\}, & \text{if } (s,t) \in \left\{ (s,t) \in \mathcal{T} : \frac{\Phi(t)}{\Phi(s)} = \frac{\Psi(t)}{\Psi(s)} \right\} \\[2mm] \{0, z_3\}, & \text{if } (s,t) \in \left\{ \frac{\Phi(t)}{\Phi(s)} \neq \frac{\Psi(t)}{\Psi(s)}, D(s,t) < 0 \right\} \\[2mm] \{0, z_3, (x_*, y_*)\}, & \text{if } (s,t) \in \left\{ \frac{\Phi(t)}{\Phi(s)} \neq \frac{\Psi(t)}{\Psi(s)}, D(s,t) = 0 \right\} \\[2mm] \{0, z_3, (x_\pm, y_\pm)\}, & \text{if } (s,t) \in \left\{ \frac{\Phi(t)}{\Phi(s)} \neq \frac{\Psi(t)}{\Psi(s)}, D(s,t) > 0 \right\}, \end{cases}$$

where $0 = (0,0)$, $z_1 = (0, \frac{\Phi(t)}{\Phi(s)})$, $z_2 = (\frac{\Phi(t)}{\Phi(s)}, 0)$, $z_3 = (\frac{\Phi(t)}{\Phi(s)}, \frac{\Phi(t)}{\Phi(s)})$, $D(s,t) = \frac{\Phi(t)}{\Phi(s)} \left(2\frac{\Psi(t)}{\Psi(s)} - \frac{\Phi(t)}{\Phi(s)} \right)$. The explicit formulas of $x_*, y_*, x_\pm$ and $y_\pm$ are given below. The sets $\left\{ \frac{\Phi(t)}{\Phi(s)} = \frac{\Psi(t)}{\Psi(s)} \right\}$, $\left\{ \frac{\Phi(t)}{\Phi(s)} = 2\frac{\Psi(t)}{\Psi(s)} \right\}$ are critical (boundary) sets of the idempotent elements transition.

(4) We have $\mathcal{I}d\left(E_5^{[s,t]}\right) =$

$$\begin{cases} \{0, z_1, z_2, z_3\}, & \text{if } (s,t) \in \{(s,t) \in \mathcal{T} : s \le t < b,\ \Phi(t) = \Phi(s)\} \\ \{0, z_3\}, & \text{if } (s,t) \in \{s \le t < b, \Phi(t) \ne \Phi(s),\ D(s,t) < 0\} \\ \{0, z_3, (x_*, y_*)\}, & \text{if } (s,t) \in \{s \le t < b, \Phi(t) \ne \Phi(s),\ D(s,t) = 0\} \\ \{0, z_3, (x_\pm, y_\pm)\}, & \text{if } (s,t) \in \{s \le t < b,\ \Phi(t) \ne \Phi(s),\ D(s,t) > 0\}, \\ \{0, z_3\}, & \text{if } t \ge b, \end{cases}$$

where z_i as in (3), $D(s,t) = \frac{\Phi(t)}{\Phi(s)}\left(2 - \frac{\Phi(t)}{\Phi(s)}\right)$.

(5) Algebras $E_i^{[s,t]}$, $i = 6, 11, 13, 15, 19$ have two idempotent elements for any time (s,t) with $s \le t < a$ and a unique idempotent for time (s,t) with $t \ge a$. The critical line of the transition is $t = a$.

(6) We have $\mathcal{I}d\left(E_7^{[s,t]}\right) =$

$$\begin{cases} \{0, z_1, z_2, z_3\}, & \text{if } (s,t) \in \{(s,t) \in \mathcal{T} : s \le t < a,\ \Psi(t) = \Psi(s)\} \\ \{0, z_3\}, & \text{if } (s,t) \in \{s \le t < a, \Psi(t) \ne \Psi(s),\ d(s,t) < 0\}, \\ \{0, z_3, (x_*, y_*)\}, & \text{if } (s,t) \in \{s \le t < a, \Psi(t) \ne \Psi(s),\ d(s,t) = 0\}, \\ \{0, z_3, (x_\pm, y_\pm)\}, & \text{if } (s,t) \in \{s \le t < a, \Psi(t) \ne \Psi(s),\ d(s,t) > 0\}, \\ 0, & \text{if } t \ge a, \end{cases}$$

where $d(s,t) = \frac{2\Psi(t)}{\Psi(s)} - 1$. The critical sets are $t = a$, $\Psi(t) = \Psi(s)$, $\Psi(s) = 2\Psi(t)$.

(7) For $a \le b$ we have $\mathcal{I}d\left(E_8^{[s,t]}\right) =$

$$\begin{cases} \{(0,0), (0,1), (1,0), (1,1)\}, & \text{if } (s,t) \in \{(s,t) \in \mathcal{T} : s \le t < a\} \\ (0,0), & \text{if } t \ge a. \end{cases}$$

For $a > b$ we have $\mathcal{I}d\left(E_8^{[s,t]}\right) =$

$$\begin{cases} \{(0,0), (0,1), (1,0), (1,1)\}, & \text{if } (s,t) \in \{(s,t) \in \mathcal{T} : s \le t < b\} \\ \{(0,0), (1,1)\}, & \text{if } (s,t) \in \{(s,t) \in \mathcal{T} : b \le t < a\} \\ (0,0), & \text{if } t \ge a. \end{cases}$$

The lines $t = a$ and $t = b$ are critical for the transition.

(8) The algebra $E_9^{[s,t]}$, has three idempotent elements $(0,0), (1,0), (0,1)$ for any time (s,t) with $t = s + 2\pi n$; has three idempotent elements $(0,0), (-1,0), (0,-1)$ for any time (s,t) with $t = s + (2n+1)\pi$ and at least one idempotent for time (s,t) with $t \ne s + \pi n$, $n \in \mathbb{Z}$.

(9) We have

$$
\mathcal{I}d\left(E_{17}^{[s,t]}\right) = \begin{cases} \{(0,0), z_2\}, & \text{if } D(s,t) < 0 \\ \{(0,0), z_2, (\frac{\Phi(s)}{2\Phi(t)}, \frac{\Psi(s)}{\Psi(t)})\}, & \text{if } D(s,t) = 0 \\ \{(0,0), z_2, (x_\pm, y_\pm)\}, & \text{if } D(s,t) > 0, \end{cases}
$$

where $D(s,t) = \frac{4\Phi^2(t)\Psi(s)}{\Phi(s)\Psi^2(t)}(g(t) - g(s)) - 1$.

(10) We have

$$
\mathcal{I}d\left(E_{18}^{[s,t]}\right) = \begin{cases} \{(0,0), (1,0)\}, & \text{if } s \le t < a, D(s,t) < 0, \\ \{(0,0), (1,0), (\frac{1}{2}, \frac{\Psi(s)}{\Psi(t)})\}, & \text{if } s \le t < a, D(s,t) = 0, \\ \{(0,0), (1,0), (x_\pm, \frac{\Psi(s)}{\Psi(t)})\}, & \text{if } s \le t < a, D(s,t) > 0, \\ \{(0,0), (\frac{h(t)\Psi(s)}{\Psi^2(t)}, \frac{\Psi(s)}{\Psi(t)})\}, & \text{if } t \ge a, \end{cases}
$$

where $D(s,t) = 1 - \frac{4\Psi(s)(h(t)-h(s))}{\Psi^2(t)}$.

(11) We have

$$
\mathcal{I}d\left(E_{20}^{[s,t]}\right) = \begin{cases} \{(0,0), z_2\}, & \text{if } s \le t < b, D(s,t) < 0, \\ \{(0,0), z_2, (\frac{\Phi(s)}{2\Phi(t)}, 1)\}, & \text{if } s \le t < b, D(s,t) = 0, \\ \{(0,0), z_2, (x_\pm, 1)\}, & \text{if } s \le t < a, D(s,t) > 0, \\ \{(0,0), z_2\} & \text{if, } t \ge b, \end{cases}
$$

where $D(s,t) = 1 - \frac{4\Phi^2(t)(v(t)-v(s))}{\Phi(s)}$.

(12) We have $\mathcal{I}d\left(E_{21}^{[s,t]}\right) =$

$$
\begin{cases} \{(0,0), (1,0)\}, & \text{if } s \le t < \min\{a,b\}, D(s,t) < 0, \\ \{(0,0), (1,0), (\frac{1}{2}, 1)\}, & \text{if } s \le t < \min\{a,b\}, D(s,t) = 0, \\ \{(0,0), (1,0), (x_\pm, 1)\}, & \text{if } s \le t < \min\{a,b\}, D(s,t) > 0, \\ \{(0,0), (1,0)\}, & \text{if } b < a, b \le t < a, \\ \{(0,0), (v(t), 1)\}, & \text{if } b > a, a \le t < b, \\ (0,0), & \text{if } t \ge \max\{a,b\}, \end{cases}
$$

where $D(s,t) = 1 - 4(v(t) - v(s))$.

(13) We have

$$
\mathcal{I}d\left(E_{23}^{[s,t]}(0,\mu)\right) = \begin{cases} \{(0,0), (0,1)\}, & \text{if } D(s,t) < 0 \\ \{(0,0), (0,1), (\frac{\theta(s)}{\theta(t)}, \frac{1}{2})\}, & \text{if } D(s,t) = 0 \\ \{(0,0), (0,1), (\frac{\theta(s)}{\theta(t)}, y_\pm)\}, & \text{if } D(s,t) > 0; \end{cases}
$$

$$\mathcal{I}d\left(E_{23}^{[s,t]}(2\mu,\mu)\right) = \begin{cases} \{(0,0),(1,0)\}, & \text{if } D(s,t) < 0 \\ \{(0,0),(1,0),(\frac{1}{2},\frac{\theta(s)}{\theta(t)})\}, & \text{if } D(s,t) = 0 \\ \{(0,0),(1,0),(x_\pm,\frac{\theta(s)}{\theta(t)})\}, & \text{if } D(s,t) > 0, \end{cases}$$

where $D(s,t) = 1 - \frac{4\theta(s)}{\theta(t)}\left(\frac{\theta(s)}{\theta(t)} - 1\right).$

(14) We have $\mathcal{I}d\left(E_{24}^{[s,t]}\right) =$

$$\begin{cases} \{(0,0),(0,1),(1,0),(1,1)\}, & \text{if } (s,t) \in \{(s,t) \in \mathcal{T} : s \le t < a\} \\ \{(0,0),\left(\frac{g(t)}{(1-g(t))^2+g^2(t)},\frac{1-g(t)}{(1-g(t))^2+g^2(t)}\right)\}, & \text{if } t \ge a. \end{cases}$$

Proof. The proof contains detailed analysis of solutions of the system (5.59) for each $E_i^{[s,t]}$. We shall give here proof of the assertion (3) which is more substantial. In case of $E_4^{[s,t]}$ the system (5.59) has the following form

$$\begin{cases} 2x = \left(\frac{\Phi(t)}{\Phi(s)} + \frac{\Psi(t)}{\Psi(s)}\right)x^2 + \left(\frac{\Phi(t)}{\Phi(s)} - \frac{\Psi(t)}{\Psi(s)}\right)y^2; \\ 2y = \left(\frac{\Phi(t)}{\Phi(s)} - \frac{\Psi(t)}{\Psi(s)}\right)x^2 + \left(\frac{\Phi(t)}{\Phi(s)} + \frac{\Psi(t)}{\Psi(s)}\right)y^2. \end{cases} \tag{5.60}$$

Case $\Phi(t) \equiv \Psi(t)$. One can see that in this case the system (5.60) has only four solutions $0 = (0,0)$ and

$$z_1 = z_1(s,t) = \left(0,\frac{\Phi(s)}{\Phi(t)}\right), \; z_2 = z_2(s,t) = \left(\frac{\Phi(s)}{\Phi(t)},0\right),$$

$$z_3 = z_3(s,t) = \left(\frac{\Phi(s)}{\Phi(t)},\frac{\Phi(s)}{\Phi(t)}\right).$$

Case $\Phi(t) \ne \Psi(t)$. In this case the solutions 0 and z_3 still exist.

Subcase $x = 0$. For $x = 0$ we have only solution $(0,0)$ if $\frac{\Phi(t)}{\Phi(s)} \ne \frac{\Psi(t)}{\Psi(s)}$ and there are two solutions $(0,0)$ and z_1 if $\frac{\Phi(t)}{\Phi(s)} = \frac{\Psi(t)}{\Psi(s)}$.

Subcase $y = 0$. For $y = 0$ we have only solution $(0,0)$ if $\frac{\Phi(t)}{\Phi(s)} \ne \frac{\Psi(t)}{\Psi(s)}$ and there are two solutions $(0,0)$ and z_2 if $\frac{\Phi(t)}{\Phi(s)} = \frac{\Psi(t)}{\Psi(s)}$.

Subcase $xy \ne 0$. Set $u = \frac{x}{y}$. From system (10.3) we get

$$(u-1)\left(\left(\frac{\Phi(t)}{\Phi(s)} - \frac{\Psi(t)}{\Psi(s)}\right)u^2 - 2\frac{\Psi(t)}{\Psi(s)}u + \left(\frac{\Phi(t)}{\Phi(s)} - \frac{\Psi(t)}{\Psi(s)}\right)\right) = 0. \tag{5.61}$$

This equation has unique solution $u = 1$ if $\frac{\Phi(t)}{\Phi(s)} = \frac{\Psi(t)}{\Psi(s)}$ or $\frac{\Phi(t)}{\Phi(s)} \ne \frac{\Psi(t)}{\Psi(s)}$ and $D(s,t) = \frac{\Phi(t)}{\Phi(s)}\left(2\frac{\Psi(t)}{\Psi(s)} - \frac{\Phi(t)}{\Phi(s)}\right) < 0$. For $\frac{\Phi(t)}{\Phi(s)} \ne \frac{\Psi(t)}{\Psi(s)}$ it has two solutions

$u = 1$ and $u = u_*(s,t)$ if $D(s,t) = 0$ and three solutions $u = 1$, $u = u_\pm(s,t)$ if $D > 0$.

Now one can describe x, y corresponding to the solutions of (5.61). The case $u = 1$, i.e. $x = y$ does not give any new solution. For $u = u_*, u_\pm$ we have $x = u_* y$ and $x = u_\pm y$, substituting these in the second equation of (5.60) after simple calculations we get the following non-zero solutions to (5.60):

$$x_* = \frac{\Phi(s)}{\Phi(t)}, \quad y_* = \frac{\Psi(s)}{\Psi(t)} - \frac{\Phi(s)}{\Phi(t)};$$

$$x_\pm = \frac{1 \pm \frac{\Psi(s)}{\Psi(t)}\sqrt{D(s,t)}}{\frac{\Phi(t)}{\Phi(s)} \pm \sqrt{D(s,t)}}, \quad y_\pm = \frac{\frac{\Phi(t)}{\Phi(s)} - \frac{\Psi(t)}{\Psi(s)}}{\frac{\Psi(t)}{\Psi(s)}\left(\frac{\Phi(t)}{\Phi(s)} \pm \sqrt{D(s,t)}\right)}.$$

Note that $x_\pm$, $y_\pm$ are well defined for any (s,t) with $\frac{\Phi(t)}{\Phi(s)} \neq \frac{\Psi(t)}{\Psi(s)}$. Thus the critical (boundary) times of the transition of idempotent elements are points (s,t) which satisfy $\frac{\Phi(t)}{\Phi(s)} = \frac{\Psi(t)}{\Psi(s)}$ or $\frac{\Phi(t)}{\Phi(s)} = \frac{2\Psi(t)}{\Psi(s)}$.

Proofs of the assertions (i), $i = 1, 2, 4 - 14$ are similar analysis of the system (5.59) for each $E_i^{[s,t]}$. $\qquad\square$

This theorem gives a very interesting "a fixed set of idempotent elements" property transition.

Remark 5.6. There are exactly solvable models in statistical mechanics, here an imprecise notion of "exactly solvable" as meaning: "The solutions can be expressed explicitly in terms of some previously known functions" is also sometimes used [16]. In such models, for example, the critical temperature can be expressed explicitly. Comparing this with examples of a property transition we also can say that a property transition of a CEA is exactly solvable if the critical time can be found exactly.

5.2.4 *Dynamics of CEAs given by matrix (5.18)*

In this subsection we consider CEAs constructed by matrix (5.18). To avoid many special cases we assume

$$\det(\mathcal{M}^{[s,t]}) = a_1^{[s,t]} a_4^{[s,t]} a_6^{[s,t]} \neq 0 \quad \text{for all } (s,t). \tag{5.62}$$

Following [197] we give the following

Theorem 5.11. *If a CEA, $E^{[s,t]}$ is defined by a matrix (5.18) satisfying (5.62) then*

1. *The algebra $E^{[s,t]}$ is baric for any $(s,t) \in \mathcal{T}$.*
2. *The algebra $E^{[s,t]}$ has a unique absolute nilpotent element $x = 0$ for any $(s,t) \in \mathcal{T}$.*
3. $\mathcal{I}d\left(E^{[s,t]}\right) =$

$$\{\lambda_1, \lambda_2\} \bigcup \begin{cases} \{\lambda_3\}, & \text{if } (s,t) \in \{(s,t) \in \mathcal{T} : D_1(s,t) = 0\} \\ \{\lambda_4, \lambda_5\}, & \text{if } (s,t) \in \{(s,t) \in \mathcal{T} : D_1(s,t) > 0\} \end{cases}$$

$$\bigcup \begin{cases} \{\lambda_6\}, & \text{if } (s,t) \in \{(s,t) \in \mathcal{T} : D_2(s,t) = D_3(s,t) = 0\} \\ \{\lambda_7, \lambda_8\}, & \text{if } (s,t) \in \{(s,t) \in \mathcal{T} : D_2(s,t) = 0, D_3(s,t) > 0\} \\ \{\lambda_9\}, & \text{if } (s,t) \in \{(s,t) \in \mathcal{T} : D_2(s,t) > 0, D_4(s,t) = 0\} \\ \{\lambda_{10}, \lambda_{11}\}, & \text{if } (s,t) \in \{(s,t) \in \mathcal{T} : D_2(s,t) > 0, D_4(s,t) > 0\} \\ \{\lambda_{12}\}, & \text{if } (s,t) \in \{(s,t) \in \mathcal{T} : D_2(s,t) > 0, D_5(s,t) = 0\} \\ \{\lambda_{13}, \lambda_{14}\}, & \text{if } (s,t) \in \{(s,t) \in \mathcal{T} : D_2(s,t) > 0, D_5(s,t) > 0\}, \end{cases}$$

where $\lambda_1 = (0,0,0)$, $\lambda_2 = \left(0,0,1/a_6^{[s,t]}\right)$, $\lambda_3 = \left(0,1/a_4^{[s,t]}, 1/(2a_6^{[s,t]})\right)$,

$$D_1(s,t) = 1 - \frac{4a_5^{[s,t]}a_6^{[s,t]}}{(a_4^{[s,t]})^2}, \quad \lambda_{4,5} = \left(0, \frac{1}{a_4^{[s,t]}}, \frac{1 \pm \sqrt{D_1(s,t)}}{2a_6^{[s,t]}}\right),$$

$$D_2(s,t) = 1 - \frac{4a_2^{[s,t]}a_4^{[s,t]}}{(a_1^{[s,t]})^2}, \quad D_3(s,t) = 1 - 4a_6^{[s,t]}\left(\frac{a_3^{[s,t]}}{(a_1^{[s,t]})^2} + \frac{a_5^{[s,t]}}{(2a_4^{[s,t]})^2}\right),$$

$$\lambda_6 = \left(\frac{1}{a_1^{[s,t]}}, \frac{1}{2a_4^{[s,t]}}, \frac{1}{2a_6^{[s,t]}}\right), \quad \lambda_{7,8} = \left(\frac{1}{a_1^{[s,t]}}, \frac{1}{2a_4^{[s,t]}}, \frac{1 \pm \sqrt{D_3(s,t)}}{2a_6^{[s,t]}}\right),$$

$$D_{4,5}(s,t) = 1 - 4a_6^{[s,t]}\left(\frac{a_3^{[s,t]}}{(a_1^{[s,t]})^2} + a_5^{[s,t]}y_{4,5}^2\right), \quad \text{with } y_{4,5} = \frac{1 \pm \sqrt{D_2(s,t)}}{2a_4^{[s,t]}},$$

$$\lambda_9 = \left(\frac{1}{a_1^{[s,t]}}, y_4, \frac{1}{2a_6^{[s,t]}}\right), \quad \lambda_{10,11} = \left(\frac{1}{a_1^{[s,t]}}, y_4, \frac{1 \pm \sqrt{D_4(s,t)}}{2a_6^{[s,t]}}\right),$$

$$\lambda_{12} = \left(\frac{1}{a_1^{[s,t]}}, y_5, \frac{1}{2a_6^{[s,t]}}\right), \quad \lambda_{13,14} = \left(\frac{1}{a_1^{[s,t]}}, y_5, \frac{1 \pm \sqrt{D_5(s,t)}}{2a_6^{[s,t]}}\right).$$

Proof. 1. Under condition (5.62) this assertion follows from Theorem 3.7.

2. From condition (5.62) and system (5.53) we conclude that $x = (0,0,0)$ is the unique solution.

3. For matrix (5.18) the system (5.59) can be written as

$$x_1 = a_1^{[s,t]} x_1^2$$

$$x_2 = a_2^{[s,t]} x_1^2 + a_4^{[s,t]} x_2^2 \qquad (5.63)$$

$$x_3 = a_3^{[s,t]} x_1^2 + a_5^{[s,t]} x_2^2 + a_6^{[s,t]} x_3^2.$$

Hence for x_1 there is two possibilities $x_1 = 0$ and $x_1 = 1/a_1^{[s,t]}$. For each value of x_1 the x_2 also has up to two possible values. For fixed values of x_1 and x_2 we have two possible values for x_3. Carefully computing the values gives the result. $\qquad\square$

Corollary 5.3. *The CEA considered in Theorem 5.11 has the following dynamics:*

1. *It has not "baric" property transition.*
2. *It has not "uniqueness absolute nilpotent element" property transition.*
3. *It has "a fixed number of idempotent elements" property transition. The transition critical sets are given by $\{(s,t) : D_i(s,t) = 0\}$, $i = 1,2,3,4,5$, i.e. the number of idempotent elements changes depending on time, when the time point crosses these sets.*

5.2.5 *Classification dynamics of CEAs*

Following [188] and [189] define new CEA $E_{27}^{[s,t]}$, given by the following matrix

$$\mathcal{M}_{27}^{[s,t]} = \begin{cases} \begin{pmatrix} 0 & 0 \\ 0 & 0 \end{pmatrix}, & \text{if} \quad s < t \le C; \\[2ex] \begin{pmatrix} 0 & \frac{\Phi(t)}{\Phi(s)} \\ 0 & 0 \end{pmatrix}, & \text{if} \quad t > C, \end{cases}$$

where $C > 0$ and Φ is an arbitrary function, with $\Phi(t) \ne 0$.

Recall that evolution algebras E and E' (denoted $E \simeq E'$) are called isomorphic if there is a linear, bijective map $\phi \colon E \longrightarrow E'$ such that $\phi(xy) = \phi(x)\phi(y)$ and the set $\{\phi(e_i)\}$ is a subset of a natural basis of E'.

The following theorem gives the time depending isomorphism dynamics of the chains of evolution algebras $E_i^{[s,t]}, i = 0, \ldots, 24$ and $E_{27}^{[s,t]}$.

Theorem 5.12. *Let E_i, $i = 1, 2, \ldots, 7$ be the evolution algebras defined in Theorem 3.13. Then*
$E_1^{[s,t]}$ *is isomorphic to E_3 for any $0 \le s \le t$.*

$$E_2^{[s,t]} \simeq \begin{cases} E_3 & \text{for all } (s,t) \in \{(s,t) : s \le t < b\}; \\ \\ E_0 & \text{for all } (s,t) \in \{(s,t) : t \ge b\}. \end{cases}$$

$E_3^{[s,t]} \simeq E_2$ *for any $s, t \in \mathcal{T}$.*

$$E_4^{[s,t]} \simeq \begin{cases} E_6\left(\frac{\Phi(t)\Psi(s)-\Psi(t)\Phi(s)}{\Phi(t)\Psi(s)+\Psi(t)\Phi(s)}, \frac{\Phi(t)\Psi(s)-\Psi(t)\Phi(s)}{\Phi(t)\Psi(s)+\Psi(t)\Phi(s)}\right) \\ \quad \text{for all } (s,t) \in \left\{(s,t) : \frac{\Phi(t)}{\Phi(s)} \ne -\frac{\Psi(t)}{\Psi(s)}\right\}; \\ \\ E_7(0) \quad \text{for all } (s,t) \in \left\{(s,t) : \frac{\Phi(t)}{\Phi(s)} = -\frac{\Psi(t)}{\Psi(s)}\right\}. \end{cases}$$

$$E_5^{[s,t]} \simeq \begin{cases} E_6\left(\frac{\Phi(t)-\Phi(s)}{\Phi(t)+\Phi(s)}, \frac{\Phi(t)-\Phi(s)}{\Phi(t)+\Phi(s)}\right) \\ \quad \text{for all } (s,t) \in \{(s,t) : s \le t < b, \quad \Phi(t) \ne -\Phi(s)\}; \\ \\ E_7(0) \quad \text{for all } (s,t) \in \{(s,t) : s \le t < b, \quad \Phi(t) = -\Phi(s)\}; \\ \\ E_2 \quad \text{for all } (s,t) \in \{(s,t) : t \ge b\}. \end{cases}$$

$$E_6^{[s,t]} \simeq \begin{cases} E_2 & \text{for all } (s,t) \in \{(s,t) : s \le t < a\}; \\ \\ E_0 & \text{for all } (s,t) \in \{(s,t) : t \ge a\}. \end{cases}$$

$$E_7^{[s,t]} \simeq \begin{cases} E_6\left(\frac{\Psi(s)-\Psi(t)}{\Psi(s)+\Psi(t)}, \frac{\Psi(s)-\Psi(t)}{\Psi(s)+\Psi(t)}\right) \\ \quad \text{for all } (s,t) \in \{(s,t) : s \le t < a, \quad \Psi(t) \ne -\Psi(s)\}; \\ \\ E_7(0) \quad \text{for all } (s,t) \in \{(s,t) : s \le t < a, \quad \Psi(t) = -\Psi(s)\}; \\ \\ E_3 \quad \text{for all } (s,t) \in \{(s,t) : t \ge a\}. \end{cases}$$

$$E_8^{[s,t]} \simeq \begin{cases} E_6(0,0) & \text{for all} \ (s,t) \in \{(s,t) : s \le t < \min\{a,b\}\}; \\[2mm] E_3 & \text{for all} \ (s,t) \in \{(s,t) : a \le t < b, \ a < b\}; \\[2mm] E_2 & \text{for all} \ (s,t) \in \{(s,t) : b \le t < a, \ b < a\}; \\[2mm] E_0 & \text{for all} \ (s,t) \in \{(s,t) : t \ge \max\{a,b\}\}. \end{cases}$$

$$E_9^{[s,t]} \simeq \begin{cases} E_6(\tan(t-s), -\tan(t-s)) \\ \quad \text{for all} \ (s,t) \in \{(s,t) : t \ne s + \frac{\pi}{2} + \pi k, \ k \in \mathbb{Z}\}; \\[2mm] E_7(0) & \text{for all} \ (s,t) \in \{(s,t) : t = s + \frac{\pi}{2} + \pi k, \ k \in \mathbb{Z}\}. \end{cases}$$

$E_{10}^{[s,t]} \simeq E_2$ *for any* $s,t \in \mathcal{T}$.

$$E_{11}^{[s,t]} \simeq \begin{cases} E_2 & \text{for all} \ (s,t) \in \{(s,t) : s \le t < a\}; \\[2mm] E_0 & \text{for all} \ (s,t) \in \{(s,t) : t \ge a\}. \end{cases}$$

$$E_{12}^{[s,t]} \simeq \begin{cases} E_1 & \text{for all} \ (s,t) \in \{(s,t) : t \ge a, \ h(s)g(s) = \pm 1\}; \\[2mm] E_2 & \text{for all} \ (s,t) \in \{(s,t) : s \le t < a, \ h^2(s)g^2(s) < 1\}; \\[2mm] E_3 & \text{for all} \ (s,t) \in \{(s,t) : s \le t < a, \ h^2(s)g^2(s) > 1\}. \end{cases}$$

$$E_{13}^{[s,t]} \simeq \begin{cases} E_1 & \text{for all} \ (s,t) \in \{(s,t) : s \le t < a, \ \psi(s) = \pm 1\}; \\[2mm] E_2 & \text{for all} \ (s,t) \in \{(s,t) : s \le t < a, \ \psi^2(s) < 1\}; \\[2mm] E_3 & \text{for all} \ (s,t) \in \{(s,t) : s \le t < a, \ \psi^2(s) > 1\}; \\[2mm] E_0 & \text{for all} \ (s,t) \in \{(s,t) : t \ge a\}. \end{cases}$$

$$E_{14}^{[s,t]} \simeq \begin{cases} E_1 & \text{for all} \ (s,t) \in \{(s,t) : \ \psi(s) = 0\}; \\[2mm] E_2 & \text{for all} \ (s,t) \in \{(s,t) : \ \Phi(s)\psi(s) > 0\}; \\[2mm] E_5 & \text{for all} \ (s,t) \in \{(s,t) : \ \Phi(s)\psi(s) < 0\}. \end{cases}$$

$$E_{15}^{[s,t]} \simeq \begin{cases} E_1 & \text{for all } (s,t) \in \{(s,t) : s \leq t < a, \quad \psi(s) = 0\}; \\[2mm] E_2 & \text{for all } (s,t) \in \{(s,t) : s \leq t < a, \quad \psi(s) > 0\}; \\[2mm] E_5 & \text{for all } (s,t) \in \{(s,t) : s \leq t < a, \quad \psi(s) < 0\}; \\[2mm] E_0 & \text{for all } (s,t) \in \{(s,t) : t \geq a\}. \end{cases}$$

$E_{16}^{[s,t]} \simeq E_1$ for any $s,t \in \mathcal{T}$.

$E_{17}^{[s,t]} \simeq E_6\left(\frac{\Phi^2(t)\psi(s)(g(t)-g(s))}{\Phi(s)\psi^2(t)}, 0\right)$ for any $s,t \in \mathcal{T}$.

$$E_{18}^{[s,t]} \simeq \begin{cases} E_6\left(\frac{\psi(s)(h(t)-h(s))}{\psi^2(t)}, 0\right) & \text{for all } (s,t) \in \{(s,t) : s \leq t < a\}; \\[2mm] E_1 & \text{for all } (s,t) \in \{(s,t) : t \geq a\}. \end{cases}$$

$$E_{19}^{[s,t]} \simeq \begin{cases} E_1 & \text{for all } (s,t) \in \{(s,t) : s \leq t < b\}; \\[2mm] E_0 & \text{for all } (s,t) \in \{(s,t) : t \geq b\}. \end{cases}$$

$$E_{20}^{[s,t]} \simeq \begin{cases} E_6\left(\frac{\Phi^2(t)(v(t)-v(s))}{\Phi(s)}, 0\right) & \text{for all } (s,t) \in \{(s,t) : s \leq t < b\}; \\[2mm] E_1 & \text{for all } (s,t) \in \{(s,t) : t \geq b, \quad w(s) = 0\}; \\[2mm] E_2 & \text{for all } (s,t) \in \left\{(s,t) : t \geq b, \frac{\Phi^2(t)w(s)}{\Phi(s)} > 0\right\}; \\[2mm] E_5 & \text{for all } (s,t) \in \left\{(s,t) : t \geq b, \frac{\Phi^2(t)w(s)}{\Phi(s)} < 0\right\}. \end{cases}$$

$$E_{21}^{[s,t]} \simeq \begin{cases} E_6(v(t) - v(s), 0) & \text{for all } (s,t) \in \{(s,t) : s \leq t < \min\{a,b\}\}; \\[2mm] E_1 & \text{for all } (s,t) \in \{(s,t) : a \leq t < b, \quad a < b\} \cup \\ & \qquad \{(s,t) : b \leq t < a, \quad a > b, \quad v(s) = 0\}; \\[2mm] E_2 & \text{for all } (s,t) \in \{(s,t) : b \leq t < a, \quad a > b, \quad v(s) > 0\}; \\[2mm] E_5 & \text{for all } (s,t) \in \{(s,t) : b \leq t < a, \quad a > b, \quad v(s) < 0\}; \\[2mm] E_0 & \text{for all } (s,t) \in \{(s,t) : t \geq \max\{a,b\}\}. \end{cases}$$

$E_{22}^{[s,t]} \simeq E_2$ for any $s,t \in \mathcal{T}$.

$$E_{23}^{[s,t]}(\lambda, \mu) \simeq$$

$$
\begin{cases}
E_6\left(\frac{2\theta(s)(\theta(s)-\theta(t))}{(\theta(s)+\theta(t))^2}, 0\right) \quad \text{for all} \quad \lambda = 2\mu \quad (s,t) \in \{(s,t) : s \leq t < a\}; \\[2ex]
E_6\left(\frac{\xi\zeta}{(1-\zeta)^2}, \frac{(1-\xi)(1-\zeta)}{\xi^2}\right) \quad \text{for all} \quad \lambda \neq 2\mu \\[1ex]
\quad s,t) \in \left\{(s,t) : \theta(t) \neq \frac{2\lambda}{2\mu-\lambda}\theta(s), \quad \theta(t) \neq \frac{2\mu-\lambda}{\lambda}\theta(s)\right\}; \\[2ex]
E_7\left(\frac{1-\zeta}{\sqrt[3]{\zeta^2}}\right) \quad \text{for all} \quad \lambda \neq 2\mu, \ \lambda \neq 0 \\[1ex]
\quad (s,t) \in \left\{(s,t) : \theta(t) = \frac{2\lambda}{2\mu-\lambda}\theta(s), \quad \theta(t) \neq \theta(s), \quad \theta(t) \neq \frac{2\mu-\lambda}{\lambda}\theta(s)\right\}; \\[2ex]
E_7\left(\frac{\xi}{\sqrt[3]{(1-\xi)^2}}\right) \quad \text{for all} \quad \lambda \neq 2\mu \\[1ex]
\quad (s,t) \in \left\{(s,t) : \theta(t) \neq \frac{2\lambda}{2\mu-\lambda}\theta(s), \quad \theta(t) \neq \theta(s), \quad \theta(t) = \frac{2\mu-\lambda}{\lambda}\theta(s)\right\}; \\[2ex]
E_7(0) \quad \text{for all} \quad \lambda \neq 2\mu, \\[1ex]
\quad s,t) \in \left\{(s,t) : \theta(t) = \frac{2\lambda}{2\mu-\lambda}\theta(s), \quad \theta(t) = \frac{2\mu-\lambda}{\lambda}\theta(s)\right\},
\end{cases}
$$

where $\xi = 1 - \frac{\lambda-2\mu}{2(\lambda-\mu)}\left(1 - \frac{\theta(t)}{\theta(s)}\right)$, $\zeta = \frac{\lambda}{2(\lambda-\mu)}\left(1 - \frac{\theta(t)}{\theta(s)}\right)$.

$$
E_{24}^{[s,t]} \simeq
\begin{cases}
E_6(0,0) \quad \text{for all} \quad (s,t) \in \{(s,t) : s \leq t < a\}; \\[2ex]
E_2 \quad \text{for all} \quad (s,t) \in \{(s,t) : t \geq a\}.
\end{cases}
$$

$$
E_{27}^{[o,t]} \simeq
\begin{cases}
E_0 \quad \text{for all} \quad (s,t) \in \{(s,t) : s < t \leq C\}, \\[2ex]
E_4 \quad \text{for all} \quad (s,t) \in \{(s,t) : t > C\}.
\end{cases}
$$

Proof.   The proof is based to following simple lemma:

Lemma 5.2. *An evolution algebra corresponding to the matrix*

$$
(i) \quad \begin{pmatrix} \lambda & -\lambda \\ -\lambda & \lambda \end{pmatrix} \quad \text{is isomorphic to}
\begin{cases}
E_0, \quad \text{if} \ \lambda = 0; \\[2ex]
E_3, \quad \text{if} \ \lambda \neq 0;
\end{cases}
$$

$$
(ii) \quad \begin{pmatrix} \lambda & \mu \\ \mu & \lambda \end{pmatrix} \quad \text{is isomorphic to}
\begin{cases}
E_0, \quad \text{if} \ \lambda = \mu = 0; \\[2ex]
E_2, \quad \text{if} \ \lambda = \mu \neq 0; \\[2ex]
E_6\left(\frac{\mu}{\lambda}, \frac{\mu}{\lambda}\right), \ \text{if} \ \lambda \neq \mu, \lambda \neq 0, \mu \in \mathbb{R}; \\[2ex]
E_7(0), \quad \text{if} \ \lambda = 0, \mu \neq 0;
\end{cases}
$$

$$(iii) \quad \begin{pmatrix} \lambda & \mu \\ -\mu & \lambda \end{pmatrix} \quad \text{is isomorphic to} \begin{cases} E_0, & \text{if } \lambda = \mu = 0; \\ E_6(\frac{\mu}{\lambda}, -\frac{\mu}{\lambda}), & \text{if } \lambda \neq 0, \mu \in \mathbb{R}; \\ E_7(0), & \text{if } \lambda = 0, \mu \neq 0; \end{cases}$$

$$(iv) \quad \begin{pmatrix} \lambda & \mu \\ \lambda & \mu \end{pmatrix} \quad \text{is isomorphic to} \begin{cases} E_0, & \text{if } \lambda = \mu = 0; \\ E_2, & \text{otherwise;} \end{cases}$$

$$(v) \quad \begin{pmatrix} \lambda & \lambda \\ \mu & \mu \end{pmatrix} \quad \text{is isomorphic to} \begin{cases} E_0, & \text{if } \lambda = \mu = 0; \\ E_1, & \text{if } \lambda\mu = 0, \lambda^2 + \mu^2 \neq 0; \\ E_2, & \text{if } \lambda \neq 0, \ \mu \neq 0; \end{cases}$$

$$(vi) \quad \begin{pmatrix} \lambda & 0 \\ \mu & 0 \end{pmatrix} \quad \text{is isomorphic to} \begin{cases} E_0, & \text{if } \lambda = \mu = 0; \\ E_1, & \text{if } \lambda \neq 0, \mu = 0; \\ E_2, & \text{if } \lambda\mu > 0; \\ E_4, & \text{if } \lambda = 0, \mu \neq 0; \\ E_5, & \text{if } \lambda\mu < 0; \end{cases}$$

$$(vii) \quad \begin{pmatrix} \lambda & 0 \\ \mu & \eta \end{pmatrix} \quad \text{is isomorphic to} \begin{cases} E_0, & \text{if } \lambda = \mu = \eta = 0; \\ E_1, & \text{if } \{\lambda = 0, \eta \neq 0\} \cup \{\lambda \neq 0, \mu = \eta = 0\}; \\ E_2, & \text{if } \lambda \neq 0, \mu \neq 0, \eta = 0; \\ E_4, & \text{if } \lambda = \eta = 0, \mu \neq 0; \\ E_6(\frac{\lambda\mu}{\eta^2}, 0), & \text{if } \lambda \neq 0, \eta \neq 0, \mu \in \mathbb{R}; \end{cases}$$

$$(viii) \quad \begin{pmatrix} \lambda & 1-\lambda \\ \mu & 1-\mu \end{pmatrix} \quad \text{is isomorphic to}$$

$$\begin{cases} E_2, & \text{if } \lambda = \mu; \\[2mm] E_6\left(\frac{\lambda\mu}{(1-\mu)^2}, \frac{(1-\lambda)(1-\mu)}{\lambda^2}\right), & \text{if } \lambda \neq \mu \text{ and } \lambda \neq 0, \mu \neq 1; \\[2mm] E_7\left(\frac{1-\mu}{\sqrt[3]{\mu^2}}\right), & \text{if } \lambda = 0, \mu \neq 0, \mu \neq 1; \\[2mm] E_7\left(\frac{\lambda}{\sqrt[3]{(1-\lambda)^2}}\right), & \text{if } \lambda \neq 0, \lambda \neq 1, \mu = 1; \\[2mm] E_7(0), & \text{if } \lambda = 0, \mu = 1, \end{cases}$$

where E_0 is the trivial evolution algebra (i.e. with zero multiplication) and the evolution algebras E_i, $i = 1, \ldots, 7$, are given in Theorem 3.13.

The following are proofs of each items of the theorem:

1. By the lemma one can see that $E_1^{[s,t]}$ is isomorphic to E_3, $\forall s, t \in \mathcal{T}$, by the change of basis $e_1' = \frac{1}{2}\frac{\Psi(s)}{\Psi(t)}e_1$, $e_2' = -\frac{1}{2}\frac{\Psi(s)}{\Psi(t)}e_2$.

2. $E_2^{[s,t]}$ is isomorphic to E_3, $\forall s, t \in \mathcal{T}, s \leq t < b$, by the change of basis $e_1' = \frac{1}{2}e_1$, $e_2' = -\frac{1}{2}e_2$, in time when $t \geq b$ will be E_0.

3. $E_3^{[s,t]}$ is isomorphic to E_2, $\forall s, t \in \mathcal{T}$, by the change of basis

$$e_1' = \frac{1}{4}\frac{\Phi(s)}{\Phi(t)}e_1 + \frac{\Phi(s)}{\Phi(t)}e_2, \quad e_2' - -\frac{1}{4}\frac{\Phi(s)}{\Phi(t)}c_1 + \frac{\Phi(s)}{\Phi(t)}c_2.$$

4. $E_4^{[s,t]}$ is isomorphic to $E_6\left(\frac{\Phi(t)\Psi(s)-\Psi(t)\Phi(s)}{\Phi(t)\Psi(s)+\Psi(t)\Phi(s)}, \frac{\Phi(t)\Psi(s)-\Psi(t)\Phi(s)}{\Phi(t)\Psi(s)+\Psi(t)\Phi(s)}\right)$, when $\frac{\Phi(t)}{\Phi(s)} \neq -\frac{\Psi(t)}{\Psi(s)}$, $\forall s, t \in \mathcal{T}$, by the change of basis

$$e_1' = \frac{1}{2}\frac{\Phi(s)\Psi(s)}{\Phi(t)\Psi(s) + \Psi(t)\Phi(s)}e_1, \quad e_2' = \frac{1}{2}\frac{\Phi(s)\Psi(s)}{\Phi(t)\Psi(s) + \Psi(t)\Phi(s)}e_2,$$

   and it is isomorphic to $E_7(0)$, when $\frac{\Phi(t)}{\Phi(s)} = \frac{-\Psi(t)}{\Psi(s)}$, $\forall s, t \in \mathcal{T}$, by the change of basis $e_1' = \frac{1}{2}\frac{\Phi(s)\Psi(s)}{\Phi(t)\Psi(s)-\Psi(t)\Phi(s)}e_1$, $e_2' = \frac{1}{2}\frac{\Phi(s)\Psi(s)}{\Phi(t)\Psi(s)-\Psi(t)\Phi(s)}e_2$.

5. $E_5^{[s,t]}$ is isomorphic to $E_6\left(\frac{\Phi(t)-\Phi(s)}{\Phi(t)+\Phi(s)}, \frac{\Phi(t)-\Phi(s)}{\Phi(t)+\Phi(s)}\right)$, when $\Phi(t) \neq -\Phi(s)$, $\forall s, t \in \mathcal{T}, s \leq t < b$, by the change of basis

$$e_1' = \frac{1}{2}\frac{\Phi(s)}{\Phi(t) + \Phi(s)}e_1, \quad e_2' = \frac{1}{2}\frac{\Phi(s)}{\Phi(t) + \Phi(s)}e_2,$$

and it is isomorphic to $E_7(0)$, when $\Phi(t) = -\Phi(s)$, $\forall s, t \in \mathcal{T}, s \leq t < b$, by the change of basis $e_1' = -e_1$, $e_2' = -e_2$. It is isomorphic to E_2, $\forall s, t \in \mathcal{T}, t \geq b$, by the change of basis

$$e_1' = \frac{1}{2}\frac{\Phi(s)}{\Phi(t)}e_1 + \frac{\Phi(s)}{\Phi(t)}e_2, \quad e_2' = -\frac{1}{2}\frac{\Phi(s)}{\Phi(t)}e_1 + \frac{\Phi(s)}{\Phi(t)}e_2.$$

6. $E_6^{[s,t]}$ is isomorphic to E_2, $\forall s, t \in \mathcal{T}, s \leq t < a$, by the change of basis $e_1' = \frac{1}{4}e_1 + \frac{1}{4}e_2, e_2' = -\frac{1}{4}e_1 + \frac{1}{4}e_2$, and in time when $t \geq a$ this algebra will be E_0.

7. $E_7^{[s,t]}$ is isomorphic to $E_6\left(\frac{\Psi(s)-\Psi(t)}{\Psi(t)+\Psi(s)}, \frac{\Psi(t)-\Psi(s)}{\Psi(t)+\Psi(s)}\right)$, when $\Psi(t) \neq -\Psi(s)$, $\forall s, t \in \mathcal{T}, s \leq t < b$, by the change of basis

$$e_1' = \frac{1}{2}\frac{\Psi(s)}{\Psi(t) + \Psi(s)}e_1, \quad e_2' = \frac{1}{2}\frac{\Psi(s)}{\Psi(t) + \Psi(s)}e_2,$$

and it is isomorphic to $E_7(0)$, when $\Psi(t) = -\Psi(s)$, $\forall s, t \in \mathcal{T}, s \leq t < b$, by the change of basis $e_1' = -e_1$, $e_2' = -e_2$. Isomorphic to E_3, for all $s, t \in \mathcal{T}, t \geq b$, by change of basis $e_1' = \frac{1}{2}\frac{\Psi(s)}{\Psi(t)}e_1, e_2' = -\frac{\Psi(s)}{\Psi(t)}e_2$.

8. $E_8^{[s,t]}$ is isomorphic to $E_6(0,0)$, $\forall s, t \in \mathcal{T}, s \leq t < \min\{a, b\}$, by the change of basis $e_1' = e_1$, $e_2' = e_2$, and it is isomorphic to E_3 $\forall s, t \in \mathcal{T}$, $a \leq t < b$, $a < b$, by the change of basis $e_1' = \frac{1}{2}e_1$, $e_2' = -\frac{1}{2}e_2$. By the change of basis $e_1' = \frac{1}{4}e_1 + \frac{1}{4}e_2$, $e_2' = -\frac{1}{4}e_1 + \frac{1}{2}e_2$, this algebra will be isomorphic to E_2, $\forall s, t \in \mathcal{T}$, $b \leq t < a$, $b < a$. And this algebra will be E_0, $\forall s, t \in \mathcal{T}, t \geq \max\{a, b\}$.

9. $E_9^{[s,t]}$ is isomorphic to $E_6(\tan(t-s), -\tan(t-s))$, $\forall s, t \in \mathcal{T}$, when $t \neq s + \frac{\pi}{2} + \pi k, k \in \mathbb{Z}$, by the change of basis

$$e_1' = \frac{1}{\cos(t-s)}e_1, \quad e_2' = \frac{1}{\cos(t-s)}e_2,$$

and it is isomorphic to $E_7(0)$ $\forall s, t \in \mathcal{T}$, when $t = s + \frac{\pi}{2} + \pi k, k \in \mathbb{Z}$, by the change of basis $e_1' = -e_1$, $e_2' = e_2$.

10. $E_{10}^{[s,t]}$ is isomorphic to E_2, $\forall s, t \in \mathcal{T}$, by the change of basis:

$$e_1' = \frac{1}{4}\left(\frac{h(s)(h(t) + g(t))}{h^2(t) + g^2(t)}e_1 + \frac{h(s)(h(t) - g(t))}{h^2(t) + g^2(t)}e_2\right),$$

$$e_2' = \frac{1}{4}\left(\frac{h(s)(h(t) + g(t))}{h^2(t) + g^2(t)}e_1 - \frac{h(s)(h(t) - g(t))}{h^2(t) + g^2(t)}e_2\right).$$

11. $E_{11}^{[s,t]}$ is isomorphic to E_2, $\forall s,t \in \mathcal{T}, s \leq t < a$, by the change of basis

$$e_1' = \frac{1}{4}\left(\frac{1+\psi(t)}{1+\psi^2(t)}e_1 + \frac{1-\psi(t)}{1+\psi^2(t)}e_2 \right),$$

$$e_2' = \frac{1}{4}\left(\frac{1+\psi(t)}{1+\psi^2(t)}e_1 - \frac{1-\psi(t)}{1+\psi^2(t)}e_2 \right),$$

in time when $t \geq 0$ will be E_0.

12. $E_{12}^{[s,t]}$ is isomorphic to E_1, $\forall s,t \in \mathcal{T}$, when $g(s)h(s) = \pm 1$. In the case when $g(s)h(s) = 1$, the change of basis will be $e_1' = \frac{1}{h(t)}e_1 + \frac{1}{h(t)}e_2$, $e_2' = e_2$, and when $g(s)h(s) = -1$, the change of basis will be $e_1' = -\frac{1}{h(t)}e_1 - \frac{1}{h(t)}e_2$, $e_2' = e_1$.

When $g^2(s)h^2(s) < 1$, $E_{12}^{[s,t]}$ is isomorphic to E_2, for all $s,t \in \mathcal{T}$, by the change of basis

$$e_1' = \frac{1}{2}\left(\frac{h(s)}{2h(t)}e_1 + \frac{h(s)}{2h(t)}e_2 \right),$$

$$e_2' = \frac{1}{4}\left(\frac{h(s)\sqrt{1-g(s)h(s)}}{2h(t)\sqrt{1+g(s)h(s)}}e_1 - \frac{h(s)\sqrt{1+g(s)h(s)}}{2h(t)\sqrt{1-g(s)h(s)}}e_2 \right).$$

When $g^2(s)h^2(s) > 1$, $E_{12}^{[s,t]}$ is isomorphic to E_3, $\forall s,t \in \mathcal{T}$, by the change of basis

$$e_1' - \frac{1}{4}\left(\frac{1+\psi(t)}{1+\psi^2(t)}e_1 + \frac{1-\psi(t)}{1+\psi^2(t)}e_2 \right),$$

$$e_2' = \frac{1}{4}\left(\frac{1+\psi(t)}{1+\psi^2(t)}e_1 - \frac{1-\psi(t)}{1+\psi^2(t)}e_2 \right),$$

in time when $t \geq 0$ will be E_0.

13. $E_{13}^{[s,t]}$ is isomorphic to E_1, $\forall s,t \in \mathcal{T}, s \leq t < a$, when $\psi(s) = \pm 1$. In the case when $\psi(s) = 1$, the change of basis will be $e_1' = e_1 + e_2$, $e_2' = e_2$, and when $\psi(s) = -1$, the change of basis will be $e_1' = e_1 + e_2$, $e_2' = e_1$.

When $\psi^2(s) < 1$, $E_{13}^{[s,t]}$ is isomorphic to E_2, $\forall s,t \in \mathcal{T}$, by the change of basis

$$e_1' = \frac{1}{2}\left(\frac{h(s)}{2h(t)}e_1 + \frac{h(s)}{2h(t)}e_2 \right),$$

$$e_2' = \frac{1}{2}\left(\frac{\sqrt{1-\psi(s)}}{2h(t)\sqrt{1+\psi(s)}}e_1 - \frac{h(s)\sqrt{1+\psi(s)}}{2h(t)\sqrt{1-\psi(s)}}e_2 \right).$$

When $\psi^2(s) > 1$, $E_{13}^{[s,t]}$ is isomorphic to E_3, $\forall s,t \in \mathcal{T}$, by the change of basis

$$e_1' = \frac{1}{4}\left(\frac{1+\psi(t)}{1+\psi^2(t)}e_1 + \frac{1-\psi(t)}{1+\psi^2(t)}e_2\right),$$

$$e_2' = \frac{1}{4}\left(\frac{1+\psi(t)}{1+\psi^2(t)}e_1 - \frac{1-\psi(t)}{1+\psi^2(t)}e_2\right),$$

in time when $t \geq 0$ will be E_0.

14. When $\psi(s) = 0$, $E_{14}^{[s,t]}$ is isomorphic to E_1, $\forall s,t \in \mathcal{T}$, by the change of basis $e_1' = \frac{\Phi(s)}{\Phi(t)}e_1$, $e_2' = e_2$. When $\Phi(s)\psi(s) > 0$, $E_{14}^{[s,t]}$ is isomorphic to E_2, $\forall s,t \in \mathcal{T}$, by the change of basis $e_1' = \frac{\Phi(s)}{\Phi(t)}e_1$, $e_2' = \frac{\sqrt{|\Phi(s)|}}{\Phi(t)\sqrt{|\psi(s)|}}e_2$, and when $\Phi(s)\psi(s) < 0$ it is isomorphic to E_5, $\forall s,t \in \mathcal{T}$, by the change of basis $e_1' = \frac{\Phi(s)}{\Phi(t)}e_1$, $e_2' = \frac{\sqrt{|\Phi(s)|}}{\Phi(t)\sqrt{|\psi(s)|}}e_2$.

15. When $\psi(s) = 0$, $E_{15}^{[s,t]}$ is isomorphic to E_1, $\forall s,t \in \mathcal{T}, s \leq t < a$, by the change of basis $e_1' = e_1$, $e_2' = e_2$. When $\psi(s) > 0$, $E_{15}^{[s,t]}$ is isomorphic to E_2, $\forall s,t \in \mathcal{T}, s \leq t < a$, by the change of basis $e_1' = e_1$, $e_2' = \frac{1}{\sqrt{|\psi(s)|}}e_2$, and in time $\forall s,t \in \mathcal{T}, s \leq t < a$, when $\psi(s) < 0$ it is isomorphic to E_5, by the change of basis $e_1' = e_1$, $e_2' = \frac{1}{\sqrt{|\psi(s)|}}e_2$, and in time when $t \geq a$ this algebra will be E_0.

16. $E_{16}^{[s,t]}$ is isomorphic to E_1, $\forall s,t \in \mathcal{T}$, by the change of basis

$$e_1' = \frac{g(t)\psi(s)}{\psi^2(t)}e_1 + \frac{\psi(s)}{\psi(t)}e_2, \quad e_2' = e_1.$$

17. $E_{17}^{[s,t]}$ is isomorphic to $E_6\left(\frac{\Phi^2(t)\psi(s)}{\Phi(s)\psi^2(t)}(g(t) - g(s)), 0\right)$, $\forall s,t \in \mathcal{T}$, by the change of basis $e_1' = \frac{\psi(s)}{\psi(t)}e_2$, $e_2' = \frac{\Phi(s)}{\Phi(t)}e_1$.

18. $E_{18}^{[s,t]}$ is isomorphic to $E_6\left(\frac{\psi(s)}{\psi^2(t)}(h(t) - h(s)), 0\right)$, $\forall s,t \in \mathcal{T}, s \leq t < a$, by the change of basis $e_1' = \frac{\psi(s)}{\psi(t)}e_2$, $e_2' = e_1$ and it is isomorphic to E_1 $\forall s,t \in \mathcal{T}, t \geq a$, by the change of basis $e_1' = \frac{h(t)\psi(s)}{\psi^2(t)}e_1 + \frac{\psi(s)}{\psi(t)}e_2$, $e_2' = e_1$.

19. $E_{19}^{[s,t]}$ is isomorphic to E_1, $\forall s,t \in \mathcal{T}, s \leq t < b$, by the change of basis $e_1' = h(t)e_1 + e_2$, $e_2' = e_1$, in time when $t \geq b$ will be E_0.

20. $E_{20}^{[s,t]}$ is isomorphic to $E_6\left(\frac{\Phi^2(t)}{\Phi(s)}(v(t) - v(s)), 0\right)$, $\forall s,t \in \mathcal{T}, s \leq t < b$, by the change of basis $e_1' = e_2$, $e_2' = \frac{\Phi(s)}{\Phi(t)}e_1$.

When $w(s) = 0$, $E_{20}^{[s,t]}$ is isomorphic to E_1, $\forall s,t \in \mathcal{T}, t \geq b$, by the change of basis $e_1' = \frac{\Phi(s)}{\Phi(t)}e_1$, $e_2' = e_2$. And when $\Phi(s)w(s) > 0$, $E_{20}^{[s,t]}$

is isomorphic to E_2, $\forall s,t \in \mathcal{T}, t \geq b$, by the change of basis

$$e_1' = \frac{\Phi(s)}{\Phi(t)}e_1, \quad e_2' = \frac{\sqrt{|\Phi(s)|}}{\Phi(t)\sqrt{|w(s)|}}e_2,$$

and when $\Phi(s)\psi(s) < 0$ it is isomorphic to E_5, $\forall s,t \in \mathcal{T}, t \geq b$, by the change of basis $e_1' = \frac{\Phi(s)}{\Phi(t)}e_1$, $e_2' = \frac{\sqrt{|\Phi(s)|}}{\Phi(t)\sqrt{|w(s)|}}e_2$.

21. $E_{21}^{[s,t]}$ is isomorphic to $E_6(v(t) - v(s), 0)$, $\forall s,t \in \mathcal{T}, s \leq t < \min\{a,b\}$, by the change of basis $e_1' = e_2$, $e_2' = e_1$.
When $v(s) = 0$, $E_{21}^{[s,t]}$ is isomorphic to E_1, $\forall s,t \in \mathcal{T}, b \leq t < a, a > b$, by the change of basis $e_1' = e_1$, $e_2' = e_2$. And when $v(s) > 0$, $E_{21}^{[s,t]}$ is isomorphic to E_2, $\forall s,t \in \mathcal{T}, b \leq t < a, a > b$, by the change of basis $e_1' = e_1$, $e_2' = \frac{1}{\sqrt{|v(s)|}}e_2$, and when $v(s) < 0$, it is isomorphic to E_5, $\forall s,t \in \mathcal{T}, b \leq t < a, a > b$, by the change of basis $e_1' = e_1$, $e_2' = \frac{1}{\sqrt{|v(s)|}}e_2$.
$E_{21}^{[s,t]}$ is isomorphic to E_1, $\forall s,t \in \mathcal{T}, b \leq t < a, a < b$, by the change of basis $e_1' = v(t)e_1 + e_2$, $e_2' = e_1$. And this algebra will be E_0, $\forall s,t \in \mathcal{T}, t \geq \max\{a,b\}$.

22. $E_{22}^{[s,t]}$ is isomorphic to E_2, $\forall s,t \in \mathcal{T}$, by the change of basis

$$e_1' = \frac{f(t)}{2f^2(t) - 2f(t) + 1}e_1 + \frac{1 - f(t)}{2f^2(t) - 2f(t) + 1}e_2$$

$$e_2' = -\frac{1 - f(t)}{2f^2(t) - 2f(t) + 1}e_1 + \frac{f(t)}{2f^2(t) - 2f(t) + 1}e_2.$$

23. $E_{23}^{[s,t]}$ is isomorphic to $E_6\left(\frac{2\theta(s)(\theta(s) - \theta(t))}{(\theta(s) + \theta(t))^2}, 0\right)$ for any $s,t \in \mathcal{T}$, when $\lambda = 2\mu$, by the change of basis $e_1' = \frac{2\theta(s)}{\theta(s) + \theta(t)}e_2$, $e_2' = e_1$.
$E_{23}^{[s,t]}$ is isomorphic to $E_6\left(\frac{\xi\zeta}{(1-\zeta)^2}, \frac{(1-\xi)(1-\zeta)}{\xi^2}\right)$ for all $\lambda \neq 2\mu$,

$$(s,t) \in \left\{(s,t) : \theta(t) \neq \frac{2\lambda}{2\mu - \lambda}\theta(s), \theta(t) \neq \frac{2\mu - \lambda}{\lambda}\theta(s)\right\},$$

by the change of basis $e_1' = \frac{1}{1-\zeta}e_2$, $e_2' = \frac{1}{\xi}e_1$.
$E_{23}^{[s,t]}$ is isomorphic to $E_7\left(\frac{1-\zeta}{\sqrt[3]{\zeta^2}}\right)$ for all $\lambda \neq 2\mu$,

$$(s,t) \in \left\{(s,t) : \theta(t) = \frac{2\lambda}{2\mu - \lambda}\theta(s), \lambda \neq 0, \theta(t) \neq \theta(s),\right.$$

$$\left. \theta(t) \neq \frac{2\mu - \lambda}{\lambda}\theta(s)\right\},$$

by the change of basis $e_1' = \frac{1}{\sqrt[3]{\zeta}}e_1$, $e_2' = \frac{1}{\sqrt[3]{\zeta^2}}e_2$.

$E_{23}^{[s,t]}$ is isomorphic to $E_7\left(\frac{1-\zeta}{\sqrt[3]{\zeta^2}}\right)$ for all $\lambda \neq 2\mu$,

$$(s,t) \in \left\{(s,t): \ \theta(t) = \frac{2\lambda}{2\mu-\lambda}\theta(s), \ \lambda \neq 0, \ \theta(t) \neq \theta(s),\right.$$

$$\left.\theta(t) \neq \frac{2\mu-\lambda}{\lambda}\theta(s)\right\},$$

by the change of basis $e'_1 = \frac{1}{\sqrt[3]{1-\xi}}e_1$, $e'_2 = \frac{1}{\sqrt[3]{(1-\xi)^2}}e_2$.

$E_{23}^{[s,t]}$ is isomorphic to $E_7(0)$ for all $\lambda \neq 2\mu$,

$$(s,t) \in \left\{(s,t): \theta(t) = \frac{2\lambda}{2\mu-\lambda}\theta(s), \theta(t) = \frac{2\mu-\lambda}{\lambda}\theta(s)\right\},$$

by the change of basis $e'_1 = e_1, e'_2 = e_2$.

24. $E_{24}^{[s,t]}$ is isomorphic to $E_6(0,0)$, $\forall s,t \in \mathcal{T}, s \leq t < a$, by the change of basis $e'_1 = e_1$, $e'_2 = e_2$, and it is isomorphic to E_2, $\forall s,t \in \mathcal{T}, t \geq a$, by the change of basis

$$e'_1 = \frac{g(t)}{2g^2(t) - 2g(t) + 1}e_1 + \frac{1 - g(t)}{2g^2(t) - 2g(t) + 1}e_2,$$

$$e'_2 = -\frac{1 - g(t)}{2g^2(t) - 2g(t) + 1}e_1 + \frac{g(t)}{2g^2(t) - 2g(t) + 1}e_2.$$

25. $E_{27}^{[s,t]}$ is isomorphic to E_4, $\forall s,t \in \mathcal{T}, t > C$, by the change of basis $e'_1 = \frac{\Phi(s)}{\Phi(t)}e_1, e'_2 = e_2$, in period of time $s < t \leq C$, it will be isomorphic to the trivial evolution algebra E_0.

$\square$

5.3 Chains of evolution algebras of chicken population

See Definition 4.1 for an evolution algebra of a "chicken" population (EACP). In this section following [217] we construct chains of EACP.

Recall that an algebra EACP, $\mathcal{E}$, is defined by a rectangular $n \times (n+1)$-matrix

$$M = \begin{pmatrix} a_{11} & a_{12} & \ldots & a_{1n} & b_1 \\ a_{21} & a_{22} & \ldots & a_{2n} & b_2 \\ \vdots & \vdots & \ldots & \vdots & \vdots \\ a_{n1} & a_{n2} & \ldots & a_{nn} & b_n \end{pmatrix},$$

which is called the matrix of structural constants of the algebra $\mathcal{E}$.

Write the matrix M in the form $M = A \oplus \mathbf{b}$, where $A = (a_{ij})_{i,j=1,\ldots,n}$ and $\mathbf{b}^T = (b_1,\ldots,b_n)$.

For two rectangular $n \times (n+1)$-matrices $M = A \oplus \mathbf{b}$ and $H = B \oplus \mathbf{c}$ define multiplication by

$$MH = AB \oplus A\mathbf{c}, \quad HM = BA \oplus B\mathbf{b}. \tag{5.64}$$

One can see that this multiplication agrees with usual multiplication of $(n+1) \times (n+1)$-matrices with zero $(n+1)$-th row.

Consider a family $\left\{ \mathcal{E}^{[s,t]} : s,t \in \mathbb{R}, \ 0 \le s \le t \right\}$ of $(n+1)$-dimensional EACP over the field $\mathbb{R}$, with basis $\{h_1,\ldots,h_n,r\}$ and multiplication table

$$h_i r = r h_i = \frac{1}{2}\left(\sum_{j=1}^{n} a_{ij}^{[s,t]} h_j + b_i^{[s,t]} r \right); \quad h_i h_j = 0, \quad 1 \le i,j \le n; \quad rr = 0. \tag{5.65}$$

Here parameters s,t are considered as time.

Denote by $M^{[s,t]} = A^{[s,t]} \oplus \mathbf{b}^{[s,t]} = \left(a_{ij}^{[s,t]} \right)_{i,j=1,\ldots,n} \oplus (b_i^{[s,t]})_{i=1,\ldots,n}$-the matrix of structural constants.

Definition 5.7. A family $\left\{ \mathcal{E}^{[s,t]} : s,t \in \mathbb{R}, \ 0 \le s \le t \right\}$ of $(n+1)$-dimensional EACP over the field $\mathbb{R}$ is called a chain of EACP (CEACP) if the matrix $M^{[s,t]}$ of structural constants satisfies the Chapman–Kolmogorov equation

$$M^{[s,t]} = M^{[s,\tau]} M^{[\tau,t]}, \quad \text{for any} \ \ s < \tau < t. \tag{5.66}$$

By the rule (5.64) of multiplication we get from (5.66) the following

$$A^{[s,t]} = A^{[s,\tau]} A^{[\tau,t]}, \quad \mathbf{b}^{[s,t]} = A^{[s,\tau]} \mathbf{b}^{[\tau,t]} \quad \text{for any} \ \ s < \tau < t. \tag{5.67}$$

Definition 5.8. A CEACP is called a time-homogeneous if the matrix $M^{[s,t]}$ depends only on $t - s$. In this case we write $M^{[t-s]}$.

Definition 5.9. A CEACP is called periodic if its matrix $M^{[s,t]}$ is periodic with respect to at least one of the variables s, t, i.e. (periodicity with respect to t) $M^{[s,t+P]} = M^{[s,t]}$ for all values of t. The constant P is called the period, and is required to be nonzero.

Remark 5.7. To construct a CEACP one has to solve the system (5.67). Note that the first equation of the system does not depend on $\mathbf{b}^{[s,t]}$, therefore, one can solve it firstly and then, for each solution, from the second equation one must find corresponding $\mathbf{b}^{[s,t]}$. Note that the first equation is usual Chapman–Kolmogorov equation (for quadratic matrices) and a wide class of their solutions are known see previous sections and [31], [218].

In the next section we shall construct several examples of CEACPs.

5.3.1 *Examples of CEACPs*

5.3.1.1 *The CEACP corresponding to a Markov process*

We recall that a left (right) stochastic matrix is a square matrix of nonnegative real numbers, with each column (row) summing to 1. A stochastic vector is a vector whose elements are nonnegative real numbers which sum to 1. Thus, each row of a right stochastic matrix (or column of a left stochastic matrix) is a stochastic vector.

Let $\left\{A^{[s,t]},\ \ 0 \le s \le t\right\}$ be a family of stochastic matrices which satisfies the first equation of the system (5.67), then it defines a Markov process. We are going to solve the second equation of (5.67).

Lemma 5.3. *If* $\left\{A^{[s,t]},\ \ 0 \le s \le t\right\}$ *is a family of left stochastic matrices then* $\sum_{i=1}^{n} b_i^{[s,t]}$ *does not depend on* s. *Moreover if* $\mathbf{b}^{[s,t]}$ *is a stochastic vector for some time* (s_0, t_0) *then it is a stochastic vector for any time* (s, t_0) *with* $s < t_0$.

Proof. The second equation of (5.67) for the coordinates has the following form

$$b_i^{[s,t]} = \sum_{j=1}^{n} a_{ij}^{[s,\tau]} b_j^{[\tau,t]}, \quad \text{for any}\ \ s < \tau < t. \tag{5.68}$$

From this we get

$$\sum_{i=1}^{n} b_i^{[s,t]} = \sum_{j=1}^{n} \left(\sum_{i=1}^{n} a_{ij}^{[s,\tau]} \right) b_j^{[\tau,t]} = \sum_{j=1}^{n} b_j^{[\tau,t]}. \tag{5.69}$$

Now if $\mathbf{b}^{[s,t]}$ is a stochastic vector for some time (s_0, t_0) then $b_i^{[s_0,t_0]} \ge 0$, $\sum_{i=1}^{n} b_i^{[s_0,t_0]} = 1$ this by (5.69) gives $\sum_{i=1}^{n} b_i^{[s,t_0]} = 1$ for any $s < t_0$. Write the formula (5.68) for $\tau = s_0$ and $t = t_0$, then we get $b_i^{[s,t_0]} \ge 0$ for any $s < t_0$. Thus the vector $\mathbf{b}^{[s,t_0]}$ is a stochastic vector for any $s < t_0$. $\square$

Consider matrix M_0 which is the $(n+1) \times (n+1)$-matrix constructed from M by adding $(n+1)$-th zero-row. The following theorem follows from Lemma 5.3.

Theorem 5.13. *For each Markov process given by* M_0, *there is a CEACP whose structural constants are transition probabilities of the process, and whose generator set (basis) is the state space of the Markov process.*

5.3.1.2 *The CEACP which does not define a process*

Now we consider several concrete examples.

Example 1. Here we describe all 2-dimensional CEACP with basis $\{h, r\}$. Such chains are given by vector $(a^{[s,t]}, b^{[s,t]})$ for which the equation (5.66) has the following form

$$a^{[s,\tau]} a^{[\tau,t]} = a^{[s,t]}, \quad a^{[s,\tau]} b^{[\tau,t]} = b^{[s,t]}. \tag{5.70}$$

The first equation of the system (5.70) is known as Cantor's second equation which has very rich family of solutions:

a) $a^{[s,t]} = \frac{\Phi(t)}{\Phi(s)}$, where Φ is an arbitrary function with $\Phi(s) \neq 0$;

b)

$$a^{[s,t]} = \begin{cases} 1, & \text{if } s \le t < \theta, \\ 0, & \text{if } t \ge \theta, \end{cases} \quad \text{where } \theta > 0.$$

In the case a) from the second equation of (5.70) we get $b^{[s,t]} = \alpha(t)/\Phi(s)$, where α is an arbitrary function.

In the case b) from the second equation of (5.70) we get

$$b^{[s,t]} = \begin{cases} b^{[\tau,t]}, & \text{if } s < \tau < \min\{\theta, t\}, \\ 0, & \text{if } \theta \le \tau < t. \end{cases}$$

This has the following solution

$$b^{[s,t]} = \begin{cases} \beta(t), & \text{if } s < \min\{\theta, t\}, \\ 0, & \text{if } \theta \le s, \end{cases}$$

where β is an arbitrary function. Thus we obtain two CEACP:

$$\mathcal{E}_1^{[s,t]}, \quad \text{with } hr = rh = \frac{1}{2\Phi(s)}(\Phi(t)h + \alpha(t)r), \ h^2 = r^2 = 0; \tag{5.71}$$

$$\mathcal{E}_2^{[s,t]}, \quad \text{with } hr = rh = \frac{1}{2}\begin{cases} h + \beta(t)r, & \text{if } s < \min\{\theta, t\}, \\ 0 & \text{if } \theta \le s, \end{cases} \quad h^2 = r^2 = 0. \tag{5.72}$$

Example 2. Consider

$$A^{[t]} = \frac{1}{2}\begin{pmatrix} \lambda^t + \mu^t & \lambda^t - \mu^t \\ \lambda^t - \mu^t & \lambda^t + \mu^t \end{pmatrix},$$

where $\lambda, \mu > 0$.

This matrix satisfies the first equation of (5.67) with $A^{[s,t]} = A^{[t-s]}$. Now we shall find solutions of the second equation corresponding to $A^{[t-s]}$. The equation has the following form

$$(\lambda^{\tau-s} + \mu^{\tau-s})b_1^{[\tau,t]} + (\lambda^{\tau-s} - \mu^{\tau-s})b_2^{[\tau,t]} = 2b_1^{[s,t]}$$

$$(\lambda^{\tau-s} - \mu^{\tau-s})b_1^{[\tau,t]} + (\lambda^{\tau-s} + \mu^{\tau-s})b_2^{[\tau,t]} = 2b_2^{[s,t]}.$$

Denoting $\psi(s,t) = b_1^{[s,t]} + b_2^{[s,t]}$ from the last system we obtain

$$\lambda^{\tau}\psi(\tau,t) = \lambda^{s}\psi(s,t).$$

Hence $\psi(s,t)$ has the form $\psi(s,t) = \lambda^{-s}\alpha(t)$, where α is an arbitrary function. Consequently $b_2^{[s,t]} = \lambda^{-s}\alpha(t) - b_1^{[s,t]}$. Then for function $b_1^{[s,t]}$ we have

$$(\lambda^{\tau-s} + \mu^{\tau-s})b_1^{[\tau,t]} + (\lambda^{\tau-s} - \mu^{\tau-s})(\lambda^{-\tau}\alpha(t) - b_1^{[\tau,t]}) = 2b_1^{[s,t]},$$

i.e.

$$2\mu^{\tau}b_1^{[\tau,t]} + (\lambda^{\tau-s} - \mu^{\tau-s})\lambda^{-\tau}\mu^{s}\alpha(t) = 2\mu^{s}b_1^{[s,t]}.$$

Denote $f(s,t) = 2\mu^{s}b_1^{[s,t]}$. Then from the last equation we get

$$f(\tau,t) - f(s,t) = ((\mu/\lambda)^{\tau} - (\mu/\lambda)^{s})\alpha(t).$$

This equation has the following solution

$$f(s,t) = (\mu/\lambda)^{s}\alpha(t) + \beta(t),$$

where β is an arbitrary function. Hence

$$b_1^{[s,t]} = \frac{1}{2}(\lambda^{-s}\alpha(t) + \mu^{-s}\beta(t)), \quad b_2^{[s,t]} = \frac{1}{2}(\lambda^{-s}\alpha(t) - \mu^{-s}\beta(t)).$$

Thus we get the following solution of the system (5.67)

$$M^{[s,t]} = \frac{1}{2}\begin{pmatrix} \lambda^{t-s} + \mu^{t-s} & \lambda^{t-s} - \mu^{t-s} & \lambda^{-s}\alpha(t) + \mu^{-s}\beta(t) \\ \lambda^{t-s} - \mu^{t-s} & \lambda^{t-s} + \mu^{t-s} & \lambda^{-s}\alpha(t) - \mu^{-s}\beta(t) \end{pmatrix}.$$

Let $\mathcal{E}^{[s,t]}$, $0 \le s \le t$ be the CEACP corresponding to this matrix. In spite of $A^{[t-s]}$ is a time-homogeneous, note that this CEACP is not a time-homogeneous, in general. But if $\alpha(t) = \lambda^{t}$, $\beta(t) = \mu^{t}$ then the CEACP is a time-homogeneous.

Depending on parameters λ, μ and parameter-functions α, β we get distinct behavior of $\mathcal{E}^{[s,t]}$ for $t - s \to +\infty$. In case if one of the following limits does not exist

$$\lim_{t-s\to+\infty} (\lambda^{-s}\alpha(t) + \mu^{-s}\beta(t)) = b_1, \quad \lim_{t-s\to+\infty} (\lambda^{-s}\alpha(t) + \mu^{-s}\beta(t)) = b_2$$

then a limiting EACP does not exist. If the limits do exist, then we have

$$\lim_{t-s\to+\infty} \mathcal{E}^{[s,t]} = \begin{cases} \mathcal{E}_0 & \text{if } 0 < \lambda, \mu < 1, \\ \mathcal{E}_1 & \text{if } \lambda = \mu = 1, \\ \mathcal{E}_{1/2} & \text{if } \lambda = 1, 0 \le \mu < 1, \\ \mathcal{E}_{-1/2} & \text{if } \mu = 1, 0 \le \lambda < 1, \\ \mathcal{E}_\infty & \text{otherwise,} \end{cases}$$

where $\mathcal{E}_0$ is an EACP with multiplication (omitted multiplications are zero):

$$h_1 r = b_1 r, \quad h_2 r = b_2 r;$$

$\mathcal{E}_1$ is an EACP with multiplication table

$$h_1 r = h_1 + b_1 r, \quad h_2 r = h_2 + b_2 r;$$

$\mathcal{E}_{1/2}$ is an EACP with

$$h_1 r = \frac{1}{2}(h_1 + h_2) + b_1 r, \quad h_2 r = \frac{1}{2}(h_1 + h_2) + b_2 r;$$

$\mathcal{E}_{-1/2}$ is an EACP with

$$h_1 r = \frac{1}{2}(h_1 - h_2) + b_1 r, \quad h_2 r = -\frac{1}{2}(h_1 - h_2) + b_2 r;$$

and $\mathcal{E}_\infty$ is a vector space which has "infinity multiplication", or we can say that in $\mathcal{E}_\infty$ an algebra structure is not defined. This example shows that a limit of a CEACP can be non evolution algebra.

Example 3. Consider a solution $A^{[s,t]} = \left(a_{ij}^{[s,t]}\right)_{i,j=1,2}$ to the first equation of (5.67) given by

$$\begin{aligned} a_{11}^{[s,t]} &= \tfrac{1}{2}\left(1 + \Phi(t)(\Psi(t) - \Psi(s)) + \tfrac{\Phi(t)}{\Phi(s)}\right), \\ a_{12}^{[s,t]} &= \tfrac{1}{2}\left(1 - \Phi(t)(\Psi(t) - \Psi(s)) - \tfrac{\Phi(t)}{\Phi(s)}\right), \\ a_{21}^{[s,t]} &= \tfrac{1}{2}\left(1 + \Phi(t)(\Psi(t) - \Psi(s)) - \tfrac{\Phi(t)}{\Phi(s)}\right), \\ a_{22}^{[s,t]} &= \tfrac{1}{2}\left(1 - \Phi(t)(\Psi(t) - \Psi(s)) + \tfrac{\Phi(t)}{\Phi(s)}\right), \end{aligned} \tag{5.73}$$

where $\Phi \neq 0$ and Ψ are arbitrary functions. Now we shall find solutions of the second equation of (5.67) corresponding to $A^{[s,t]}$. Denoting $f(s,t) = b_1^{[s,t]} + b_2^{[s,t]}$, $g(s,t) = b_1^{[s,t]} - b_2^{[s,t]}$ the second equation can be written as

$$f(\tau,t) + \Phi(\tau)(\Psi(\tau) - \Psi(s))g(\tau,t) = f(s,t),$$

$$(\Phi(\tau)/\Phi(s))g(\tau,t) = g(s,t).$$

From the second equation of this system we get $\Phi(s)g(s,t) = \alpha(t)$, where α is an arbitrary function. Substituting $g(\tau,t) = \alpha(t)/\Phi(\tau)$ in the first equation of the last system we get $f(s,t) = \beta(t) - \Psi(s)\alpha(t)$, where β is an arbitrary function. Finally we obtain the following

$$
\begin{aligned}
b_1^{[s,t]} &= \tfrac{1}{2}\left(\beta(t) + (\tfrac{1}{\Phi(s)} - \Psi(s))\alpha(t)\right), \\
b_2^{[s,t]} &= \tfrac{1}{2}\left(\beta(t) - (\tfrac{1}{\Phi(s)} + \Psi(s))\alpha(t)\right).
\end{aligned}
\tag{5.74}
$$

Thus we showed that matrix $A^{[s,t]}$ given by (5.73) and vector $\mathbf{b}^{[s,t]}$ given by (5.74) generate a CEACP, $\mathcal{E}^{[s,t]}$, $0 \leq s \leq t$. This CEACP varies by four parameter-functions, for example, if Φ, Ψ, α and β such that

$$\lim_{t \to +\infty} \Phi(t)\Psi(t) = \lim_{t \to +\infty} \Phi(t) = \lim_{t \to +\infty} \alpha(t) = \lim_{t \to +\infty} \beta(t) = 0$$

then for a fixed s we have $\lim_{t \to \infty} \mathcal{E}^{[s,t]} = \mathcal{E}_{1/2}$, where $\mathcal{E}_{1/2}$ is an EACP with multiplication table

$$h_1 r = h_2 r = \frac{1}{2}(h_1 + h_2).$$

Example 4. For any n we shall give an example of time non-homogeneous $n+1$-dimensional CEACP. Let $\{T^{[t]}, t \geq 0\}$ be a family of invertible (for all t), $n \times n$ matrices. Define the following matrix

$$A^{[s,t]} = T^{[s]}(T^{[t]})^{-1}, \tag{5.75}$$

where $(T^{[t]})^{-1}$ is the inverse of $T^{[t]}$.

Note that a construction of a family of invertible $n \times n$ matrices $T^{[t]}$ is not difficult, for example, one can take $T^{[t]}$ as a lower or upper triangular $n \times n$ matrix. Then the matrices are invertible iff $\det(T^{[t]}) \neq 0$ for all t, i.e. the diagonal elements of the triangular matrix are non-zero.

Let $\mathbf{c}^{[t]}$, $t > 0$ be an arbitrary family of vectors. Define

$$\mathbf{b}^{[s,t]} = T^{[s]}\mathbf{c}^{[t]}. \tag{5.76}$$

Proposition 5.11. *The matrix* $M^{[s,t]} = A^{[s,t]} \oplus \mathbf{b}^{[s,t]}$ *given by (5.75) and (5.76) generates a* $n+1$-*dimensional CEACP.*

Proof. We shall show that the matrix $M^{[s,t]}$ given in the statement of the Proposition satisfies the equation (5.66):

$$M^{[s,\tau]}M^{[\tau,t]} = (A^{[s,\tau]} \oplus \mathbf{b}^{[s,\tau]})(A^{[\tau,t]} \oplus \mathbf{b}^{[\tau,t]}) = A^{[s,\tau]}A^{[\tau,t]} \oplus A^{[s,\tau]}\mathbf{b}^{[\tau,t]}$$

$$= T^{[s]}\left((T^{[\tau]})^{-1}T^{[\tau]}\right)(T^{[t]})^{-1} \oplus T^{[s]}\left((T^{[\tau]})^{-1}T^{[\tau]}\right)\mathbf{c}^{[t]}$$

$$= T^{[s]}(T^{[t]})^{-1} \oplus T^{[s]}\mathbf{c}^{[t]} = M^{[s,t]}. \qquad \square$$

Thus each family (with one parameter) of invertible $n \times n$ matrices together with a family of vectors (with one parameter) define a CEACP $\mathcal{E}^{[s,t]}$ which is time non-homogeneous, in general.

Example 5. One can see that the matrix

$$A^{[s,t]} = \begin{pmatrix} \cos(t-s) & \sin(t-s) \\ -\sin(t-s) & \cos(t-s) \end{pmatrix} \tag{5.77}$$

satisfies the equation (5.66). We are going to find corresponding solutions of the second equation of (5.67). This equation has the following form

$$\cos(\tau - s)b_1^{[\tau,t]} + \sin(\tau - s)b_2^{[\tau,t]} = b_1^{[s,t]}$$
$$-\sin(\tau - s)b_1^{[\tau,t]} + \cos(t - s)b_2^{[\tau,t]} = b_2^{[s,t]}. \tag{5.78}$$

One can check that this system has the following solution

$$b_1^{[s,t]} = c_1 \sin(t-s) - c_2 \cos(t-s), \quad b_2^{[s,t]} = c_1 \cos(t-s) + c_2 \sin(t-s),$$

where c_1 and c_2 are arbitrary numbers. Thus the following matrix satisfies (5.66):

$$M^{[s,t]} = \begin{pmatrix} \cos(t-s) & \sin(t-s) & c_1 \sin(t-s) - c_2 \cos(t-s) \\ -\sin(t-s) & \cos(t-s) & c_1 \cos(t-s) + c_2 \sin(t-s) \end{pmatrix}. \tag{5.79}$$

Since this matrix is periodic with period $P = 2\pi$, the corresponding CEACP $\mathcal{E}^{[t]}$ is also periodic. Moreover, for arbitrary 3-dimensional EACP $\mathcal{E}_a^+$, or $\mathcal{E}_a^-$, $a \in [-1, 1]$ with structural constants matrix

$$M_a^{\pm} = \begin{pmatrix} a & \pm\sqrt{1-a^2} & \pm c_1\sqrt{1-a^2} - c_2 a \\ \mp\sqrt{1-a^2} & a & c_1 a \pm c_2\sqrt{1-a^2} \end{pmatrix}$$

respectively, there is a sequence $t_n = t_n(a)$ of times such that $\lim_{n\to\infty} \mathcal{E}^{[t_n]} = \mathcal{E}_a^+$ or $\mathcal{E}_a^-$. We have $\mathcal{E}_a^{\pm} \neq \mathcal{E}_b^{\pm}$ if $a \neq b$. Moreover the following is true

Proposition 5.12. *1) For any $a, b \in [-1, 1]$, $a \neq b$, the algebras $\mathcal{E}_a^+$ and $\mathcal{E}_b^+$ are not isomorphic.*

2) For any $a, b \in [-1, 1]$, $a \neq b$, the algebras $\mathcal{E}_a^-$ and $\mathcal{E}_b^-$ are not isomorphic.

Proof. 1) Let $\varphi = (\alpha_{ij})_{i,j=1,2,3}$ be an isomorphism of the evolution algebra $\mathcal{E}_a^+$ to the evolution algebra $\mathcal{E}_b^+$. Here $\det(\varphi) \neq 0$. Since $\det(A^{[s,t]}) = 1$, from $\varphi(h_1 h_2) = \varphi(h_1)\varphi(h_2) = 0$, $\varphi(h_1^2) = 0$, $\varphi(h_2^2) = 0$ and $\varphi(rr) = 0$ we get

$$\alpha_{11}\alpha_{23} + \alpha_{21}\alpha_{13} = 0, \quad \alpha_{12}\alpha_{23} + \alpha_{22}\alpha_{13} = 0;$$

$$\alpha_{11}\alpha_{13} = 0, \quad \alpha_{12}\alpha_{13} = 0; \quad \alpha_{21}\alpha_{23} = 0, \quad \alpha_{22}\alpha_{23} = 0;$$

$$\alpha_{31}\alpha_{33} = 0, \quad \alpha_{32}\alpha_{33} = 0.$$

Since $\det(\varphi) \neq 0$ the last equations give the following possibilities to φ:

$$F_1 = \left\{ \varphi = \begin{pmatrix} \alpha_{11} & \alpha_{12} & 0 \\ \alpha_{21} & \alpha_{22} & 0 \\ 0 & 0 & \alpha_{33} \end{pmatrix} : (\alpha_{11}\alpha_{22} - \alpha_{12}\alpha_{21})\alpha_{33} \neq 0 \right\},$$

$$F_2 = \left\{ \varphi = \begin{pmatrix} \alpha_{11} & \alpha_{12} & 0 \\ 0 & 0 & \alpha_{23} \\ \alpha_{31} & \alpha_{32} & 0 \end{pmatrix} : (\alpha_{12}\alpha_{31} - \alpha_{11}\alpha_{32})\alpha_{23} \neq 0 \right\},$$

$$F_3 = \left\{ \varphi = \begin{pmatrix} 0 & 0 & \alpha_{13} \\ \alpha_{21} & \alpha_{22} & 0 \\ \alpha_{31} & \alpha_{32} & 0 \end{pmatrix} : (\alpha_{21}\alpha_{32} - \alpha_{22}\alpha_{31})\alpha_{13} \neq 0 \right\}.$$

Note that the classes F_i are the same up to renumbering of indexes. Therefore, we consider only class F_1. Take $\varphi \in F_1$ then we get the following relation between matrices M_a^+ and

$$M_b^+ = \varphi(M_a^+) = \begin{pmatrix} m_{11} & m_{12} & m_{13} \\ m_{21} & m_{22} & m_{23} \end{pmatrix}, \tag{5.80}$$

where m_{ij} is function of a given by the following formulas

$$m_{11} = \frac{(\alpha_{11}a - \alpha_{12}\sqrt{1-a^2})\alpha_{22} - (\alpha_{11}\sqrt{1-a^2} + \alpha_{12}a)\alpha_{21}}{\alpha_{11}\alpha_{22} - \alpha_{12}\alpha_{21}}$$

$$= a - \frac{(\alpha_{11}\alpha_{21} + \alpha_{12}\alpha_{22})\sqrt{1-a^2}}{\alpha_{11}\alpha_{22} - \alpha_{12}\alpha_{21}},$$

$$m_{12} = \frac{-(\alpha_{11}a - \alpha_{12}\sqrt{1-a^2})\alpha_{12} + (\alpha_{11}\sqrt{1-a^2} + \alpha_{12}a)\alpha_{11}}{\alpha_{11}\alpha_{22} - \alpha_{12}\alpha_{21}}$$

$$= \frac{(\alpha_{11}^2 + \alpha_{12}^2)\sqrt{1-a^2}}{\alpha_{11}\alpha_{22} - \alpha_{12}\alpha_{21}},$$

$$m_{13} = \frac{\alpha_{11}(c_1\sqrt{1-a^2} - c_2 a) + \alpha_{12}(c_1 a + c_2\sqrt{1-a^2})}{\alpha_{33}},$$

$$m_{21} = \frac{(\alpha_{21}a - \alpha_{22}\sqrt{1-a^2})\alpha_{22} - (\alpha_{21}\sqrt{1-a^2} + \alpha_{22}a)\alpha_{21}}{\alpha_{11}\alpha_{22} - \alpha_{12}\alpha_{21}}$$

$$= -\frac{(\alpha_{21}^2 + \alpha_{22}^2)\sqrt{1-a^2}}{\alpha_{11}\alpha_{22} - \alpha_{12}\alpha_{21}},$$

$$m_{22} = \frac{-(\alpha_{21}a - \alpha_{22}\sqrt{1-a^2})\alpha_{12} + (\alpha_{21}\sqrt{1-a^2} + \alpha_{22}a)\alpha_{11}}{\alpha_{11}\alpha_{22} - \alpha_{12}\alpha_{21}}$$

$$= a + \frac{(\alpha_{11}\alpha_{21} + \alpha_{12}\alpha_{22})\sqrt{1-a^2}}{\alpha_{11}\alpha_{22} - \alpha_{12}\alpha_{21}},$$

$$m_{23} = \frac{\alpha_{21}(c_1\sqrt{1-a^2} - c_2 a) + \alpha_{22}(c_1 a + c_2\sqrt{1-a^2})}{\alpha_{33}}.$$

By (5.80) we should have $m_{11} = m_{22} = b$ and $m_{12} = -m_{21} = \sqrt{1-b^2}$. These equalities for $a \neq \pm 1$ give

$$\alpha_{11}\alpha_{21} + \alpha_{12}\alpha_{22} = 0, \quad \alpha_{11}^2 + \alpha_{12}^2 = \alpha_{21}^2 + \alpha_{22}^2. \tag{5.81}$$

Consequently, if $a \neq \pm 1$ then $b = a$. Moreover, if $a = \pm 1$ then again we get $m_{11} = m_{22} = b = a = \pm 1$. Hence isomorphisms of the class F_1 can not give an isomorphism between $\mathcal{E}_a^+$ and $\mathcal{E}_b^+$ if $a \neq b$. Similar argument works for classes F_2 and F_3.

2) The proof of 2) is similar to the proof of 1). $\qquad\square$

Consider now discrete time n, $n \in \mathbb{N}$ and the CEACP $\{\mathcal{E}^{[n]}, n \in \mathbb{N}\}$ given by matrix (5.79) with $t - s = n$.

Proposition 5.13. *The discrete time CEACP $\mathcal{E}^{[n]}$, $n \in \mathbb{N}$, is dense in the set $\{\mathcal{E}_a^{\pm}, a \in [-1,1]\}$ of EACP, i.e. for an arbitrary EACP $\mathcal{E}_a^{\pm}$ there exists a sequence $\{n_k\}_{k=1,2,\ldots}$ of natural numbers such that $\lim_{k\to\infty} \mathcal{E}^{[n_k]} = \mathcal{E}_a^+$ or $\mathcal{E}_a^-$.*

Proof.   It is known that the sequences $\{\sin n\}$ and $\{\cos n\}, n \in \mathbb{N}$, are dense in $[-1, 1]$ (see e.g. [57]). Hence for any $a \in [-1, 1]$ there is a sequence $\{n_k\}_{k=1,2,\dots}$ of natural numbers such that $\lim\limits_{k \to \infty} \cos(n_k) = a$. The same sequence can be used to get $\lim\limits_{k \to \infty} \mathcal{E}^{[n_k]} = \mathcal{E}_a^+$ or $\mathcal{E}_a^-$. $\qquad\square$

5.3.2   *Time depending dynamics of CEACP*

In this section for 2- and 3-dimensional algebras of CEACP we shall study the time depending dynamics of isomorphic EACPs in the chain.

Let $\mathcal{E}$ be a 2-dimensional EACP and $\{h, r\}$ be a basis of this algebra.

It is evident that if $\dim \mathcal{E}^2 = 0$ then $\mathcal{E}$ is an abelian algebra, i.e. an algebra with all products equal to zero.

Recall the following

Proposition 5.14. [142] *Any 2-dimensional, non-trivial EACP $\mathcal{E}$ is isomorphic to one of the following pairwise non isomorphic algebras:*

$$
\begin{aligned}
E_1: \quad & rh = hr = h, \quad h^2 = r^2 = 0, \\
E_2: \quad & rh = hr = \tfrac{1}{2}(h + r), \quad h^2 = r^2 = 0.
\end{aligned}
$$

The following proposition gives time dynamics of (5.71) and (5.72).

Proposition 5.15. *We have*

$$
\mathcal{E}_1^{[s,t]} \cong
\begin{cases}
E_1, & \text{if } t \in \{t : \alpha(t) = 0\}, \\
E_2, & \text{if } t \in \{t : \alpha(t) \neq 0\}.
\end{cases}
$$

$$
\mathcal{E}_2^{[s,t]} \cong
\begin{cases}
E_1, & \text{if } (s,t) \in \{(s,t) : \beta(t) = 0, s < \min\{\theta, t\}\}, \\
E_2, & \text{if } (s,t) \in \{(s,t) : \beta(t) \neq 0, s < \min\{\theta, t\}\}, \\
E_0, & \text{if } (s,t) \in \{(s,t) : s \geq 0\},
\end{cases}
$$

where E_0 is the algebra with zero multiplication.

Proof.   Since $\Phi(t) \neq 0$ if $\alpha(t) = 0$ then by change of basis $h' = h$ and $r' = \frac{2\Phi(s)}{\Phi(t)} r$ we get the algebra E_1. In case $\alpha(t) \neq 0$ the change $h' = \frac{\Phi(s)}{\Phi(t)} r$, and $r' = \frac{\Phi(s)}{\alpha(t)} h$ implies the algebra E_2.

For $\mathcal{E}_2^{[s,t]}$ the proof is similar. $\qquad\square$

To give some illustration of Proposition 5.15 we consider the following example.

Example 6. Take

$$\alpha(t) = \begin{cases} (8-t)(12-t)(2019-t), & \text{if } 0 \le t \le 2019, \\ 0, & \text{if } t > 2019. \end{cases}$$

By the proposition the corresponding CEACP will be isomorphic to E_1 if time $t \in \{8, 12\} \cup [2019, +\infty)$. So there are three critical times 8, 12, 2019 at which the chain changes the algebras and remain with the same algebra (up to isomorphism) between critical times. This is like a phase transition property of physical systems [77]. There parameter is the temperature and the system changes its phase (state) at critical temperatures.

Let now $\mathcal{E}$ be a 3-dimensional EACP and $\{h_1, h_2, r\}$ be a basis of this algebra. Note that any 3-dimensional EACP $\mathcal{E}$ with $\dim(\mathcal{E}^2) = 1$ is isomorphic to one of the following pairwise non isomorphic algebras:

$$\begin{aligned} \mathcal{E}_1: \quad & h_1 r = r; \\ \mathcal{E}_2: \quad & h_1 r = h_2; \\ \mathcal{E}_3: \quad & h_1 r = h_1 + r. \end{aligned}$$

In each algebra $r h_i = h_i r$, $i = 1, 2$ and all omitted products are zero.

In order to use this fact, first we give a class of 3-dimensional CEACP $\mathcal{E}^{[s,t]}$ with $\dim\left(\left(\mathcal{E}^{[s,t]}\right)^2\right) = 1$ for any (s, t). Such a CEACP has matrix $M^{[s,t]}$ in the following form

$$M^{[s,t]} = \begin{pmatrix} a_1^{[s,t]} & a_2^{[s,t]} & b^{[s,t]} \\ ca_1^{[s,t]} & ca_2^{[s,t]} & cb^{[s,t]} \end{pmatrix},$$

where $c = c(s, t)$ is a given function.

For this operator the equation (5.67) can be written as

$$a_1^{[\tau,t]} \left(a_1^{[s,\tau]} + c(\tau, t) a_2^{[s,\tau]} \right) = a_1^{[s,t]},$$

$$a_2^{[\tau,t]} \left(a_1^{[s,\tau]} + c(\tau, t) a_2^{[s,\tau]} \right) = a_2^{[s,t]},$$

$$b^{[\tau,t]} \left(a_1^{[s,\tau]} + c(\tau, t) a_2^{[s,\tau]} \right) = b^{[s,t]},$$

$$c(s, \tau) a_1^{[\tau,t]} \left(a_1^{[s,\tau]} + c(\tau, t) a_2^{[s,\tau]} \right) = c(s, t) a_1^{[s,t]},$$

$$c(s, \tau) a_2^{[\tau,t]} \left(a_1^{[s,\tau]} + c(\tau, t) a_2^{[s,\tau]} \right) = c(s, t) a_2^{[s,t]},$$

$$c(s, \tau) b^{[\tau,t]} \left(a_1^{[s,\tau]} + c(\tau, t) a_2^{[s,\tau]} \right) = c(s, t) b^{[s,t]}.$$

$$(5.82)$$

Comparing the first and the fourth equations we get $c(s,\tau) = c(s,t)$, i.e. c should not depend on t. So denote this function as $c(s)$. Then from system (5.82) it remains the following system

$$a_1^{[\tau,t]}\left(a_1^{[s,\tau]} + c(\tau)a_2^{[s,\tau]}\right) = a_1^{[s,t]},$$

$$a_2^{[\tau,t]}\left(a_1^{[s,\tau]} + c(\tau)a_2^{[s,\tau]}\right) = a_2^{[s,t]}, \tag{5.83}$$

$$b^{[\tau,t]}\left(a_1^{[s,\tau]} + c(\tau)a_2^{[s,\tau]}\right) = b^{[s,t]}.$$

Denote $\gamma(s,t) = a_1^{[s,t]} + c(t)a_2^{[s,t]}$ and $\delta(s,t) = a_1^{[s,t]} - c(t)a_2^{[s,t]}$. Then from first and second equations of (5.82) we get

$$\gamma(s,t) = \gamma(s,\tau)\gamma(\tau,t), \quad \delta(s,t) = \gamma(s,\tau)\delta(\tau,t), \ s \le \tau \le t.$$

The first equation is Cantor's second equation which has a solution $\gamma(s,t) = \frac{\phi(t)}{\phi(s)}$, where ϕ is an arbitrary function with $\phi(s) \ne 0$. Using the solution from the second equation we find $\delta(s,t) = \frac{\psi(t)}{\phi(s)}$, where ψ is an arbitrary function. Then we obtain

$$a_1^{[s,t]} = \begin{cases} \frac{\phi(t)+\psi(t)}{2\phi(s)}, & \text{if } c(t) \ne 0, \\[2mm] \frac{\phi(t)}{\phi(s)}, & \text{if } c(t) = 0, \end{cases}$$

$$a_2^{[s,t]} = \begin{cases} \frac{\phi(t)-\psi(t)}{2c(t)\phi(s)}, & \text{if } c(t) \ne 0 \\[2mm] \frac{\psi(t)}{\phi(s)}, & \text{if } c(t) = 0, \end{cases} \qquad b^{[s,t]} = \frac{\alpha(t)}{\phi(s)}.$$

Consequently, for t such that $c(t) \ne 0$ we have

$$M^{[s,t]} = \frac{1}{2\phi(s)}\begin{pmatrix} \phi(t) + \psi(t) & \frac{\phi(t)-\psi(t)}{c(t)} & 2\alpha(t) \\ c(t)(\phi(t)+\psi(t)) & \phi(t) - \psi(t) & 2c(t)\alpha(t) \end{pmatrix} \tag{5.84}$$

and for t such that $c(t) = 0$ we have

$$M^{[s,t]} = \frac{1}{\phi(s)}\begin{pmatrix} \phi(t) & \psi(t) & \alpha(t) \\ 0 & 0 & 0 \end{pmatrix}. \tag{5.85}$$

Thus we proved the following

Theorem 5.14. *The matrix (5.84) and (5.85) generates a CEACP $\mathcal{E}^{[s,t]}$ with $\dim\left(\left(\mathcal{E}^{[s,t]}\right)^2\right) = 1$ for any (s,t).*

Consider a 3-dimensional EACP $\mathcal{E}$ with $\dim(\mathcal{E}^2) = 1$ then it has the following form

$$h_1 r = r h_1 = \frac{1}{2}(ah_1 + bh_2 + Ar), \quad h_2 r = r h_2 = \frac{c}{2}(ah_1 + bh_2 + Ar),$$

$$h_1^2 = h_2^2 = h_1 h_2 = r^2 = 0.$$

Note that non-zero coefficients of $h_1 r$ can be taken 1. Indeed, if $abA \neq 0$ then the change of basis $h_1' = \frac{2}{A}h_1$, $h_2' = \frac{2b}{aA}h_2$, $r' = \frac{2}{a}r$ makes all coefficients of $h_1 r$ equal 1, i.e. we obtain

$$h_1 r = r h_1 = h_1 + h_2 + r,$$

$$h_2 r = r h_2 = \frac{bc}{a}(h_1 + h_2 + r), \tag{5.86}$$

$$h_1^2 = h_2^2 = h_1 h_2 = r^2 = 0.$$

In case some a, b, A is equal 0 then one can choose a suitable change of basis to make non-zero coefficients of $h_1 r$ equal to 1. But we do not consider these particular cases here, because they are simpler than the case $abA \neq 0$.

Proposition 5.16. *Assume there is no zero in the first row of matrices (5.84) and (5.85). Then for the corresponding CEACP $\mathcal{E}^{[s,t]}$ we have $\mathcal{E}^{[s,t]} \cong \mathcal{E}_3$ for any (s,t).*

Proof. By the above-mentioned change of the basis the algebra $\mathcal{E}^{[s,t]}$ can be written as (5.86). In this case we have

$$\frac{bc}{a} = \begin{cases} \frac{\phi(t)-\psi(t)}{\phi(t)+\psi(t)}, & \text{if } c(t) \neq 0, \\ 0, & \text{if } c(t) = 0. \end{cases}$$

By the change

$$h_1' = \frac{\phi(t)+\psi(t)}{2\phi(t)}(h_1 + h_2),$$

$$h_2' = \frac{\psi(t)-\phi(t)}{2\phi(t)}h_1 + \frac{\psi(t)+\phi(t)}{2\phi(t)}h_2,$$

$$r' = \frac{\phi(t)+\psi(t)}{2\phi(t)}r$$

we get the algebra $\mathcal{E}_3$. $\qquad\square$

5.4 Flows of finite dimensional algebras

In this section following [140] and [141] define flow of algebras.

5.4.1   *Definitions and interpretations*

Define first quadratic stochastic processes, with continuous time, which are related with QSOs as well as Markov processes with linear operators [183], [237], [238].

Definition 5.10. A family $\left\{ P_{ij,k}^{[s,t]} : i, j, k \in E,\ s, t \in \mathbb{R}_+,\ t - s \geq 1 \right\}$ of functions $P_{ij,k}^{[s,t]}$ with initial state $x^{(0)} \in S^{m-1}$ is called a quadratic stochastic process (QSP) if, for fixed $s, t \in \mathbb{R}_+$, it satisfies the following conditions:

(i) $P_{ij,k}^{[s,t]} = P_{ji,k}^{[s,t]}$ for any $i, j, k \in E$;

(ii) $P_{ij,k}^{[s,t]} \geq 0$ and $\sum_{k \in E} P_{ij,k}^{[s,t]} = 1$ for any $i, j, k \in E$;

(iii) An analogue of the Kolmogorov–Chapman equation: there are two types: for the initial point $x^{(0)}$ and $s < r < t$ such that $t - r \geq 1$, $r - s \geq 1$ one has

(iii_A)

$$P_{ij,k}^{[s,t]} = \sum_{m,l \in E} P_{ij,m}^{[s,r]} P_{ml,k}^{[r,t]} x_l^{(r)},$$

where $x_k^{(r)}$ is defined as

$$x_k^{(r)} = \sum_{i,j \in E} P_{ij,k}^{[0,r]} x_i^{(0)} x_j^{(0)};$$

(iii_B)

$$P_{ij,k}^{[s,t]} = \sum_{m,l,g,h \in E} P_{im,l}^{[s,r]} P_{jg,h}^{[s,r]} P_{lh,k}^{[r,t]} x_m^{(s)} x_g^{(s)}.$$

The QSP is called of type (A) (resp. type (B)) if it satisfies the equation (iii_A) (resp. (iii_B)). The equations (iii_A) and (iii_B) can be interpreted as different laws of the behavior of the 'offspring'. Let us give some known examples of different types of QSPs.

Example 5.2. Let $E = \{1, 2\}$ and $(x, 1 - x)$ be an initial distribution on E, $0 \leq x \leq 1$. Consider the quadratic stochastic operator, $V \colon S^1 \to S^1$, defined as

$$x_1' = (1 - 2\epsilon)x_1^2 + (1 - 2\epsilon)x_1 x_2,$$

$$x_2' = 2\epsilon x_1^2 + (1 + 2\epsilon)x_1 x_2 + x_2^2$$

and the following system of transition probabilities:

$$P^{[s,t]}_{11,1} = \frac{(1-2\epsilon)^{t-s}}{2^{t-s-1}} \left[\left(2^{t-s-1} - 1\right) (1 - 2\epsilon)^s x + 1 \right];$$

$$P^{[s,t]}_{12,1} = P^{[s,t]}_{21,1} = \frac{(1-2\epsilon)^{t-s}}{2^{t-s-1}} \left[\left(2^{t-s-1} - 1\right) (1 - 2\epsilon)^s x + \tfrac{1}{2} \right];$$

$$P^{[s,t]}_{22,1} = \frac{(1-2\epsilon)^{t-s}}{2^{t-s-1}} \left(2^{t-s-1} - 1\right) x;$$

$$P^{[s,t]}_{ij,2} = 1 - P^{[s,t]}_{ij,1}, \qquad\qquad i,j = 1,2.$$

For $\epsilon \in [0, 1/2]$ these transition probabilities generate a QSP which is of type (A) and type (B), simultaneously. In this case we have

$$x_1^{(t)} = (1 - 2\epsilon)^t x, \qquad x_2^{(t)} = 1 - (1 - 2\epsilon)^t x.$$

Example 5.3. Let $E = \{1, 2\}$ and $(x, 1 - x)$ be an initial distribution on E, $0 \le x \le 1$. Consider the following system of transition functions:

$$P^{[s,t]}_{11,1} = \frac{\epsilon^{t-s}}{2^{t-s-1}} \cdot \frac{s+1}{t+1} \cdot \left[\left(2^{t-s-1} - 1\right) \cdot \frac{\epsilon^s}{s+1} x + 1 \right];$$

$$P^{[s,t]}_{12,1} = P^{[s,t]}_{21,1} = \frac{\epsilon^{t-s}}{2^{t-s-1}} \cdot \frac{s+1}{t+1} \cdot \left[\left(2^{t-s-1} - 1\right) \cdot \frac{\epsilon^s}{s+1} x + \tfrac{1}{2} \right];$$

$$P^{[s,t]}_{22,1} = \frac{2^{t-s-1}-1}{2^{t-s-1}} \cdot \frac{\epsilon^t}{t+1} x;$$

$$P^{[s,t]}_{ij,2} = 1 - P^{[s,t]}_{ij,1}, \qquad\qquad i,j = 1,2.$$

These functions generate a QSP of type (A) and (B) for any $\epsilon \in [0, 1]$. Moreover,

$$x_1^{(t)} = \frac{\epsilon^t}{t+1} x, \qquad x_2^{(t)} = 1 - \frac{\epsilon^t}{t+1} x, \qquad t \ge 1.$$

Example 5.4. Let $E = \{1, 2, 3\}$ and $(x_1, x_2, 1 - x_1 - x_2)$ be an initial distribution on E, where $x_1 \ge 0$, $x_2 \ge 0$, $x_1 + x_2 \le 1$. Consider the following system of transition probabilities:

$$P^{[s,t]}_{11,1} = 2\epsilon^{t-s} + \frac{2^{t-s-1}-1}{2^{t-s-1}} x_1^{(t+1)};$$

$$P^{[s,t]}_{12,1} = P^{[s,t]}_{13,1} = \epsilon^{t-s} + \frac{2^{t-s-1}-1}{2^{t-s-1}} x_1^{(t+1)};$$

$$P^{[s,t]}_{22,1} = P^{[s,t]}_{23,1} = P^{[s,t]}_{33,1} = \frac{2^{t-s-1}-1}{2^{t-s-1}} x_1^{(t+1)};$$

$$P^{[s,t]}_{11,2} = P^{[s,t]}_{13,2} = P^{[s,t]}_{33,2} = \frac{2^{t-s-1}-1}{2^{t-s-1}} x_2^{(t+1)};$$

$$P^{[s,t]}_{22,2} = 2\epsilon^{t-s} + \frac{2^{t-s-1}-1}{2^{t-s-1}} x_2^{(t+1)};$$

$$P^{[s,t]}_{12,2} = P^{[s,t]}_{23,2} = \epsilon^{t-s} + \frac{2^{t-s-1}-1}{2^{t-s-1}} x_2^{(t+1)};$$

$$P^{[s,t]}_{ij,3} = 1 - P^{[s,t]}_{ij,1} - P^{[s,t]}_{ij,2}, \qquad\qquad i,j = 1, 2, 3,$$

where $x_1^{(t)} = (2\epsilon)^t x_1$, $x_2^{(t)} = (2\epsilon)^t x_2$. For $\epsilon \in [0, 1/2]$, these transition probabilities generate a QSP which is of type (A), but is not of type (B).

Consider a family $\left\{\mathcal{A}^{[s,t]} : s,t \in \mathbb{R},\ 0 \le s \le t\right\}$ of arbitrary m-dimensional algebras over the field $\mathbb{R}$, with basis $e_1, \ldots, e_m$ and multiplication table

$$e_i e_j = \sum_{k=1}^{m} c_{ijk}^{[s,t]} e_k, \quad i,j = 1, \ldots, m. \tag{5.87}$$

Here parameters s, t are considered as time.

Denote by $\mathcal{M}^{[s,t]} = \left(c_{ijk}^{[s,t]}\right)_{i,j,k=1,\ldots,m}$ the matrix of structural constants of $\mathcal{A}^{[s,t]}$.

Definition 5.11. Define the following flow of algebras (FA):

- A family $\left\{\mathcal{A}^{[s,t]} : s,t \in \mathbb{R},\ 0 \le s \le t\right\}$ of m-dimensional algebras over the field $\mathbb{R}$ is called a stochastic flow of algebras (SFA) if the matrices $\mathcal{M}^{[s,t]}$ of structural constants satisfy the conditions of Definition 5.10. The SFA is called of type (A) (resp. type (B)) if it corresponds to a QSP of type (A) (resp. (B)).
- A family $\left\{\mathcal{A}^{[s,t]} : s,t \in \mathbb{R},\ 0 \le s \le t\right\}$ of m-dimensional algebras over the field $\mathbb{R}$ is called an FA if the matrices $\mathcal{M}^{[s,t]}$ of structural constants satisfy the Kolmogorov–Chapman equation (for cubic matrices):

$$\mathcal{M}^{[s,t]} = \mathcal{M}^{[s,\tau]} \mathcal{M}^{[\tau,t]}, \qquad \text{for all } 0 \le s < \tau < t. \tag{5.88}$$

The FA is called of type (C) (resp. type (D), (E)) if the equation (5.88) is considered with respect to multiplication of type (C) (resp. (D), (E)) mentioned above.

Remark 5.8. Note that

- SFA corresponds to a family of stochastic matrices (i.e. satisfy condition (ii) of Definition 5.10). But in general for FAs we do not assume that the corresponding matrices are stochastic, they only satisfy (5.88).
- Since there are several different kinds of multiplications of cubic matrices, Definition 5.11 gives several types of FA. To construct an FA one has to fix a multiplication rule first and then consider equation (5.88) with respect to the fixed multiplication.

Remark 5.9. Interpretations: The multiplication $e_i e_j$ given by the equality (5.87) can be considered as action of the particle e_i on the particle e_j at time s, as an interaction process, then with rate $c_{ijk}^{[s,t]}$ a particle e_k appears at time t. Kolmogorov–Chapman equation (5.88) gives time depending evolution law of the interacting process (dynamical system).

Comparison with the deformations of algebras: In case of finite-dimensional algebras a deformation is the transformation of a given algebra to another algebra (with the same dimension) [78]. This is equivalent to change (transform) a given multiplication table (structural constants) to another one. Often, the algebra at $t = 0$ is considered as an initial algebra and the algebra at the current time t as the current algebra. In [78] deformations of an algebra A are given by a bilinear function f_t, i.e., one considers the algebra A_t with multiplication f_t as the generic element of a "one-parameter family of deformations of A". Thus deformations of an algebra are given by the rule f_t (which has an explicit form), similarly a flow of algebras is given by $\mathcal{M}^{[s,t]}$ with the rule (5.88). Some relations of an FA can be also seen in the following example: a plane deformation [194] is restricted to the plane described by the basis vectors e_1, e_2. It is known that the deformation gradient of this plane deformation has the form

$$
\mathbf{F} = \begin{pmatrix} \cos\theta & \sin\theta & 0 \\ -\sin\theta & \cos\theta & 0 \\ 0 & 0 & 1 \end{pmatrix} \begin{pmatrix} \lambda_1 & 0 & 0 \\ 0 & \lambda_2 & 0 \\ 0 & 0 & 1 \end{pmatrix},
$$

where θ is the angle of rotation and λ_1, λ_2 are the principal stretches. Comparing this matrix with matrix (5.92) (see below) one can see that the corresponding FA and the plane deformation have similar matrices.

5.4.2 *Examples of FAs*

Now we shall give several concrete examples of FAs.

Note that all above-mentioned examples of QSP generate stochastic flow of algebras (SFAs). The following is another example:

Example 5.5. Let $E = \{1, 2, \ldots, m\}$. Consider a family of m-dimensional stochastic vectors: $a(t) = (a_1(t), a_2(t), \ldots, a_m(t))$, i.e. $a_i(t) \geq 0$, $\sum_i a_i(t) = 1$ for any $t \geq 0$. For each pair s, t define a stochastic matrix $Q^{[s,t]} = (q_{il}^{[s,t]})_{i,l \in E}$, where $q_{il}^{[s,t]} = a_l(t)$ for all $i \in E$, i.e. it does not depend on s. One can see that this matrix satisfies the Kolmogorov–Chapman equation:

$$
Q^{[s,t]} = Q^{[s,\tau]} Q^{[\tau,t]}, \qquad \text{for all } 0 \leq s < \tau < t.
$$

Now define functions

$$
P_{ij,k}^{[s,t]} = q_{ik}^{[s,t]} = a_k(t).
$$

One can check that the defined family $\{P_{ij,k}^{[s,t]}\}$ is a QSP. Thus it generates a SFA.

Define the following (Maksimov's) multiplications for cubic matrices (see (2.2) for general multiplications):

Type (C):

$$E_{ijk} \cdot E_{lnr} = \begin{cases} E_{ijr}, & \text{if } k = l, \text{ and } j = n, \\ 0, & \text{otherwise.} \end{cases}$$

Extending this multiplication to arbitrary cubic matrices

$$A = (a_{ijk})_{i,j,k=1}^m, \quad B = (b_{ijk})_{i,j,k=1}^m, \quad C = (c_{ijk})_{i,j,k=1}^m,$$

we get that the entries of $C = AB$ can be written as

$$c_{ijr} = \sum_{k=1}^m a_{ijk} b_{kjr}. \tag{5.89}$$

Type (D):

$$E_{ijk} \cdot E_{lnr} = \begin{cases} E_{ijr}, & \text{if } k = l, \\ 0, & \text{otherwise.} \end{cases}$$

In this case

$$c_{ijr} = \sum_{k,n=1}^m a_{ijk} b_{knr}. \tag{5.90}$$

Type (E):

$$E_{ijk} \cdot E_{lnr} = \begin{cases} E_{inr}, & \text{if } k = l, \\ 0, & \text{otherwise.} \end{cases}$$

Consequently

$$c_{inr} = \sum_{j,k=1}^m a_{ijk} b_{knr}.$$

Now we take $m = 2$, i.e. $E = \{1, 2\}$ and construct examples of a flow of two-dimensional algebras of type (C). In this case equation (5.88) has the following form:

$$c_{ijr}^{[s,t]} = c_{ij1}^{[s,\tau]} c_{1jr}^{[\tau,t]} + c_{ij2}^{[s,\tau]} c_{2jr}^{[\tau,t]}, \qquad i, j, r = 1, 2.$$

This is a quadratic system of eight equations with eight unknown functions $c_{ijr}^{[s,t]}$ of two variables s, t, $0 \le s < t$. One can see that the equations for $j = 1$ and $j = 2$ are independent. Therefore it suffices to solve the system only for $j = 1$. Denoting $a_{ir}^{[s,t]} = c_{i1r}^{[s,t]}$ we get

$$
\begin{aligned}
a_{11}^{[s,t]} &= a_{11}^{[s,\tau]} a_{11}^{[\tau,t]} + a_{12}^{[s,\tau]} a_{21}^{[\tau,t]}, \\
a_{12}^{[s,t]} &= a_{11}^{[s,\tau]} a_{12}^{[\tau,t]} + a_{12}^{[s,\tau]} a_{22}^{[\tau,t]}, \\
a_{21}^{[s,t]} &= a_{21}^{[s,\tau]} a_{11}^{[\tau,t]} + a_{22}^{[s,\tau]} a_{21}^{[\tau,t]}, \\
a_{22}^{[s,t]} &= a_{21}^{[s,\tau]} a_{12}^{[\tau,t]} + a_{22}^{[s,\tau]} a_{22}^{[\tau,t]}.
\end{aligned}
\tag{5.91}
$$

The full set of solutions to the system (5.91) is not known yet. But there is a very wide class of its solutions see Section 5.1.2 and [31, 197, 218]. One of these known solutions is the following (non-stochastic, time-homogeneous) matrix:

$$
\begin{pmatrix} a_{11}^{[s,t]} & a_{12}^{[s,t]} \\ a_{21}^{[s,t]} & a_{22}^{[s,t]} \end{pmatrix} = \begin{pmatrix} \cos(t-s) & \sin(t-s) \\ -\sin(t-s) & \cos(t-s) \end{pmatrix}.
\tag{5.92}
$$

Thus the following is a wide class of examples:

Example 5.6. Any pair $(a_{ij}^{[s,t]}), (\bar{a}_{ij}^{[s,t]})$ of solutions of the system (5.91) generates an FA of type (C) corresponding to the matrix $\mathcal{M}^{[s,t]}$ with entries

$$
c_{i1r}^{[s,t]} = a_{ir}^{[s,t]}, \qquad c_{i2r}^{[s,t]} = \bar{a}_{ir}^{[s,t]}.
$$

In the next section we shall give general arguments to solve the equation (5.88) and after that we shall give examples of FAs of type (D) and (E).

Note that for each fixed s, t the matrix $\mathcal{M}^{[s,t]}$ of a SFA defines an algebra $\mathcal{A}^{[s,t]}$ over $\mathbb{R}$. This algebra can be considered as an *evolution algebra of a free population* (see page 15 and Chapter 3 in [155]), but Lyubich used for this algebra the notion *evolutionary algebra*.

In [155] for the evolution algebra of a free population it is proven that there is a character $\chi(x) = \sum_i x_i$, therefore that algebra is baric.

The following proposition is known (see [269]):

Proposition 5.17. *A finite-dimensional commutative real algebra A is a baric algebra if and only if A has a basis $e_1, \ldots, e_m$ such that the constants c_{ijk} defined by*

$$
e_i e_j = \sum_{k=1}^{m} c_{ijk} e_k
$$

satisfy the relation

$$\sum_{k=1}^{m} c_{ijk} = 1.$$

In the next sections we shall study the dynamics (depending on time) of properties of algebras in FAs.

5.4.3   *Reduction of Kolmogorov–Chapman's equations from cubic matrices to square ones*

In this section for each multiplication (C), (D), (E) we show that equation (5.88) written for cubic matrices (not necessarily stochastic) can be reduced to similar equations for square matrices (different for different multiplications).

Case (C). Let $\mathcal{M}_j^{[s,t]} = (c_{ijk}^{[s,t]})_{i,k=1}^{m}$ be the jth layer of the matrix $\mathcal{M}^{[s,t]}$. The following proposition is very useful:

Proposition 5.18. *Any solution of the equation (5.88) for multiplication of type (C) is a direct sum of solutions of the following m independent equations:*

$$\mathcal{M}_j^{[s,t]} = \mathcal{M}_j^{[s,\tau]} \mathcal{M}_j^{[\tau,t]}, \quad \text{for all} \ \ 0 \le s < \tau < t, \quad j = 1, \ldots, m.$$

Proof. This is consequence of the equality (5.89), which shows that layers are independent in the multiplication of type (C).   □

Case (D). For the cubic matrix $\mathcal{M}^{[s,t]}$ consider the square matrix $\overline{\mathcal{M}}^{[s,t]} = (\bar{c}_{ik}^{[s,t]})$ with

$$\bar{c}_{ik}^{[s,t]} = \sum_{j=1}^{m} c_{ijk}^{[s,t]}, \qquad i, k = 1, \ldots, m. \tag{5.93}$$

Proposition 5.19. *Any solution of equation (5.88) for the multiplication of type (D) can be given by a solution of the system (5.93) with a matrix* $\overline{\mathcal{M}}^{[s,t]} = (\bar{c}_{ik}^{[s,t]})$ *which satisfies*

$$\overline{\mathcal{M}}^{[s,t]} = \overline{\mathcal{M}}^{[s,\tau]} \, \overline{\mathcal{M}}^{[\tau,t]}, \quad \text{for all} \ \ 0 \le s < \tau < t. \tag{5.94}$$

Proof. Using the multiplication rule (5.90), we write the equation (5.88) in the following form:

$$c_{ijr}^{[s,t]} = \sum_{k=1}^{m} \left(c_{ijk}^{[s,\tau]} \sum_{n=1}^{m} c_{knr}^{[\tau,t]} \right) = \sum_{k=1}^{m} c_{ijk}^{[s,\tau]} \bar{c}_{kr}^{[\tau,t]}. \tag{5.95}$$

Taking sum over j, from the last equality we get

$$\bar{c}_{ir}^{[s,t]} = \sum_{k=1}^{m} \bar{c}_{ik}^{[s,\tau]}\bar{c}_{kr}^{[\tau,t]},$$

this is the equality (5.94). $\qquad\square$

To illustrate Proposition 5.19 we give the following example:

Example 5.7. Take $m = 2$ and an arbitrary function $\Psi(t)$, $t \geq 0$, with $\Psi(t) \neq 0$. Consider the following square matrix:

$$\overline{\mathcal{M}}^{[s,t]} = \frac{1}{2}\begin{pmatrix} \frac{\Psi(t)}{\Psi(s)} & -\frac{\Psi(t)}{\Psi(s)} \\ -\frac{\Psi(t)}{\Psi(s)} & \frac{\Psi(t)}{\Psi(s)} \end{pmatrix}. \tag{5.96}$$

One can check that this matrix satisfies the equality (5.94). Write cubic matrix $\mathcal{M}^{[s,t]}$ for $m = 2$ in the following convenient form:

$$\mathcal{M}^{[s,t]} = \begin{pmatrix} c_{111}^{[s,t]} \ c_{112}^{[s,t]} & c_{211}^{[s,t]} \ c_{212}^{[s,t]} \\ c_{121}^{[s,t]} \ c_{122}^{[s,t]} & c_{221}^{[s,t]} \ c_{222}^{[s,t]} \end{pmatrix}. \tag{5.97}$$

Using Proposition 5.19 we shall construct $\mathcal{M}^{[s,t]}$ which corresponds to the taken matrix (5.96): from (5.88) (i.e. (5.95)) we get

$$c_{ij1}^{[s,t]} = \frac{\Psi(t)}{2\Psi(\tau)}c_{ij1}^{[s,\tau]} - \frac{\Psi(t)}{2\Psi(\tau)}c_{ij2}^{[s,\tau]}, \qquad i,j = 1,2,$$

$$c_{ij2}^{[s,t]} = -\frac{\Psi(t)}{2\Psi(\tau)}c_{ij1}^{[s,\tau]} + \frac{\Psi(t)}{2\Psi(\tau)}c_{ij2}^{[s,\tau]}, \qquad i,j = 1,2.$$

Consequently $c_{ij1}^{[s,t]} = -c_{ij2}^{[s,t]}$. Therefore

$$c_{ij1}^{[s,t]} = \frac{\Psi(t)}{\Psi(\tau)}c_{ij1}^{[s,\tau]} \Leftrightarrow \frac{c_{ij1}^{[s,t]}}{\Psi(t)} = \frac{c_{ij1}^{[s,\tau]}}{\Psi(\tau)}.$$

It follows from the last equality that $\frac{c_{ij1}^{[s,t]}}{\Psi(t)}$ does not depend on t, i.e. there exists a function $\kappa_{ij}(s)$ such that $\frac{c_{ij1}^{[s,t]}}{\Psi(t)} = \kappa_{ij}(s)$. Thus

$$c_{ij1}^{[s,t]} = \kappa_{ij}(s)\Psi(t), \qquad c_{ij2}^{[s,t]} = -\kappa_{ij}(s)\Psi(t). \tag{5.98}$$

By (5.93) for (5.96) we should have

$$c_{111}^{[s,t]} + c_{121}^{[s,t]} = -(c_{112}^{[s,t]} + c_{122}^{[s,t]}) = -(c_{211}^{[s,t]} + c_{221}^{[s,t]}) = c_{212}^{[s,t]} + c_{222}^{[s,t]} = \frac{\Psi(t)}{2\Psi(s)}.$$

By using (5.98), we get from these equalities that

$$\kappa_{11}(s) + \kappa_{12}(s) = -(\kappa_{21}(s) + \kappa_{22}(s)) = \frac{1}{2\Psi(s)}.$$

Consequently the matrix (5.97) has the following form:

$$\mathcal{M}^{[s,t]} = \Psi(t) \begin{pmatrix} \kappa_{11}(s) & -\kappa_{11}(s) \\ \frac{1}{2\Psi(s)} - \kappa_{11}(s) & \kappa_{11}(s) - \frac{1}{2\Psi(s)} \end{pmatrix}$$
$$\left. \begin{matrix} \kappa_{21}(s) & -\kappa_{21}(s) \\ -\frac{1}{2\Psi(s)} - \kappa_{21}(s) & \kappa_{21}(s) + \frac{1}{2\Psi(s)} \end{matrix} \right), \qquad (5.99)$$

where $\Psi \neq 0$, κ_{11} and κ_{21} are arbitrary functions of $t \geq 0$. This is a $2 \times 2 \times 2$ cubic matrix which satisfies (5.88) and has three parameter functions. Thus we proved the following:

Proposition 5.20. *For any fixed functions* $\Psi \neq 0$, κ_{11} *and* κ_{21} *the matrix* (5.99) *generates an FA of type (D).*

Now we shall construct a periodic FA.

Example 5.8. Again consider $m = 2$ and using Proposition 5.19 we shall construct $\mathcal{M}^{[s,t]}$ corresponding to the matrix (5.92): from (5.88) we get

$$c_{ij1}^{[s,t]} = \cos(t - \tau)c_{ij1}^{[s,\tau]} - \sin(t - \tau)c_{ij2}^{[s,\tau]}, \qquad i,j = 1,2,$$

$$c_{ij2}^{[s,t]} = \sin(t - \tau)c_{ij1}^{[s,\tau]} + \cos(t - \tau)c_{ij2}^{[s,\tau]}, \qquad i,j = 1,2.$$

One can check that this system has a solution in the form

$$c_{ij1}^{[s,t]} = A_{ij}\cos(t - s) - B_{ij}\sin(t - s),$$
$$\qquad (5.100)$$
$$c_{ij2}^{[s,t]} = B_{ij}\cos(t - s) + A_{ij}\sin(t - s),$$

where A_{ij} and B_{ij} are arbitrary numbers.

By (5.93) for (5.92) we should have

$$c_{111}^{[s,t]} + c_{121}^{[s,t]} = \cos(t - s), \qquad c_{112}^{[s,t]} + c_{122}^{[s,t]} = \sin(t - s),$$

$$c_{211}^{[s,t]} + c_{221}^{[s,t]} = -\sin(t - s), \qquad c_{212}^{[s,t]} + c_{222}^{[s,t]} = \cos(t - s).$$

By using (5.100), we get from these equalities that

$$(A_{11} + A_{12})\cos(t - s) - (B_{11} + B_{12})\sin(t - s) = \cos(t - s),$$

$$(B_{11} + B_{12})\cos(t - s) + (A_{11} + A_{12})\sin(t - s) = \sin(t - s),$$

$$(A_{21} + A_{22})\cos(t - s) - (B_{21} + B_{22})\sin(t - s) = -\sin(t - s),$$

$$(B_{21} + B_{22})\cos(t - s) + (A_{21} + A_{22})\sin(t - s) = \cos(t - s).$$

Consequently

$$A_{12} = 1 - A_{11}, \quad B_{12} = -B_{11}, \quad A_{22} = -A_{21}, \quad B_{22} = 1 - B_{21}.$$

Now denoting $a = A_{11}$, $b = B_{11}$, $c = A_{21}$, $d = B_{21}$ and $\hat{t} = t - s$, we get the cubic matrix (5.97) as

$$\mathcal{M}^{[s,t]} = \mathcal{M}^{[\hat{t}]} = \left(\begin{array}{cc} a\cos\hat{t} - b\sin\hat{t} & b\cos\hat{t} + a\sin\hat{t} \\ (1-a)\cos\hat{t} + b\sin\hat{t} & -b\cos\hat{t} + (1-a)\sin\hat{t} \\ c\cos\hat{t} - d\sin\hat{t} & d\cos\hat{t} + c\sin\hat{t} \\ -c\cos\hat{t} - (1-d)\sin\hat{t} & (1-d)\cos\hat{t} - c\sin\hat{t} \end{array}\right), \quad (5.101)$$

this is a $2 \times 2 \times 2$ cubic matrix which satisfies (5.88) and has four arbitrary parameters $a, b, c, d \in \mathbb{R}$. Thus we proved the following:

Proposition 5.21. *For any fixed parameters $a, b, c, d \in \mathbb{R}$ the matrix (5.101) generates a homogeneous, periodic FA of type (D).*

A m-dimensional time non-homogeneous FA of type (D). For arbitrary m we shall give an example of time non-homogeneous FA of type (D), i.e. we shall proof the following:

Theorem 5.15. *Let $\{A^{[t]} = (a_{ij}^{[t]}),\ t \geq 0\}$ be a family of invertible (for all t), $m \times m$ square matrices[2] and let $(A^{[t]})^{-1} = (b_{ij}^{[t]})$ denote the inverse of $A^{[t]}$. Assume that $\beta_{ijk}^{(s)}$, $i, j, k - 1, \ldots, m$, are arbitrary functions such that*

$$\sum_{j=1}^{m} \beta_{ijk}^{(s)} = a_{ik}^{[s]}, \qquad \textit{for any } i, k \textit{ and } s.$$

Then cubic matrix

$$\mathcal{M}^{[s,t]} = \left(\sum_{k=1}^{m} \beta_{ijk}^{(s)} b_{kr}^{[t]}\right)_{i,j,r=1}^{m}$$

generates an FA of type (D).

Proof. Define the following matrix:

$$\overline{\mathcal{M}}^{[s,t]} = A^{[s]}(A^{[t]})^{-1}.$$

---

[2] Construction of a family of invertible $m \times m$ matrices $A^{[t]}$ is not difficult, for example, one can take a lower or an upper triangular matrix with non-zero entries in the diagonal. Thus this theorem describes a very rich class of FAs.

We know from previous chapters that this matrix satisfies (5.94). We shall use it to construct a cubic matrix which satisfies (5.88). For each i, j define an m-dimensional vector

$$\alpha_{ij}^{[s,t]} = \left(c_{ij1}^{[s,t]}, \dots, c_{ijm}^{[s,t]}\right).$$

By multiplication rule (D) we have from (5.88) that

$$\alpha_{ij}^{[s,t]} = \alpha_{ij}^{[s,\tau]} A^{[\tau]} (A^{[t]})^{-1}.$$

This can be rewritten as

$$\alpha_{ij}^{[s,t]} A^{[t]} = \alpha_{ij}^{[s,\tau]} A^{[\tau]}.$$

It follows from the last equality that the vector $\alpha_{ij}^{[s,t]} A^{[t]}$ should not depend on t, i.e. there is a vector $\beta_{ij}^{(s)} = (\beta_{ijk}^{(s)})_{k=1}^m$ such that $\alpha_{ij}^{[s,t]} A^{[t]} = \beta_{ij}^{(s)}$. Then

$$\alpha_{ij}^{[s,t]} = \beta_{ij}^{(s)} (A^{[t]})^{-1},$$

i.e.

$$c_{ijr}^{[s,t]} = \sum_{k=1}^m \beta_{ijk}^{(s)} b_{kr}^{[t]}.$$

By (5.93) we should have

$$\sum_{j=1}^m c_{ijr}^{[s,t]} = \sum_{k=1}^m \sum_{j=1}^m \beta_{ijk}^{(s)} b_{kr}^{[t]} = (A^{[s]}(A^{[t]})^{-1})_{ir} = \sum_{k=1}^m a_{ik}^{[s]} b_{kr}^{[t]},$$

i.e.,

$$\sum_{k=1}^m b_{kr}^{[t]} \left(\sum_{j=1}^m \beta_{ijk}^{(s)} - a_{ik}^{[s]} \right) = 0,$$

the last equality follows from the condition of theorem. $\qquad\square$

Case (E). Comparing (D) and (E) one can see that these multiplications are very similar: in case (D) the right hand side of the equation (5.88) has sum over two indexes of the matrix $\mathcal{M}^{[\tau,t]}$, but in the case (E) such sum is with respect to matrix $\mathcal{M}^{[s,\tau]}$. With respect to the time these multiplications interchange the role of s and t, i.e. multiplication (D) gives a forward flow, but multiplication (E) corresponds to a backward flow. Therefore one can show that Proposition 5.19 is true in the case (E) too. Here we shall give an example of an FA of type (E), which corresponds to the matrix (5.96).

Example 5.9. We shall construct a cubic matrix $\mathcal{M}^{[s,t]}$ for $m = 2$ which satisfies (5.88) in the case of the multiplication (E). Find such a matrix $\mathcal{M}^{[s,t]}$ corresponding to the matrix (5.96): from (5.88) we get (compare with case (D) to see the interchange of s and t)

$$c_{1ij}^{[s,t]} = \frac{\Psi(\tau)}{2\Psi(s)} c_{1ij}^{[\tau,t]} - \frac{\Psi(\tau)}{2\Psi(s)} c_{2ij}^{[\tau,t]}, \quad i,j = 1,2,$$

$$c_{2ij}^{[s,t]} = -\frac{\Psi(\tau)}{2\Psi(s)} c_{1ij}^{[\tau,t]} + \frac{\Psi(\tau)}{2\Psi(s)} c_{2ij}^{[\tau,t]}, \quad i,j = 1,2.$$

Consequently $c_{1ij}^{[s,t]} = -c_{2ij}^{[s,t]}$. Therefore

$$c_{1ij}^{[s,t]} = \frac{\Psi(\tau)}{\Psi(s)} c_{1ij}^{[\tau,t]} \Leftrightarrow \Psi(s) c_{1ij}^{[s,t]} = \Psi(\tau) c_{1ij}^{[\tau,t]}.$$

Hence $\Psi(s) c_{1ij}^{[s,t]}$ does not depend on s, i.e. there exists a function $\gamma_{ij}(t)$ such that $\Psi(s) c_{1ij}^{[s,t]} = \gamma_{ij}(t)$. Thus

$$c_{1ij}^{[s,t]} = \frac{\gamma_{ij}(t)}{\Psi(s)}, \qquad c_{2ij}^{[s,t]} = -\frac{\gamma_{ij}(t)}{\Psi(s)}. \tag{5.102}$$

We should have

$$c_{111}^{[s,t]} + c_{121}^{[s,t]} = -(c_{112}^{[s,t]} + c_{122}^{[s,t]}) = -(c_{211}^{[s,t]} + c_{221}^{[s,t]}) = c_{212}^{[s,t]} + c_{222}^{[s,t]} = \frac{\Psi(t)}{2\Psi(s)}.$$

By using (5.102), we get from this equalities that

$$\gamma_{11}(t) + \gamma_{21}(t) = -(\gamma_{12}(t) + \gamma_{22}(t)) = \frac{\Psi(t)}{2}.$$

Now the matrix (5.97) has the following form:

$$\mathcal{M}^{[s,t]} = \frac{1}{\Psi(s)} \begin{pmatrix} \gamma_{11}(t) & \gamma_{12}(t) & -\gamma_{11}(t) & -\gamma_{12}(t) \\ \frac{\Psi(t)}{2} - \gamma_{11}(t) & -\gamma_{12}(t) - \frac{\Psi(t)}{2} & -\frac{\Psi(t)}{2} + \gamma_{11}(t) & \gamma_{12}(t) + \frac{\Psi(t)}{2} \end{pmatrix}, \tag{5.103}$$

where $\Psi \neq 0$, γ_{11} and γ_{12} are arbitrary functions. This is a $2 \times 2 \times 2$ cubic matrix which satisfies (5.88) and has three parameter functions. Thus we proved the following:

Proposition 5.22. *For any fixed* $\Psi \neq 0$, γ_{11} *and* γ_{12} *the matrix* (5.103) *generates an FA of type (E).*

The following theorem is an analogue of Theorem 5.15 for an FA of type (E):

Theorem 5.16. *Let $\{A^{[t]} = (a_{ij}^{[t]}),\ t \geq 0\}$ be a family of invertible (for all t), $m \times m$ square matrices and let $(A^{[t]})^{-1} = (b_{ij}^{[t]})$ denote the inverse of $A^{[t]}$. Assume that $\gamma_{ijk}^{(t)},\ i, j, k = 1, \dots, m$, are arbitrary functions such that*

$$\sum_{j=1}^{m} \gamma_{ijk}^{(t)} = b_{ik}^{[t]}, \qquad for\ any\ \ i, k\ \ and\ \ t.$$

Then the cubic matrix

$$\mathcal{M}^{[s,t]} = \left(\sum_{k=1}^{m} a_{ik}^{[s]} \gamma_{kjr}^{(t)} \right)_{i,j,r=1}^{m}$$

generates an FA of type (E).

Proof. The proof is similar to the proof of Theorem 5.15. Here instead of $\alpha_{ij}^{[s,t]}$ one considers

$$v_{jr}^{[s,t]} = \left(c_{1jr}^{[s,t]}, \dots, c_{mjr}^{[s,t]} \right)^{T}.$$

By the multiplication rule (E) we have from (5.88) that

$$v_{jr}^{[s,t]} = A^{[s]} (A^{[\tau]})^{-1} v_{jr}^{[\tau,t]}.$$

Now repeating similar calculations in the proof of Theorem 5.15, we obtain the result. $\qquad\square$

5.4.4 *Constructions and time dynamics of FAs*

Let $\{\mathcal{A}^{[s,t]} : s, t \in \mathbb{R},\ 0 \leq s \leq t\}$ be a family of arbitrary m-dimensional algebras over the field F, with basis $e_1, \dots, e_m$ and multiplication table

$$e_i e_j = \sum_{k=1}^{m} c_{ijk}^{[s,t]} e_k, \quad i, j = 1, \dots, m.$$

Denote by $\mathcal{M}^{[s,t]} = \left(c_{ijk}^{[s,t]} \right)_{i,j,k=1}^{m}$ the matrix of structural constants of $\mathcal{A}^{[s,t]}$.

Fix an arbitrary multiplication (not necessarily the Maksimov's multiplication) of cubic matrices, say $*_\mu$.

Recall that a family $\{\mathcal{A}^{[s,t]} : s,t \in \mathbb{R},\ 0 \leq s \leq t\}$ of m-dimensional algebras over the field F is called an FA of type μ if the matrices $\mathcal{M}^{[s,t]}$ of structural constants satisfy the Kolmogorov–Chapman equation (for cubic matrices):

$$\mathcal{M}^{[s,t]} = \mathcal{M}^{[s,\tau]} *_\mu \mathcal{M}^{[\tau,t]}, \qquad \text{for all } 0 \leq s < \tau < t. \tag{5.104}$$

An FA is called a discrete-time FA if the matrix $\mathcal{M}^{[s,t]}$ depends only on $t, s \in \mathbb{N}$. In this case we write $\mathcal{M}^{[n,m]}$, $n, m \in \mathbb{N}$.

Now following [141] we construct FAs, i.e. find solutions to the equation (5.104).

5.4.4.1 *Power-associative multiplications*

Recall that an algebra is called *power-associative* if for each element a the subalgebra generated by a is associative, that is $a^n a^m = a^{n+m}$, for all nonnegative integers n, m.

Fix an arbitrary multiplication rule between cubic matrices, say $*_\mu$. For a cubic matrix Q denote

$$Q^{*_\mu n} = \underbrace{Q *_\mu Q *_\mu \cdots *_\mu Q}_{n \ \text{times}}, \qquad n = 1, 2, \ldots.$$

Condition 1. Assume that the algebra $\mathfrak{C}_\mu$ of cubic matrices is power-associative.

Proposition 5.23.

1. *If Condition 1 is satisfied for the multiplication $*_\mu$ then $\mathcal{M}^{[n,m]} = Q^{*_\mu(m-n)}$ generates a discrete time FA of type μ.*
2. *If ACM $\mathfrak{C}_\mu$ has an idempotent, i.e. there exists $\mathcal{I} \in \mathfrak{C}_\mu$ such that $\mathcal{I} *_\mu \mathcal{I} = \mathcal{I}$ then $\mathcal{M}^{[s,t]} = \mathcal{I}$, for all $0 \leq s < t$, generates an FA of type μ.*

Proof. 1. Using power-associativity of the ACM for any natural numbers $n < k < m$ we get

$$Q^{*_\mu(m-n)} = Q^{*_\mu(k-n)} *_\mu Q^{*_\mu(m-k)}.$$

Thus $\mathcal{M}^{[n,m]} = Q^{*_\mu(m-n)}$ is a solution to (5.88) and therefore it generates a discrete-time FA of type μ, denote it by $\mathcal{A}_1^{[n,m]}$.

2. Straightforward. Denote the corresponding FA by $\mathcal{A}_2^{[s,t]}$. $\qquad\square$

Remarks on time dynamics. The time dynamics of $\mathcal{A}_1^{[n,m]}$ depends on the fixed matrix Q. This FA is a time homogeneous, and periodic iff the powers (with respect to multiplication μ) of the cubic matrix Q have periods, i.e. if there is $p \in \mathbb{N}$ such that $Q^{*\mu p} = Q$; if Q is such that $\lim_{n \to \infty} Q^{*\mu n} = \tilde{Q}$ exists then the FA has a limit algebra:

$$\tilde{\mathcal{A}}_1 = \lim_{m-n \to \infty} \mathcal{A}_1^{[n,m]}.$$

The time dynamics of $\mathcal{A}_2^{[s,t]}$ is trivial: it does not depend on time.

5.4.4.2 *Exponential solutions*

In the theory of continuous-time Markov chains, the matrix exponent of square matrices plays a crucial role (see e.g. Chapter 3 of [22], Chapter 2 [251]). Here we adapt this notion of matrix exponent for cubic matrices.

Recall that an algebra is *unital* if it has an element u with $ux = x = xu$ for all x in the algebra. The element u is called *unit* element.

Condition 2. Assume $*_\mu$ is such that the corresponding ACM, $\mathfrak{C}_\mu = (\mathfrak{C}, *_\mu)$ is unital, i.e. it has a unit matrix denoted by $\mathbb{I}$.

In the unital ACM $\mathfrak{C}_\mu$ for each cubic matrix Q define $Q^{*\mu 0} = \mathbb{I}$.
A *matrix exponent* $\exp_\mu(tQ)$ is defined by

$$\exp_\mu(tQ) = \mathbb{I} + \sum_{n \geq 1} \frac{(tQ)^{*\mu n}}{n!} = \sum_{n \geq 0} \frac{(tQ)^{*\mu n}}{n!}, \qquad (5.105)$$

i.e.

$$\left(\exp_\mu(tQ) \right)_{ijk} = \sum_{n \geq 0} \frac{t^n (Q^{*\mu n})_{ijk}}{n!}.$$

For a cubic matrix Q, the parameter t in (5.105) can be any real number. We consider the case $t \geq 0$.

Condition 3. Assume that $\mathfrak{C}_\mu$ is a normed space with norm $\|\cdot\|$ such that for any two cubic matrices A, B, we have

$$\|A *_\mu B\| \leq \|A\| \|B\|. \qquad (5.106)$$

Condition 4. Assume $\mathfrak{C}_\mu$ is an associative algebra.

Proposition 5.24. *Assume Conditions 1–4 are satisfied. Let Q be a cubic matrix. Then:*

(a) The series in (5.105) converges.

(b) *The function* $\exp_\mu(tQ)$ *is differentiable with respect to t with*

$$\frac{d}{dt}\exp_\mu(tQ) = Q *_\mu \exp_\mu(tQ) = \exp_\mu(tQ) *_\mu Q, \quad t \in \mathbb{R}.$$

(c) *The exponent matrices satisfy the semigroup property:*

$$\exp_\mu((s+t)Q) = \exp_\mu(sQ) *_\mu \exp_\mu(tQ), \quad \text{for all } s,t \in \mathbb{R}. \qquad (5.107)$$

(d) *The function* $\exp_\mu(tQ)$ *is the only solution to*

$$\frac{d}{dt}Y(t) = Y(t) *_\mu Q = Q *_\mu Y(t), \quad \text{with } Y(0) = \mathbb{I}. \qquad (5.108)$$

These are called the Kolmogorov's forward and backward equations.

Proof.   (a) Using properties of the norm and (5.106) we get

$$\|\exp_\mu(tQ)\| = \Big\| \sum_{n\geq 0} \frac{(tQ)^{*_\mu n}}{n!} \Big\| \leq \sum_{n\geq 0} \frac{|t|^k \|Q\|^n}{n!} = \exp(|t|\,\|Q\|) < +\infty.$$

(b) Write

$$\frac{d}{dt}\exp_\mu(tQ) = \frac{d}{dt} \sum_{n\geq 0} \frac{(tQ)^{*_\mu n}}{n!}$$

$$= \sum_{n\geq 1} \frac{t^{n-1}Q^{*_\mu n}}{(n-1)!} = Q *_\mu \sum_{n\geq 1} \frac{(tQ)^{*_\mu n-1}}{(n-1)!} = Q *_\mu \exp_\mu(tQ).$$

It is obvious that

$$Q *_\mu \exp_\mu(tQ) = Q *_\mu \sum_{n\geq 1} \frac{(tQ)^{*_\mu n-1}}{(n-1)!}$$

$$= \sum_{n\geq 1} \frac{(tQ)^{*_\mu n-1}}{(n-1)!} *_\mu Q = \exp_\mu(tQ) *_\mu Q.$$

(c) From (5.105) we have

$$\exp_\mu(sQ) *_\mu \exp_\mu(tQ) = \Big(\sum_{n\geq 0} \frac{(sQ)^{*_\mu n}}{n!} \Big) *_\mu \Big(\sum_{k\geq 0} \frac{(tQ)^{*_\mu k}}{k!} \Big)$$

$$= \sum_{n\geq 0}\sum_{k\geq 0} \frac{Q^{*_\mu (n+k)} s^n t^k}{n!\,k!}.$$

Denoting $j = n + k$ from the last equality, we get

$$\exp_\mu(sQ) *_\mu \exp_\mu(tQ) = \sum_{j \geq 0} \sum_{k \geq 0} \frac{Q^{*_\mu j} s^{j-k} t^k}{(j-k)!\, k!}$$

$$= \sum_{j \geq 0} \frac{Q^{*_\mu j}}{j!} \sum_{k \geq 0} \frac{j!}{(j-k)!\, k!} s^{j-k} t^k = \sum_{j \geq 0} \frac{Q^{*_\mu j} (s+t)^j}{j!} = \exp_\mu((s+t)Q).$$

(d) From part (b) it follows that $Y(t) = \exp_\mu(tQ)$ is a solution to (5.108). We show its uniqueness. Assume $Z(t)$ is another solution. Then take

$$\frac{d}{dt}\left(Z(t) *_\mu \exp_\mu(-tQ)\right) = \frac{d}{dt} Z(t) *_\mu \exp_\mu(-tQ) + Z(t) *_\mu \frac{d}{dt}\exp_\mu(-tQ)$$

$$= Q *_\mu Z(t) *_\mu \exp_\mu(-tQ) - Z(t) *_\mu Q *_\mu \exp_\mu(-tQ) = 0.$$

Thus $Z(t) *_\mu \exp_\mu(-tQ)$ is independent on t. Consequently, since at $t = 0$ we have this function equal to $\mathbb{I}$ it follows that $Z(t) *_\mu \exp_\mu(-tQ) = \mathbb{I}$, consequently by part (c) we get $Z(t) = \exp_\mu(tQ)$. $\qquad\square$

Using (5.107) one easily checks that $\mathcal{M}^{[s,t]} = \exp_\mu((t-s)Q)$ satisfies (5.104).

Summarizing, we get the following.

Theorem 5.17. *Let μ be a multiplication such that Conditions 1–4 are satisfied. Let Q be a cubic matrix. Then the family of matrices $\{\mathcal{M}^{[s,t]} = \exp_\mu((t-s)Q)\}$, $0 \leq s < t$, generates a time-homogeneous FA of type μ, denoted by $\mathcal{A}_3^{[s,t]}$.*

Remarks on time dynamics. To study time dynamics of FA $\mathcal{A}_3^{[s,t]}$ one needs to compute the matrix exponent $\exp_\mu(tQ)$. Note that even in case of square matrices finding methods to compute the matrix exponent is difficult, and this is still a topic of considerable current research. But for some simple Q one can compute the matrix exponent. For example, if Q is nilpotent, i.e. there exists $q \in \mathbb{N}$ such that $Q^{*_\mu q} = 0$. Then we have

$$\exp_\mu(tQ) = \mathbb{I} + tQ + \frac{t^2}{2} Q^{*_\mu 2} + \cdots + \frac{t^{q-1}}{(q-1)!} Q^{*_\mu(q-1)}.$$

For the corresponding $\mathcal{A}_3^{[s,t]}$ we see that if Q is nilpotent then it does not exist a limit algebra at $t \to \infty$.

5.4.5 *Time non-homogeneous FAs*

Let $\mathbb{I} \in \mathfrak{C}_\mu$ be a unit matrix (under Condition 2). Matrix $A^{-1} \in \mathfrak{C}_\mu$ is called inverse of an $A \in \mathfrak{C}_\mu$ if

$$A *_\mu A^{-1} = A^{-1} *_\mu A = \mathbb{I}.$$

The following theorem gives an example of m-dimensional time non-homogeneous FA.

Theorem 5.18. *Assume $*_\mu$ such that Condition 2 and 4 are satisfied. Let $\{A^{[t]}, t \geq 0\} \subset \mathfrak{C}_\mu$ be a family of invertible (for all t) m^3-matrices. Then the matrix*

$$\mathcal{M}^{[s,t]} = A^{[s]} *_\mu (A^{[t]})^{-1},$$

generates an FA of type μ, where $(A^{[t]})^{-1}$ is the inverse of $A^{[t]}$.

Proof. By associativity of the multiplication $*_\mu$ of matrices we get

$$\mathcal{M}^{[s,\tau]} *_\mu \mathcal{M}^{[\tau,t]} = A^{[s]} *_\mu \left((A^{[\tau]})^{-1} *_\mu A^{[\tau]} \right) *_\mu (A^{[t]})^{-1}$$

$$= A^{[s]} *_\mu (A^{[t]})^{-1} = \mathcal{M}^{[s,t]}.$$

Thus $\mathcal{M}^{[s,t]}$ satisfies (5.104), consequently each family (with one parameter) of invertible cubic matrices defines an FA, denoted by $\mathcal{A}_4^{[s,t]}$, which is time non-homogeneous, in general. But it will be a time homogeneous FA, for example, if $A^{[t]}$ is equal to tth power of an invertible matrix A. $\square$

5.4.6 *Constructions for Maksimov's multiplications*

Recall that $\mathcal{O}_m$ denotes the set of all associative binary operations on $I = \{1, 2, \ldots, m\}$.

Definition 5.12. An operation $a \in \mathcal{O}_m$ is called uniformly distributed if

$$\sum_{l,n:a(l,n)=j} 1 = m,$$

independently on $j \in I$.

For example, an operation a with $a(i,j) = j$, for all $i, j \in I$, is uniformly distributed. But $a(i,j) = 1$, for all $i, j \in I$, is not uniformly distributed.

Theorem 5.19. *Consider the Maksimov's multiplication (2.3) with respect to a uniformly distributed a. Take arbitrary functions $f_i(t)$ and $g_i(t)$, $i = 1, \ldots m$, of $t \in \mathbb{R}$ such that*

$$\sum_{k=1}^{m} f_k(t) g_k(t) = \frac{1}{m}, \quad \text{for any } t \in \mathbb{R}. \tag{5.109}$$

Then the family of cubic matrices $\mathcal{M}^{[s,t]} = (f_i(s) g_k(t))_{i,j,k=1}^{m}$ generates an FA of type a, denoted by $\mathcal{A}_5^{[s,t]}$.

Notice that the entries of these cubic matrices do not depend on the middle index j.

Proof. Let $\mathcal{M}^{[s,t]} = (f_i(s) g_k(t))_{i,j,k=1}^{m}$ then the equation (5.88) has the form

$$f_i(s) g_r(t) = \sum_{l,n:\, a(l,n)=j} \sum_{k} f_i(s) g_k(\tau) f_k(\tau) g_r(t). \tag{5.110}$$

Simplify the system (5.110) to get

$$f_i(s) g_r(t) \left[\sum_{l,n:\, a(l,n)=j} \sum_{k} f_k(\tau) g_k(\tau) - 1 \right] = 0.$$

By conditions of the theorem we have

$$\sum_{l,n:\, a(l,n)=j} \sum_{k} f_k(\tau) g_k(\tau) - 1 = 0, \quad \text{for all } \tau \in \mathbb{R}.$$

This completes the proof. $\qquad\square$

Now we construct a very rich class of FAs of type a_0, where a_0 is an operation in $\mathcal{O}_m$ such that $a_0(j,n) = j$ for any $j, n \in I$. Note that this operation is uniformly distributed. Then the corresponding Maksimov's multiplication for arbitrary cubic matrices

$$A = (a_{ijk})_{i,j,k=1}^{m}, \quad B = (b_{ijk})_{i,j,k=1}^{m}, \quad C = (c_{ijk})_{i,j,k=1}^{m},$$

is $C = A *_{a_0} B$ where

$$c_{ijr} = \sum_{k,n=1}^{m} a_{ijk} b_{knr}. \tag{5.111}$$

Theorem 5.20. *Let $g_i(t)$ and $\gamma_{ij}(t)$, $i, j \in I$, be arbitrary functions of $t \in \mathbb{R}$ such that $g_i(t) \neq 0$ and*

$$m \sum_{j=1}^{m} \gamma_{ij}(s) = g_i(s), \quad \text{for all } i \in I, \ s \in \mathbb{R}. \tag{5.112}$$

Then the cubic matrix

$$\mathcal{M}^{[s,t]} = \left(\frac{\gamma_{ij}(s)}{g_r(t)} \right)^{m}_{i,j,r=1}$$

generates an FA of type a_0 denoted by $\mathcal{A}_6^{[s,t]}$.

Proof. For $*_{a_0}$, using (5.111), the equation (5.104) can be written as

$$M_{ijr}^{[s,t]} = \sum_{k,n=1}^{m} M_{ijk}^{[s,\tau]} M_{knr}^{[\tau,t]}, \quad \text{for all } i,j,r \in I. \tag{5.113}$$

Denote

$$f_{ir}(s,t) = \sum_{j=1}^{m} M_{ijr}^{[s,t]}. \tag{5.114}$$

Then from (5.113) we get

$$f_{ir}(s,t) = \sum_{k=1}^{m} f_{ik}(s,\tau) f_{kr}(\tau,t), \quad \text{for all } i,r \in I. \tag{5.115}$$

Consider arbitrary functions $g_i(t)$, $i = 1, \ldots, m$, such that $g_i(t) \neq 0$. Then one can see that the system of functional equations (5.115) has the following solution

$$f_{ij}(s,t) = \frac{g_i(s)}{m \, g_j(t)}.$$

Using this equality, by (5.114) and (5.113), we get

$$M_{ijr}^{[s,t]} g_r(t) = \frac{1}{m} \sum_{k=1}^{m} M_{ijk}^{[s,\tau]} g_k(\tau), \quad \text{for all } i,j,r \in I.$$

From this equality it is clear that the quantity $M_{ijr}^{[s,t]} g_r(t)$ should not depend on t and r, i.e. $M_{ijr}^{[s,t]} g_r(t) = \gamma_{ij}(s)$, for some function $\gamma_{ij}(s)$. Thus

$$M_{ijr}^{[s,t]} = \frac{\gamma_{ij}(s)}{g_r(t)}, \quad \text{for all } i,j,r \in I.$$

Now the equality (5.114) gives the condition (5.112) on γ_{ij}. $\square$

Remarks on time dynamics. In general the behavior of $\mathcal{A}_5^{[s,t]}$ depends on given functions $f_i(t)$ and $g_i(t)$. One can choose these function suitable to an expected property of the dynamics. Here we consider one example to make $\mathcal{A}_5^{[s,t]}$ a periodic FA. Consider

$$f_k(t) = 2 + \sin(kt), \quad g_k(t) = \frac{1}{m^2(2 + \sin(kt))}, \quad k = 1, \ldots, m.$$

Then the condition (5.109) is satisfied. Consequently, the matrix

$$\mathcal{M}^{[s,t]} = \left(\frac{2 + \sin(is)}{m^2 \left(2 + \sin(kt) \right)} \right)^m_{i,j,k=1}$$

generates a periodic FA $\mathcal{A}_5^{[s,t]}$.

The time behavior of $\mathcal{A}_6^{[s,t]}$ (is similar to the behavior of $\mathcal{A}_5^{[s,t]}$) depends on its parameter-functions $g_i(t)$ and $\gamma_{ij}(t)$, $i, j \in I$.

Let S_m be the group of permutations on I.

Take $a \in \mathcal{O}_m$ and define an action of $\pi \in S_m$ on a (denoted by πa) as

$$\pi a(i, j) = \pi a(\pi^{-1}(i), \pi^{-1}(j)), \qquad \text{for all} \ \ i, j \in I.$$

Theorem 5.21. *Let $a \in \mathcal{O}_m$ and $*_a$ be a Maksimov's multiplication (see (2.3)). If there exists a permutation $\pi \in S_m$ such that $\pi b = a$, that is*

$$a(j, n) = \pi^{-1}(b(\pi(j), \pi(n))), \qquad \text{for all} \ \ j, n \in I, \tag{5.116}$$

then there is an FA of type a iff there is an FA of type b.

Proof. Assume there is an FA of type a, i.e. the equation (5.104) has a solution, say $\mathcal{M}^{[s,t]} = \left(M_{ijr}^{[s,t]} \right)$, with respect to the multiplication $*_a$. That is

$$M_{ijr}^{[s,t]} = \sum_{l,n:\, a(l,n)=j} \sum_k M_{ilk}^{[s,\tau]} M_{knr}^{[\tau,t]}. \tag{5.117}$$

Denote $\pi(i) = i'$, then using (5.116), since π is one-to-one, from (5.117) we get

$$M^{[s,t]}_{\pi^{-1}(i')\pi^{-1}(j')\pi^{-1}(r')}$$

$$= \sum_{l',n':\, b(l',n')=j'} \sum_{k'} M^{[s,\tau]}_{\pi^{-1}(i')\pi^{-1}(l')\pi^{-1}(k')} M^{[\tau,t]}_{\pi^{-1}(k')\pi^{-1}(n')\pi^{-1}(r')}.$$

Thus $\tilde{\mathcal{M}}^{[s,t]} = \left(\tilde{M}_{ijr}^{[s,t]} \right)$ with $\tilde{M}_{ijr}^{[s,t]} = M^{[s,t]}_{\pi^{-1}(i)\pi^{-1}(j)\pi^{-1}(r)}$ is a solution of the equation (5.104) with respect to the multiplication $*_b$. This completes the proof. $\qquad\square$

5.4.7 *Multiplication of solutions*

In this subsection we assume the following

Condition 5. Assume $\mathfrak{C}_\mu$ is a commutative algebra.

Theorem 5.22. *Assume $*_\mu$ such that Conditions 4 and 5 are satisfied. Let $\{\mathcal{M}^{[s,t]}\}, \{\mathcal{N}^{[s,t]}\} \subset \mathfrak{C}_\mu$ be two families of m^3-matrices which generate two FAs of type μ. Then the matrix*

$$\mathcal{B}^{[s,t]} = \mathcal{M}^{[s,t]} *_\mu \mathcal{N}^{[s,t]},$$

generates an FA of type μ.

Proof. By associativity and commutativity of the multiplication $*_\mu$ of matrices we get

$$\mathcal{B}^{[s,\tau]} *_\mu \mathcal{B}^{[\tau,t]} = \left(\mathcal{M}^{[s,\tau]} *_\mu \mathcal{N}^{[s,\tau]}\right) *_\mu \left(\mathcal{M}^{[\tau,t]} *_\mu \mathcal{N}^{[\tau,t]}\right)$$

$$= \left(\mathcal{M}^{[s,\tau]} *_\mu \mathcal{M}^{[\tau,t]}\right) *_\mu \left(\mathcal{N}^{[s,\tau]} *_\mu \mathcal{N}^{[\tau,t]}\right) = \mathcal{M}^{[s,t]} *_\mu \mathcal{N}^{[s,t]} = \mathcal{B}^{[s,t]}. \qquad \square$$

Remark 5.10. Note that Theorem 5.22 can be generalized, i.e. under Conditions 4 and 5, let $\{\mathcal{M}_i^{[s,t]}\} \subset \mathfrak{C}_\mu$, $i = 1, \ldots, k$, be k families of m^3-matrices which generate k FAs of type μ. Then the matrix

$$\mathcal{B}^{[s,t]} = \mathcal{M}_1^{[s,t]} *_\mu \mathcal{M}_2^{[s,t]} *_\mu \cdots *_\mu \mathcal{M}_k^{[s,t]},$$

generates an FA of type μ.

This formula is useful to construct new FAs by known ones (using above mentioned examples of FAs and other examples given in [138]).

Remark 5.11. To check the Conditions 1–5 for an algebra $\mathfrak{C}_\mu$ one has to check known conditions ([116]) on $\mu = \left(C_{ijk,lnr}^{uvw}\right)$ of the matrix of structural constants.[3] Indeed, one can numerate the elements of the set $\{ijk : i, j, k = 1, \ldots, m\}$ to write it in the form $\mathcal{J} = \{1, 2, \ldots, m^3\}$ then the matrix μ can be written as usual form: $\mu = (c_{ij}^k)$ where $i, j, k \in \mathcal{J}$. Then the following conditions are known for an algebra with matrix μ:

- Commutative iff:

$$c_{ij}^k = c_{ji}^k, \quad \text{for all } i, j, k \in \mathcal{J}.$$

---

[3] Since there is one-to-one correspondence between multiplications and the matrices of structural constants we identify the multiplication μ with the corresponding matrix of structural constants.

- Associative iff:

$$\sum_{r=1}^{n} c_{ij}^{r} c_{rk}^{l} = \sum_{r=1}^{n} c_{ir}^{l} c_{jk}^{r}, \quad \text{for all } i,j,k,l \in \mathcal{J}.$$

- Existence of an idempotent element: this problem is equivalent to the existence of a fixed point of the map $V \colon \mathbb{R}^{m^3} \longrightarrow \mathbb{R}^{m^3}$, $x \mapsto V(x) = x'$, given by

$$V \colon x_k' = \sum_{i,j=1}^{m^3} c_{ij}^{k} x_i x_j, \quad k = 1, \ldots, m^3.$$

For example, if μ is a stochastic cubic matrix in the sense that

$$c_{ij}^{k} \geq 0, \quad \sum_{k=1}^{m^3} c_{ij}^{k} = 1, \quad \text{for all } i,j,$$

then from known theorems about fixed points it follows that the corresponding operator V has at least one fixed point (i.e. the algebra $\mathfrak{C}_\mu$ has at least one idempotent element).

Let us give an example of ACM for which all above mentioned conditions can be easily checked.

Example 5.10. Let $\alpha : \mathcal{J} \times \mathcal{J} \to \mathcal{J}$ be a binary operation on $\mathcal{J} = \{1, 2, \ldots, m^3\} \equiv \{ijk : i,j,k = 1, \ldots, m\}$ and assume that $(\mathcal{J}, \alpha)$ is a group with the operation α, i.e., the operation satisfies axioms: associativity, has identity element (denoted by $i_0 j_0 k_0$) and each element has an inverse (for each ijk its inverse is denoted by $\overline{ijk}$).

Define the following multiplication for basis matrices E_{ijk}:

$$E_{ijk} *_\alpha E_{lnr} = E_{\alpha(ijk, lnr)}. \tag{5.118}$$

Since $(\mathcal{J}, \alpha)$ is a group one can see that the ACM $\mathfrak{C}_\alpha$ is associative, with unit matrix $\mathbb{I} = E_{i_0 j_0 k_0}$ and each basis element E_{ijk} has its inverse denoted by $E_{\overline{ijk}}$. Indeed, since $\alpha(ijk, \overline{ijk}) = i_0 j_0 k_0$ we have

$$E_{ijk} *_\alpha E_{\overline{ijk}} = E_{\alpha(ijk, \overline{ijk})} = E_{i_0 j_0 k_0}.$$

Moreover, if the group is commutative then the algebra $\mathfrak{C}_\alpha$ is also commutative.

5.4.7.1 *Differential equations for flows of algebras*

For a continuous-time Markov process, Kolmogorov derived forward equations and backward equations, which are a pair of systems of differential equations that describe the time-evolution of the probability transition probabilities $P_{ij}^{[s,t]}$ giving the process [132]. For quadratic stochastic processes partial differential equations with delaying argument were derived. These equations then were used to describe some processes (see [183]).

In this section we shall derive partial differential equations for the matrices $\mathcal{M}^{[s,t]}$.

Let $\mathcal{M}^{[s,t]} = \left(M_{ijk}^{[s,t]} \right)_{i,j,k=1}^{m}$ generate an FA of type μ. Take a small $h > 0$ such that $s + h < \tau < t$ and using the equation (5.88), we write

$$\mathcal{M}^{[s+h,t]} - \mathcal{M}^{[s,t]} = \mathcal{M}^{[s+h,\tau]} *_\mu \mathcal{M}^{[\tau,t]} - \mathcal{M}^{[s,\tau]} *_\mu \mathcal{M}^{[\tau,t]}$$

$$= \left(\mathcal{M}^{[s+h,\tau]} - \mathcal{M}^{[s,\tau]} \right) *_\mu \mathcal{M}^{[\tau,t]}.$$

Dividing this expression by h and assuming the existence of the limits we obtain

$$\lim_{h \to 0} \frac{\mathcal{M}^{[s+h,t]} - \mathcal{M}^{[s,t]}}{h} = \lim_{h \to 0} \frac{\mathcal{M}^{[s+h,\tau]} - \mathcal{M}^{[s,\tau]}}{h} *_\mu \mathcal{M}^{[\tau,t]},$$

i.e.

$$\frac{\partial}{\partial s} \mathcal{M}^{[s,t]} = \left(\frac{\partial}{\partial s} \mathcal{M}^{[s,\tau]} \right) *_\mu \mathcal{M}^{[\tau,t]}, \tag{5.119}$$

here $\frac{\partial}{\partial s} \mathcal{M}^{[s,t]} = \left(\frac{\partial}{\partial s} M_{ijk}^{[s,t]} \right)_{i,j,k=1}^{m}$.

Similarly with respect t we get

$$\frac{\partial}{\partial t} \mathcal{M}^{[s,t]} = \mathcal{M}^{[s,\tau]} *_\mu \left(\frac{\partial}{\partial t} \mathcal{M}^{[\tau,t]} \right). \tag{5.120}$$

Summarize this in the following theorem.

Theorem 5.23. *If $\mathcal{M}^{[s,t]}$ generates an FA of type μ then it satisfies the partial differential equations (5.119) and (5.120).*

Remark 5.12. The equation (5.119) is called forward equation and (5.120) is called backward equation. Note that the equations (5.108) are particular cases of these equations. As it was shown above the solution of the equation (5.108) gives only time-homogeneous FAs. Each matrix of the FAs mentioned in the previous section satisfy equations (5.119), (5.120).

5.4.8 *Time dependent dynamics of algebraic properties in an FA*

By definitions, an FA is a continuous-time dynamical system which in a fixed pair of time (s, t) is an algebra. Therefore properties of the algebras in an FA depend on time.

5.4.8.1 *Dynamics of a baric property*

The following theorem gives dynamical baric properties:

Theorem 5.24.

(i) Let $\{\mathcal{A}^{[s,t]},\ 0 \le s \le t\}$ be a SFA. Then the algebra $\mathcal{A}^{[s,t]}$ is baric for any $(s, t) \in \mathcal{T}$.

(ii) An algebra from an FA of type (C) may or may not be a baric algebra.

(iii) Let $\{\mathcal{A}^{[s,t]},\ 0 \le s < t\}$ be a flow of m-dimensional algebras of type (D) or (E). Then $\mathcal{A}^{[s,t]}$ is not baric if $m \ge 2$, i.e. $\mathcal{T}_{baric} = \emptyset$ if $m \ge 2$.

Proof. We use Proposition 5.17:

(i) By definition of a SFA the corresponding matrix $\mathcal{M}^{[s,t]}$ is stochastic, i.e. the condition of Proposition 5.17 is satisfied.

(ii) In the case of multiplication (C), if the matrix $\mathcal{M}^{[s,t]} = (c_{ijk}^{[s,t]})$ satisfies conditions

$$c_{ijk}^{[s,t]} = c_{jik}^{[s,t]}, \qquad \sum_{k=1}^{m} c_{ijk}^{[s,t]} = 1, \quad \text{for some } (s,t), \tag{5.121}$$

then the corresponding $\mathcal{A}^{[s,t]}$ is baric, otherwise it is not baric.

(iii) We should check (5.121) for the multiplication (D) (the case (E) is similar). Assume $\mathcal{A}^{[s,t]}$ is baric for any $(s,t) \in \mathcal{T}$. Then from (5.88) we have

$$c_{ijr}^{[s,t]} = \sum_{k=1}^{m} c_{ijk}^{[s,\tau]} \sum_{n=1}^{m} c_{knr}^{[\tau,t]},$$

and, by using (5.121) for any (s, t) we get

$$1 = \sum_{r=1}^{m} c_{ijr}^{[s,t]} = \sum_{k=1}^{m} c_{ijk}^{[s,\tau]} \sum_{n=1}^{m} \sum_{r=1}^{m} c_{knr}^{[\tau,t]} = m.$$

This is a contradiction to the assumption that $m \ge 2$. $\qquad\square$

Remark 5.13. Since the matrix of an FA is not stochastic in general, using the above mentioned examples of FAs one can give several baric property diagrams. For example, in case (C) one can take matrices mentioned in Example 5.6 which satisfy condition (5.121). But in cases (D) and (E) baric FAs arise only for $m = 1$.

5.4.8.2 *Limit algebras of flows*

As it was mentioned above each flow of algebras can be considered as a continuous time dynamical system such that in each fixed pair (s, t) of time its state is a finite-dimensional algebra. The main problem in such dynamical systems is to know what will be the limit algebra (if it exists):

$$\lim_{t-s \to +\infty} \mathcal{A}^{[s,t]} = \mathcal{A}.$$

In case the limit does not exist then one asks about possible limit algebras (see Proposition 2.7 of [31] for an interesting example).

Case Example 5.2. Simple calculations show that the limit algebra (depending on parameter ϵ of the example) is $\mathcal{A}_\epsilon = \langle e_1, e_2 \rangle$ with multiplication table:

$$\begin{cases} e_1^2 = e_1 e_2 = e_2 e_1 = e_2^2 = e_2, & \text{if } \epsilon \in (0, \tfrac{1}{2}], \\ e_1^2 = e_1 e_2 = e_2 e_1 = e_2^2 = x e_1 + (1 - x) e_2, & \text{if } \epsilon = 0. \end{cases}$$

Case Example 5.3. In this case the limit algebra (independently on parameter ϵ of the example) is $\mathcal{A} = \langle e_1, e_2 \rangle$ with multiplication table:

$$e_1^2 = e_1 e_2 = e_2 e_1 = e_2^2 = e_2.$$

Case Example 5.4. The limit algebra (depending on parameter ϵ and initial point $(x_1, x_2, 1 - x_1 - x_2)$ of the example) is $\mathcal{A}_\epsilon = \langle e_1, e_2, e_3 \rangle$ with multiplication table:

$$e_i e_j = \begin{cases} e_3, & \text{if } \epsilon \in [0, \tfrac{1}{2}), \\ x_1 e_1 + x_2 e_2 + (1 - x_1 - x_2) e_3, & \text{if } \epsilon = \tfrac{1}{2}, \end{cases} \quad \text{for any } i, j = 1, 2, 3.$$

Note that these examples present SFAs of different types, but their limit algebras may coincide.

Case Example 5.5. The limit algebra depends on the given stochastic vector $a(t)$. If $\lim_{t \to \infty} a(t)$ exists then the limit algebra also exists. But one can choose $a(t)$ such that the limit does not exist. For example take $a(t) = (\sin^2(t), \cos^2(t), 0, \ldots, 0)$ then the limit does not exist and the set of all possible limit algebras is a continuum set. We do this point more clear in the case of Example 5.8.

Case Example 5.8. It is clear that the limit algebra does not exist in Example 5.8 either. Here we shall show that the set of all limit algebras in discrete time is dense: Consider discrete time n, $n \in \mathbb{N}$, and the FA $\{\mathcal{A}^{[n]}, n \in \mathbb{N}\}$ given by the matrix (5.101). Denote by $\mathcal{A}_a$, $a \in [-1, 1]$, the algebra which is given by the matrix of structural constants (5.101) with $\sin(\hat{t}) = a$.

Proposition 5.25. *The discrete time FA $\mathcal{A}^{[n]}$, $n \in \mathbb{N}$, is dense in the set $\{\mathcal{A}_a, \ a \in [-1, 1]\}$ of algebras, i.e. for an arbitrary algebra $\mathcal{A}_a$ there exists a sequence $\{n_k\}_{k=1,2,\ldots}$ of natural numbers such that $\lim_{k \to \infty} \mathcal{A}^{[n_k]} = \mathcal{A}_a$.*

Proof. It is known (see [213]) that the sequences $\{\sin n\}$ and $\{\cos n\}$, $n \in \mathbb{N}$, are dense in $[-1, 1]$. Hence for any $a \in [-1, 1]$ there is a sequence $\{n_k\}_{k=1,2,\ldots}$ of natural numbers such that $\lim_{k \to \infty} \sin(n_k) = a$. The same sequence can be used to get $\lim_{k \to \infty} \mathcal{A}^{[n_k]} = \mathcal{A}_a$. $\qquad\square$

Since all other above-constructed FAs depend on arbitrary functions $(\psi, \kappa, \gamma, \ldots)$ their limit algebras depend on these functions. One can control their limit by choosing these parameter functions.

5.4.8.3 *Dynamics of commutativity*

It is clear that a cubic matrix (c_{ijk}) generates a commutative algebra iff $c_{ijk} = c_{jik}$. The algebras in FAs given by Examples 5.2–5.5 are commutative for any time $(s, t) \in \mathcal{T}$.

In the case of Example 5.6 an algebra $\mathcal{A}^{[s,t]}$ of the FA of type (C) is commutative iff $a_{2r}^{[s,t]} = \bar{a}_{1r}^{[s,t]}$. For example, take for the matrix $(a_{ij}^{[s,t]})$ the matrix (5.92) and for $(\bar{a}_{ij}^{[s,t]})$ its transpose. Then as a particular case of Example 5.6 the following cubic matrix generates an FA of type (C):

$$\mathcal{M}^{[s,t]} = \left(\begin{array}{cc|cc} \cos(t-s) & \sin(t-s) & -\sin(t-s) & \cos(t-s) \\ \cos(t-s) & -\sin(t-s) & \sin(t-s) & \cos(t-s) \end{array}\right).$$

The algebra $\mathcal{A}^{[s,t]}$ of this FA is commutative iff $\cos(t-s) = -\sin(t-s)$, i.e. $t = s + \frac{3\pi}{4} + \pi n$, $n = 0, 1, 2, \ldots$. Since the FA is a time-homogeneous, we can consider its dynamics with respect to one time parameter $\hat{t} = t - s$. What was shown above is that the commutativity of the algebras in this FA holds only at values $\frac{3\pi}{4} + \pi n$, $n = 0, 1, 2, \ldots$ of the time $\hat{t}$. Thus dynamics of the commutativity is as follows: starting from $\hat{t} = 0$ one first meets commutativity at $\hat{t} = \frac{3\pi}{4}$, then the commutativity appears only after each adding time π, i.e. with period π. Therefore this is an example of almost non-commutative (with respect to Lebesgue measure on $\mathcal{T}$) FA of type (C).

In the case of Example 5.7 one can see that the duration of the commutativity is the set

$$\left\{ (s,t) \in \mathcal{T} : \kappa_{21}(s) = \frac{1}{2\Psi(s)} - \kappa_{11}(s) \right\}.$$

Since κ_{11}, κ_{21} and Ψ are arbitrary one can choose these functions to control this set (the case of Example 5.9 is similar).

The following proposition shows that algebras of the FA in Example 5.8 will stay always commutative (or always non-commutative):

Proposition 5.26. *The FA corresponding to Example 5.8 (i.e. to the matrix (5.101)) is always commutative iff $a = 1 - c$ and $b = -d$, otherwise they are always non-commutative.*

Proof. Commutativity of the algebra $\mathcal{A}^{[t]}$ is reduced to the system

$$(1 - a - c)\cos\hat{t} + (b + d)\sin\hat{t} = 0,$$
$$-(1 - a - c)\sin\hat{t} + (b + d)\cos\hat{t} = 0. \tag{5.122}$$

Since $\det \begin{pmatrix} \cos\hat{t} & \sin\hat{t} \\ -\sin\hat{t} & \cos\hat{t} \end{pmatrix} = 1$ for any $\hat{t} \geq 0$, the system (5.122) holds only in the case $a = 1 - c$ and $b = -d$. $\square$

The following proposition is a corollary of Theorem 5.15 and Theorem 5.16:

Proposition 5.27. *The commutativity duration for the FA given in Theorem 5.15 (resp. Theorem 5.16) is*

$$\left\{ (s,t) \in \mathcal{T} : \quad \beta^{(s)}_{ijk} = \beta^{(s)}_{jik}, \quad \text{for any} \ \ i,j,k = 1,\dots,m \right\}.$$

$$\left(resp. \ \left\{ (s,t) \in \mathcal{T} : \quad \gamma^{(t)}_{ijk} = \gamma^{(t)}_{jik}, \quad \text{for any} \ \ i,j,k = 1,\dots,m \right\} \right).$$

Proof. From the cubic matrix $\mathcal{M}^{[s,t]}$ mentioned in Theorem 5.15 we get that the commutativity of the corresponding algebra $\mathcal{A}^{[s,t]}$ holds for (s,t) such that

$$\sum_{k=1}^{m} \left[\beta^{(s)}_{ijk} - \beta^{(s)}_{jik} \right] b^{(t)}_{kr} = 0, \quad \text{for all} \ i,j.$$

Since for any t, by the condition of Theorem 5.15, we have $\det((A^{[t]})^{-1}) = \det((b^{(t)}_{kr})^{m}_{k,r=1}) \neq 0$, and from the last system of equations we get $\beta^{(s)}_{ijk} = \beta^{(s)}_{jik}$ for any i,j and k. $\square$

5.4.8.4 *Duration to be an evolution algebra*

The following proposition is for stochastic FAs:

Proposition 5.28. *Any SFA $\{\mathcal{A}^{[s,t]} : s, t \in \mathbb{R}\}$ does not contain an evolution algebra, i.e. $\mathcal{A}^{[s,t]} \neq EA$ for any $(s,t) \in \mathcal{T}$.*

Proof. Since in a stochastic flow of algebras we have the condition that

$$\sum_{k=1}^{m} P_{ij,k}^{[s,t]} = 1, \text{ for all } i, j \text{ and for all } (s,t) \in \mathcal{T},$$

for each (s,t) and i, j, there exists $k_0 = k_0(i, j, s, t)$ such that $P_{ij,k_0}^{[s,t]} > 0$. Consequently, $e_i e_j \neq 0$ for each $i \neq j$, i.e. the algebra is not an EA. $\square$

Note that a non-stochastic FA may contain EAs: in the case of Example 5.7 one can see that the duration to be an EA in the corresponding FA is the set

$$\mathcal{E} = \left\{ (s,t) \in \mathcal{T} : \kappa_{21}(s) = \frac{1}{2\Psi(s)} - \kappa_{11}(s) = 0 \right\}.$$

Since κ_{11}, κ_{21} and Ψ are arbitrary for each given subset $\widetilde{\mathcal{T}} \subset \mathcal{T}$ one can choose these functions in such a way that $\mathcal{E} = \widetilde{\mathcal{T}}$ (the case of Example 5.9 is similar).

5.4.8.5 *Dynamics of associativity*

It is known [116] that a finite-dimensional algebra A with a matrix of structural constants $(c_{ijk})_{i,j,k=1}^{n}$ is associative iff

$$\sum_{r=1}^{n} c_{ijr} c_{rkl} = \sum_{r=1}^{n} c_{irl} c_{jkr}, \qquad \text{for all } i, j, k, l. \tag{5.123}$$

In this subsection we check the associativity condition for FAs of Theorems 5.15 and 5.16.

Theorem 5.25. *The duration of associativity for the FA $\{\mathcal{A}^{[s,t]}\}$ given in Theorem 5.15 (resp. Theorem 5.16) is*

$$\left\{ (s,t) \in \mathcal{T} : \sum_{p,r} \left(\beta_{ijp}^{(s)} \beta_{rkq}^{(s)} - \beta_{irq}^{(s)} \beta_{jkp}^{(s)} \right) b_{pr}^{[t]} = 0, \text{ for all } i, j, k, q \right\}.$$

$$\left(resp. \left\{ (s,t) \in \mathcal{T} : \sum_{r,q} \left(a_{rq}^{[s]} \gamma_{pjr}^{(t)} \gamma_{qkl}^{(t)} - a_{jq}^{[s]} \gamma_{prl}^{(t)} \gamma_{qkr}^{(t)} \right) = 0, for\, all\, p, j, k, l \right\} \right).$$

Proof. For $\mathcal{A}^{[s,t]}$ the condition (5.123) can be written as

$$\sum_{p,r,q} \beta^{(s)}_{ijp}\beta^{(s)}_{rkq} b^{[t]}_{pr} b^{[t]}_{ql} = \sum_{p,r,q} \beta^{(s)}_{irq}\beta^{(s)}_{jkp} b^{[t]}_{pr} b^{[t]}_{ql}, \qquad \text{for all } i,j,k,l.$$

Rewrite it in the following form:

$$\sum_{q} X_{ijk,q} b^{[t]}_{ql} = 0, \qquad \text{for all } i,j,k,l, \tag{5.124}$$

where

$$X_{ijk,q}(s,t) = \sum_{p,r} \left(\beta^{(s)}_{ijp}\beta^{(s)}_{rkq} - \beta^{(s)}_{irq}\beta^{(s)}_{jkp} \right) b^{[t]}_{pr}.$$

Since $\det((A^{[t]})^{-1}) \neq 0$ from (5.124) we get $X_{ijk,q}(s,t) \equiv 0$, i.e.

$$\sum_{p,r} \left(\beta^{(s)}_{ijp}\beta^{(s)}_{rkq} - \beta^{(s)}_{irq}\beta^{(s)}_{jkp} \right) b^{[t]}_{pr} = 0.$$

$\square$

Remark 5.14. An interesting problem is to construct an FA which contains (depending on time) as many known finite-dimensional algebras as possible.

Bibliographical notes. In this chapter we presented recently introduced flows of algebras. The chapter is based on [29], [31], [40], [57], [60], [77], [88], [132], [139]–[142], [155], [158], [183], [188], [189], [197], [217], [218], [237], [251]. See [3], [115] and [164] for flow and algebra relations.

PART 2
Probabilistic approach

Chapter 6

Markov processes of cubic stochastic matrices

In the first section of this chapter we consider Maksimov's associative multiplications of cubic matrices. The notion of stochastic cubic matrix varies depending on the real models of application. The second section is devoted to a Markov interaction process which generalizes a Markov process. In the third section we define Markov processes of cubic stochastic (in a fixed sense) matrices which also called a quadratic stochastic process (QSP). Note that a QSP is a continuous-time dynamical system whose states are stochastic cubic matrices satisfying an analogue of the Kolmogorov–Chapman equation (KCE) considered with respect to a fixed multiplication of cubic matrices. The existence of a stochastic (at each time) solution to the KCE provides the existence of a QSP. For some specially chosen notions of stochastic cubic matrices and their multiplications we construct corresponding QSPs. Some time-dependent behavior (dynamics) of such processes are studied. The fourth section gives applications to the biology, constructing a QSP which describes the dynamics of a population with the possibility of twin births.

6.1 Maksimov's cubic stochastic matrices

6.1.1 *Multiplications*

Denote $I = \{1, 2, \ldots, m\}$. Let $\mathfrak{C}$ be the set of all m^3-dimensional cubic matrices over the field of real numbers. Recall that E_{ijk}, $i, j, k \in I$ are basis cubic matrices in $\mathfrak{C}$.

Following [158] define the following multiplications for basis matrices E_{ijk}:

$$E_{ijk} *_0 E_{lnr} = \delta_{kl}\delta_{jn}E_{ijr}, \tag{6.1}$$

where δ_{kl} is the Kronecker symbol.

227

Then for any two cubic matrices $A = (a_{ijk}), B = (b_{ijk}) \in \mathfrak{C}$ the matrix $A *_0 B = (c_{ijk})$ is defined by

$$c_{ijr} = \sum_{k=1}^{m} a_{ijk} b_{kjr}. \tag{6.2}$$

Proposition 6.1. *The algebra of cubic matrices* $(\mathfrak{C}, *_0)$ *is a direct sum of algebras of square matrices.*

Proof.   Denote by Q_{ij} an m^2-square matrix whose (i, j)th entry is equal to 1 and all other entries are equal to 0. By (6.1) we have $E_{ijk} *_0 E_{kjr} = E_{ijr}$ that is for fixed j, the collection of basis matrices E_{ijk} is a semigroup which under the mapping $E_{ijk} \to Q_{ik}$ is isomorphic to a semigroup generated by the basis Q_{ik} with respect to multiplication

$$Q_{ij} * Q_{km} = \delta_{jk} Q_{im}.$$

Therefore, the jth section in the algebra of cubic matrices with respect to multiplication (6.1) is isomorphic to the algebra of square matrices with respect to multiplication $*$. Moreover, $E_{ijk} *_0 E_{lnr} = 0$ if $\delta_{kl}\delta_{jn} = 0$, i.e. multiplications of cross-sections are equal to zero. $\square$

Recall also the multiplication given by (2.2):

$$E_{ijk} *_a E_{lnr} = \delta_{kl} E_{ia(j,n)r}, \tag{6.3}$$

where $a\colon I \times I \to I$, $(j, n) \mapsto a(j, n) \in I$, is an arbitrary associative binary operation.

Note that (6.1) is *not* a particular case of (6.3).

Denote by $\mathcal{O}_m$ the set of all associative binary operations on I.

The general formula for the multiplication is the extension of (6.3) by bilinearity, i.e. for any two cubic matrices $A = (a_{ijk}), B = (b_{ijk}) \in \mathfrak{C}$ the matrix $A *_a B = (c_{ijk})$ is defined by

$$c_{ijr} = \sum_{l,n:\, a(l,n)=j} \sum_{k} a_{ilk} b_{knr}.$$

Note that $c_{ijr} = 0$ for j such that $\{l, n : a(l, n) = j\} = \emptyset$.

Denote by $\mathfrak{C}_a \equiv \mathfrak{C}_a^m = (\mathfrak{C}, *_a)$, $a \in \mathcal{O}_m$, the ACM given by the multiplication $*_a$.

Lemma 6.1. *The multiplication (6.3) is associative for each associative* $a \in \mathcal{O}_m$.

Proof. For any a we have

$$[E_{ijk} *_a E_{knr}] *_a E_{rst} = E_{ia(j,n)r} *_a E_{rst} = E_{ia(a(j,n),s)t}.$$

$$E_{ijk} *_a [E_{knr} *_a E_{rst}] = E_{ijk} *_a E_{ka(n,s)t} = E_{ia(j,a(n,s))t}.$$

Associativity of a completes the proof. $\qquad\square$

If the equation $a(x, u) = v$ (resp. $a(u, x) = v$) is uniquely solvable for any $u, v \in I$ then the operation a on I has right (resp. left) unique solvability.

Lemma 6.2. *If the operation a on I has right or left unique solvability, then*

$$\sum_{d \in I} \sum_{\substack{j,m: \\ a(j,m)=d}} \gamma_{j,m} = \sum_{j \in I} \sum_{m \in I} \gamma_{j,m}.$$

Proof. Assume the operation a have the left unique solvability (the right case is similar). Then for any m and $d \in I$, one can uniquely define the element j_d as a solution of the equation $a(j_d, m) = d$. Moreover, the elements j_d are different for different d. Thus for any fixed $m \in I$, the set $\{j_d, \; d \in I\}$ is a permutation of the set I. Consequently,

$$\sum_{d \in I} \sum_{\substack{j,m: \\ a(j,m)=d}} \gamma_{j,m} = \sum_{m \in I} \sum_{a(j_d,m),\, d \in I} \gamma_{j,m}$$

$$= \sum_{m \in I} \sum_{a(j,m),\, j \in I} \gamma_{j,m} - \sum_{j \in I} \sum_{m \in I} \gamma_{j,m}. \qquad\square$$

6.1.2 *Stochasticity*

Define several kinds of cubic stochastic matrices (see [158, 183]): a cubic matrix $P = (p_{ijk})_{i,j,k=1}^{m}$ is called

$(1, 2)$-*stochastic* if

$$p_{ijk} \geq 0, \qquad \sum_{i,j=1}^{m} p_{ijk} = 1, \quad \text{for all } k.$$

$(1, 3)$-*stochastic* if

$$p_{ijk} \geq 0, \qquad \sum_{i,k=1}^{m} p_{ijk} = 1, \quad \text{for all } j.$$

$(2, 3)$-*stochastic* if

$$p_{ijk} \geq 0, \qquad \sum_{j,k=1}^{m} p_{ijk} = 1, \quad \text{for all } i.$$

3-*stochastic* if

$$p_{ijk} \geq 0, \qquad \sum_{k=1}^{m} p_{ijk} = 1, \quad \text{for all } i, j.$$

The last one can be also given with respect to first and second index.

Maksimov [158] also defined a twice stochastic matrix: a (2,3)-stochastic cubic matrix is called *twice stochastic* if

$$\sum_{i=1}^{m} p_{ijk} = \frac{1}{m}, \qquad \text{for all } j, k.$$

Proposition 6.2. $(2, 3)$-*stochastic (and twice) stochastic cubic matrices form a convex semigroup*[1] *with respect to multiplication (6.3).*

Proof. *Case:* $(2, 3)$-*stochastic.* Let $P = (p_{ijk})$ and $Q = (q_{ijk})$ are two $(2, 3)$-stochastic cubic matrices. By (6.3) for the $t_{ia(j,m)r}$ element of $P *_a Q$, we have

$$t_{idr} = \sum_{\substack{j,m: \\ a(j,m)=d}} \sum_{k \in I} p_{ijk} q_{kmr}.$$

Using Lemma 6.2 we get

$$\sum_{d,r \in I} t_{idr} = \sum_{r \in I} \sum_{d=a(j,m)} \sum_{k \in I} p_{ijk} q_{kmr}$$

$$= \sum_{k \in I} \sum_{r \in I} \sum_{j \in I} \sum_{m \in I} p_{ijk} q_{kmr} = \sum_{j,k \in I} p_{ijk} \sum_{m,r \in I} q_{kmr} = 1.$$

Convexity: Take $\lambda \in [0, 1]$ then $\lambda P + (1 - \lambda)Q = (\lambda p_{ijk} + (1 - \lambda)q_{ijk})$. Therefore, it is clear that if P and Q are $(2, 3)$-stochastic then $\lambda P + (1 - \lambda)Q$ is $(2, 3)$-stochastic too.

Case: twice stochastic. We have

$$\sum_{i \in I} t_{idr} = \sum_{i \in I} \sum_{\substack{j,s \in I: \\ a(j,s)=d}} \sum_{k \in I} p_{ijk} q_{ksr}$$

$$= \sum_{\substack{j,s \in I: \\ a(j,s)=d}} \frac{1}{m} \sum_{k \in I} q_{ksr} = \frac{1}{m^2} \sum_{\substack{j,s \in I: \\ a(j,s)=d}} 1 = \frac{m}{m^2} = \frac{1}{m}$$

the last sum equals to m because of a unique (left or right) solvability for any fixed $d \in I$. The convexity is straightforward. $\qquad\square$

---

[1]A semigroup is an algebraic structure consisting of a set together with an associative binary operation.

Remark 6.1. One also can show that $(1,2)$-stochastic cubic matrices form a convex semigroup. But the collection of $(1,3)$-stochastic matrices does not form a semigroup with respect to multiplication (6.3).

6.1.3 *Probabilistic interpretations of stochasticity*

Let us give Maksimov's interpretations of stochastic cubic matrices with multiplications (6.1) and (6.3).

1. **Case** (6.1): Let $P = (p_{ijk})$ be a stochastic matrix then p_{ijk} gives the transition probability of a particle (or a biological element like allele, gene, species etc.) of the type j from the state i to the state k. If P_j denotes the jth layer of the matrix P, then by the multiplication (6.1), the jth layer of the matrix P^n is equal to P_j^n. Therefore one associates to the matrix P the direct sum of independent Markov walks on I with transition matrices $P_1, P_2, \ldots, P_s$. This walk may be considered as a walk of one particle on the states I^s with a transition matrix which is the Kronecker product of matrices P_j, $j \in I$.

Consider a particular case of the multiplication (6.3):

$$E_{ijk} *_1 E_{mnr} = \delta_{km} E_{ijr}. \tag{6.4}$$

2. **Case** (6.4): Let us consider the probabilistic interpretation of multiplication (6.4).

2.1. For the $(2,3)$-stochastic matrix $P = (p_{ijk})$ the entry p_{ijk} may be interpreted as a probability of appearance of a particle of the type k per unit of time as a result of interaction of particles of types i and j. Then the ijr-th entry $p_{ijr}^{(n)}$ of the matrix P^n may be interpreted as a probability of appearance of a particle of type r from the particle of the type i per n units of time under the condition that on the first step the particle of the type i interacts with the particle of the type j.

2.2. For the case of $(1,2)$-stochastic matrix $P = (p_{ijk})$ the entry p_{ijk} one may interpret as a probability for a particle of the type k to have an ordered pair of parents of types i and j. Then the ijr-th entry $p_{ijr}^{(n)}$ of the matrix P^n may be interpreted as a possibility for a particle of the type r obtained per n units of time to have an initial ordered pair of parents of the particles of types i and j.

2.3. Let the matrix $P = (p_{ijk})$ be twice stochastic (with respect to the last index), with multiplication (6.4). Using $\sum_{k \in I} p_{ijk} = 1/m$, one can assume that a particle of the type j will appear with probability $1/m$ per unit of time and during the same time a particle of the type i will appear with probability mp_{ijk} which interacting with the particle j generates a particle of the type k. The ijr-th entry $p_{ijr}^{(n)}$ of the matrix P^n may be considered as a probability of appearance of a particle of the type r from a particle of the type i per r units of time under the condition that a particle of the type j appears at the first step.

The value $\sum_{s \in I} \frac{1}{m} mp_{ksr} = p_{kr}$ may be interpreted as a probability of transition of a particle of the type k into a particle of the type r in one step under the interaction with any particle of the type s since $\sum_{r \in I} p_{kr} = 1$. This process may be represented as follows. At the first step, we choose with probability $1/m$ a particle of any of m types, say of the type j, which interacting with a particle of type i generates a particle of type k. Further appearance of particles occurs by the scheme of a Markov chain with transition matrix (p_{kr}).

3. **Case** (6.3): Consider two $(2,3)$-stochastic matrices $P = (p_{ijk})$ and $Q = (q_{kmr})$ and let the multiplication be defined by (6.3). Then the idr-th entry c_{idr} of the product $P *_a Q$ is equal to $c_{idr} = \sum_{a(j,s)=d} \sum_k p_{ijk} q_{ksr}$.

Now consider p_{ijk} as a probability of transforming a particle of the type i into a particle of the type k per unit of time under the action of the factor j and if we assume that the superposition of the acting factors is again a factor defined by the operation a, then c_{ijk} gives a probability of transforming a particle of the type i into a particle of the type r in two units of time under the action of a superposition of any two factors j and s for which $a(j,s) = d$. Similarly one can treat elements of any power P^n.

Note that the associative property of the multiplication gives a unique definiteness of the factor per n units of time by factors which act at each step. Moreover, the associative property guarantees the preservation of recurrent relations for transition probabilities as in ordinary Markov chains.

6.2   Markov interaction process

Following [158] note that a Markov process which is not actually connected with the multiplication of cubic matrices, and uses only $(1,2)$-stochastic or $(2,3)$-stochastic matrices. In this process the multiplication (6.3) will be used to describe events connected with the process and describing its limit behavior.

Let us give a strict formal description of the process.

Definition 6.1. The Markov interaction process (MIP) is given on I by three elements:

(a) an initial distribution $\mu = \{p_i\}$ on I, i.e.

$$p_i \geq 0, \ \forall i \in I, \ \sum_{i \in I} p_i = 1.$$

(b) a stochastic square matrix (p_{ij}) on I, i.e.

$$p_{ij} \geq 0, \ \forall i, j \in I, \ \sum_{j \in I} p_{ij} = 1, \ \forall i \in I.$$

(c) a 3-stochastic cubic matrix (q_{ijk}), i.e.,

$$q_{ijk} \geq 0, \qquad \sum_{k \in I} q_{ijk} = 1, \quad \text{for all } i, j \in I.$$

The cubic matrices in point (c) will be called probabilistic and numbers q_{ijk} will be called transition probabilities.

Assume at the initial time we have a particle of the type i with probability p_i. Then the MIP proceeds in the following way. A particle of the type j arises per unit of time with probability p_{ij} and a particle of the type i_1 arises with probability q_{iji_1} as a result of interaction of a particle of the type j with a particle of the type i. This also can be considered as a result of the interaction, a particle of the type i is transformed into a particle of the type i_1. At the next moment of time with probability $p_{i_1 j_1}$, a particle of the type j_1 arises; as a result of the interaction of this particle with a particle of the type i_1, a particle of the type i_2 arises, and so on.

Remark 6.2. The MIP is proceeded for a physical system of particles, but one can adapt the notations to biological systems. In particular, a MIP can be interpreted as a population dynamics.

A trajectory (dynamics) of the MIP can be described by a two-row matrix:

$$\begin{pmatrix} i, & i_1, & i_2, & \cdots \\ j, & j_1, & j_2, & \cdots \end{pmatrix},$$

where the first row shows the types of evolving particles at sequential times and the second row shows the types of particles arising for interaction. This is interpreted either as the marking of a particle of the type itself or as the marking of transition of a particle of the type i_k into a particle of the type j_k. Therefore the probability of transition from i to i_1 under interaction of i with j is equal to

$$p_{iji_1} = p_{ij} q_{iji_1} \tag{6.5}$$

and the matrix (p_{iji_1}) is $(2,3)$-stochastic. Moreover the product

$$p_{iji_1}p_{i_1j_1i_2}\cdots p_{i_{n-1}j_{n-1}i_n}$$

gives the probability of the trajectory

$$\begin{pmatrix} i, & i_1, & \cdots & i_{n-1}, & i_n \\ j, & j_1, & \cdots & j_{n-1}, & \bullet \end{pmatrix}.$$

Proposition 6.3. *There exists a one-to-one correspondence between matrices of transition probabilities of the MIP and cubic $(2,3)$-stochastic matrices.*

Proof. Note that the matrix with entries (6.5) is $(2,3)$-stochastic. Conversely, for $(2,3)$-stochastic matrix (p_{iji_1}) setting $p_{ij} = \sum_{i_1} p_{iji_1}$ we obtain the stochastic square matrix (p_{ij}). Moreover, setting $q_{iji_1} = \frac{p_{iji_1}}{p_{ij}}$, we obtain the probabilistic matrix (q_{iji_1}). $\qquad\square$

Note that the ordinary Markov chain is a particular case of the MIP when $q_{iji_1} = q_{ii_1}$ for all $j \in I$. However, in contrast to a Markov chain the limit behavior of probabilities of having a particle of the type i at n-th step essentially depends on the set of particles $j, j_1, \ldots, j_{n-1}$ providing the interaction.

Proposition 6.4. *Let $P = (p_{ijk})$ be the $(2,3)$-stochastic matrix defining MIP by Proposition 6.3. The entry $p_{idk}^{(n)}$ of the matrix $P^n = \left(p_{idk}^{(n)}\right)$ with multiplication (6.3) gives the probability of transition in the MIP under consideration of particles of the type i per n units of time into a particle of the type k under an arbitrary set $j, j_1, \ldots, j_{n-1}$ satisfying the condition $a(j_{n-1}, a(j_{n-2}, \ldots, a(j, j_1)) \ldots) = d$.*

Proof. By multiplication (6.3) we have

$$p_{idk}^{(n)} = \sum_{a(j_{n-1},a(j_{n-2},\ldots,a(j,j_1))\ldots)=d} \ \sum_{i_1,\ldots,i_{n-1}\in I} p_{iji_1}p_{i_1j_1i_2}\cdots p_{i_{n-1}j_{n-1}k}.$$

Using formula (6.5) we get

$$p_{idk}^{(n)} = \sum_{a(j_{n-1},a(j_{n-2},\ldots,a(j,j_1))\ldots)=d} \ \sum_{i_1,\ldots,i_{n-1}\in I} A(i_1,\ldots,i_{n-1};j,j_1,\ldots,j_{n-1}),$$

where

$$A(i_1,\ldots,i_{n-1};j,j_1,\ldots,j_{n-1}) = p_{ij}q_{iji_1}p_{i_1j_1}q_{i_1j_1i_2}\cdots p_{i_{n-1}j_{n-1}}q_{i_{n-1}j_{n-1}k}.$$

This A is equal to the probability of transition of a particle of the type i to a particle of the type j per n units of time in the MIP, if the particle of the type i sequentially generates descendants $i_1, \ldots, i_{n-1}, i_n = k$ with the help of particles types $j, j_1, \ldots, j_{n-1}$.

Thus stationarity of the MIP depends on the choice of multiplication in the sense of which the powers P^n of matrix P converge. $\qquad\square$

6.2.1 *Ergodic property of MIP*

For the MIP a limit behavior of probabilities depends on the sequences of particle types involved in the interactions and is closely connected with a notion of being stationary for cubic matrices.

Definition 6.2. The MIP is called ergodic with respect to multiplication (6.3) if under this multiplication, the powers P^n of the cubic matrix P (corresponding to the MIP) tend to the matrix Π with positive entries.

Note that the limit matrix Π is stationary, i.e., $P\Pi = \Pi$ and $\Pi^2 = \Pi$.

Hence the convergence of transition probabilities of a particle of the type i into a particle of the type j for increasing n is naturally defined with respect to the set $(j, j_1, j_2, \ldots, j_{n-1})$ of particle types appearing for interaction for which $a(j_{n-1}, a(j_{n-2}, \ldots, a(j, j_1)) \ldots) = d$.

Any $(2,3)$-stochastic matrix P can be uniquely represented in the form

$$P = \sum_{i,j,k} p_{ijk} E_{ijk} = \sum_{i,k} q_{ik} \varepsilon_{ik}, \qquad (6.6)$$

where

$$\varepsilon_{ik} = \sum_{t \in I} c_{ilk} E_{itk}, \quad c_{itk} \geq 0, \quad \sum_{t \in I} c_{itk} = 1,$$

and $q_{ik} = p_{ijk}/c_{ijk}$.

The representation (6.6) is called the Markov decomposition of the cubic stochastic matrix P.

Let a be an associate operation on I, if it is a semigroup operation I, then it defines a semigroup G on I. In what follows, if we need to emphasize the semigroup structure of I, we shall write G instead of I.

By $M^+(G)$ denote a collection of various linear combinations of elements of G with nonnegative coefficients. The operation a naturally induces on $M^+(G)$ the semigroup operation. Moreover the semigroup $M^+(G)$ includes a semigroup $M^1(G)$ of all probability measures on G.

Define a mapping

$$\tau : \varepsilon_{ik} = \sum_{t \in I} c_{itk} E_{itk} \rightarrow \tau(\varepsilon_{ik}) = \sum_{t \in G} c_{itk} t.$$

Note that τ is one-to-one.

Proposition 6.5. *The following equality holds*

$$\tau(a(\varepsilon_{ik}, \varepsilon_{km})) = a(\tau(\varepsilon_{ik}), \tau(\varepsilon_{km})).$$

Proof. Using multiplication (6.3) we obtain

$$a(\varepsilon_{ik}, \varepsilon_{km}) = a\left(\sum_{t \in I} c_{itk} E_{itk}, \sum_{u \in I} c_{kum} E_{kum}\right)$$

$$= \sum_{t,u \in I} c_{itk} c_{kum} E_{ia(t,u)m}$$

$$= \sum_{t \in I} \left(\sum_{\substack{a(u,v)=t: \\ u,v \in I}} c_{iuk} c_{kvm}\right) E_{itm}.$$

Therefore

$$\tau(a(\varepsilon_{ik}, \varepsilon_{km})) = \sum_{t \in G} \left(\sum_{\substack{a(u,v)=t: \\ u,v \in G}} c_{iuk} c_{kvm}\right) t$$

$$= a\left(\sum_{u \in G} c_{iuk} u, \sum_{v \in G} c_{kvm} v\right) = a(\tau(\varepsilon_{ik}), \tau(\varepsilon_{km})).$$

$$\square$$

Using (6.6), the mapping τ can be linearly and homomorphically extended onto a semigroup of $(2,3)$-stochastic cubic matrices. This leads to the consideration of square matrices with entries from $M^+(G)$ with ordinary multiplication.

Definition 6.3. The square matrix (δ_{ij}), $\delta_{ij} \in M^+(G)$, is called a stochastic matrix over $M^+(G)$, if $\sum_{k \in I} \delta_{ik} \in M^1(G)$ for all $i \in I$.

Note that these matrices form a convex semigroup. The following proposition will be useful.

Proposition 6.6. *A semigroup of cubic $(2,3)$-stochastic matrices with multiplication (6.3) is isomorphic to a semigroup of stochastic matrices over $M^+(G)$.*

Proof. Let P be a $(2,3)$-stochastic matrix with the Markov decomposition:

$$P = \sum_{i,k \in I} p_{ik} \varepsilon_{ik}.$$

Clearly, the mapping $\psi : P \to (\tau(p_{ik})\tau(\varepsilon_{ik}))$ is one-to-one, convex, and holomorphic in view of Proposition 6.5. Moreover, it is clear that the matrix $(\tau(p_{ik})\varepsilon_{ik})$ is stochastic in the sense of Definition 6.3. This completes the proof. □

Consider the following particular case of (6.3):

$$E_{ijk} *_2 E_{mnr} = \delta_{km} E_{inr}. \tag{6.7}$$

Proposition 6.6 permits us to describe stationary matrices for multiplications (6.4) and (6.7).

Indeed, for distributions μ_1 and μ_2 with multiplication (6.4), the equality $\mu_1 *_1 \mu_2 = \mu_1$ holds, and for multiplication (6.7), we have the equality $\mu_1 *_2 \mu_2 = \mu_2$. Therefore, for both multiplications (6.4) and (6.7) we have $\mu * \mu = \mu$.

Proposition 6.7. *For the general cubic stationary ergodic matrix P with multiplication (6.4), the mapping ψ is of the form $\Pi = (\Pi_{ik})$, where $\Pi_{ik} = p_k \varepsilon_i$, $p_k \geq 0$, $\sum_k p_k = 1$, and $\varepsilon_i \in M^1(G)$.*

Proof. Let $\psi(P) = (\tau(p_{ik})\tau(\varepsilon_{ik}))$ be the image of the cubic stochastic matrix P under the mapping ψ. One can see that the mapping of P into $(\tau(p_{ik}))$ is homomorphic with respect to multiplication. Consequently, since P is stationary we get that $(\tau(p_{ik}))$ is stationary and ergodic. Hence we have $\tau(p_{ik}) = p_k$, $\sum_{k \in I} p_k = 1$ ([248]). Therefore,

$$(\tau(p_{ik})\tau(\varepsilon_{ik})) = (p_{ik}\tau(\varepsilon_{ik})).$$

Since this matrix is idempotent because it is stationary, squaring it, we obtain in view of the equality $\mu_1 * \mu_2 = \mu_1$ in the ith row

$$p_1 \sum_{j=1}^{m} p_j \tau(\varepsilon_{ij}), \quad \cdots \quad , p_m \sum_{j=1}^{m} p_j \tau(\varepsilon_{ij}).$$

Now comparing this row with the ith row of the matrix $(p_k \tau(\varepsilon_{ij}))$, one can see that

$$\tau(\varepsilon_{i1}) = \tau(\varepsilon_{i2}) = \cdots = \tau(\varepsilon_{im}) = \varepsilon_i, \quad \forall i \in I.$$

Conversely, we have that $(p_k \varepsilon_i)^2 = (p_k \varepsilon_i)$. This completes the proof. □

The following proposition can be similarly proved.

Proposition 6.8. *For the general cubic stationary ergodic matrix P with multiplication (6.7), the mapping ψ is of the form $\Pi = (\Pi_{ik})$, where $\Pi_{ik} = p_i\varepsilon_i$, $p_i \geq 0$, $\sum_i p_i = 1$, and $\varepsilon_i \in M^1(G)$.*

Consider the case, when a two-sided invariant measure ε exists in a semigroup G, i.e., there exists a measure for which $a(\mu, \varepsilon) = a(\varepsilon, \mu) = \varepsilon$ for any $\mu \in M^1(G)$. The measure ε must always have a support coinciding with G since otherwise its support would have to be a proper ideal of G, the presence of which contradicts the property of a unique solvability of the operation a.

For semigroups with operation (6.4) or (6.7), the two-sided invariant distribution does not exist. Therefore, this case is in direct contrast to the case under consideration. For an arbitrary finite group G, the invariant distribution on G is a unique two-sided invariant measure on G.

Let us represent stochastic matrices over $M^1(G)$ as $(p_{ij}\Delta_{ij})$, where Δ_{ij} are probabilistic distributions on G and (p_{ij}) a stochastic matrix. It will be convenience, to write $(p_{ij}\Delta_{ij}) = (P, \Delta)$, where $P = (p_{ij})$ and $\Delta = (\Delta_{ij})$. In particular, let $\varepsilon = (\varepsilon)$ be a matrix each entry of which is the two-sided invariant measure ε and let $T = (p_k)$ be a stationary stochastic square matrix.

Lemma 6.3. *Let $P = (p_{ij})$, $Q = (q_{jm})$ and $\Delta = (\Delta_{ij})$ be stochastic matrices with $\Delta_{ij} \in M^1(G)$. The following equalities hold:*

$$(P, \Delta)(Q, \varepsilon) = (PQ, \varepsilon), \quad (Q, \varepsilon)(P, \varepsilon) = (QP, \varepsilon).$$

Proof. Since the measure ε is two-sided invariant, we obtain

$$\Delta_{ij}\varepsilon = \varepsilon, \quad \varepsilon\Delta_{ij} = \varepsilon.$$

Consequently,

$$(p_{ij}\Delta_{ij})(q_{jm}\varepsilon) = \left(\sum_j p_{ij}q_{jm}\Delta_{ij}\varepsilon\right) = \left(\left(\sum_j p_{ij}q_{jm}\right)\varepsilon\right) = (PQ, \varepsilon).$$

The second equality is proved similarly. $\square$

Theorem 6.1. *Let $P = (p_{ink})$ be a cubic $(2, 3)$-stochastic matrix. Then the powers P^n converge to a stationary matrix with positive elements (ergodic) if and only if all elements P^{n_0} are positive for some n_0.*

Proof. The necessity is evident. Let us prove the sufficiency. Assume that all entries of the cubic matrix P are positive. Then its ψ-image is equal to $(\tilde{p}_{ij}\Delta_{ij})$, where $\tilde{p}_{ij} = \sum_{n \in I} p_{inj}$ are entries of a stochastic matrix constructed by the matrix P. By (6.6) and the definition of the mapping ψ, all $\tilde{p}_{ij}$ are positive, and a support of each distribution Δ_{ij} coincides with G. Therefore, for some ϵ, $0 < \epsilon < 1$, the following decomposition holds:

$$(\Delta_{ij}) = \epsilon(\varepsilon) + (1 - \epsilon)(\varepsilon_{ij}), \quad \varepsilon_{ij} \in M^1(G).$$

Therefore

$$(\tilde{p}_{ij}\Delta_{ij}) = \epsilon(\tilde{p}_{ij}\varepsilon) + (1 - \epsilon)(\tilde{p}_{ij}\varepsilon_{ij}). \tag{6.8}$$

Raising both sides of (6.8) to the nth power and using Lemma 6.5, we get

$$\psi(P^n) = (\tilde{p}_{ij}\Delta_{ij})^n = \{1 - (1 - \epsilon)^n\}(\tilde{p}_{ij}^{(n)}\varepsilon) + (1 - \epsilon)^n(\tilde{p}_{ij}\varepsilon_{ij})^n,$$

where $(\tilde{p}_{ij}^{(n)})$ is the nth power of the matrix $(\tilde{p}_{ij})$, and $(\tilde{p}_{ij}\varepsilon_{ij})^n$ is some stochastic matrix over $M^+(G)$. Since $\tilde{p}_{ij} > 0$, it follows that $(\tilde{p}_{ij}^{(n)})$ converges to some stochastic stationary matrix T with positive elements. Therefore, $\psi(P^n)$ converges to the matrix (T, ε). Thus P^n converges to a stochastic matrix that may be represented as a tensor product of the stationary matrix T and the two-sided invariant distribution ε. Since P^{n_0} has positive elements, then as has been proved above,

$$\lim_{t \to \infty} \psi(P^{n_0 t}) = (T, \epsilon).$$

Consequently,

$$\psi(P^n) = \psi(P^r P^{n_0 t}) = \psi(P^r)\psi(P^{n_0 t}), \quad 0 \le r < n_0, \quad \forall n.$$

Then

$$\psi(P^r)\psi(P^{n_0 t}) = \psi(P^r)[(T, \epsilon) + \delta_n],$$

where δ_n tends to zero. Since T is a stationary matrix, by Lemma 6.5 we get $\psi(P^r)(T, \varepsilon) = (T, \varepsilon)$. The theorem is proved. $\qquad\square$

Remark 6.3. Note that the MIP being uniquely defined by a cubic $(2, 3)$-stochastic matrix is not generally speaking connected with the multiplication of cubic matrices. A multiplication only helps to define the set of particles $\{j, j_1, j_2, \dots\}$ with respect to which the final probabilities of appearance of descendants exist. Depending on the cubic matrix P, it may turn out that the MIP is ergodic under one multiplication and is not ergodic under another. Therefore, a question of interest for applications arises: Under which conditions on P can one always define a multiplication such that the MIP defined by P is ergodic? This gives a strategy for the choice of particles $\{j_s\}$. Theorem 6.1 gives one such case: when all entries of the matrix P are positive, the MIP will be ergodic under any choice of the group operation a. Then the natural question is on the choice of the simplest set $J_d^{(n)} = \{j, j_1, j_2, \ldots, j_{n-1}\}$ for which $a(j_{n-1}, a(j_{n-2}, \ldots, a(j, j_1)) \ldots) = d$ arises. This choice can be done by the following way: if I has m elements, then one has to choose the simplest commutative groups of m elements. For example, we can consider the structure of a commutative additive group modulo m. If $m = 2^s$, then it is better to consider the direct product of s groups of second order.

6.2.2 *Markov chains induced by cubic matrices*

Operations (6.3) do the transformation of the set I into itself by the rule $i \star E_{ljk} = k$ if $i = l$ and zero otherwise, which interprets the killing. Verification of the consistency of the operation $\star$ with the associativity of any of these multiplications is evident. The transformation of I^2 into itself which is consistent with the multiplication of cubic matrices is also defined. Recall that by Q_{ij} we denote an m^2- square matrix whose (i, j)th entry is equal to 1 and all other entries are equal to 0.

For the operation (6.3), we define

$$Q_{hl} \star E_{kmr} = \delta_{lk} Q_{a(h,m)r}. \tag{6.9}$$

Then we have

$$[Q_{hk} \star E_{kmr}] \star E_{rst} = Q_{a(h,m)r} \star E_{rst} = Q_{a(a(h,m),s)t}.$$

On the other hand,

$$Q_{hk} \star [E_{kmr} *_a E_{rst}] = Q_{hk} \star E_{ka(m,s)t} = Q_{a(h,a(m,s))t}.$$

Since a is associative we get

$$Q_{a(a(h,m),s)t} = Q_{a(h,a(m,s))t}.$$

In particular, for multiplications (6.4) and (6.7), we have, respectively,

$$Q_{ij} \star E_{tjk} = Q_{ik}, \quad Q_{it} \star E_{tjk} = Q_{jk}, \tag{6.10}$$

in other cases we obtain zero. Define the following symmetric operations

$$Q_{it} \star E_{ijk} = Q_{kt}, \quad Q_{it} \star E_{ijk} = Q_{kj}, \tag{6.11}$$

Using (6.9), the $(2,3)$-stochastic matrix P naturally defines a stochastic matrix on the set I^2. For $(h, l) \in I^2$, define

$$Q_{hl} \star P = Q_{hl} \star \left[\sum_{k,m,r} p_{kmr} E_{kmr} \right] = \sum_{m,r} p_{lmr} Q_{a(h,m)r}. \tag{6.12}$$

Assume that the operation a is uniquely solvable from the right. Therefore, for $s \in I$ and fixed h, the element $a(h, s)$ runs over all elements of I. On the other hand, $\sum_{m,r} p_{lmr} = 1$ for any $l \in I$.

Thus on I^2 we defined the twice-stochastic matrix $\sigma(P)$, whose entries are composed of elements $P = (p_{ijk})$. By (6.12), the entry $P_{(h,l),(d,r)}$ of the matrix $\sigma(P)$ is equal to $P_{lm_d r}$, where $a(h, m_d) = d$. Consequently for any fixed numeration of the pairs (h, l), we have for the columns c composed from them

$$c \star P = \sigma(P)c.$$

Proposition 6.9. *The mapping σ is one-to-one, homomorphic, and linearly convex.*

Proof. The homomorphism of the mapping σ follows from the consistency of the operation $\star$ with the associative property of multiplication of cubic matrices. Indeed, by definition we have

$$c \star PQ = \sigma(PQ)c.$$

On the other hand, by the consistency of $\star$ with the associative property of multiplication, we have

$$c \star PQ = \sigma(P)c \star Q = \sigma(P)\sigma(Q)c.$$

Consequently

$$\sigma(PQ) = \sigma(P)\sigma(Q).$$

The mutual uniqueness and linearity of the mapping σ are evident. Therefore, the mapping σ can be considered as a representation of the algebra of cubic matrices with the help of square matrices. $\qquad \square$

Finally, define another Markov process on I^2 which is naturally generated by the MIP. If

$$\begin{pmatrix} i, & i_1, & i_2, & \cdots \\ j, & j_1, & j_2, & \cdots \end{pmatrix}$$

is a trajectory of the MIP, then the probability of transition of the pair (i_m, j_m) to the pair (i_{m+1}, j_{m+1}) is equal to

$$P_{i_m j_m, \, i_{m+1} j_{m+1}} = \overset{\circ}{P}_{i_m j_m, \, i_{m+1}} P_{i_{m+1} j_{m+1}}. \tag{6.13}$$

By the properties (b) and (c) of the definition of the MIP (Definition 6.1) and by (6.13), we have

$$\sum_{i_{m+1}, j_{m+1}} P_{i_m j_m, \, i_{m+1} j_{m+1}} = 1.$$

Therefore the probabilities in (6.13) define the Markov process on I^2. We know that (Proposition 6.3), the MIP is uniquely defined by the cubic matrix (p_{ijk}), which defines matrix $(\tilde{p}_{ik})$ by

$$\tilde{p}_{ik} = \sum_{j \in I} p_{ijk}.$$

It is clear that $\tilde{p}_{ik}$ are the probabilities of transition of particle i into particle k independent of intermediate interacting particles j. Thus the ordinary Markov process arises, which is remarkable since its transition probabilities in n steps completely define the respective transition probabilities for the process on I^2. More precisely, let $(\tilde{p}_{kt}^{(n)}) = (\tilde{p}_{kt})^n$ and let $P_{ij,th}$ be the probability of transition of the pair (i, j) into the pair (t, h) in n steps. Then we have

$$P_{ij,th}^{(n)} = \left(\sum_{u \in I} p_{iju} \tilde{p}_{ut}^{(n-1)} \right) p_{th}.$$

6.3   Markov process as a quadratic stochastic process

6.3.1   *Preliminaries, problems and motivations*

6.3.1.1   *Markov process of square matrices*

Let us recall first the notion of Markov process for square stochastic matrices. This will be useful to compare with Markov processes of cubic matrices.

A square matrix $\mathcal{U} = (U_{ij})_{i,j=1}^{m}$ is called *right stochastic* if

$$U_{ij} \geq 0, \quad \forall i, j = 1, \ldots, m; \qquad \sum_{j=1}^{m} U_{ij} = 1, \quad \forall i = 1, \ldots, m.$$

Similarly one can define a *left stochastic* matrix being a non-negative real square matrix, with each column summing to 1 and a *doubly stochastic* matrix being a square matrix of non-negative real numbers with each row and column summing to 1.

A family of stochastic matrices $\{\mathcal{U}^{[s,t]} : s, t \geq 0\}$ is called a *Markov process* if it satisfies the Kolmogorov–Chapman equation:

$$\mathcal{U}^{[s,t]} = \mathcal{U}^{[s,\tau]}\mathcal{U}^{[\tau,t]}, \qquad \text{for all } 0 \leq s < \tau < t. \tag{6.14}$$

Let $I = \{1, 2, \ldots, m\}$. A *distribution* (or *state*) of the set I is a probability measure $x = (x_1, \ldots, x_m)$, where x_i is a probability of $i \in I$. The set of all such vectors is called a simplex and denoted by

$$S^{m-1} = \left\{ x \in \mathbb{R}^m : x_i \geq 0, \sum_{i=1}^{m} x_i = 1 \right\}.$$

Let $x^{(0)} = (x_1^{(0)}, \ldots, x_m^{(0)}) \in S^{m-1}$ be an initial distribution on I. Denote by $x^{(t)} = (x_1^{(t)}, \ldots, x_m^{(t)}) \in S^{m-1}$ the distribution of the system at the moment t. For arbitrary moments of time s and t with $s < t$ the matrix $\mathcal{U}^{[s,t]} = \left(U_{ij}^{[s,t]} \right)$ gives the transition probabilities from the distribution $x^{(s)}$ to the distribution $x^{(t)}$. Moreover $x^{(t)}$ depends linearly from $x^{(s)}$:

$$x_k^{(t)} = \sum_{i=1}^{m} U_{ik}^{[s,t]} x_i^{(s)}, \qquad k = 1, \ldots, m.$$

A Markov chain is a type of Markov process that has either discrete state space or discrete time, but the precise definition of a Markov chain varies (see e.g. [12, 248, 251] for the theory of Markov process).

6.3.1.2 *Markov process as a quadratic stochastic process*

Following [32] and [140] define a quadratic stochastic process.

Denote by $\mathcal{S}$ the set of all possible kinds of stochasticity and denote by $\mathbb{M}$ the set of all possible multiplication rules of cubic matrices.

Let parameters $s \geq 0$, $t \geq 0$, are considered as time.

Denote by $\mathcal{M}^{[s,t]} = \left(P_{ijk}^{[s,t]} \right)_{i,j,k=1}^{m}$ a cubic matrix with two parameters.

Definition 6.4. [140] A family $\{ \mathcal{M}^{[s,t]} : s, t \in \mathbb{R}_+ \}$ is called a Markov process of cubic matrices (or a quadratic stochastic process (QSP)) of type $(\sigma | \mu)$ if for each time s and t the cubic matrix $\mathcal{M}^{[s,t]}$ is stochastic in sense $\sigma \in \mathcal{S}$ and satisfies the Kolmogorov–Chapman equation (for cubic matrices):

$$\mathcal{M}^{[s,t]} = \mathcal{M}^{[s,\tau]} *_\mu \mathcal{M}^{[\tau,t]}, \qquad \text{for all } 0 \leq s < \tau < t \tag{6.15}$$

with respect to the multiplication $\mu \in \mathbb{M}$.

Note that this definition of QSP gives an alternative of Definition 5.10 and a natural generalization of the Markov process of Subsection 6.3.1.1. Comparing Definition 5.10 with Definition 6.4 one can see that in the Definition 6.4 of QSP we do not assume $t - s \geq 1$ and it does not depend on an initial point $x^{(0)}$. For the theory of QSPs in the sense of Definition 5.10 see [183].

Definition 6.5. A QSP is called a (time) homogeneous if the matrix $\mathcal{M}^{[s,t]}$ depends only on $t - s$. In this case we write $\mathcal{M}^{[t-s]}$.

Definition 6.6. A QSP is called periodic if the matrix $\mathcal{M}^{[s,t]}$ depends on time s or/and t periodically, i.e. periodicity with respect to s (resp. t): there is $S > 0$ (resp. $T > 0$), such that $\mathcal{M}^{[s+S,t]} = \mathcal{M}^{[s,t]}$ for all $0 \le s < s+S < t$ (resp. $\mathcal{M}^{[s,t+T]} = \mathcal{M}^{[s,t]}$, for all $0 \le s < t$).

6.3.1.3 *Motivations and interpretations*

Quadratic stochastic processes arise naturally in the study of biological and physical systems with interactions (see [126]).

To give such a model, assume there are m different types of particles, denoted by $I = \{1, 2, \ldots, m\}$ the set of all types and the main aim is to study the asymptotic behavior of the variables $\xi_i^{(t)} =$ number of particles of type i at the time t. The initial state is taken to be fixed and described by $\xi_i^{(0)}$, the numbers of particles of type $i \in I$ in the initial (zero) time. These numbers are assumed finite. Denote by

$$x_i^{(t)} = P^{(t)}(i) = \frac{\xi_i^{(t)}}{\sum_{j=1}^{m} \xi_j^{(t)}},$$

fraction of particles of type i at the time t. Thus the vector $x^{(t)} = (x_1^{(t)}, \ldots, x_m^{(t)}) \in S^{m-1}$ is a distribution of the system (i.e. the vector describing fractions of all types of particles) at the moment t.

Let $x^{(0)} = (x_1^{(0)}, \ldots, x_m^{(0)}) \in S^{m-1}$ be an initial distribution on I. For arbitrary moments of time $s \ge 0$ and $t > 0$, with $s < t$, the matrix $\mathcal{M}^{[s,t]}$ gives the transition probabilities from the distribution $x^{(s)}$ to the distribution $x^{(t)}$. To use the matrix $\mathcal{M}^{[s,t]} = \left(P_{ijk}^{[s,t]} \right)$ assume that a particle of type $i \in I$ and a particle of type $j \in I$ have interaction at time s, as an interaction process, then with probability $P_{ijk}^{[s,t]}$ a particle of type k appears at time t. The Kolmogorov–Chapman equation (6.15) gives the time-dependent evolution law of the interacting process (dynamical system).

Since $x^{(t)} \in S^{m-1}$, one can consider the following models:

- Consider $P_{ijk}^{[s,t]}$ as the conditional probability $P^{[s,t]}(k|i,j)$ that ith and jth particles (physics) or species (biology) interbred successfully at time s, then they produce an individual k at time t.

  Assume the "parents" ij are independent for any moment of time s, that is

$$P^{(s)}(i,j) = P^{(s)}(i)P^{(s)}(j) = x_i^{(s)} x_j^{(s)},$$

  and we assume that the matrix $\left(P_{ijk}^{[s,t]} \right)$ is 3-stochastic, then the probability distribution $x^{(t)}$ can be found by the formula of the total probability as

$$x_k^{(t)} = \sum_{i,j=1}^{m} P^{(s)}(i,j) P^{[s,t]}(k|i,j) = \sum_{i,j=1}^{m} P_{ijk}^{[s,t]} x_i^{(s)} x_j^{(s)},$$

where $k = 1, \ldots, m, \ \ 0 \le s < t$.

For 1-stochastic and 2-stochastic it can be defined similarly, by replacing the corresponding indices.

- Consider now a physical (biological, chemical) system where there are m types of "particles" or molecules, the set of types is denoted by $I = \{1, \ldots, m\}$, and each particle may split to two new ones having types from I. Consider $P_{ijk}^{[s,t]}$ as the conditional probability $P^{[s,t]}(i,j|k)$ that a particle of type k starts splitting at time s and finishes splitting at time t and the result is two particles with ith and jth types. For a biological model see Section 6.4.

Assume $\left(P_{ijk}^{[s,t]} \right)$ is (1,2)-stochastic then $x^{(t)}$ can be defined by

$$x_k^{(t)} = \frac{1}{2} \sum_{i,j=1}^{m} \left(P_{kij}^{[s,t]} + P_{ikj}^{[s,t]} \right) x_j^{(s)}, \qquad k = 1, \ldots, m, \ \ 0 \le s < t.$$

$$(6.16)$$

For (1,3)-stochastic and (2,3)-stochastic cases one can define similarly by replacing the indices.

Thus finding $P_{ijk}^{[s,t]}$ from the equation (6.15) (at a fixed (σ, μ)) and studying the time-dependent behavior of $P_{ijk}^{[s,t]}$ we can describe the time-dependent evolution of $x^{(t)}$.

6.3.1.4 *The main problems*

A QSP of type (σ, μ) corresponds to a solution of (6.15). In this section we study QSPs, for the following two cases

- $\sigma = 3$-stochastic and μ is the Maksimov's multiplication $\mu = 0$ given by

$$E_{ijk} *_0 E_{lnr} = \delta_{kl} \delta_{jn} E_{ijr}. \tag{6.17}$$

Then for any two cubic matrices $A = (a_{ijk}), B = (b_{ijk}) \in \mathfrak{C}$ the matrix $A *_0 B = (c_{ijk})$ is defined by

$$c_{ijr} = \sum_{k=1}^{m} a_{ijk} b_{kjr}. \tag{6.18}$$

Call this QSP of type $(3|0)$. Under these conditions the equation (6.15) has the following form

$$P_{ijr}^{[s,t]} = \sum_{k=1}^{m} P_{ijk}^{[s,\tau]} P_{kjr}^{[\tau,t]}, \quad \forall i,j,r \in I, \qquad (6.19)$$

where

$$P_{ijk}^{[s,t]} \geq 0, \quad \forall i,j,k \in I; \qquad \sum_{k=1}^{n} P_{ijk}^{[s,t]} = 1, \quad \text{for all} \quad i,j \in I, \quad 0 \leq s < t.$$
$$(6.20)$$

Thus a QSP of type $(3|0)$ is a solution to the system (6.19) and (6.20).

- $\sigma = (1,2)$-stochastic and μ is the Maksimov's multiplication with the operation $a = a_0$ such that $a_0(i,j) = i$ for any $i,j \in I$. Call this QSP of type $(12|a_0)$. Under these conditions the equation (6.15) has the following form

$$P_{ijr}^{[s,t]} = \sum_{k,n=1}^{m} P_{ijk}^{[s,\tau]} P_{knr}^{[\tau,t]}, \quad \forall i,j,r \in I, \qquad (6.21)$$

where

$$P_{ijr}^{[s,t]} \geq 0, \quad \forall i,j,r \in I; \qquad \sum_{i,j=1}^{m} P_{ijr}^{[s,t]} = 1, \quad \forall r \in I, \quad 0 \leq s < t. \quad (6.22)$$

Hence a QSP of type $(12|a_0)$ is a solution to the system (6.21) and (6.22).

As mentioned in the previous subsection, each QSP (in particular, QSP corresponding to two cases described in this subsection) is useful to construct a Markov process of cubic matrices.

For the multiplication (6.18) one can see that if two cubic matrices, say A and B, are 3-stochastic then their multiplication $A *_0 B$ is 3-stochastic too.

Remark 6.4. The multiplications considered in this section chosen for simplicity to be able to solve (6.15). Below we show that the type $(3|0)$ is related to natural multiplications of square matrices. While the type $(12|a_0)$ also has natural applications (see Section 6.4), the example given in Section 6.4 can not be modeled as a Markov process of square matrices.

6.3.2 *QSPs of type* $(3|0)$

Let $\mathcal{M}_j^{[s,t]} = (P_{ijk}^{[s,t]})_{i,k=1}^m$ be the jth layer of the matrix $\mathcal{M}^{[s,t]}$. The following proposition characterizes all QSPs of type $(3|0)$.

Proposition 6.10 ([140]). *Any solution of the equation* (6.15) *for the multiplication* (6.17) *is a direct sum of solutions of the following m independent equations:*

$$\mathcal{M}_j^{[s,t]} = \mathcal{M}_j^{[s,\tau]} \mathcal{M}_j^{[\tau,t]}, \qquad \text{for all} \quad 0 \le s < \tau < t, \quad j = 1, \ldots, m.$$

The following lemma is obvious.

Lemma 6.4. *The matrix* $\mathcal{M}^{[s,t]}$ *is 3-stochastic if and only if the square matrix* $\mathcal{M}_j^{[s,t]}$ *is right stochastic for any* $j = 1, \ldots, m$.

As corollary of Proposition 6.10 and Lemma 6.4 we have the following.

Theorem 6.2. *Any QSP of type* $(3|0)$ *is a direct sum of m right stochastic square matrices* $\mathcal{M}_j^{[s,t]} = (P_{ijk}^{[s,t]})_{i,k=1}^m$ *satisfying equation* (6.14). *Consequently, any QSP of type* $(3|0)$ *consists m independent collection of usual Markov processes (see Subsection 6.3.1.1).*

Remark 6.5. Denote by $\mathfrak{M}$ the set of all usual Markov processes. The independence mentioned in Theorem 6.2 allows us to say that there is one-to-one correspondence between the set of all QSPs of type $(3|0)$ and the set $\underbrace{\mathfrak{M} \times \mathfrak{M} \times \cdots \times \mathfrak{M}}_{m \text{ times}}$. Since the basic theory of Markov process of square matrices is well developed, this theory can be adapted (by Theorem 6.2) to QSPs of type $(3|0)$.

Here we give examples of QSPs of type $(3|0)$. These examples also will be used to construct QSPs of type $(12|a_0)$.

Example 6.1. In Subsection 5.1.2 to construct chains of some algebras, for $m = 2$, a wide class of solutions of (6.14) is presented, many of them are non-stochastic matrices, in general. Here we recall the following families of (left, right, doubly) stochastic square matrices, which satisfy the equation (6.14), i.e. they generate independently interesting Markov processes:

$$\mathbb{Q}_1^{[s,t]} = \begin{pmatrix} g(s) & g(s) \\ 1 - g(s) & 1 - g(s) \end{pmatrix},$$

where $g(s) \in [0,1]$ is an arbitrary function;

$$Q_2^{[s,t]} = \frac{1}{2} \begin{pmatrix} 1 + \frac{\Psi(t)}{\Psi(s)} & 1 - \frac{\Psi(t)}{\Psi(s)} \\ 1 - \frac{\Psi(t)}{\Psi(s)} & 1 + \frac{\Psi(t)}{\Psi(s)} \end{pmatrix},$$

where $\Psi(t) > 0$ is an arbitrary decreasing function of $t \geq 0$;

$$Q_3^{[s,t]} = \begin{cases} \begin{pmatrix} 1 & 0 \\ 0 & 1 \end{pmatrix}, & \text{if } s \leq t < b, \\[2em] \frac{1}{2} \begin{pmatrix} 1 & 1 \\ 1 & 1 \end{pmatrix}, & \text{if } t \geq b, \end{cases}$$

where $b > 0$;

$$Q_4^{[s,t]} = \begin{pmatrix} 1 & 0 \\ 1 - \frac{\psi(t)}{\psi(s)} & \frac{\psi(t)}{\psi(s)} \end{pmatrix},$$

where $\psi(t) > 0$ is a decreasing function of $t \geq 0$;

$$Q_5^{[s,t]} = \begin{pmatrix} f(t) & 1 - f(t) \\ f(t) & 1 - f(t) \end{pmatrix},$$

where $f(t) \in [0,1]$ is an arbitrary function;

$$Q_6^{[s,t]}(\lambda, \mu) = \begin{pmatrix} 1 - \frac{\lambda - 2\mu}{2(\lambda - \mu)}\left(1 - \frac{\theta(t)}{\theta(s)}\right) & \frac{\lambda - 2\mu}{2(\lambda - \mu)}\left(1 - \frac{\theta(t)}{\theta(s)}\right) \\ \frac{\lambda}{2(\lambda - \mu)}\left(1 - \frac{\theta(t)}{\theta(s)}\right) & 1 - \frac{\lambda}{2(\lambda - \mu)}\left(1 - \frac{\theta(t)}{\theta(s)}\right) \end{pmatrix},$$

where λ, μ are real parameters such that $0 < 2\mu < \lambda$ and $\theta(t) > 0$ is an arbitrary decreasing function;

$$Q_7^{[s,t]} = \begin{cases} \begin{pmatrix} 1 & 0 \\ 0 & 1 \end{pmatrix}, & \text{if } s \leq t < a, \\[2em] \begin{pmatrix} g(t) & 1 - g(t) \\ g(t) & 1 - g(t) \end{pmatrix}, & \text{if } t \geq a, \end{cases}$$

where $g(t) \in [0,1]$ is an arbitrary function.

Using the right stochastic matrices we can construct the following QSPs of type $(3|0)$:

$$\mathcal{M}^{[s,t]} = \left(\mathcal{M}_1^{[s,t]} \,\middle|\, \mathcal{M}_2^{[s,t]}\right), \text{ with any } \mathcal{M}_1^{[s,t]}, \mathcal{M}_2^{[s,t]} \in \left\{\mathbb{Q}_2^{[s,t]}, \mathbb{Q}_3^{[s,t]}, \ldots, \mathbb{Q}_7^{[s,t]}\right\}.$$

Note that the matrices $\mathbb{Q}_i^{[s,t]}$, $i = 1, \ldots, 7$, generate interesting usual Markov processes: some of them independent on time, some depend only on t, but many of them non-homogeneously depend on both s, t. Depending on the statistical models of real-world processes one can choose parameter functions (i.e. g, Ψ, ψ, f, θ) and be able then to control the evolution (with respect to time) of such Markov processes. Then the evolution of the QSP of type $(3|0)$ will be given by the evolution of two independent Markov processes.

6.3.3 *QSPs of type* $(12|a_0)$

Let $\mathcal{M}^{[s,t]} = \left(P_{ijk}^{[s,t]}\right)$ be a cubic matrix, define the square matrix $\overline{\mathcal{M}}^{[s,t]} = (\bar{c}_{ik}^{[s,t]})$ with

$$\bar{c}_{ik}^{[s,t]} = \sum_{j=1}^{m} P_{ijk}^{[s,t]}, \qquad i, k = 1, \ldots, m. \tag{6.23}$$

Proposition 6.11 ([140]). *Any solution of equation* (6.15) *for the multiplication of type* a_0 *(equivalently equation* (6.21)*) can be given by a solution of the system* (6.23) *with a matrix* $\overline{\mathcal{M}}^{[s,t]} = (\bar{c}_{ik}^{[s,t]})$ *which satisfies* (6.14).

From this proposition it follows that the family of matrices $\overline{\mathcal{M}}^{[s,t]} = (\bar{c}_{ik}^{[s,t]})$ is a Markov process if and only if the matrices are left stochastic.

The following lemma gives a connection between left stochastic and $(1,2)$-stochastic matrices.

Lemma 6.5. *The matrix* $\mathcal{M}^{[s,t]} = \left(P_{ijk}^{[s,t]}\right)$, *with* $P_{ijk}^{[s,t]} \geq 0$, *is $(1,2)$-stochastic if and only if the corresponding matrix* $\overline{\mathcal{M}}^{[s,t]}$ *is left stochastic.*

Proof. It is consequence of the equality (6.23). $\square$

6.3.3.1 *Two-dimensional cases*

Now we construct QSPs of type $(12|a_0)$ corresponding to the left stochastic matrices mentioned in Example 6.1. Write a cubic matrix $\mathcal{M}^{[s,t]}$ for $m = 2$

in the following convenient form:

$$\mathcal{M}^{[s,t]} = \begin{pmatrix} P_{111}^{[s,t]} & P_{112}^{[s,t]} & P_{211}^{[s,t]} & P_{212}^{[s,t]} \\ P_{121}^{[s,t]} & P_{122}^{[s,t]} & P_{221}^{[s,t]} & P_{222}^{[s,t]} \end{pmatrix}. \tag{6.24}$$

Case $\mathbb{Q}_1^{[s,t]}$: Let $\overline{\mathcal{M}}^{[s,t]} = \mathbb{Q}_1^{[s,t]}$. Then from (6.21) by (6.23) we get

$$P_{ij1}^{[s,t]} = g(s)P_{ij1}^{[s,\tau]} + (1 - g(s))P_{ij2}^{[s,\tau]}, \qquad i,j = 1,2,$$

$$P_{ij2}^{[s,t]} = g(s)P_{ij1}^{[s,\tau]} + (1 - g(s))P_{ij2}^{[s,\tau]}, \qquad i,j = 1,2.$$

Consequently $P_{ij1}^{[s,t]} = P_{ij2}^{[s,t]}$. Therefore, by the last system we have $P_{ij1}^{[s,t]} = P_{ij1}^{[s,\tau]}$. Hence $P_{ij1}^{[s,t]}$ should not depend on t, i.e. there exists a function $u_{ij}(s)$ such that

$$P_{ij1}^{[s,t]} = P_{ij2}^{[s,t]} = u_{ij}(s). \tag{6.25}$$

By (6.23) and (6.25) we shall have

$$P_{111}^{[s,t]} + P_{121}^{[s,t]} = P_{112}^{[s,t]} + P_{122}^{[s,t]} = u_{11}(s) + u_{12}(s) = g(s),$$

$$P_{211}^{[s,t]} + P_{221}^{[s,t]} = P_{212}^{[s,t]} + P_{222}^{[s,t]} = u_{21}(s) + u_{22}(s) = 1 - g(s).$$

Consequently the matrix (6.24) has the following form:

$$\mathcal{M}_{(1)}^{[s,t]} = \begin{pmatrix} u_{11}(s) & u_{11}(s) & u_{21}(s) & u_{21}(s) \\ g(s) - u_{11}(s) & g(s) - u_{11}(s) & 1 - g(s) - u_{21}(s) & 1 - g(s) - u_{21}(s) \end{pmatrix}, \tag{6.26}$$

where u_{11} and u_{21} are arbitrary functions of $s \geq 0$.

Thus we proved the following.

Proposition 6.12. *Let $g(s) \in [0, 1]$ be an arbitrary function. The family of matrices (6.26), $\mathcal{M}_{(1)}^{[s,t]}$, is a QSP of type $(12|a_0)$ if and only if the functions $u_{11}(s)$ and $u_{21}(s)$ are such that*

$$0 \leq u_{11}(s) \leq g(s), \qquad 0 \leq u_{21}(s) \leq 1 - g(s).$$

For this QSP $\mathcal{M}_{(1)}^{[s,t]}$, using (6.16), let us give the time behavior of the distribution $x^{(t)} = (x_1^{(t)}, x_2^{(t)}) \in S^1$. Fix $s \geq 0$ and by taking a vector $x^{(s)} = (x_1^{(s)}, x_2^{(s)}) \in S^1$, then by formula (6.16) independently on the vector $x^{(s)}$, for any $t > s$, we get

$$x_1^{(t)} = A(s) \equiv \frac{1}{2}\big(g(s) + u_{11}(s) + u_{21}(s)\big),$$

$$x_2^{(t)} = 1 - A(s) = 1 - \frac{1}{2}\big(g(s) + u_{11}(s) + u_{21}(s)\big).$$

Thus the time behavior of $x^{(t)}$ is clear: start process at time s with an arbitrary initial distribution vector $x^{(s)}$ then as soon as the time t turns on the distribution of the system goes to the distribution $(A(s), 1 - A(s))$ and this distribution remains stable during all time $t > s$.

Case $\mathbb{Q}_2^{[s,t]}$: Let $\overline{\mathcal{M}}^{[s,t]} = \mathbb{Q}_2^{[s,t]}$. Then from (6.21) by (6.23) we get

$$P_{ij1}^{[s,t]} = \frac{1}{2} P_{ij1}^{[s,\tau]} \left(1 + \frac{\Psi(t)}{\Psi(\tau)}\right) + \frac{1}{2} P_{ij2}^{[s,\tau]} \left(1 - \frac{\Psi(t)}{\Psi(\tau)}\right), \qquad i,j = 1,2,$$

$$P_{ij2}^{[s,t]} = \frac{1}{2} P_{ij1}^{[s,\tau]} \left(1 - \frac{\Psi(t)}{\Psi(\tau)}\right) + \frac{1}{2} P_{ij2}^{[s,\tau]} \left(1 + \frac{\Psi(t)}{\Psi(\tau)}\right), \qquad i,j = 1,2.$$

Denoting $\alpha_{ij}(s,t) = P_{ij1}^{[s,t]} - P_{ij2}^{[s,t]}$ and $\beta_{ij}(s,t) = P_{ij1}^{[s,t]} + P_{ij2}^{[s,t]}$, from the last system of equations we get

$$\frac{\alpha_{ij}(s,t)}{\Psi(t)} = \frac{\alpha_{ij}(s,\tau)}{\Psi(\tau)}, \qquad \beta_{ij}(s,t) = \beta_{ij}(s,\tau).$$

It follows from the last equalities that $\frac{\alpha_{ij}(s,t)}{\Psi(t)}$ and $\beta_{ij}(s,t)$ do not depend on t, i.e. there are functions $\gamma_{ij}(s)$ and $\zeta_{ij}(s)$ such that

$$\alpha_{ij}(s,t) = \gamma_{ij}(s)\Psi(t), \quad \beta_{ij}(s,t) = \zeta_{ij}(s).$$

Consequently,

$$P_{ij1}^{[s,t]} = \frac{1}{2} \left(\gamma_{ij}(s)\Psi(t) + \zeta_{ij}(s)\right), \quad P_{ij2}^{[s,t]} = \frac{1}{2} \left(\zeta_{ij}(s) - \gamma_{ij}(s)\Psi(t)\right). \tag{6.27}$$

By these equalities from (6.23) (for $\overline{\mathcal{M}}^{[s,t]} = \mathbb{Q}_2^{[s,t]}$) we get

$$\left(\gamma_{11}(s) + \gamma_{12}(s) - \frac{1}{\Psi(s)}\right) \Psi(t) + \zeta_{11}(s) + \zeta_{12}(s) = 1,$$

$$\zeta_{11}(s) + \zeta_{12}(s) - \left(\gamma_{11}(s) + \gamma_{12}(s) - \frac{1}{\Psi(s)}\right) \Psi(t) = 1,$$

$$\left(\gamma_{21}(s) + \gamma_{22}(s) + \frac{1}{\Psi(s)}\right) \Psi(t) + \zeta_{21}(s) + \zeta_{22}(s) = 1,$$

$$\zeta_{21}(s) + \zeta_{22}(s) - \left(\gamma_{21}(s) + \gamma_{22}(s) + \frac{1}{\Psi(s)}\right) \Psi(t) = 1.$$

From this system we obtain

$$\zeta_{11}(s) + \zeta_{12}(s) = \zeta_{21}(s) + \zeta_{22}(s) = 1,$$

$$\gamma_{11}(s) + \gamma_{12}(s) = -\left(\gamma_{21}(s) + \gamma_{22}(s)\right) = \frac{1}{\Psi(s)}.$$

Using these equalities and (6.27) the matrix (6.24) can be written in the following form:

$$\mathcal{M}_{(2)}^{[s,t]} = \frac{1}{2}\begin{pmatrix} \zeta_{11}(s) + \gamma_{11}(s)\Psi(t) & \zeta_{11}(s) - \gamma_{11}(s)\Psi(t) \\ 1 - \zeta_{11}(s) + \left(\frac{1}{\Psi(s)} - \gamma_{11}(s)\right)\Psi(t) & 1 - \zeta_{11}(s) - \left(\frac{1}{\Psi(s)} - \gamma_{11}(s)\right)\Psi(t) \end{pmatrix}$$

$$\left| \begin{array}{cc} \zeta_{21}(s) + \gamma_{21}(s)\Psi(t) & \zeta_{21}(s) - \gamma_{21}(s)\Psi(t) \\ 1 - \zeta_{21}(s) - \left(\frac{1}{\Psi(s)} + \gamma_{21}(s)\right)\Psi(t) & 1 - \zeta_{21}(s) + \left(\gamma_{21}(s) + \frac{1}{\Psi(s)}\right)\Psi(t) \end{array} \right),$$

$$\tag{6.28}$$

where $\Psi(t) > 0$ is a decreasing function, γ_{11}, γ_{21}, ζ_{11} and ζ_{21} are arbitrary functions of $s \geq 0$.

Proposition 6.13. *The family of matrices* (6.28), $\mathcal{M}_{(2)}^{[s,t]}$ *(with $\Psi(t) > 0$ a decreasing function), is a QSP of type* $(12|a_0)$ *if and only if for the functions* γ_{11}, γ_{21}, ζ_{11} *and* ζ_{21} *the following conditions hold:*

$$\frac{1}{2}\left(\frac{1}{\Psi(s)} - \frac{1}{\Psi(t)}\right) \leq \gamma_{11}(s) \leq \frac{1}{2}\left(\frac{1}{\Psi(s)} + \frac{1}{\Psi(t)}\right), \quad \forall s,t,\ 0 \leq s < t;$$

$$-\frac{1}{2}\left(\frac{1}{\Psi(s)} + \frac{1}{\Psi(t)}\right) \leq \gamma_{21}(s) \leq \frac{1}{2}\left(\frac{1}{\Psi(s)} - \frac{1}{\Psi(t)}\right), \quad \forall s,t,\ 0 \leq s < t;$$

$$0 \leq \zeta_{11}(s) \leq 1, \qquad 0 \leq \zeta_{21}(s) \leq 1, \quad \forall s \geq 0.$$

Proof. One can see that the matrix $\mathcal{M}_{(2)}^{[s,t]}$ satisfies $\sum_{i,j} P_{ijk}^{[s,t]} = 1$, for $k = 1, 2$. Therefore we shall show that $P_{ijk}^{[s,t]} \geq 0$. The system of inequalities $P_{1ij}^{[s,t]} \geq 0$, $i, j = 1, 2$, is equivalent to

$$0 \leq \zeta_{11}(s) + \gamma_{11}(s)\Psi(t) \leq 1 + \frac{\Psi(t)}{\Psi(s)},$$

$$0 \leq \zeta_{11}(s) - \gamma_{11}(s)\Psi(t) \leq 1 - \frac{\Psi(t)}{\Psi(s)}.$$

Solving this system of inequalities with respect to $\zeta_{11}(s)$ and $\gamma_{11}(s)$ we get the conditions mentioned in the proposition. The conditions for $\zeta_{21}(s)$ and $\gamma_{21}(s)$ can be obtained similarly from the system of inequalities $P_{2ij}^{[s,t]} \geq 0$, $i, j = 1, 2$. $\qquad\square$

Remark 6.6. If $\Psi(t) > 0$ is a bounded function, say $\Psi(0) \leq \Psi(t) \leq \Psi(\infty)$, then the condition of Proposition 6.13 can be given uniformly with respect to t, i.e. one gets

$$\frac{1}{2}\left(\frac{1}{\Psi(s)} - \frac{1}{\Psi(\infty)}\right) \le \gamma_{11}(s) \le \frac{1}{2}\left(\frac{1}{\Psi(s)} + \frac{1}{\Psi(\infty)}\right), \qquad \text{for all } s \ge 0;$$

$$-\frac{1}{2}\left(\frac{1}{\Psi(s)} + \frac{1}{\Psi(\infty)}\right) \le \gamma_{21}(s) \le \frac{1}{2}\left(\frac{1}{\Psi(s)} - \frac{1}{\Psi(0)}\right), \qquad \text{for all } s \ge 0.$$

Now let us give, for the QSP $\mathcal{M}_{(2)}^{[s,t]}$, the time behavior of the distribution $x^{(t)} = (x_1^{(t)}, x_2^{(t)}) \in S^1$. Fix $s \ge 0$ and by taking an initial distribution $x^{(s)} = (x_1^{(s)}, x_2^{(s)}) \in S^1$, then by formula (6.16) for *any* $t > s$, we get

$$x_1^{(t)} = \frac{1}{4}\left(1 + \zeta_{11}(s) + \zeta_{21}(s) + \left\{\gamma_{11}(s) + \gamma_{21}(s) + \frac{1}{\Psi(s)}\right\}\Psi(t)\right) x_1^{(s)}$$

$$+ \frac{1}{4}\left(1 + \zeta_{11}(s) + \zeta_{21}(s) - \left\{\gamma_{11}(s) + \gamma_{21}(s) + \frac{1}{\Psi(s)}\right\}\Psi(t)\right) x_2^{(s)},$$

$$x_2^{(t)} = \frac{1}{4}\left(3 - \zeta_{11}(s) - \zeta_{21}(s) - \left\{\gamma_{11}(s) + \gamma_{21}(s) + \frac{1}{\Psi(s)}\right\}\Psi(t)\right) x_1^{(s)}$$

$$+ \frac{1}{4}\left(3 - \zeta_{11}(s) - \zeta_{21}(s) + \left\{\gamma_{11}(s) + \gamma_{21}(s) + \frac{1}{\Psi(s)}\right\}\Psi(t)\right) x_2^{(s)}.$$

Thus the time behavior of $x^{(t)}$ depends on the function $\Psi(t)$:

- If $\Psi(t)$ has a limit, say $\Psi(\infty)$, then depending on the initial vector $x^{(s)}$ we have the following limit distribution:

$$\lim_{t\to\infty} x^{(t)} = \left(x_1^{(\infty)}(s), 1 - x_1^{(\infty)}(s)\right),$$

  where

$$x_1^{(\infty)}(s) = \frac{1}{4}\left(1 + \zeta_{11}(s) + \zeta_{21}(s) + \left\{\gamma_{11}(s) + \gamma_{21}(s) + \frac{1}{\Psi(s)}\right\}\Psi(\infty)\right) x_1^{(s)}$$

$$+ \frac{1}{4}\left(1 + \zeta_{11}(s) + \zeta_{21}(s) - \left\{\gamma_{11}(s) + \gamma_{21}(s) + \frac{1}{\Psi(s)}\right\}\Psi(\infty)\right) x_2^{(s)}.$$

- If $\Psi(t)$ is a periodic function, then for any $t > s$ the behavior of $x^{(t)}$ will be periodic.

Thus if one starts the process at time s with an arbitrary initial distribution vector $x^{(s)}$ then the distribution of the system goes to a limit distribution if Ψ has a limit, otherwise, the set of limit points of $x^{(s)}$ is equivalent to the set of limit points of $\Psi(t)$. Thus, choosing the parameter functions Ψ, ζ_{11}, ζ_{21}, γ_{11}, and γ_{21}, one can control the behavior of the distribution $x^{(t)}$ during all time $t > s$.

Case $\mathbb{Q}_3^{[s,t]}$: Let $\overline{\mathcal{M}}^{[s,t]} = \mathbb{Q}_3^{[s,t]}$. Then from (6.21) by (6.23) for $t < b$ we get

$$P_{ij1}^{[s,t]} = P_{ij1}^{[s,\tau]}, \qquad P_{ij2}^{[s,t]} = P_{ij2}^{[s,\tau]}, \quad i,j = 1,2.$$

Consequently, there are $\eta_{ij}(s)$ and $\xi_{ij}(s)$ such that

$$P_{ij1}^{[s,t]} = \eta_{ij}(s), \qquad P_{ij2}^{[s,t]} = \xi_{ij}(s), \quad i,j = 1,2.$$

Moreover, by (6.23), for any $s \geq 0$, we shall have

$$\eta_{11}(s) + \eta_{12}(s) = 1, \qquad \eta_{21}(s) + \eta_{22}(s) = 0.$$
$$\xi_{11}(s) + \xi_{12}(s) = 0, \qquad \xi_{21}(s) + \xi_{22}(s) = 1.$$

In case $t \geq b$ the solution of the equation can be reduced to the case $\mathbb{Q}_1^{[s,t]}$ with $g(s) \equiv \frac{1}{2}$. Therefore, we get

$$\mathcal{M}_{(3)}^{[s,t]} = \begin{cases} \left(\begin{array}{cc|cc} \eta_{11}(s) & \xi_{11}(s) & \eta_{21}(s) & \xi_{21}(s) \\ 1 - \eta_{11}(s) & -\xi_{11}(s) & -\eta_{21}(s) & 1 - \xi_{21}(s) \end{array} \right), & \text{if } s \leq t < b, \\[2ex] \left(\begin{array}{cc|cc} \kappa_{11}(s) & \kappa_{11}(s) & \kappa_{21}(s) & \kappa_{21}(s) \\ \frac{1}{2} - \kappa_{11}(s) & \frac{1}{2} - \kappa_{11}(s) & \frac{1}{2} - \kappa_{21}(s) & \frac{1}{2} - \kappa_{21}(s) \end{array} \right), & \text{if } t \geq b. \end{cases}$$

The following proposition is obvious.

Proposition 6.14. *The family of matrices $\mathcal{M}_{(3)}^{[s,t]}$ is a QSP of type $(12|a_0)$ if and only if*

$$\eta_{11}(x), \xi_{21}(s) \in [0,1], \quad \eta_{21}(s) = \xi_{11}(s) \equiv 0, \quad \kappa_{11}(x), \kappa_{21}(s) \in [0,1/2].$$

6.3.3.2 *m-dimensional case*

For arbitrary m the following theorem gives an example of time non-homogeneous QSP of type $(12|a_0)$:

Theorem 6.3. *Let $\{A^{[t]} = (a_{ij}^{[t]}), t \geq 0\}$ be a family of invertible $m \times m$ square matrices (for all t), and let $(A^{[t]})^{-1} = (b_{ij}^{[t]})$ denote the inverse of $A^{[t]}$. Assume that*

(i) The square matrix $\overline{\mathcal{M}}^{[s,t]} = A^{[s]}(A^{[t]})^{-1}$ is left stochastic for any $s < t$.
(ii) Take arbitrary functions $\beta_{ijk}^{(s)}$, $i,j,k = 1,\ldots,m$, such that

$$\sum_{j=1}^{m} \beta_{ijk}^{(s)} = a_{ik}^{[s]}, \qquad \text{for any } i,k \text{ and } s,$$

$$\sum_{k=1}^{m} \beta_{ijk}^{(s)} b_{kr}^{[t]} \geq 0, \qquad \text{for any } i,j,r \text{ and } s < t.$$

Then the cubic matrix

$$\mathcal{M}^{[s,t]} = \left(\sum_{k=1}^{m} \beta_{ijk}^{(s)} b_{kr}^{[t]} \right)_{i,j,r=1}^{m} \tag{6.29}$$

generates a QSP of type $(12|a_0)$.

Proof. By Theorem 5.15 it is known that the matrix (6.29) satisfies the equation (6.15) for the multiplication a_0. By the conditions (i), (ii) and Lemma 6.5 we conclude that the matrix (6.29) is (1,2)-stochastic. Thus this matrix satisfies conditions (6.21) and (6.22), i.e. is a QSP of type $(12|a_0)$. $\square$

Since the inverse of a stochastic matrix may not be stochastic, one wants to have an example of a family of matrices $A^{[t]}$ satisfying conditions of Theorem 6.3. The following proposition gives such an example.

Proposition 6.15. *Let* $m = 2$ *and suppose that the matrix* $A^{[t]}$, $t \geq 0$, *has the form*

$$A^{[t]} = \begin{pmatrix} a(t) & 1 - b(t) \\ 1 - a(t) & b(t) \end{pmatrix},$$

where $a(t), b(t) \in (0, 1)$ *are arbitrary increasing (resp. decreasing) functions such that* $a(t) + b(t) > 1$ *(resp.* $a(t) + b(t) < 1$*),* $\forall t \in (0, 1)$. *Then the matrix* $A^{[t]}$ *satisfies condition (i) of Theorem 6.3.*

Proof. From condition $\det(A^{[t]}) = a(t) + b(t) - 1 \neq 0$ it follows that $A^{[t]}$ is invertible for any $t \in (0, 1)$. We shall prove that it satisfies the condition (i) of Theorem 6.3. We have

$$(A^{[t]})^{-1} = \frac{1}{a(t) + b(t) - 1} \begin{pmatrix} b(t) & -1 + b(t) \\ -1 + a(t) & a(t) \end{pmatrix}.$$

Using this equality we get

$$A^{[s]}(A^{[t]})^{-1} = (\mathcal{A}_{ij}^{[s,t]})_{i,j=1,2} = \frac{1}{a(t) + b(t) - 1}$$

$$\times \begin{pmatrix} a(s)b(t) - (1 - b(s))(1 - a(t)) & a(t)(1 - b(s)) - a(s)(1 - b(t)) \\ b(t)(1 - a(s)) & b(s)(1 - a(t)) & a(t)b(s) - (1 - a(s))(1 - b(t)) \end{pmatrix}.$$

One can see that $\mathcal{A}_{1j}^{[s,t]} + \mathcal{A}_{2j}^{[s,t]} = 1$, $j = 1, 2$. Therefore it remains to check that $\mathcal{A}_{ij}^{[s,t]} \geq 0$. We assume $a(t) + b(t) > 1$, $\forall t \in (0, 1)$ (the case $a(t) + b(t) < 1$ can be considered similarly). Then, since $0 < a(t), b(t), 1 - a(t), 1 - b(t) < 1$ for all $t \geq 0$, we have

- the inequality $\mathcal{A}_{11}^{[s,t]} \geq 0$ is equivalent to

$$a(s)b(t) - \big(1 - b(s)\big)\big(1 - a(t)\big) \geq 0$$

which is true since $a(s) > 1 - b(s)$ and $b(t) > 1 - a(t)$.
- the inequality $\mathcal{A}_{21}^{[s,t]} \geq 0$ is equivalent to

$$\frac{b(t)}{1 - a(t)} \geq \frac{b(s)}{1 - a(s)},$$

for all $s < t$. The last inequality follows from the condition that $a(t) > a(s)$ and $b(t) > b(s)$, for all $t > s$ (increasing functions).
- the inequality $\mathcal{A}_{12}^{[s,t]} \geq 0$ is equivalent to

$$\frac{a(t)}{1 - b(t)} \geq \frac{a(s)}{1 - b(s)},$$

for all $s < t$. The last inequality follows again from the condition that a and b are increasing functions.
- the inequality $\mathcal{A}_{22}^{[s,t]} \geq 0$ is equivalent to

$$a(t)b(s) - \big(1 - a(s)\big)\big(1 - b(t)\big) \geq 0$$

which is true since by the assumption we have $a(t) > 1 - b(t)$ and $b(s) > 1 - a(s)$.

This completes the proof. $\hspace{2cm}\square$

Denote

$$\alpha(s) = \beta_{111}^{(s)}, \qquad \beta(s) = \beta_{112}^{(s)}, \qquad \gamma(s) = \beta_{211}^{(s)}, \qquad \delta(s) = \beta_{212}^{(s)}.$$

Using Theorem 6.3 and Proposition 6.15 we construct the following cubic matrix

$$\mathcal{N}^{[s,t]} = \left(P_{ijk}^{[s,t]} \right) = \frac{1}{a(t) + b(t) - 1}$$

$$\times \left(\begin{array}{cc} \alpha(s)b(t) + \beta(s)(a(t) - 1) & \alpha(s)(b(t) - 1) + \beta(s)a(t) \\ A(s,t) & B(s,t) \\ \gamma(s)b(t) + \delta(s)(a(t) - 1) & \gamma(s)(b(t) - 1) + \delta(s)a(t) \\ C(s,t) & D(s,t) \end{array} \right).$$

where

$$A(s,t) = (a(s) - \alpha(s))b(t) + (1 - b(s) - \beta(s))(a(t) - 1),$$

$$B(s,t) = (a(s) - \alpha(s))(b(t) - 1) + (1 - b(s) - \beta(s))a(t),$$

$$C(s,t) = (1 - a(s) - \gamma(s))b(t) + (b(s) - \delta(s))(a(t) - 1),$$

$$D(s,t) = (1 - a(s) - \gamma(s))(b(t) - 1) + (b(s) - \delta(s))a(t).$$

The following proposition illustrates Theorem 6.3.

Proposition 6.16. *Let $a(t), b(t) \in (0,1)$ be functions such that $a(t)+b(t)-1 > 0$ for any $t \geq 0$. The family of matrices $\mathcal{N}^{[s,t]}$ is a QSP of type $(12|a_0)$ if and only if the functions $\alpha(s)$, $\beta(s)$, $\gamma(s)$ and $\delta(s)$ satisfy the following*

$$0 \leq \alpha(s) \leq a(s), \qquad 0 \leq \beta(s) \leq 1 - b(s), \quad \forall s \geq 0; \qquad (6.30)$$
$$0 \leq \gamma(s) \leq 1 - a(s), \qquad 0 \leq \delta(s) \leq b(s), \qquad \forall s \geq 0. \qquad (6.31)$$

Proof. One can see that the matrix $\mathcal{N}^{[s,t]}$ satisfies $\sum_{i,j} P_{ijk}^{[s,t]} = 1$, for $k = 1, 2$. Therefore we shall show that $P_{ijk}^{[s,t]} \geq 0$. The system of inequalities $P_{1ij}^{[s,t]} \geq 0$, $i, j = 1, 2$, is equivalent to the following system of inequalities with respect to $\alpha(s)$ and $\beta(s)$:

$$0 \leq b(t)\alpha(s) - (1 - a(t))\beta(s) \leq a(s)b(t) - (1 - a(t))(1 - b(s)),$$
$$0 \leq -(1 - b(t))\alpha(s) + a(t)\beta(s) \leq -a(s)(1 - b(t)) + a(t)(1 - b(s)).$$

By dividing the first inequalities by $1 - a(t) > 0$ and by dividing the second inequalities by $a(t) > 0$ and by summing the resulting inequalities, we get

$$0 \leq \frac{a(t) + b(t) - 1}{(1 - a(t))a(t)}\alpha(s) \leq a(s)\frac{a(t) + b(t) - 1}{(1 - a(t))a(t)}.$$

Since $\frac{a(t)+b(t)-1}{(1-a(t))a(t)} > 0$, we get the first inequalities of (6.30). The second inequalities of (6.30) can be obtained similarly.

Now the system of inequalities $P_{2ij}^{[s,t]} \geq 0$, $i, j = 1, 2$, is equivalent to the following system of inequalities with respect to $\gamma(s)$ and $\delta(s)$:

$$0 \leq b(t)\gamma(s) - (1 - a(t))\delta(s) \leq (1 - a(s))b(t) - (1 - a(t))b(s),$$
$$0 \leq -(1 - b(t))\gamma(s) + a(t)\delta(s) \leq -(1 - a(s))(1 - b(t)) + a(t)b(s).$$

By dividing the first (resp. second) inequalities by $1 - a(t) > 0$ (resp. $a(t) > 0$) and by summing the resulting inequalities, we get

$$0 \leq \frac{a(t) + b(t) - 1}{(1 - a(t))a(t)}\gamma(s) \leq (1 - a(s))\frac{a(t) + b(t) - 1}{(1 - a(t))a(t)}.$$

Again since $\frac{a(t)+b(t)-1}{(1-a(t))a(t)} > 0$, we get the first inequalities of (6.31). The second inequalities of (6.31) can be obtained similarly. $\square$

6.4   A population with possibility of twin birth

For better biological interpretation, we renumber the set I, starting at 0 instead of 1. Consider the set $I = \{0, 1, 2\}$ as the set of types in a population. The element 0 will play the role of an "empty body", the element 1 represents a "female", while the element 2 is a "male".

Consider $P_{ijk}^{[s,t]}$ as the conditional probability $P^{[s,t]}(i,j|k)$ that a member of type $k \in I$ starts its "pregnancy" period at time s and finishes at time t with zero (in case $i = j = 0$), one (in case $i = 0$ or $j = 0$, $i + j \neq 0$) or two (in case $i \neq 0$ and $j \neq 0$, i.e. with a twin) offspring of ith and jth types.

Then it is natural to define $P_{ijk}^{[s,t]}$ as follows:

$$P_{ij0}^{[s,t]} = \begin{cases} 1, & \text{if } i = j = 0, \\ 0, & \text{otherwise;} \end{cases} \tag{6.32}$$

$$P_{ij1}^{[s,t]} \geq 0, \quad \text{for all } i, j \in I, \tag{6.33}$$

here $P_{001}^{[s,t]}$ can be strictly positive, which corresponds, for example, to the case in which the female 1 cannot have a child, because "she" is ill.

$$P_{ij2}^{[s,t]} = \begin{cases} 1, & \text{if } i = j = 0, \\ 0, & \text{otherwise.} \end{cases} \tag{6.34}$$

Remark 6.7. Note that in (6.32) and (6.34) the types 0 and 2 play the same role. But in (6.33) the types 0 and 2 have different outcome. For example,

$P_{001}^{[s,t]}$ means the probability of nobody born,

$P_{221}^{[s,t]}$ means the probability of male-male twin born.

Hence the cubic matrix $\mathcal{M}^{[s,t]} = \left(P_{ijk}^{[s,t]} \right)_{i,j,k=0}^{2}$ has the following form:

$$\mathcal{M}^{[s,t]} = \begin{pmatrix} 1 \ a^{[s,t]} \ 1 & 0 \ \alpha^{[s,t]} \ 0 & 0 \ u^{[s,t]} \ 0 \\ 0 \ b^{[s,t]} \ 0 & 0 \ \beta^{[s,t]} \ 0 & 0 \ v^{[s,t]} \ 0 \\ 0 \ c^{[s,t]} \ 0 & 0 \ \gamma^{[s,t]} \ 0 & 0 \ w^{[s,t]} \ 0 \end{pmatrix}. \tag{6.35}$$

One can check that the functions $P_{ijk}^{[s,t]}$, for $k = 0$ and $k = 2$, satisfy equation (6.21) and condition (6.22), where one should use sums for $i, j = 0, 1, 2$. The equation (6.21) for $P_{ij1}^{[s,t]}$, taking into account the matrix (6.35), can be written as the following system of nine equations

$$a^{[s,t]} = a^{[\tau,t]} + b^{[\tau,t]} + c^{[\tau,t]}$$
$$+ a^{[s,\tau]}\left(\alpha^{[\tau,t]} + \beta^{[\tau,t]} + \gamma^{[\tau,t]}\right) + u^{[\tau,t]} + v^{[\tau,t]} + w^{[\tau,t]}, \tag{6.36}$$

$$b^{[s,t]} = b^{[s,\tau]}\left(\alpha^{[\tau,t]} + \beta^{[\tau,t]} + \gamma^{[\tau,t]}\right), c^{[s,t]} = c^{[s,\tau]}\left(\alpha^{[\tau,t]} + \beta^{[\tau,t]} + \gamma^{[\tau,t]}\right); \tag{6.37}$$

$$\alpha^{[s,t]} = \alpha^{[s,\tau]}\left(\alpha^{[\tau,t]} + \beta^{[\tau,t]} + \gamma^{[\tau,t]}\right), \beta^{[s,t]} = \beta^{[s,\tau]}\left(\alpha^{[\tau,t]} + \beta^{[\tau,t]} + \gamma^{[\tau,t]}\right),$$

$$\gamma^{[s,t]} = \gamma^{[s,\tau]}\left(\alpha^{[\tau,t]} + \beta^{[\tau,t]} + \gamma^{[\tau,t]}\right); \tag{6.38}$$

$$u^{[s,t]} = u^{[s,\tau]}\left(\alpha^{[\tau,t]} + \beta^{[\tau,t]} + \gamma^{[\tau,t]}\right), v^{[s,t]} = v^{[s,\tau]}\left(\alpha^{[\tau,t]} + \beta^{[\tau,t]} + \gamma^{[\tau,t]}\right),$$

$$w^{[s,t]} = w^{[s,\tau]}\left(\alpha^{[\tau,t]} + \beta^{[\tau,t]} + \gamma^{[\tau,t]}\right). \tag{6.39}$$

This is a system of two-variable-functional equations. We should only consider non-negative solutions (see the condition (6.22)) which for any $0 \le s < t$ satisfy

$$a^{[s,t]} + b^{[s,t]} + c^{[s,t]} + \alpha^{[s,t]} + \beta^{[s,t]} + \gamma^{[s,t]} + u^{[s,t]} + v^{[s,t]} + w^{[s,t]} = 1. \tag{6.40}$$

Denoting

$$f(s,t) = \alpha^{[s,t]} + \beta^{[s,t]} + \gamma^{[s,t]}$$

from system (6.38) we get

$$f(s,t) = f(s,\tau)f(\tau,t).$$

This equation has a very rich family of solutions:

(a) $f(s,t) \equiv 0$;
(b) $f(s,t) = \frac{\Phi(t)}{\Phi(s)}$, where Φ is an arbitrary function with $\Phi(s) \ne 0$;
(c)

$$f(s,t) = \begin{cases} 1, & \text{if } s \le t < a, \\ 0, & \text{if } t \ge a, \end{cases} \qquad \text{where } a > 0.$$

Case of the solution (a): In this case we get from system (6.36)–(6.39) and (6.40) that

$$a^{[s,t]} \equiv 1, \quad b^{[s,t]} = c^{[s,t]} = \alpha^{[s,t]} = \beta^{[s,t]} = \gamma^{[s,t]} = u^{[s,t]} = v^{[s,t]} = w^{[s,t]} \equiv 0.$$

Thus we constructed a QSP of type $(12|a_0)$. To give a biological interpretation, let us compute distributions $x^{(t)} = (x_0^{(t)}, x_1^{(t)}, x_2^{(t)})$. By using formula (6.16) we get

$$x_0^{(t)} = \frac{1}{2}\sum_{i,j=0}^{2}\left(P_{0ij}^{[s,t]} + P_{i0j}^{[s,t]}\right)x_j^{(s)} = x_0^{(s)} + x_1^{(s)} + x_2^{(s)} = 1,$$

$$x_1^{(t)} = x_2^{(t)} = 0, \qquad 0 \le s < t.$$

Since $x_0^{(s)} = 1$ is the probability to have 0 type, the biological interpretation of the process is clear: independently on initial distribution $x^{(s)}$, the population will die as soon as the time $t > s$ turns on.

Case of the solution (b): In this case from system (6.37) we get

$$b^{[s,t]} = b^{[s,\tau]}\left(\alpha^{[\tau,t]} + \beta^{[\tau,t]} + \gamma^{[\tau,t]}\right) = b^{[s,\tau]}f(\tau,t) = b^{[s,\tau]}\frac{\Phi(t)}{\Phi(\tau)},$$

consequently,

$$\frac{b^{[s,t]}}{\Phi(t)} = \frac{b^{[s,\tau]}}{\Phi(\tau)}.$$

From the last equality it follows that the function $\frac{b^{[s,t]}}{\Phi(t)}$ should not depend on t, i.e. there exists a function, say $b(s)$, such that

$$\frac{b^{[s,t]}}{\Phi(t)} = b(s) \quad \Rightarrow \quad b^{[s,t]} = b(s)\Phi(t).$$

Similarly one can prove that there are functions $c(s), \alpha(s), \beta(s), \ldots, w(s)$ such that

$$c^{[s,t]} = c(s)\Phi(t), \quad \alpha^{[s,t]} = \alpha(s)\Phi(t), \quad \beta^{[s,t]} = \beta(s)\Phi(t), \quad \gamma^{[s,t]} = \gamma(s)\Phi(t),$$

$$u^{[s,t]} = u(s)\Phi(t), \quad v^{[s,t]} = v(s)\Phi(t), \quad w^{[s,t]} = w(s)\Phi(t). \tag{6.41}$$

By definition of $f(s,t)$ we shall have

$$f(s,t) = \alpha^{[s,t]} + \beta^{[s,t]} + \gamma^{[s,t]} = \Phi(t)(\alpha(s) + \beta(s) + \gamma(s)) = \frac{\Phi(t)}{\Phi(s)},$$

i.e.

$$\alpha(s) + \beta(s) + \gamma(s) = \frac{1}{\Phi(s)}. \tag{6.42}$$

By using equalities (6.41) and (6.42) from (6.36) we get

$$a^{[s,t]} = a^{[\tau,t]} + a^{[s,\tau]}\frac{\Phi(t)}{\Phi(\tau)} + \Phi(t)\big(b(\tau) + c(\tau) + u(\tau) + v(\tau) + w(\tau)\big).$$

Denoting $g(s,t) = \frac{a^{[s,t]}}{\Phi(t)}$ from the last equation we get

$$g(s,t) = g(s,\tau) + g(\tau,t) + b(\tau) + c(\tau) + u(\tau) + v(\tau) + w(\tau).$$

This equation has the following solution[2]

$$g(s,t) = \kappa(t) - \kappa(s) - b(s) - c(s) - u(s) - v(s) - w(s),$$

---

[2] Recall that the equation $g(s,t) = g(s,\tau) + g(\tau,t)$ is known as Cantor's first equation. One can check that this equation has very rich class of solutions, i.e. $g(s,t) = \kappa(t) - \kappa(s)$ is a solution for an *arbitrary* function κ.

where $\kappa(t)$ is an arbitrary function. Consequently, we get

$$a^{[s,t]} = \Phi(t)\left(\kappa(t) - \kappa(s) - b(s) - c(s) - u(s) - v(s) - w(s)\right).$$

Thus we obtained a solution of the system (6.36)–(6.39), for which the condition (6.40) has the form

$$\Phi(t)\left(\kappa(t) - \kappa(s) + \frac{1}{\Phi(s)}\right) = 1,$$

i.e.

$$\kappa(s) - \frac{1}{\Phi(s)} = \kappa(t) - \frac{1}{\Phi(t)}.$$

This equality says that the function $\kappa(t) - \frac{1}{\Phi(t)}$ should not depend on t, i.e. there is a constant K such that

$$\kappa(t) = K + \frac{1}{\Phi(t)}.$$

Thus

$$a^{[s,t]} = \Phi(t)\left(\frac{1}{\Phi(t)} - \frac{1}{\Phi(s)} - b(s) - c(s) - u(s) - v(s) - w(s)\right).$$

Now we are ready to write an explicit formula for the corresponding cubic matrix:

$$\mathcal{M}^{[s,t]} = \Phi(t)\begin{pmatrix} 1\,\frac{1}{\Phi(t)} - \frac{1}{\Phi(s)} - b(s) - c(s) - u(s) - v(s) - w(s) & 1 \\ 0 & b(s) & 0 \\ 0 & c(s) & 0 \end{pmatrix}$$

$$\begin{vmatrix} 0 & \alpha(s) & 0 & 0 & u(s) & 0 \\ 0 & \beta(s) & 0 & 0 & v(s) & 0 \\ 0 & \frac{1}{\Phi(s)} - \alpha(s) - \beta(s) & 0 & 0 & w(s) & 0 \end{vmatrix}. \qquad (6.43)$$

Thus we have proved the following.

Proposition 6.17. *The family of matrices* $\mathcal{M}^{[s,t]}$, *(6.43), is a QSP of type* $(12|a_0)$ *if and only if* $b(t), c(t), \alpha(t), \beta(t), u(t), v(t), w(t) \in [0,1]$, $\Psi(t) > 0$, *are arbitrary functions such that*

$$\frac{1}{\Phi(t)} - \frac{1}{\Phi(s)} \geq b(s) + c(s) + u(s) + v(s) + w(s),$$

$$\alpha(t) + \beta(t) \leq \frac{1}{\Phi(t)}, \quad \textit{for all } 0 \leq s < t.$$

Time behavior of this QSP depends on fixed functions. Let us give some interesting interpretations:

- Assume the following limit exists

$$\lim_{t \to \infty} \Phi(t) = \Phi(\infty) > 0.$$

  In this case, if, for example, $\beta(s) > 0$ then the population has a positive probability, $P_{111}^{[s,\infty]} = \Phi(\infty)\beta(s) > 0$ to have twins (i.e. female-female twins). In this case, the condition $\gamma(s) + v(s) > 0$ gives positive probability $\Phi(\infty)(\gamma(s) + v(s))$ of having female-male twins, and similarly if $w(s) > 0$ then $\Phi(\infty)w(s) > 0$ is the positive probability of male-male twins.

- If at the initial time s some probability is positive, for example, $\Psi(t)\beta(s) > 0$, then during all fixed time t, with $t > s$, this probability remains positive. For this example, it means that if initially, the population had possibility, to have a female-female twin, then it will have this possibility always, although this might not be true in the limiting case.

- If $\Phi(\infty) = 0$ then the population asymptotically dies.

- It is known that the human twin birth rate is about 2 percent of standard one-child birth. This can be used to choose the parameter functions. For example, one can take $\beta(t) = 0.02\alpha(t)$, etc.

Case of the solution (c): In this case for $t < a$ we have

$$\alpha^{[\tau,t]} + \beta^{[\tau,t]} + \gamma^{[\tau,t]} = 1, \qquad 0 \le \tau < t < a.$$

Then from the system (6.37)–(6.39) we get that there are functions $b_0(s), c_0(s), \ldots, w_0(s)$ such that

$$b^{[s,t]} = b_0(s), \ c^{[s,t]} = c_0(s), \ \alpha^{[s,t]} = \alpha_0(s), \ \beta^{[s,t]} = \beta_0(s), \ \gamma^{[s,t]} = \gamma_0(s),$$

$$u^{[s,t]} = u_0(s), \ v^{[s,t]} = v_0(s), \ w^{[s,t]} = w_0(s), \ 0 \le \tau < t < a.$$

Then by equation (6.36) we get that there is $\kappa_0(t)$ such that

$$a^{[s,t]} = \kappa_0(t) - \kappa_0(s) - b_0(s) - c_0(s) - u_0(s) - v_0(s) - w_0(s).$$

Then by condition (6.40) we get that $\kappa_0(t) = \kappa_0(s)$. To make the corresponding matrix a (1,2)-stochastic we need $b_0(s), c_0(s), u_0(s), v_0(s), w_0(s) \in [0,1]$ and by previous results we get

$$a^{[s,t]} = -b_0(s) - c_0(s) - u_0(s) - v_0(s) - w_0(s),$$

which is non-negative if and only if

$$b_0(s) = c_0(s) = u_0(s) = v_0(s) = w_0(s) \equiv 0.$$

Thus, for $t < a$, the cubic matrix has the following form

$$\mathcal{M}_1^{[s,t]} = \begin{pmatrix} 1\,0\,1 & 0 & \alpha_0(s) & 0 & 0\,0\,0 \\ 0\,0\,0 & 0 & \beta_0(s) & 0 & 0\,0\,0 \\ 0\,0\,0 & 0\ 1-\alpha_0(s)-\beta_0(s)\ 0 & 0\,0\,0 \end{pmatrix}.$$

The case $t \geq a$ is simpler, because in this case $f(s,t) = 0$ and the solution of the system is

$$a^{[s,t]} = 1, \qquad b^{[s,t]} = \cdots = w^{[s,t]} = 0.$$

Thus, for $t \geq a$, the cubic matrix has the following form

$$\mathcal{M}_0^{[s,t]} = \begin{pmatrix} 1\,1\,1 & 0\,0\,0 & 0\,0\,0 \\ 0\,0\,0 & 0\,0\,0 & 0\,0\,0 \\ 0\,0\,0 & 0\,0\,0 & 0\,0\,0 \end{pmatrix}.$$

Define now

$$\mathcal{M}^{[s,t]} = \begin{cases} \mathcal{M}_1^{[s,t]}, & \text{if } s \leq t < a, \\ \mathcal{M}_0^{[s,t]}, & \text{if } t \geq a, \end{cases} \qquad \text{where } a > 0. \qquad (6.44)$$

Thus we have proved the following.

Proposition 6.18. *The family of matrices* $\mathcal{M}^{[s,t]}$, *(6.44), is a QSP of type* $(12|a_0)$ *if and only if* $\alpha_0(t), \beta_0(t) \in [0,1]$, *are arbitrary functions such that*

$$\alpha_0(t) + \beta_0(t) \leq 1, \quad \text{for all } 0 \leq s < t < a.$$

The time behavior of this QSP depends on fixed functions. But it is simpler than the previous case. Let us give some interesting interpretations:

- Start the process at time s, then for any $t < a$, the probabilities $\{P_{101}^{[s,t]}, P_{111}^{[s,t]}, P_{121}^{[s,t]}\} = \{\alpha_0(s), \beta_0(s), 1 - \alpha_0(s) - \beta_0(s)\}$ are independent on time $t < a$. While all the other probabilities independent on both s and t. Thus the process is stable for any $t < a$, i.e. until $t = a$.
- Start the process at time s as soon as $t \geq a$ then the population immediately dies. This phenomenon reminds a cataclysm (catastrophe): "everything is going good, good, $\ldots$, died".

Bibliographical notes. The chapter is based on [31], [32], [139]–[142], [155], [158], [183]. We refer to [113], [133], [183], [201]–[205], [235], [252] for further reading about a nonlinear Markov evolution, which is a dynamical system generated by a measure-valued ordinary differential equation with the specific feature of preserving positivity. These works devoted to the theory of nonlinear Markov processes: existence, uniqueness, constructions, approximation schemes, regularity, law of large numbers and probabilistic interpretations. The book [134] devoted to the dynamical systems when all relevant variables of the system are nonnegative. Such dynamical systems arise is in biology, demography or economics, where the levels of populations or prices of goods are positive.

Chapter 7

Cubic stochastic operators and processes

In this chapter analogically as quadratic stochastic operators and processes, we define cubic stochastic operator (CSO) and cubic stochastic processes (CSP). We give a construction of a CSO and show that dynamical systems generated by such a CSO can be studied by studying of the behavior of trajectories of a Volterra CSO given on a finite dimensional simplex. We define a CSP and drive differential equations for such CSPs with continuous time.

7.1 Construction of CSO for finite set

7.1.1 *Definitions*

Let $E = \{1, 2, \ldots, m\}$ be a fined set and S^{m-1} be $(m-1)$-dimensional simplex.

A cubic stochastic operator CSO arises as follows: consider a population consisting of m species. Let $x^0 = (x_1^0, \ldots, x_m^0)$ be the probability distribution (where $x_i^0 = P(i)$ is the probability of $i \in E$) of species in the initial generation, and $P_{ijk,l}$ the probability that individuals in the ith, jth and kth species interbreed to produce an individual l, more precisely $P_{ijk,l}$ is the conditional probability $P(l|i,j,k)$ that ith, jth and kth species interbred successfully, then they produce an individual l. Assume the population is free, i.e., in any generation the "parents" ijk are independent i.e., $P(i,j,k) = P(i)P(j)P(k) = x_i x_j x_k$. Then the probability distribution $x' = (x_1', \ldots, x_m')$ of the species in the first generation can be found by the total probability

$$x_l' = \sum_{i,j,k=1}^{m} P(l|i,j,k)P(i,j,k) = \sum_{i,j,k=1}^{m} P_{ijk,l} x_i^0 x_j^0 x_k^0, \quad l = 1, \ldots, m. \quad (7.1)$$

Thus the association $x^0 \to x'$ defines a map W, called cubic stochastic operator, where $P_{ijk,l}$ are coefficients of 'heredity' and

$$P_{ijk,l} \geq 0, \quad \sum_{l=1}^{m} P_{ijk,l} = 1, \quad i,j,k,l = 1,...,m \qquad (7.2)$$

In [130], [128] and [216] the behavior of trajectories of some CSOs were studied.

Note that each CSO can be uniquely defined by a stochastic matrix $\mathbb{P} = \{P_{ijk,l}\}_{i,j,k,l=1}^{m}$. In [63] a constructive description of quadratic stochastic operator is given. Some particular cases of this construction were defined in [64].

Following [215] we give a similar construction for the CSOs.

Let $G = (\mathcal{L}, L)$ be a finite graph without loops and multiple edges, where $\mathcal{L}$ is the set of vertexes and L is the set of edges of the graph.

Let Φ be a finite set, called the set of alleles. The function $\sigma : \mathcal{L} \to \Phi$ is called a cell.

Denote by Ω the set of all cells. Let $S(\mathcal{L}, \Phi)$ be the set of all probability measures defined on the finite set Ω.

Let $\{\mathcal{L}_i, i = 1, ..., N\}$ be the set of maximal connected subgraphs (components) of the graph G.

For $\sigma \in \Omega$ denote by $\sigma(M)$ its "projection" (or "restriction") to $M \subset \mathcal{L}$, i.e., $\sigma(M) = \{\sigma(x)\}_{x \in M}$.

Any $\sigma \in \Omega$ has the form $\sigma = (\sigma_1, \ldots, \sigma_N)$, where $\sigma_i = \sigma(\mathcal{L}_i)$. The set $\sigma(M)$ is a called a subcell iff M is a maximal connected subgraph of G.

Fix three cells $\sigma, \varphi, \psi \in \Omega$, and put

$$\Omega(\sigma, \varphi, \psi) = \{\tau = (\tau_1, \ldots, \tau_N) \in \Omega : \tau_i \in \{\sigma_i, \varphi_i, \psi_i\}, \forall i = 1, \ldots, N\}.$$

Remark 7.1. The set $\Omega(\sigma, \varphi, \psi)$ can be interpreted as the set of all possible 'children' of the 'parents' $\theta = (\sigma, \varphi, \psi)$. A child τ can be born from θ if it only consists the subcells of its parents θ.

Now let $\mu \in S(\mathcal{L}, \Phi)$ be a probability measure defined on Ω such that $\mu(\sigma) > 0$ for any cell $\sigma \in \Omega$;. The heredity coefficients $P_{\sigma\varphi\psi,\tau}$ are defined as

$$P_{\sigma\varphi\psi,\tau} = \begin{cases} \dfrac{\mu(\tau)}{\mu(\Omega(\sigma,\varphi,\psi))}, & \text{if } \tau \in \Omega(\sigma,\varphi,\psi), \\ 0 & \text{otherwise.} \end{cases} \qquad (7.3)$$

Obviously, $P_{\sigma\varphi\psi,\tau} \geq 0$, and $\sum_{\tau\in\Omega} P_{\sigma\varphi\psi,\tau} = 1$ for all $\sigma, \varphi, \psi \in \Omega$.

The CSO $W \equiv W_\mu$ acting on the simplex $S(\mathcal{L}, \Phi)$ and determined by coefficients (7.3) is defined as follows: for an arbitrary measure $\lambda \in S(\mathcal{L}, \Phi)$, the measure $W(\lambda) = \lambda' \in S(\mathcal{L}, \Phi)$ is defined by the equality

$$\lambda'(\tau) = \sum_{\sigma,\varphi,\psi\in\Omega} P_{\sigma\varphi\psi,\tau}\lambda(\sigma)\lambda(\varphi)\lambda(\psi) \tag{7.4}$$

for any cell $\tau \in \Omega$.

The CSO construction is also closely related to the graph structure on the set $\mathcal{L}$.

A CSO is called Volterra if the coefficients $P_{ijk,l}$ may be nonzero only when $l \in \{i, j, k\}$ and vanish in all the remaining cases (see [130], [128]).

One can see that any Volterra CSO has the following form

$$W : x_l' = x_l\left(x_l^2 + x_l\sum_{\substack{i=1\\i\neq l}}^{m} a_{i,l}x_i + \sum_{\substack{i,j=1\\i\neq l,\,j\neq l}}^{m} b_{ij,l}x_ix_j\right), \quad l = 1, ..., m \tag{7.5}$$

where $a_{i,l}$ and $b_{ij,l}$ are some coefficients depending on $P_{ijk,l}$.

Theorem 7.1. *The CSO (7.4) is Volterra if and only if the graph G is connected.*

Proof. Let G be connected then $\Omega(\sigma, \varphi, \psi) = \{\sigma, \varphi, \psi\}$. Consequently, by (7.3) it follows that the corresponding operator is Volterra. Conversely, if (7.3) satisfies $P_{\sigma\varphi\psi,\tau} = 0$, for $\tau \not\subset \{\sigma, \varphi, \psi\}$ then by condition $\mu(\sigma) > 0$ it follows that G is connected. $\square$

7.1.2 *Non-Volterra CSOs*

In this section we describe a condition on measure μ under which the CSO W_μ generated by μ (using the construction described in the previous section) can be studied using the theory of Volterra CSO.

Denote by $\Omega_i = \Phi^{\mathcal{L}_i}$ the set of all cells defined on component $\mathcal{L}_i$, $i = 1, ..., N$. Let μ_i be a probability measure defined on Ω_i, such that $\mu_i(\sigma) > 0$ for any $\sigma \in \Omega_i$, $i = 1, ..., N$.

Consider probability measure μ on $\Omega = \Omega_1 \times \cdots \times \Omega_N$ defined as

$$\mu(\sigma) = \prod_{i=1}^{N} \mu_i(\sigma_i), \tag{7.6}$$

where $\sigma = (\sigma_1, ..., \sigma_N)$, with $\sigma_i \in \Omega_i, i = 1, ..., N$.

By Theorem 7.1 if $N = 1$ then QSO constructed on G is Volterra QSO.

Theorem 7.2. *The CSO constructed by (7.3) with measure (7.6) is reducible to N separate Volterra CSOs.*

Proof. For any $\sigma = (\sigma_1, ..., \sigma_N)$, $\varphi = (\varphi_1, ..., \varphi_N)$, $\psi = (\psi_1, ..., \psi_N) \in \Omega$ we have

$$\mu(\Omega(\sigma, \varphi, \psi)) = \sum_{\substack{\tau_1, ..., \tau_N: \\ \tau_i \in \{\sigma_i, \varphi_i, \psi_i\}, i=1,...,N}} \prod_{i=1}^{N} \mu_i(\tau_i) = \prod_{i=1}^{N} \left(\mu_i(\sigma_i) + \mu_i(\varphi_i) + \mu_i(\psi_i) \right).$$

Using this equality by (7.3) we get

$$P_{\sigma\varphi\psi,\tau} = \begin{cases} \prod_{i=1}^{N} \dfrac{\mu_i(\tau_i)}{\mu_i(\sigma_i) + \mu_i(\varphi_i) + \mu_i(\psi_i)}, & \text{if } \tau \in \Omega(\sigma, \varphi, \psi), \\ 0 & \text{otherwise.} \end{cases} \tag{7.7}$$

Thus CSO generated by measure (7.6) can be written as

$$\lambda'(\tau) = \lambda'(\tau_1, ..., \tau_N) =$$

$$\sum_{\substack{\sigma = (\sigma_1, ..., \sigma_N) : \sigma_i \in \Omega_i \\ \varphi = (\varphi_1, ..., \varphi_N) : \varphi_i \in \Omega_i \\ \psi = (\psi_1, ..., \psi_N) : \psi_i \in \Omega_i}} \prod_{i=1}^{N} \frac{\mu_i(\tau_i) \mathbf{1}_{(\tau_i \in \{\sigma_i, \varphi_i, \psi_i\})}}{\mu_i(\sigma_i) + \mu_i(\varphi_i) + \mu_i(\psi_i)} \lambda(\sigma)\lambda(\varphi)\lambda(\psi). \tag{7.8}$$

Denote

$$X_{i,\omega} = \sum_{\substack{\tau \in \Omega: \\ \tau_i = \omega}} \lambda(\tau) = \sum_{\substack{\tau_1, ..., \tau_{i-1}, \tau_{i+1}, ..., \tau_N \\ \tau_k \in \Omega_k, k \neq i}} \lambda(\tau_1, ..., \tau_{i-1}, \omega, \tau_{i+1}, ..., \tau_N). \tag{7.9}$$

From (7.8) we have

$$X'_{i,\omega} = \sum_{\substack{\tau \in \Omega: \\ \tau_i = \omega}} \lambda'(\tau) = \sum_{\substack{\tau \in \Omega: \\ \tau_i = \omega}} \left[\sum_{\substack{\sigma_1, ..., \sigma_{i-1}, \sigma_{i+1}, ..., \sigma_N \\ \varphi, \psi \in \Omega}} \frac{\mu_i(\omega)}{\mu_i(\omega) + \mu_i(\varphi_i) + \mu_i(\psi_i)} \right.$$

$$\times \prod_{\substack{j=1 \\ j \neq i}}^{N} \frac{\mu_j(\tau_j) \mathbf{1}_{(\tau_j \in \{\sigma_i, \varphi_j, \psi_j\})}}{\mu_j(\sigma_j) + \mu_j(\varphi_j) + \mu_j(\psi_j)} \lambda(\sigma_1, ..., \sigma_{i-1}, \omega, \sigma_{i+1}, ..., \sigma_N)\lambda(\varphi)\lambda(\psi)$$

$$+ \sum_{\substack{\varphi_1, ..., \varphi_{i-1}, \varphi_{i+1}, ..., \varphi_N \\ \sigma, \psi \in \Omega}} \frac{\mu_i(\omega)}{\mu_i(\sigma_i) + \mu_i(\omega) + \mu_i(\psi_i)}$$

$$\times \prod_{\substack{j=1 \\ j\neq i}}^{N} \frac{\mu_j(\tau_j)\mathbf{1}_{(\tau_j\in\{\sigma_i,\varphi_j,\psi_j\})}}{\mu_j(\sigma_j)+\mu_j(\varphi_j)+\mu_j(\psi_j)}\lambda(\sigma)\lambda(\varphi_1,...,\varphi_{i-1},\omega,\varphi_{i+1},...,\varphi_N)\lambda(\psi)$$

$$+\sum_{\substack{\psi_1,...,\psi_{i-1},\psi_{i+1},...,\psi_N \\ \sigma,\varphi\in\Omega}} \frac{\mu_i(\omega)}{\mu_i(\sigma_i)+\mu_i(\varphi_i)+\mu_i(\omega)}$$

$$\times \prod_{\substack{j=1 \\ j\neq i}}^{N} \frac{\mu_j(\tau_j)\mathbf{1}_{(\tau_j\in\{\sigma_i,\varphi_j,\psi_j\})}}{\mu_j(\sigma_j)+\mu_j(\varphi_j)+\mu_j(\psi_j)}\lambda(\sigma)\lambda(\varphi)\lambda(\psi_1,...,\psi_{i-1},\omega,\psi_{i+1},...,\psi_N)\Bigg]$$

$$= 3 \sum_{\substack{\sigma_1,...,\sigma_{i-1},\sigma_{i+1},...,\sigma_N \\ \varphi,\psi\in\Omega}} \frac{\mu_i(\omega)}{\mu_i(\omega)+\mu_i(\varphi_i)+\mu_i(\psi_i)}$$

$$\times \sum_{\substack{\tau\in\Omega: \\ \tau_i=\omega}} \prod_{\substack{j=1 \\ j\neq i}}^{N} \frac{\mu_j(\tau_j)\mathbf{1}_{(\tau_j\in\{\sigma_j,\varphi_j,\psi_j\})}}{\mu_j(\sigma_j)+\mu_j(\varphi_j)+\mu_j(\psi_j)}\lambda(\sigma_1,...,\sigma_{i-1},\omega,\sigma_{i+1},...,\sigma_N)\lambda(\varphi)\lambda(\psi).$$

$$(7.10)$$

One can see that

$$\sum_{\tau_1,...,\tau_{i-1},\tau_{i+1},...,\tau_N} \prod_{\substack{j=1 \\ j\neq i}}^{N} \frac{\mu_j(\tau_j)\mathbf{1}_{(\tau_j\in\{\sigma_j,\varphi_j,\psi_j\})}}{\mu_j(\sigma_j)+\mu_j(\varphi_j)+\mu_j(\psi_j)} = 1.$$

Thus from (7.10) we have

$$\text{RHS of } (7.10)$$

$$= 3 \sum_{\substack{\sigma_1,...,\sigma_{i-1},\sigma_{i+1},...,\sigma_N \\ \varphi,\psi\in\Omega}} \frac{\mu_i(\omega)}{\mu_i(\omega)+\mu_i(\varphi_i)+\mu_i(\psi_i)}$$

$$\times \lambda(\sigma_1,...,\sigma_{i-1},\omega,\sigma_{i+1},...,\sigma_N)\lambda(\varphi)\lambda(\psi)$$

$$= \sum_{\substack{\sigma,\varphi,\psi \\ \sigma_i=\varphi_i=\psi_i=\omega}} \lambda(\sigma)\lambda(\varphi)\lambda(\psi) + 6 \sum_{\psi_i\in\Omega_i\setminus\omega} \frac{\mu_i(\omega)}{2\mu_i(\omega)+\mu_i(\psi_i)}$$

$$\times \sum_{\substack{\sigma_1,...,\sigma_{i-1},\sigma_{i+1},...,\sigma_N \\ \varphi_1,...,\varphi_{i-1},\varphi_{i+1},...,\varphi_N \\ \psi_1,...,\psi_{i-1},\psi_{i+1},...,\psi_N}} \lambda(\sigma_1,...,\sigma_{i-1},\omega,\sigma_{i+1},...,\sigma_N)$$

$$\times \lambda(\varphi_1, ..., \varphi_{i-1}, \omega, \varphi_{i+1}, ..., \varphi_N)\lambda(\psi)$$

$$3 \sum_{\varphi_i, \psi_i \in \Omega_i \setminus \omega} \frac{\mu_i(\omega)}{\mu_i(\omega) + \mu_i(\varphi_i) + \mu_i(\psi_i)}$$

$$+ \sum_{\substack{\sigma_1, ..., \sigma_{i-1}, \sigma_{i+1}, ..., \sigma_N \\ \varphi_1, ..., \varphi_{i-1}, \varphi_{i+1}, ..., \varphi_N \\ \psi_1, ..., \psi_{i-1}, \psi_{i+1}, ..., \psi_N}} \lambda(\sigma_1, ..., \sigma_{i-1}, \omega, \sigma_{i+1}, ..., \sigma_N)\lambda(\varphi)\lambda(\psi)$$

$$= X_{i,\omega}^3 + \sum_{\psi \in \Omega_i \setminus \omega} \frac{6\mu_i(\omega)}{2\mu_i(\omega) + \mu_i(\psi)} X_{i,\omega}^2 X_{i,\psi}$$

$$+ \sum_{\varphi, \psi \in \Omega_i \setminus \omega} \frac{3\mu_i(\omega)}{\mu_i(\omega) + \mu_i(\varphi) + \mu_i(\psi)} X_{i,\omega} X_{i,\varphi} X_{i,\psi}.$$

Thus operator (7.8) can be rewritten as

$$X_{i,\omega}' = X_{i,\omega} \left(X_{i,\omega}^2 + \sum_{\psi \in \Omega_i \setminus \omega} \frac{6\mu_i(\omega)}{2\mu_i(\omega) + \mu_i(\psi)} X_{i,\omega} X_{i,\psi} \right.$$

$$\left. + \sum_{\varphi, \psi \in \Omega_i \setminus \omega} \frac{3\mu_i(\omega)}{\mu_i(\omega) + \mu_i(\varphi) + \mu_i(\psi)} X_{i,\varphi} X_{i,\psi} \right), \tag{7.11}$$

where $X_{i,\omega}$ is defined by (7.9), $\omega \in \Omega_i, i = 1, ..., N$.

Note that $\sum_{\omega \in \Omega_i} X_{i,\omega} = 1$ for any $i = 1, ..., N$. One can see that for each fixed i ($i = 1, ..., N$) the operator (7.11) is a Volterra CSO $W^{(i)} : S^{|\Omega_i|-1} \rightarrow S^{|\Omega_i|-1}$. $\qquad\square$

This theorem allows us to use the theory of Volterra CSO to describe the behavior of trajectories of non-Volterra CSO (7.8).

If for each $i \in \{1, ..., N\}$ the asymptotical behavior of trajectories of CSO $W^{(i)}$ is known, say $X_{i,\omega}^{(n)} \rightarrow X_{i,\omega}^*$, $n \rightarrow \infty$, then asymptotical behavior of W (i.e. (7.8)), say $\lambda^{(n)}(\tau) \rightarrow \lambda^*(\tau)$, $n \rightarrow \infty$, can be found from the following system of linear equations

$$\sum_{\tau \in \Omega : \tau_i = \omega} \lambda^*(\tau) = X_{i,\omega}^*, \quad \omega \in \Omega_i, i = 1, ..., N. \tag{7.12}$$

7.1.3 *A four-dimensional case*

In this section we shall illustrate the restriction of a non-Volterra CSO to two Volterra operators and study the trajectory of the non-Volterra operator by these two Volterra operators.

Consider graph $G = (\mathcal{L}, L)$ with $\mathcal{L} = \{1, 2\}$ and $L = \emptyset$. Take $\Phi = \{1, 2\}$. Then non-Volterra CSO (7.8) has the form

$$x_1' = x_1^3 + 3\beta_1(x_1^2 x_2 + x_1 x_2^2) + 3\alpha_1(x_1^2 x_3 + x_1 x_3^2)$$
$$+ 3\alpha_1\beta_1[x_1^2 x_4 + x_1 x_4^2 + x_2^2 x_3 + x_2 x_3^2$$
$$+ 2(x_1 x_2 x_3 + x_1 x_2 x_4 + x_1 x_3 x_4 + x_2 x_3 x_4)]$$

$$x_2' = x_2^3 + 3\beta_2(x_1^2 x_2 + x_1 x_2^2) + 3\alpha_1(x_2^2 x_4 + x_2 x_4^2)$$
$$+ 3\alpha_1\beta_2[x_1^2 x_4 + x_1 x_4^2 + x_2^2 x_3 + x_2 x_3^2$$
$$+ 2(x_1 x_2 x_3 + x_1 x_2 x_4 + x_1 x_3 x_4 + x_2 x_3 x_4)]$$

$$x_3' = x_3^3 + 3\alpha_2(x_1 x_3^2 + x_1^2 x_3) + 3\beta_1(x_3^2 x_4 + x_3 x_4^2)$$
$$+ 3\alpha_2\beta_1[x_1^2 x_4 + x_1 x_4^2 + x_2^2 x_3 + x_2 x_3^2$$
$$+ 2(x_1 x_2 x_3 + x_1 x_2 x_4 + x_1 x_3 x_4 + x_2 x_3 x_4)]$$

$$x_4' = x_4^3 + 3\alpha_2(x_2 x_4^2 + x_2^2 x_4) + 3\beta_2(x_3^2 x_4 + x_3 x_4^2)$$
$$+ 3\alpha_2\beta_2[x_1^2 x_4 + x_1 x_4^2 + x_2^2 x_3 + x_2 x_3^2$$
$$+ 2(x_1 x_2 x_3 + x_1 x_2 x_4 + x_1 x_3 x_4 + x_2 x_3 x_4)],$$

$$(7.13)$$

where $\mu_1 = (\alpha_1, \alpha_2)$, $\alpha_j > 0$, $\alpha_1 + \alpha_2 = 1$; $\mu_2 = (\beta_1, \beta_2)$, $\beta_j \geq 0$, $\beta_1 + \beta_2 = 1$.

Putting $x_1 + x_2 = X_{1,1}$, $x_3 + x_4 = X_{1,2}$ and $x_1 + x_3 = X_{2,1}$, $x_2 + x_4 = X_{2,2}$ we get the Volterra cubic operators:

$$X_{1,1}' = X_{1,1}\left(X_{1,1}^2 + 3\alpha_1 X_{1,2}(X_{1,1} + X_{1,2})\right)$$
$$X_{1,2}' = X_{1,2}\left(X_{1,2}^2 + 3\alpha_2 X_{1,1}(X_{1,1} + X_{1,2})\right)$$

$$(7.14)$$

and

$$X_{2,1}' = X_{2,1}\left(X_{2,1}^2 + 3\beta_1 X_{2,2}(X_{2,1} + X_{2,2})\right)$$
$$X_{2,2}' = X_{2,2}\left(X_{2,2}^2 + 3\beta_2 X_{2,1}(X_{2,1} + X_{2,2})\right).$$

$$(7.15)$$

Since $X_{i,1} + X_{i,2} = 1$, $i = 1, 2$ the study of both operators (7.14) and (7.15) can be reduced to the study of a dynamical system given by the function $f_\alpha(x) = x(x^2 + 3\alpha(1 - x))$, $x \in [0, 1]$. This is an increasing function of $x \in [0, 1]$ for each parameter $\alpha \in [0, 1]$.

We have

$$\text{Fix}(f_\alpha) = \{x \in [0,1] : f_\alpha(x) = x\} = \begin{cases} \{0,1\} & \text{if } \alpha \in [0,1/3] \cup [2/3,1] \\ \{0, 3\alpha - 1, 1\} & \text{if } \alpha \in (1/3, 2/3). \end{cases}$$

Using above-mentioned properties of the function $f_\alpha(x)$ and checking $|f'_\alpha(a)|$ at $a \in \text{Fix}(f_\alpha)$ one can see that the sequence $x^{(n)} = f_\alpha(x^{(n-1)})$, $n \geq 1$ for $x^{(0)} \in [0,1]$ has the following limits

$$\lim_{n\to\infty} x^{(n)} = \begin{cases} 0, & \text{for any } x^{(0)} \in [0,1), \quad \alpha \in [0,1/3] \\ 3\alpha - 1, & \text{for any } x^{(0)} \in (0,1), \quad \alpha \in (1/3,2/3) \\ 1, & \text{for any } x^{(0)} \in (0,1], \quad \alpha \in [2/3,1]. \end{cases} \tag{7.16}$$

By equalities (7.16) for operators (7.14) we get the following

$$\lim_{n\to\infty} (X_{1,1}^{(n)}, X_{1,2}^{(n)}) = \begin{cases} (0,1), & \text{for any } X_{1,1}^{(0)} \in [0,1), \quad \alpha_1 \in [0,1/3] \\ (3\alpha_1 - 1, 2 - 3\alpha_1), & \forall\, X_{1,1}^{(0)} \in (0,1), \quad \alpha_1 \in (1/3,2/3) \\ (1,0), & \text{for any } X_{1,1}^{(0)} \in (0,1], \quad \alpha_1 \in [2/3,1]. \end{cases}$$
$$\tag{7.17}$$

A similar formula is true for the operator (7.15) where α_1 replaced by β_1. Combining these formulas and using formula (7.12) one proves the following.

Proposition 7.1. *The trajectory of the non-Volterra CSO (7.13) has following limit*

$$\lim_{n\to\infty} x^{(n)} = \begin{cases} (1,0,0,0), & \text{if } \alpha_1, \beta_1 \in [2/3,1], \\ (0,1,0,0), & \text{if } \alpha_1 \in [2/3,1], \beta_1 \in [0,1/3], \\ (0,0,1,0), & \text{if } \alpha_1 \in [0,1/3], \beta_1 \in [2/3,1], \\ (0,0,0,1), & \text{if } \alpha_1, \beta_1 \in [0,1/3], \\ (0,0,3\beta_1 - 1, 2 - 3\beta_1), & \text{if } \alpha_1 \in [0,1/3], \beta_1 \in (1/3,2/3), \\ (3\beta_1 - 1, 2 - 3\beta_1, 0, 0), & \text{if } \alpha_1 \in [2/3,1], \beta_1 \in (1/3,2/3), \\ (0, 3\alpha_1 - 1, 0, 2 - 3\alpha_1), & \text{if } \alpha_1 \in (1/3,2/3), \beta_1 \in [0,1/3], \\ (3\alpha_1 - 1, 0, 2 - 3\alpha_1, 0), & \text{if } \alpha_1 \in (1/3,2/3), \beta_1 \in [2/3,1], \\ \in U, & \text{if } \alpha_1 \in (1/3,2/3), \beta_1 \in (1/3,2/3), \end{cases}$$

where

$$U = \{x \in S^3 : x_1 + x_2 = 3\alpha_1 - 1, \quad x_3 + x_4 = 2 - 3\alpha_1,$$
$$x_1 + x_3 = 3\beta_1 - 1, \quad x_2 + x_4 = 2 - 3\beta_1\}.$$

7.2 Construction of CSO for continual set

Now following [160] we shall give a construction of CSO for a continual set E. Note that for the continual set E one of the key problem is to determine the set of coefficients of heredity which is already infinite dimensional.

In [183], [237], [238] the authors introduced a continuous-time dynamical system as quadratic stochastic processes.

7.2.1 *Definitions*

Consider a measurable space $(E, \mathcal{F})$ and let $S(E, \mathcal{F})$ be the set of all probability measures on $(E, \mathcal{F})$.

Definition 7.1. A mapping $W : S(E, \mathcal{F}) \to S(E, \mathcal{F})$ is called a cubic stochastic operator (CSO) if, for an arbitrary measure $\lambda \in S(E, \mathcal{F})$, the measure $\lambda' = W\lambda$ is defined by

$$\lambda'(A) = \int_E \int_E \int_E P(x, y, z, A) d\lambda(x) d\lambda(y) d\lambda(z), \quad \forall A \in \mathcal{F},$$

where $P(x, y, z, A)$, satisfies the following conditions:

(i) $P(x, y, z, \cdot) \in S(E, \mathcal{F})$ for any fixed $x, y, z \in E$;

(ii) $P(x, y, z, A)$, regarded a function of three variables x, y, and z is measurable on $(E \times E \times E, \mathcal{F} \otimes \mathcal{F} \otimes \mathcal{F})$ for any fixed $A \in \mathcal{F}$.

When E is finite, a CSO on $S(E, \mathcal{F}) = S^{m-1}$ is as in (7.1) with $P_{ijk,l} = P(i, j, k, l)$.

Let $(E, \mathcal{F}, \mathcal{M})$ be triple, where $\mathcal{F}$ is a σ-algebra of subsets of E and $\mathcal{M} = S(E, \mathcal{F})$, i.e. the set of all probability measures on $(E, \mathcal{F})$. For any three elements $x, y, z \in E$ and a given measure $m_{t_0} \in \mathcal{M}$ at some moment t_0 of the time we assume that we know the law of probability distribution $m_{t_1} \in \mathcal{M}$ of the system E at the moment $t_1 > t_0$ of time.

Denote by $P(t_1, x, y, z, t_2, A)$ the probability of obtaining an element from the set $A \in \mathcal{F}$ at the moment t_2, provided that the elements x, y and z of E interact starting at moment t_1, where $t_2 \geq t_1 + 1$.

Thus if at the moment t_1 we start with a probability measure $m_{t_1} \in \mathcal{M}$, then $m_{t_2} \subset \mathcal{M}$ for any $t_2 \geq t_1 + 1$ is defined by

$$m_{t_2}(A) = \int_E \int_E \int_E P(t_1, x, y, z, t_2, A) m_{t_1}(dx) m_{t_1}(dy) m_{t_1}(dz). \quad (7.18)$$

Without loss of generality we assume that the process starts at the moment $t = 0$.

Definition 7.2. A family $\{P(t_1, x, y, z, t_2, A) \colon x, y, z \in E,\ A \in \mathcal{F},\ t_1, t_2 \in \mathbb{R}^+,\ t_2 - t_1 \geq 1\}$ is called cubic stochastic process (CSP) if it satisfies the following conditions

(I) $P(t, x, y, z, t+1, A) = P(0, x, y, z, 1, A)$ for any $t \geq 1$;

(II)   The value of $P(t_1, x, y, z, t_2, A)$ is independent on any permutations of variables x, y, z for all $x, y, z \in E$, and $A \in \mathcal{F}$;

(III) $P(t_1, x, y, z, t_2, A)$ is a probability measure on $(E, \mathcal{F})$ for all $x, y, z \in E$ and $t_1, t_2 \in \mathbb{R}^+,\ t_2 - t_1 \geq 1$;

(IV) $P(t_1, x, y, z, t_2, A)$ as function of the three variables x, y and z is measurable with respect to $(E \times E \times E,\ \ \mathcal{F} \otimes \mathcal{F} \otimes \mathcal{F})$ for all $A \in \mathcal{F}$;

(V)   For any $t_1 < t_2 < t_3$ such that $t_2 - t_1 \geq 1$ and $t_3 - t_2 \geq 1$ the following holds

$$P(t_1, x, y, z, t_3, A)$$

$$= \int_E \int_E \int_E P(t_1, x, y, z, t_2, du)\, P(t_2, u, \vartheta, q, t_3, A)\, m_{t_2}(d\vartheta) m_{t_2}(dq),$$

$$(7.19)$$

where m_{t_2} on $(E, \mathcal{F})$ is defined by (7.18).

Remark 7.2. Let us point out the following comments

1. The consideration of CSO and CSP are motivated by their appearance in biology (for example in gene engineering, a triple crossings for different sorts of plants to obtain another sort, and free population with ternary production) [13], in physics (for example in spin systems with ternary interactions) [159]. Note that one may have more general form of stochastic operators (resp. processes) with order $n \geq 1$, and coefficients $P_{i_1 i_2 \ldots i_n, j} \geq 0$ (resp. $P(t_1, x_1, x_2, \ldots, x_n, t_2, A)$) in biology (gene engineering) and physics (n-nary interactions). In this section, we restrict ourselves to the case $n = 3$, i.e. CSOs and CSPs. Because even for the considered setting associated dynamical systems are very complicated in comparison with $n = 1$ (i.e. Markov chains) and $n = 2$ (i.e. QSOs).

2. Note that if one defines a new process by

$$Q(s, x, y, t, A) = \int_E P(s, x, y, z, t, A) m_s(dz)$$

then one can check that the defined process is QSP (see [184, 185]). We recall that the functions $Q(s, x, y, t, A)$ denote the probability that under the interaction of the elements x and y at time s an event A comes into effect at time t. Since for physical, chemical and biological phenomena, a certain time is necessary for the realization of an interaction, it is taken the greatest such time to be equal to 1 (see the Boltzmann model [120] or the biological model [155]). Thus the probability $Q(s,x,y,t,A)$ is defined for $t-s \geq 1$. Hence, for CSP, the probabilities $P(s,x,y,z,t,A)$ are also defined for $t-s \geq 1$.

3. It should be noted that the CSPs are related to CSOs (see [120, 183]) in the same way as Markov processes are related to linear transformations.

4. The equation (7.19) in Definition 7.2 is an analogue of Chapman–Kolmogorov equation. In [183], [185], [238] such equations were extended to QSPs. Note that for QSPs there are two types of the Chapman–Kolmogorov equations: type A and type B. Similarly, one also can define (at least) two types of the Chapman–Kolmogorov equations for a CSP: The equation (7.19) corresponds to the type A of QSP, the following equation is an analogue of the type B for CSP:

$$P(s,x,y,z,t,A) = \int_E \int_E \int_E \int_E \int_E \int_E P(s,x,y_1,z_1,\tau,du)\, P(s,y,y_2,z_2,\tau,dv)$$

$$\times P(s,z,y_3,z_3,\tau,dw)\, P(\tau,u,v,w,t,A)\, m_s(dy_1)m_s(dz_1)$$

$$m_s(dy_2)m_s(dz_2)m_s(dy_3)m_s(dz_3).$$

Here for the sake of simplicity, we shall only consider CSPs which satisfy (7.19).

7.2.2 *Examples of CSP*

Let us provide some examples of CSPs.

Let $x^{(0)} = \left(x_1^{(0)}, x_2^{(0)}, \ldots, x_n^{(0)}\right)$ be an initial distribution on $E = \{1,2,...,n\}$.

Denote

$$P_{ijk,l} = P(0,i,j,k,1,\{l\}), \quad P_{ijk,l}^{[s,t]} = P(s,i,j,k,t,\{l\}).$$

By equation (7.18) at the moment $t=1$ the vector $x^{(1)} = \left(x_1^{(1)}, x_2^{(1)}, \ldots, x_n^{(1)}\right)$ is defined as follows

$$x_l^{(1)} = \sum_{i,j,k=1}^{n} P_{ijk,l}\, x_i^{(0)} x_j^{(0)} x_k^{(0)}, \quad l \in E.$$

In this case the condition (I) can be reduced to

$$P_{ijk,l}^{[t,t+1]} = P_{ijk,l}.$$

In general, from (7.18), (7.19) one finds

$$x_l^{(t)} = \sum_{i,j,k} P_{ijk,l}^{[s,t]} x_i^{(s)} x_j^{(s)} x_k^{(s)} \tag{7.20}$$

$$P_{ijk,l}^{[s,t]} = \sum_{m,\gamma,\delta} P_{ijk,m}^{[s,\tau]} P_{m\gamma\delta,l}^{[\tau,t]} x_\gamma^{(\tau)} x_\delta^{(\tau)}, \tag{7.21}$$

where $\tau - s \geq 1$ and $t - \tau \geq 1$.

Example 7.1. To define a CSP for $E = \{1,2\}$ we first define a CSO by the matrix $(P_{ijk,l})$, where

$$P_{111,1} = 1, \quad P_{112,1} = P_{121,1} = P_{211,1} = \frac{2}{3},$$

$$P_{122,1} = P_{212,1} = P_{221,1} = \frac{1}{3}, \quad P_{222,1} = 0;$$

$$P_{111,2} = 0, \quad P_{112,2} = P_{121,2} = P_{211,2} = \frac{1}{3},$$

$$P_{122,2} = P_{212,2} = P_{221,2} = \frac{2}{3}, \quad P_{222,2} = 1.$$

One can see that this CSO is the identity mapping, i.e. $x_1' = x_1$, $x_2' = x_2$. Take an initial vector $(x, 1 - x)$ and using this CSO by formulas (7.20), (7.21) one can define a CSP:

$$P_{111,1}^{[s,t]} = \frac{1}{3^{t-s-1}} \left[1 + (3^{t-s-1} - 1)x \right],$$

$$P_{112,1}^{[s,t]} = P_{121,1}^{[s,t]} = P_{211,1}^{[s,t]} = \frac{1}{3^{t-s-1}} \left[\tfrac{2}{3} + (3^{t-s-1} - 1)x \right],$$

$$P_{122,1}^{[s,t]} = P_{212,1}^{[s,t]} = P_{221,1}^{[s,t]} = \frac{1}{3^{t-s-1}} \left[\tfrac{2}{3} + (3^{t-s-1} - 1)(1 - x) \right],$$

$$P_{222,1}^{[s,t]} = \frac{3^{t-s-1} - 1}{3^{t-s-1}} x,$$

$$P_{ijk,2}^{[s,t]} = 1 - P_{ijk,1}^{[s,t]}, \quad \text{for all } i,j,k \in E = \{1,2\}.$$

Example 7.2. Let $E = \{1, 2, \ldots, n\}$. Take a family of stochastic vectors: $a(t) = (a_1(t), a_2(t), \ldots, a_n(t))$, i.e. $a_i(t) \geq 0$, $\sum_i a_i(t) = 1$ for any $t \geq 0$. For each pari s,t define a stochastic matrix $Q^{[s,t]} = (q_{il}^{[s,t]})_{i,l \in E}$, where

$q_{il}^{[s,t]} = a_l(t)$ for all $i \in E$, i.e. it does not depend on s. One can see that this matrix satisfies the Kolmogorov–Chapman equation:

$$Q^{[s,t]} = Q^{[s,\tau]}Q^{[\tau,t]}, \quad \text{for all} \ \ 0 \le s < \tau < t.$$

Now define functions

$$P(s,i,j,k,t,\{l\}) = q_{il}^{[s,t]} = a_l(t).$$

One can check that the defined family $\{P(s,i,j,k,t,\{l\})\}$ is a CSP.

Example 7.3. (cf. Example 4.2.1 of [183].) Let $(E, \mathcal{F})$ be a measurable space and m_0 be an initial measure on this space. Consider the following functions

$$P(s,x,y,z,t,A) = \frac{1}{3^{t-s-1}}\left(\frac{\delta_x(A) + \delta_y(A) + \delta_z(A)}{3} + (3^{t-s-1} - 1)m_0(A)\right),$$

where $t - s \ge 1$, $x, y, z \in E$ and $A \in \mathcal{F}$,

$$\delta_x(A) = \begin{cases} 1, & \text{if} \ x \in A \\ 0, & \text{if} \ x \notin A. \end{cases}$$

One can see that the defined family is CSP.

7.2.3 *The construction*

Let $G = (\Lambda, L)$ be a countable graph. For a finite set Φ denote by Ω the set of all functions $\sigma : \Lambda \to \Phi$. Let $S(\Omega, \Phi)$ be the set of all probability measures defined on $(\Omega, \mathcal{F})$, where $\mathcal{F}$ is the σ-algebra generated by the finite-dimensional cylindrical set. Let μ be a measure on $(\Omega, \mathcal{F})$ such that $\mu(B) > 0$ for any finite-dimensional cylindrical set $B \in \mathcal{F}$.

Let $M \subset \Lambda$ be a finite connected subgraph. Two elements $\sigma, \varphi \in \Omega$ are called equivalent if $\sigma(x) = \varphi(x)$ for any $x \in M$, i.e. $\sigma(M) = \varphi(M)$. Let $\xi = \{\Omega_i, i = 1, 2, ..., |\Phi|^{|M|}\}$, be the partition of Ω generated by this equivalent relation, where $|\cdot|$ denotes the cardinality of a set and Ω_i contains all equivalent elements.[1]

Denote $\langle ijk, l \rangle = \delta_{il} + \delta_{jl} + \delta_{kl}$, where δ is the Kronecker's symbol, i.e.

$$\delta_{ij} = \begin{cases} 1, & \text{if} \ i = j \\ 0, & \text{if} \ i \ne j. \end{cases}$$

Consider

$$P_{\sigma_1 \sigma_2 \sigma_3, \sigma} = \frac{\langle ijk, l \rangle \mu(\Omega_l)}{\mu(\Omega_i) + \mu(\Omega_j) + \mu(\Omega_k)} \quad \text{if} \ \sigma_1 \in \Omega_i, \ \sigma_2 \in \Omega_j, \ \sigma_3 \in \Omega_k, \ \sigma \in \Omega_l.$$

$$(7.22)$$

---

[1]Note that ξ depends on M, therefore all quantities which we define using ξ also depend on M. But for simplicity of formulas we will omit M from the formulas.

For arbitrary σ one can see that $P_{\sigma_1\sigma_2\sigma_3,\sigma}$ is invariant with respect to any permutations of $\sigma_1,\sigma_2,\sigma_3$.

Then the coefficients $P(\sigma_1,\sigma_2,\sigma_3,A)$ $(\sigma_1,\sigma_2,\sigma_3 \in \Omega, A \in \mathcal{F})$ are defined as

$$P(\sigma_1,\sigma_2,\sigma_3,A) = Z(\sigma_1,\sigma_2,\sigma_3) \int_A P_{\sigma_1\sigma_2\sigma_3,\sigma}d\mu(\sigma)$$

$$= Z(\sigma_1,\sigma_2,\sigma_3) \sum_{l=1}^{m} \int_{A\cap\Omega_l} P_{\sigma_1\sigma_2\sigma_3,\sigma}d\mu(\sigma),$$

where $m = |\Phi|^{|M|}$ and $Z(\sigma_1,\sigma_2,\sigma_3)$ is the normalizing factor, which is chosen by the condition that $P(\sigma_1,\sigma_2,\sigma_3,\Omega) = 1$.

One can obtain the following

$$P(\sigma_1,\sigma_2,\sigma_3,A) = \begin{cases} \frac{\mu(\Omega_i)\mu(A\cap\Omega_i)+\mu(\Omega_j)\mu(A\cap\Omega_j)+\mu(\Omega_k)\mu(A\cap\Omega_k)}{\mu^2(\Omega_i)+\mu^2(\Omega_j)+\mu^2(\Omega_k)}, \\ \qquad \text{if } \sigma_1 \in \Omega_i, \sigma_2 \in \Omega_j, \sigma_3 \in \Omega_k, i \neq k, j \neq k, i \neq j \\[4pt] \frac{2\mu(\Omega_i)\mu(A\cap\Omega_i)+\mu(\Omega_j)\mu(A\cap\Omega_j)}{2\mu^2(\Omega_i)+\mu^2(\Omega_j)}, \text{if } \sigma_1,\sigma_2 \in \Omega_i, \sigma_3 \in \Omega_j, i \neq j \\[4pt] \frac{\mu(A\cap\Omega_i)}{\mu(\Omega_i)} \text{ if } \sigma_1,\sigma_2,\sigma_3 \in \Omega_i. \end{cases}$$

$$(7.23)$$

The CSO W acting on the set $S(\Omega,\Phi)$ is determined by coefficients (7.23) is defined as follows: for an arbitrary measure $\lambda \in S(\Omega,\Phi)$, the measure $\lambda' = W\lambda$ is

$$\lambda'(A) = \int_\Omega \int_\Omega \int_\Omega P(\sigma_1,\sigma_2,\sigma_3,A)d\lambda(\sigma_1)d\lambda(\sigma_2)d\lambda(\sigma_3). \qquad (7.24)$$

Using (7.23) from (7.24) we obtain

$$\lambda'(A) = \sum_{i=1}^{m} a_i(A)\lambda^3(\Omega_i)$$

$$+ 3\sum_{i=1}^{m}\sum_{\substack{j=1\\j\neq i}}^{m} b_{ij}(A)\lambda^2(\Omega_i)\lambda(\Omega_j) + 6\sum_{1\leq i<j<k\leq m} c_{ijk}(A)\lambda(\Omega_i)\lambda(\Omega_j)\lambda(\Omega_k),$$

$$(7.25)$$

where

$$a_i(A) = \frac{\mu(A\cap\Omega_i)}{\mu(\Omega_i)},$$

$$b_{ij}(A) = \frac{2\mu(\Omega_i)\mu(A\cap\Omega_i)+\mu(\Omega_j)\mu(A\cap\Omega_j)}{2\mu^2(\Omega_i)+\mu^2(\Omega_j)}, \qquad (7.26)$$

$$c_{ijk}(A) = \frac{\mu(\Omega_i)\mu(A\cap\Omega_i)+\mu(\Omega_j)\mu(A\cap\Omega_j)+\mu(\Omega_k)\mu(A\cap\Omega_k)}{\mu^2(\Omega_i)+\mu^2(\Omega_j)+\mu^2(\Omega_k)}.$$

One can see that

$$a_i(\Omega_l) = \begin{cases} 1, & \text{if } l = i, \\ 0, & \text{if } l \neq i \end{cases} \qquad b_{ij}(\Omega_l) = \begin{cases} \dfrac{2\mu^2(\Omega_i)}{2\mu^2(\Omega_i)+\mu^2(\Omega_j)}, & l = i \\[2mm] \dfrac{\mu^2(\Omega_j)}{2\mu^2(\Omega_i)+\mu^2(\Omega_j)}, & l = j \\[2mm] 0, & l \neq i, l \neq j. \end{cases} \tag{7.27}$$

$$c_{ijk}(\Omega_l) = \begin{cases} \dfrac{\mu^2(\Omega_l)}{\mu^2(\Omega_i)+\mu^2(\Omega_j)+\mu^2(\Omega_k)}, & l \in \{i,j,k\} \\[2mm] 0, & l \notin \{i,j,k\}. \end{cases}$$

For a given measure $\lambda \in S(\Lambda, \Phi)$ the trajectory $\{\lambda^{(n)}\}, n = 1, 2, \dots$ of the operator (7.24) is defined by $\lambda^{(n+1)}(A) = W(\lambda^{(n)})(A)$, where $n = 0, 1, 2, \dots$ and $\lambda^{(0)} = \lambda$, $A \in \mathcal{F}$.

By (7.25) and (7.27) we have

$$\lambda'(\Omega_l) = \sum_{i=1}^{m} a_i(\Omega_l)\lambda^3(\Omega_i)$$

$$+ 3\sum_{i=1}^{m}\sum_{\substack{j=1 \\ j \neq i}}^{m} b_{ij}(\Omega_l)\lambda^2(\Omega_i)\lambda(\Omega_j) + 6\sum_{1 \leq i < j < k \leq m} c_{ijk}(\Omega_l)\lambda(\Omega_i)\lambda(\Omega_j)\lambda(\Omega_k).$$

$$\tag{7.28}$$

Recall that a CSO (7.1) is called Volterra if the coefficients $P_{ijk,l}$ may be nonzero only when $l \in \{i,j,k\}$ and vanish in all the remaining cases (see [130], [128]).

One can see that any Volterra CSO has the following form

$$W : \lambda'_l = \lambda_l\left(\lambda_l^2 + \lambda_l\sum_{\substack{i=1 \\ i \neq l}}^{m} a_{i,l}\lambda_i + \sum_{\substack{i,j=1 \\ i \neq l, \, j \neq l}}^{m} b_{ij,l}\lambda_i\lambda_j\right), \quad (l = 1, \dots, m) \tag{7.29}$$

where $a_{i,l}$ and $b_{ij,l}$ are some coefficients depending on $P_{ijk,l}$.

Denoting $\lambda_i = \lambda(\Omega_i)$, and $a_{i,l} = 3b_{li}(\Omega_l)$, $b_{ij,l} = 6c_{ijl}(\Omega_l)$ the operator (7.28) can be written as (7.29).

Note that the nth iteration $\lambda^{(n)} = W^{(n)}\lambda^{(0)}$ of the operator (7.24) (i.e. (7.25)) can be written as

$$\lambda^{(n+1)}(A) = \sum_{i=1}^{m} a_i(A)(\lambda_i^{(n)})^3$$

$$+ 3\sum_{i=1}^{m}\sum_{\substack{j=1 \\ j \neq i}}^{m} b_{ij}(A)(\lambda_i^{(n)})^2\lambda_j^{(n)} + 6\sum_{1 \leq i < j < k \leq m} c_{ijk}(A)\lambda_i^{(n)}\lambda_j^{(n)}\lambda_k^{(n)}, \tag{7.30}$$

where $\lambda_j^{(n)}$, $j = 1, ..., m$ are coordinates of the trajectory of operator (7.29) for the given λ.

Thus in order to study the trajectory of operator (7.24) it is enough to know the behavior of trajectories of the operator (7.29), i.e. we proved the following

Theorem 7.3. *For any finite $M \subset \Lambda$ the dynamical system generated by the CSO (7.24) is reducible to a dynamical system generated by a Volterra CSO acting on $(m-1)$-dimensional simplex.*

From Theorem 7.3 we get

Corollary 7.1. *Assume for a given measure μ and $\lambda = (\lambda_1, ..., \lambda_m) \in S^{m-1}$ for the trajectory of the Volterra operator (7.29) we have*

$$\lim_{n \to \infty} \lambda^{(n)} = (\lambda_1^*, \lambda_2^*, ..., \lambda_m^*).$$

Then the corresponding trajectory $\{\lambda^{(n)}(A)\}$ of the operator (7.24) has the following limit

$$\lambda(A) = \lim_{n \to \infty} \lambda^{(n)}(A) = \sum_{i=1}^{m} a_i(A)(\lambda_i^*)^3$$

$$+3 \sum_{i=1}^{m} \sum_{\substack{j=1 \\ j \neq i}}^{m} b_{ij}(A)(\lambda_i^*)^2 \lambda_j^* + 6 \sum_{1 \leq i < j < k \leq m} c_{ijk}(A)\lambda_i^* \lambda_j^* \lambda_k^*. \tag{7.31}$$

Remark 7.3. As it was mentioned above the theory of Volterra QSOs is well studied (see for example [68]). But Volterra CSOs were not exhaustively studied, because such cubic operators are still complicated. There are just a few articles devoted to Volterra CSOs [130], [128], [215], [216]. Therefore formula (7.31) is already helpful by using the results for the Volterra CSOs studied in these papers.

7.3 Integro-differential equations for CSP

The equations which we want to drive here were given in [161] for finite E. So we consider the continual case of E.

Consider a CSP on a measurable space $(E, \mathcal{F})$ with initial measure m_0. For $t > s + 2$ from condition (V) of Definition 7.2 we get

$$P(s, x, y, z, t + \Delta, A) - P(s, x, y, z, t, A)$$

$$= \int\limits_E \int\limits_E \int\limits_E P(s,x,y,z,t-1,du)\,\{P(t-1,u,\vartheta,q,t+\Delta,A)$$

$$-P(t-1,u,\vartheta,q,t,A)\}\,m_{t-1}(d\vartheta)m_{t-1}(dq).$$

Assume the following limit exists

$$C(t,u,\vartheta,q,A) = \lim_{\Delta\to 0} \frac{P(t-1,u,\vartheta,q,t+\Delta,A) - P(t-1,u,\vartheta,q,t,A)}{\Delta}.$$

Then taking limit $\Delta \to 0$ we obtain the *first integro-differential equation*:

$$\frac{\partial P(s,x,y,z,t,A)}{\partial t}$$

$$= \int\limits_E \int\limits_E \int\limits_E P(s,x,y,z,t-1,du)\,C(t,u,\vartheta,q,A)m_{t-1}(d\vartheta)m_{t-1}(dq). \quad (7.32)$$

Similarly, one gets the *second integro-differential equation*:

$$\frac{\partial P(s,x,y,z,t,A)}{\partial s}$$

$$= -\int\limits_E \int\limits_E \int\limits_E C(s+1,x,y,z,du)P(s+1,u,\vartheta,q,t,A)m_{s+1}(d\vartheta)m_{s+1}(dq).$$

$$(7.33)$$

Let $E = \mathbb{R}$ and $A_w = (-\infty, w]$, where $w \in \mathbb{R}$. Denote

$$F(s,x,y,z,t,w) = P(s,x,y,z,t,A_w).$$

It is clear that $F(s,x,y,z,t,w)$ as the function of w is monotone, right-continuous and

$$F(s,x,y,z,t,-\infty) = 0, \quad F(s,x,y,z,t,+\infty) = 1.$$

The condition (V) of Definition 7.2 for the function $F(s,x,y,z,t,w)$ has the following form

$$F(s,x,y,z,t,w) = \int\limits_E \int\limits_E \int\limits_E dF(s,x,y,z,\tau,u)F(\tau,u,\vartheta,q,t,w)m_\tau(d\vartheta)m_\tau(dq).$$

If the function $F(s,x,y,z,t,w)$ is absolutely continuous with respect to variable w then

$$F(s,x,y,z,t,w) = \int\limits_{-\infty}^{w} f(s,x,y,z,t,u)du,$$

where $f(s, x, y, z, t, w)$ is a non-negative function and measurable with respect to variables x, y, z, w, moreover it satisfies the following conditions

$$\int_{-\infty}^{\infty} f(s, x, y, z, t, w) dw = 1.$$

$$f(s, x, y, z, t, w) = \int_E \int_E \int_E f(s, x, y, z, \tau, u) f(\tau, u, \vartheta, q, t, w) m_\tau(d\vartheta) m_\tau(dq) du.$$

$$(7.34)$$

From (7.32) and (7.33) for $F(s, x, y, z, t, w)$ we get

$$\frac{\partial F(s, x, y, z, t, w)}{\partial t}$$

$$= \int_E \int_E \int_E \frac{\partial F(s, x, y, z, t - 1, u)}{\partial u} C(t, u, \vartheta, q, w) du\, m_{t-1}(d\vartheta) m_{t-1}(dq),$$

$$\frac{\partial F(s, x, y, z, t, w)}{\partial s}$$

$$= \int_E \int_E \int_E \frac{\partial C(s + 1, x, y, z, u)}{\partial u} F(s + 1, u, \vartheta, q, t, w) du\, m_{s+1}(d\vartheta) m_{s+1}(dq).$$

Assume the following limit exists:

$$a(t, u, \vartheta, q, w) = \lim_{\Delta \to 0} \frac{f(t - 1, u, \vartheta, q, t + \Delta, w) - f(0, u, \vartheta, q, 1, w)}{\Delta}.$$

Then we obtain the following integro-differential equations:

$$\frac{\partial f(s, x, y, z, t, w)}{\partial t}$$

$$= \int_E \int_E \int_E a(t, u, \vartheta, q, w) f(s, x, y, z, t - 1, u) du\, m_{t-1}(d\vartheta) m_{t-1}(dq) \quad (7.35)$$

$$\frac{\partial f(s, x, y, z, t, w)}{\partial s}$$

$$= -\int_E \int_E \int_E a(s + 1, x, y, z, u) f(s + 1, u, \vartheta, q, t, w) du\, m_{s+1}(d\vartheta) m_{s+1}(dq).$$

$$(7.36)$$

7.4 Reduction of the integro-differential equations to differential equations

Under some conditions the integro-differential equations (7.35) and (7.36) can be reduced to differential equations. In this subsection we illustrate this for equation (7.36).

Let $t > s + 2$, then from (7.34) we get

$$f(s, x, y, z, t, w) - f(s + \Delta, x, y, z, t, w)$$

$$= \int_E \int_E \int_E \{f(s, x, y, z, s + 1 + \Delta, u) - f(s + \Delta, x, y, z, s + 1 + \Delta, u)\}$$

$$\times f(s + 1 + \Delta, u, \vartheta, q, t, w) m_{s+1+\Delta}(d\vartheta) m_{s+1+\Delta}(dq) du. \tag{7.37}$$

We consider function f such that the decomposition of $f(s + 1 + \Delta, u, \vartheta, q, t, w)$ into Taylor's series in a neighborhood of the point (x, y, z) has the form:

$$f(s + 1 + \Delta, u, \vartheta, q, t, w)$$

$$= f(s + 1 + \Delta, x, y, z, t, w) + \frac{\partial f(s + 1 + \Delta, x, y, z, t, w)}{\partial u}(u - x)$$

$$+ \frac{\partial f(s + 1 + \Delta, x, y, z, t, w)}{\partial \vartheta}(\vartheta - y) + \frac{\partial f(s + 1 + \Delta, x, y, z, t, w)}{\partial q}(q - z)$$

$$+ \frac{1}{2}\frac{\partial^2 f(s + 1 + \Delta, x, y, z, t, w)}{\partial u^2}(u - x)^2$$

$$+ \frac{1}{2}\frac{\partial^2 f(s + 1 + \Delta, x, y, z, t, w)}{\partial \vartheta^2}(\vartheta - y)^2$$

$$+ \frac{1}{2}\frac{\partial^2 f(s + 1 + \Delta, x, y, z, t, w)}{\partial q^2}(q - z)^2$$

$$+ \frac{\partial^2 f(s + 1 + \Delta, x, y, z, t, w)}{\partial u \partial \vartheta}(u - x)(\vartheta - y)$$

$$+ \frac{\partial^2 f(s + 1 + \Delta, x, y, z, t, w)}{\partial u \partial q}(u - x)(q - z)$$

$$+ \frac{\partial^2 f(s + 1 + \Delta, x, y, z, t, w)}{\partial \vartheta \partial q}(\vartheta - y)(q - z)$$

$$+\frac{1}{2}\frac{\partial^3 f(s+1+\Delta,x,y,z,t,w)}{\partial u^2\partial\vartheta}(u-x)^2(\vartheta-y)$$

$$+\frac{1}{2}\frac{\partial^3 f(s+1+\Delta,x,y,z,t,w)}{\partial u^2\partial q}(u-x)^2(q-z)$$

$$+\frac{1}{2}\frac{\partial^3 f(s+1+\Delta,x,y,z,t,w)}{\partial\vartheta^2\partial u}(\vartheta-y)^2(u-x)$$

$$+\frac{1}{2}\frac{\partial^3 f(s+1+\Delta,x,y,z,t,w)}{\partial\vartheta^2\partial q}(\vartheta-y)^2(q-z)$$

$$+\frac{1}{2}\frac{\partial^3 f(s+1+\Delta,x,y,z,t,w)}{\partial q^2\partial u}(q-z)^2(u-x)$$

$$+\frac{1}{2}\frac{\partial^3 f(s+1+\Delta,x,y,z,t,w)}{\partial q^2\partial\vartheta}(q-z)^2(\vartheta-y)$$

$$+\frac{1}{6}\frac{\partial^3 f(s+1+\Delta,x,y,z,t,w)}{\partial q^3}(u-x)^3$$

$$+\frac{1}{6}\frac{\partial^3 f(s+1+\Delta,x,y,z,t,w)}{\partial\vartheta^2}(\vartheta-y)^3$$

$$+\frac{1}{6}\frac{\partial^3 f(s+1+\Delta,x,y,z,t,w)}{\partial q^2}(q-z)^3$$

$$+\frac{\partial^3 f(s+1+\Delta,x,y,z,t,w)}{\partial u\partial\vartheta\partial q}(u-x)(\vartheta-y)(q-z). \tag{7.38}$$

Substituting (7.38) into (7.37) we consider non-zero summands:

$$\int_E\int_E\int_E \{f(s,x,y,z,s+1+\Delta,u)-f(s+\Delta,x,y,z,s+1+\Delta,u)\}$$

$$\times\frac{\partial f(s+1+\Delta,x,y,z,t,w)}{\partial x}(u-x)m_{s+1+\Delta}(d\vartheta)m_{s+1+\Delta}(dq)du$$

$$=\frac{\partial f(s+1+\Delta,x,y,z,t,w)}{\partial x}\int_E \{f(s,x,y,z,s+1+\Delta,u)$$

$$-f(s+\Delta,x,y,z,s+1+\Delta,u)\}(u-x)du.$$

Denote

$$a(s, x, y, z, \Delta)$$

$$= \int_E \{f(s, x, y, z, s+1+\Delta, u) - f(s+\Delta, x, y, z, s+1+\Delta, u)\}\,(u-x)du.$$

Now consider the summands with second order of derivations

$$\int_E \int_E \int_E \{f(s, x, y, z, s+1+\Delta, u) - f(s+\Delta, x, y, z, s+1+\Delta, u)\}$$

$$\times \frac{1}{2}\frac{\partial^2 f(s+1+\Delta, x, y, z, w)}{\partial x^2}(u-x)^2 m_{s+1+\Delta}(d\vartheta) m_{s+1+\Delta}(dq)du$$

$$= \frac{1}{2}\frac{\partial^2 f(s+1+\Delta, x, y, z, w)}{\partial x^2}\int_E \{f(s, x, y, z, s+1+\Delta, u)$$

$$- f(s+\Delta, x, y, z, s+1+\Delta, u)\}\,(u-x)^2 du.$$

Denote

$$b^2(s, x, y, z, \Delta)$$

$$= \int_E \{f(s, x, y, z, s+1+\Delta, u) - f(s+\Delta, x, y, z, s+1+\Delta, u)\}(u-x)^2 du.$$

Then

$$\int_E \int_E \int_E \{f(s, x, y, z, s+1+\Delta, u) - f(s+\Delta, x, y, z, s+1+\Delta, u)\}$$

$$\times \frac{\partial^2 f(s+1+\Delta, x, y, z, w)}{\partial x \partial y}(u-x)(\vartheta-y) m_{s+1+\Delta}(d\vartheta) m_{s+1+\Delta}(dq)du$$

$$= \frac{\partial^2 f(s+1+\Delta, x, y, z, t, w)}{\partial x \partial y}\int_E \{f(s, x, y, z, s+1+\Delta, u)$$

$$- f(s+\Delta, x, y, z, s+1+\Delta, u)\}\,(u-x)du \int_E (\vartheta-y)m_{s+1+\Delta}(d\vartheta).$$

Denote

$$\int_E (\vartheta-y)m_{s+1+\Delta}(d\vartheta) = \alpha(s+1, y, \Delta).$$

Consequently

$$\int_E \int_E \int_E \{f(s,x,y,z,s+1+\Delta,u) - f(s+\Delta,x,y,z,s+1+\Delta,u)\}$$

$$\times \frac{\partial^2 f(s+1+\Delta,x,y,z,t,w)}{\partial x \partial z}(u-x)(q-z)m_{s+1+\Delta}(d\vartheta)m_{s+1+\Delta}(dq)du$$

$$= \frac{\partial^2 f(s+1+\Delta,x,y,z,t,w)}{\partial x \partial z}\int_E \{f(s,x,y,z,s+1+\Delta,u)$$

$$- f(s+\Delta,x,y,z,s+1+\Delta,u)\}\int_E (u-x)du\int_E (q-z)m_{s+1+\Delta}(dq).$$

$$\int_E \int_E \int_E \{f(s,x,y,z,s+1+\Delta,u) - f(s+\Delta,x,y,z,s+1+\Delta,u)\}$$

$$\times \frac{1}{2}\frac{\partial^3 f(s+1+\Delta,x,y,z,t,w)}{\partial x^2 \partial y}(u-x)^2(\vartheta-y)m_{s+1+\Delta}(d\vartheta)m_{s+1+\Delta}(dq)du$$

$$= \frac{1}{2}\frac{\partial^3 f(s+1+\Delta,x,y,z,t,w)}{\partial x^2 \partial y}\int_E \{f(s,x,y,z,s+1+\Delta,u)$$

$$- f(s+\Delta,x,y,z,s+1+\Delta,u)\}(u-x)^2 du\int_E (\vartheta-y)m_{s+1+\Delta}(d\vartheta)$$

$$= \frac{1}{2}\frac{\partial^3 f(s+1+\Delta,x,y,z,t,w)}{\partial x^2 \partial y}b^2(s,x,y,z,\Delta)\alpha(s+1,y,\Delta).$$

$$\int_E \int_E \int_E \{f(s,x,y,z,s+1+\Delta,u) - f(s+\Delta,x,y,z,s+1+\Delta,u)\}$$

$$\times \frac{1}{2}\frac{\partial^3 f(s+1+\Delta,x,y,z,t,w)}{\partial x^2 \partial z}(u-x)^2(q-z)m_{s+1+\Delta}(d\vartheta)m_{s+1+\Delta}(dq)du$$

$$= \frac{1}{2}\frac{\partial^3 f(s+1+\Delta,x,y,z,t,w)}{\partial x^2 \partial z}\int_E \{f(s,x,y,z,s+1+\Delta)$$

$$- f(s + \Delta, x, y, z, s + 1 + \Delta, u)\} (u - x)^2 du \int_E (q - z) m_{s+1+\Delta}(dq)$$

$$= \frac{1}{2} \frac{\partial^3 f(s + 1 + \Delta, x, y, z, t, w)}{\partial x^2 \partial z} b^2(s, x, y, z, \Delta) \alpha(s + 1, z, \Delta).$$

$$\int_E \int_E \int_E \{f(s, x, y, z, s + 1 + \Delta, u) - f(s + \Delta, x, y, z, s + 1 + \Delta, u)\}$$

$$\times \frac{1}{2} \frac{\partial^3 f(s + 1 + \Delta, x, y, z, t, w)}{\partial^2 y \partial x} (\vartheta - y)^2 (u - x) m_{s+1+\Delta}(d\vartheta) m_{s+1+\Delta}(dq) du$$

$$= \frac{1}{2} \frac{\partial^3 f(s + 1 + \Delta, x, y, z, t, w)}{\partial^2 y \partial x} \int_E \{f(s, x, y, z, s + 1 + \Delta, u)$$

$$- f(s + \Delta, x, y, z, s + 1 + \Delta, u)\} (u - x) du \int_E (\vartheta - y)^2 m_{s+1+\Delta}(d\vartheta)$$

$$= \frac{1}{2} a(s, x, y, z, \Delta) \frac{\partial^3 f(s + 1 + \Delta, x, y, z, t, w)}{\partial^2 y \partial x} \cdot \alpha_2(s + 1, y, \Delta),$$

where

$$\alpha_2(s + 1, y, \Delta) = \int_E (\vartheta - y)^2 m_{s+1+\Delta}(d\vartheta).$$

$$\int_E \int_E \int_E \{f(s, x, y, z, s + 1 + \Delta, u) - f(s + \Delta, x, y, z, s + 1 + \Delta, u)\}$$

$$\times \frac{1}{2} \frac{\partial^3 f(s + 1 + \Delta, x, y, z, t, w)}{\partial z^2 \partial x} (q - z)^2 (u - x) m_{s+1+\Delta}(d\vartheta) m_{s+1+\Delta}(dq) du$$

$$= \frac{1}{2} \frac{\partial^3 f(s + 1 + \Delta, x, y, z, t, w)}{\partial z^2 \partial x} \int_E \{f(s, x, y, z, s + 1 + \Delta, u)$$

$$- f(s + \Delta, x, y, z, s + 1 + \Delta, u)\} (u - x) du \int_E (q - z)^2 m_{s+1+\Delta}(dv)$$

$$= \frac{1}{2} \frac{\partial^3 f(s + 1 + \Delta, x, y, z, t, w)}{\partial z^2 \partial x} a(s, x, y, z, \Delta) \alpha_2(s + 1, z, \Delta).$$

$$\iiint\limits_{E\ E\ E} \{f(s,x,y,z,s+1+\Delta,u) - f(s+\Delta,x,y,z,s+1+\Delta,u)\}$$

$$\times \frac{\partial^3 f(s+1+\Delta,x,y,z,t,w)}{\partial z \partial y \partial z}(u-x)(\vartheta-y)(q-z)m_{s+1+\Delta}(d\vartheta)m_{s+1+\Delta}(dq)du$$

$$= \frac{\partial^3 f(s+1+\Delta,x,y,z,t,w)}{\partial x \partial y \partial z}\int\limits_{E} \{f(s,x,y,z,s+1+\Delta,u)$$

$$- f(s+\Delta,x,y,z,s+1+\Delta,u)\}(u-x)du\int\limits_{E}(\vartheta-y)m_{s+1+\Delta}(d\vartheta)$$

$$\times \int\limits_{E}(q-z)m_{s+1+\Delta}(dq)$$

$$= \frac{\partial^3 f(s+1+\Delta,x,y,z,t,w)}{\partial x \partial y \partial z}a(s,x,y,z,\Delta)\alpha(s+1,y,\Delta)\alpha(s+1,z,\Delta).$$

Since other summands are equal to zero, we get

$$f(s,x,y,z,t,w) - f(s+\Delta,x,y,z,t,w)$$

$$= a(s,x,y,z,\Delta)\frac{\partial f(s+1+\Delta,x,y,z,t,w)}{\partial x}$$

$$+b^2(s,x,y,z,\Delta)\cdot\frac{1}{2}\frac{\partial^2 f(s+1+\Delta,x,y,z,t,w)}{\partial x^2}$$

$$+\frac{1}{2}a(s,x,y,z,\Delta)\alpha(s+1,y,\Delta)\frac{\partial^2 f(s+1+\Delta,x,y,z,t,w)}{\partial x \partial y}$$

$$+a(s,x,y,z,\Delta)\frac{\partial^2 f(s+1+\Delta,x,y,z,t,w)}{\partial x \partial z}\alpha(s+1,z,\Delta)$$

$$+\frac{1}{2}\frac{\partial^3 f(s+1+\Delta,x,y,z,t,w)}{\partial x^2 \partial y}b^2(s,x,y,z,\Delta)\alpha(s+1,y,\Delta)$$

$$+\frac{1}{2}\frac{\partial^3 f(s+1+\Delta,x,y,z,t,w)}{\partial x^2 \partial z}b^2(s,x,y,z,\Delta)\alpha(s+1,z,\Delta)$$

$$+\frac{1}{2}a(s,x,y,z,\Delta)\frac{\partial^3 f(s+1+\Delta,x,y,z,t,w)}{\partial y^2 \partial x}\alpha_2(s+1,y,\Delta)$$

$$+\frac{1}{2}\frac{\partial^3 f(s+1+\Delta,x,y,z,t,w)}{\partial z^2 \partial x}a(s,x,y,z,\Delta)\alpha_2(s+1,z,\Delta)+a(s,x,y,z,\Delta)$$

$$\times\alpha(s+1,y,\Delta)\alpha(s+1,z,\Delta)\frac{\partial^3 f(s+1+\Delta,x,y,z,t,w)}{\partial x \partial y \partial z}. \qquad (7.39)$$

Dividing both sides of the equality (7.39) by Δ and passing to the limit (assuming the limits exist) as $\Delta \to 0$, and denoting

$$A(s,x,y,z)=\lim_{\Delta\to 0}\frac{a(s,x,y,z,\Delta)}{\Delta}, \quad B^2(s,x,y,z)=\lim_{\Delta\to 0}\frac{b^2(s,x,y,z,\Delta)}{2\Delta},$$

$$D(s+1,y)=\lim_{\Delta\to 0}\alpha(s+1,y,\Delta), \quad D_2(s+1,y)=\lim_{\Delta\to 0}\frac{\alpha_2(s+1,y,\Delta)}{2\Delta}$$

from (7.39) we get

$$\frac{\partial f(s,x,y,z,t,w)}{\partial s}=-A(s,x,y,z)\frac{\partial f(s+1,x,y,z,t,w)}{\partial x}$$

$$-B^2(s,x,y,z)\frac{\partial^2 f(s+1,x,y,z,t,w)}{\partial x^2}$$

$$-\frac{1}{2}A(s,x,y,z)D(s+1,y)\frac{\partial^2 f(s+1,x,y,z,t,w)}{\partial x \partial y}$$

$$-A(s,x,y,z)D(s+1,z)\frac{\partial^2 f(s+1,x,y,z,t,w)}{\partial x \partial z}$$

$$-B^2(s,x,y,z)D(s+1,y)\frac{\partial^3 f(s+1,x,y,z,t,w)}{\partial x^2 \partial y}$$

$$-B^2(s,x,y,z)D(s+1,z)\frac{\partial^3 f(s+1,x,y,z,t,w)}{\partial x^2 \partial z}$$

$$-\frac{1}{2}A(s,x,y,z)D_2(s+1,y)\frac{\partial^3 f(s+1,x,y,z,t,w)}{\partial y^2 \partial x}$$

$$-\frac{1}{2}A(s,x,y,z)D_2(s+1,z)\frac{\partial^3 f(s+1,x,y,z,t,w)}{\partial z^2 \partial x}$$

$$-A(s,x,y,z)D(s+1,y)D(s+1,z)\frac{\partial^3 f(s+1,x,y,z,t,w)}{\partial x \partial y \partial z}. \qquad (7.40)$$

Remark 7.4. Note that the equation (7.40) is known as differential equation with advanced argument. Similarly one can reduce the equation (7.35) to a differential equation which is known as a differential equation with delay argument. For theory of such kind of equations we refer to [46] and [173].

Example 7.4. Let $E = \mathbb{R}$. If $m_\tau(du) = r_\tau(u)du$ then from (7.34) we get

$$f(s, x, y, z, t, w)$$

$$= \int_{-\infty}^{\infty} \int_{-\infty}^{\infty} \int_{-\infty}^{\infty} f(s, x, y, z, \tau, u)f(\tau, u, \vartheta, q, t, w)r_\tau(\vartheta)r_\tau(q)dud\vartheta dq. \quad (7.41)$$

Let $a(s, t)$ and $b(t)$ be strictly positive functions. Consider

$$f(s, x, y, z, t, w) = (\pi a(s, t))^{-1/2} \exp\left[-\frac{(w - x - y - z)^2}{a(s, t)}\right], \quad (7.42)$$

$$r_t(u) = (\pi b(t))^{-1/2} \exp\left[-\frac{u^2}{b(t)}\right].$$

For any $A, B, C \in \mathbb{R}$, with $A > 0$ it is known[2] that

$$\int_{-\infty}^{\infty} \exp(-Ax^2 + Bx + C)dx = \sqrt{\frac{\pi}{A}} \exp\left(\frac{B^2}{4A} + C\right).$$

Using this formula (three times) one can see that the function (7.42) satisfies (7.41) iff functions $a(s, t) > 0$ and $b(t) > 0$ satisfy the following equation

$$a(s, t) = 2b(\tau) + a(s, \tau) + a(\tau, t), \quad \text{for any } s < \tau < t \text{ with } \tau - s \geq 1, \ t - \tau \geq 1. \quad (7.43)$$

Thus we obtain a family of CSP generated by functions (7.42) with $a(s, t)$, $b(t)$ satisfying (7.43). For example, take $\epsilon \in (0, 1)$, $a(s, t) = t - s - \epsilon$ and $b(t) = \epsilon/2$. These functions satisfy (7.43) and the corresponding f and r are defined by

$$f(s, x, y, z, t, w) = \frac{\exp\left[-\frac{(w - x - y - z)^2}{t - s - \epsilon}\right]}{\sqrt{(t - s - \epsilon)\pi}},$$

$$r_t(u) = \frac{\exp\left[-\frac{2u^2}{\epsilon}\right]}{\sqrt{\epsilon\pi/2}}$$

generate a CSP.

Bibliographical notes. The chapter is based on [13], [46], [63], [64], [68], [120], [130], [128], [155], [159], [160], [161], [173], [183], [184], [185], [215], [216], [216], [237], [238]. For future reading see [10], [18], [47], [106], [117], [118], [174], [179], [243], [249] and the references therein.

[2]See https://en.wikipedia.org/wiki/Gaussian_integral.

PART 3
Concrete populations dynamics

Chapter 8

Dynamics generated by quadratic stochastic operators

In this chapter we present results related to trajectories of the discrete-time dynamical systems generated by quadratic stochastic operators (QSOs). The first section devoted to Volterra QSOs. There are many works related with such operators, we will only give some of results which have clear biological interpretations. Other sections of this chapter are devoted to a wide class of non-Volterra QSOs and their dynamical systems. Note that there are many (non-)Volterra QSOs which we do not present in this book, but we will give citations to them for interested readers.

8.1 Free population: Volterra's discrete model

Recall that a quadratic stochastic operator (QSO) is a mapping of the simplex

$$S^{n-1} = \left\{ x = (x_1, ..., x_n) \in \mathbb{R}^n : x_i \geq 0, \sum_{i=1}^{n} x_i = 1 \right\}$$

into itself, of the form

$$V : x_k' = \sum_{i,j=1}^{n} p_{ij,k} x_i x_j, \quad k \in E = \{1, ..., n\}, \tag{8.1}$$

where $p_{ij,k}$ are coefficients of heredity, which satisfy

$$p_{ij,k} \geq 0, \quad p_{ij,k} = p_{ji,k} \quad \sum_{k=1}^{n} p_{ij,k} = 1, \quad i, j, k \in E. \tag{8.2}$$

The population evolves by starting from an arbitrary state (probability distribution on E) $x \in S^{n-1}$ then passing to the state $V(x)$ (in the next "generation"), then to $V^2(x)$, and so on.

For a given $x^{(0)} \in S^{n-1}$ the trajectory $\{x^{(m)}\}, m = 0, 1, 2, \ldots$ of QSO (8.1) is defined by $x^{(m+1)} = V(x^{(m)})$. By $\omega(x^0)$ we denote the set of all limit points of the trajectory $\{x^{(m)}\}$. Compactness of S^{n-1} implies that $\omega(x^0) \neq \emptyset$.

One of the main problems in mathematical biology consists in the study of asymptotical behavior of the trajectories and description of $\omega(x^0)$.

In this section following [69]–[72] we give results related to Volterra QSO.

Definition 8.1. A QSO $V : S^{n-1} \to S^{n-1}$ give by (8.1) is called *Volterra* if

$$p_{ij,k} = 0, \quad \text{if} \ \ k \notin \{i, j\}. \tag{8.3}$$

This condition biologically means that the offspring repeats the genotype of one of its parents.

8.1.1 *Canonical form of Volterra operator*

Proposition 8.1. *Let V be a Volterra QSO, then it can be represented by*

$$x'_k = x_k \left(1 + \sum_{i=1}^{n} a_{ki} x_i \right), \tag{8.4}$$

where $a_{ki} = 2p_{ik,k} - 1$ for $i \neq k$ and $a_{kk} = 0$. Moreover $a_{ki} = -a_{ik}$ and $|a_{ki}| \leq 1$.

Proof. Since $p_{kk,i} = 0$ for $i \neq k$ and $\sum_{i=1}^{n} p_{kk,i} = 1$, we have $p_{kk,k} = 1$. By condition $p_{ij,k} = p_{ji,k}$ from (8.1) we get

$$x'_k = x_k \left(x_k + 2 \sum_{\substack{i=1 \\ i \neq k}}^{n} p_{ik,k} x_i \right).$$

Using $\sum_{i=1}^{n} x_i = 1$ we obtain

$$x'_k = x_k \left(1 + \sum_{\substack{i=1 \\ i \neq k}}^{n} (2p_{ik,k} - 1) x_i \right).$$

Denoting $a_{ki} = 2p_{ik,k} - 1$ for $i \neq k$ and $a_{kk} = 0$ from the last equality we get (8.4).

From $0 \leq p_{ik,k} \leq 1$ it follows $|a_{ki}| \leq 1$. Finally using (8.3) we have $p_{ik,k} + p_{ik,i} = 1$. Hence

$$a_{ki} + a_{ik} = 2p_{ik,k} - 1 + 2p_{ki,i} - 1 = 2(p_{ik,k} + p_{ik,i} - 1) = 0,$$

i.e. $a_{ki} = -a_{ik}$. $\qquad \square$

For any $I \subset E = \{1, ..., n\}$ define *face* Γ_I of S^{n-1} by
$$\Gamma_I = \{x \in S^{n-1} : x_i = 0 \text{ for any } i \in I\}.$$
Relative inside of Γ_I is defined by
$$\text{int}\Gamma_I = \{x \in \Gamma_I : x_i > 0, \forall i \notin I\}.$$

Proposition 8.2. *Let V be a Volterra QSO. Then the following assertions hold true*

(i) *Any face of S^{n-1} is invariant set with respect to V;*
(ii) *The vertices of the simplex S^{n-1} are fixed points of V;*
(iii) *The relative inside of any face of S^{n-1} is invariant with respect to V.*

Proof. (i) If V is Volterra QSO, then from (8.4) it is clear that, if $x_i = 0$ then $x'_i = 0$. Hence $V(\Gamma_I) \subset \Gamma_I$. The assertion (ii) is an immediate corollary of (i).

(iii) Assume $x_k > 0$. Then $a_{kk} = 0$ implies that
$$x'_k = x_k(1 + a_{k1}x_1 + ... + a_{kk-1}x_{k-1} + a_{kk+1}x_{k+1} + ... + a_{kn}x_n)$$
$$\geq x_k(1 - x_1 - ... - x_{k-1} - x_{k+1} - ... - x_n) = x_k^2 > 0.$$
Hence $V(\text{int}\Gamma_I) \subset \text{int}\Gamma_I$. $\square$

Remark 8.1. As was mentioned above the biological treatment of Volterra operator is that the offspring repeats the genotype of one of its parents. The assertion (i) of Proposition 8.2 says that, any missing genotype of the initial offspring can not be appeared in the next offsprings of the evolution. The assertion (ii) says that, a present genotype of the initial offspring will be present in any future offsprings of the population.

Example 8.1. Assume an evolution operator $V : S^1 \to S^1$ has a form
$$\begin{cases} x'_1 = x_1(1 - ax_2), \\ x'_2 = x_2(1 + ax_1), \end{cases}$$
where $a = 2p_{12,2} - 1 > 0$.

If $x^0 = (x_1^0, x_2^0) \in S^1$ is an initial probability distribution then the probability distribution of offspring $m + 1$ is defined by $x^{(m+1)}$:
$$\begin{cases} x_1^{(m+1)} = x_1^{(m)}(1 - ax_2^{(m)}), \\ x_2^{(m+1)} = x_2^{(m)}(1 + ax_1^{(m)}), m = 0, 1, 2, \ldots \end{cases}$$
The relation $1 + ax_1^{(m)} \geq 1$ for any m implies that the sequence $\{x_2^{(m)}\}$ $(0 \leq x_2^{(m)} \leq 1)$ is nondecreasing, and similarly $\{x_1^{(m)}\}$ $(0 \leq x_1^{(m)} \leq 1)$ is non increasing. Therefore,
$$\lim_{m \to \infty} x_1^{(m)} = \alpha, \qquad \lim_{m \to \infty} x_2^{(m)} = \beta.$$
One can show that if $x_2^0 > 0$ then $\beta = 1$, and $\alpha = 0$.

Example 8.2. Assume an evolution operator $V : S^2 \to S^2$ has a form

$$\begin{cases} x_1' = x_1(1 - x_2 + x_3), \\ x_2' = x_2(1 + x_1 - x_3), \\ x_3' = x_3(1 - x_1 + x_2). \end{cases}$$

This example is well known (see [126]). One can see that this operator has four fixed points:

$$M_1(1, 0, 0), \ M_2(0, 1, 0), \ M_3(0, 0, 1), \ C\left(\frac{1}{3}, \frac{1}{3}, \frac{1}{3}\right).$$

Theorem 8.1. *([126], [273]) For any $x^0 \neq C$ lying inside of S^2, the trajectory $\{x^{(m)}\}$ does not converge. The set of limit points of the trajectory is an infinite subset of the boundary ∂S^2 of S^2.*

Proof. In Subsection 8.1.3 (below) we prove theorems which generalize this theorem. $\square$

Remark 8.2. A QSO V satisfies the *ergodic* theorem if the limit

$$\lim_{m \to \infty} \frac{1}{m} \sum_{j=0}^{m-1} x^{(j)}$$

exists for any $x^0 \in S^{n-1}$. On the basis of Example 8.2 and numerical calculations Ulam conjectured that the ergodic theorem holds for any QSO. In [273] it was proven that this conjecture is false in general.

8.1.2 *Lyapunov functions*

Let V be a Volterra operator on S^{n-1}. Then due to Proposition 8.1 we have

$$(V(x))_k = x_k\left(1 + \sum_{i=1}^{n} a_{ki}x_i\right), \ k = 1, \ldots, n \tag{8.5}$$

where $a_{ki} = 2p_{ik,k} - 1$ for $i \neq k$, $a_{kk} = 0$, $a_{ki} = -a_{ik}$, $|a_{ki}| \leq 1$.

Definition 8.2. A continuous functional $\varphi : S^{n-1} \to \mathbb{R}$ is called a *Lyapunov function* for V if for any initial point $x^0 \in S^{n-1}$ there exists $\lim_{m \to \infty} \varphi(x^{(m)})$.

It is clear that if $\lim_{m \to \infty} \varphi(X^{(m)}) = c$, then $\omega(x^0) \subset \varphi^{-1}(c)$.

Lemma 8.1. *If $A = (a_{ki})_{k,i=1}^{n}$ is a skew-symmetric matrix (i.e. $a_{ki} = -a_{ik}$) then*

$$P = \left\{x \in S^{n-1} : \sum_{i=1}^{n} a_{ki}x_i \geq 0, \ k = 1, \ldots, n\right\} \neq \emptyset. \tag{8.6}$$

Proof. Denote $F_k = \left\{ x \in S^{n-1} : \sum_{i=1}^{n} a_{ki} x_i \geq 0 \right\}$. For any $I \subset E$ put

$$I^c = \{1, 2, \ldots, n\} \setminus I.$$

Then we have

$$\Gamma_{I^c} \subset \bigcup_{k \in I} F_k. \tag{8.7}$$

Indeed, take $x \in \Gamma_{I^c}$. By $a_{ki} = -a_{ik}$, one gets $\sum_{i,k \in I} a_{ki} x_k x_i = 0$. Since $x_k \geq 0$, at that $x_k > 0$ for at least one $k \in I$, we have

$$\sum_{i \in I} a_{ki} x_i \geq 0 \tag{8.8}$$

otherwise for some k one has

$$\sum_{i,k \in I} a_{ki} x_k x_i = \sum_{k \in I} x_k \sum_{i \in I} a_{ki} x_i < 0.$$

Since $x \in \Gamma_{I^c}$ by (8.8) we obtain $\sum_{i=1}^{n} a_{ki} x_i = \sum_{i \in I} a_{ki} x_i \geq 0$. Consequently, $x \in \bigcup_{k \in I} F_k$. Thanks to the Sperner's combinatorial Lemma[1] from (8.7) one gets $P = \bigcap_{k=1}^{n} F_k \neq \emptyset$. $\square$

Corollary 8.1. *If $A = (a_{ki})_{k,i=1}^{n}$ is a skew-symmetric matrix then*

$$Q = \left\{ x \in S^{n-1} : \sum_{i=1}^{n} a_{ki} x_i \leq 0, \ k = 1, ..., n \right\} \neq \emptyset. \tag{8.9}$$

Proof. The proof immediately follows from Lemma 8.1 since the matrix A is skew-symmetric as well. $\square$

It is clear that P and Q are convex sets.

Denote

$$\mathrm{Fix}(V) = \{ x \in S^{n-1} : V(x) = x \}$$

the set of fixed points of V.

Corollary 8.2. *Let V be a Volterra operator, then $P \subset \mathrm{Fix}(V)$, $Q \subset \mathrm{Fix}(V)$.*

Proof. Let, for example, $x \in P$. Then from (8.5) one has

$$(V(x))_k = x_k \left(1 + \sum_{i=1}^{n} a_{ki} x_i \right) \geq x_k, \quad k = 1, \ldots, n. \tag{8.10}$$

The equality $\sum_{k=1}^{n} x_k = \sum_{k=1}^{n} (V(x))_k = 1$ with (8.10) yields that $(V(x))_k = x_k$ for every $k \in \{1, \ldots, n\}$. $\square$

[1] https://en.wikipedia.org/wiki/Sperner27s_lemma.

Let us recall Young's[2] inequality:

$$b_1^{p_1} \cdots b_n^{p_n} \le p_1 b_1 + \cdots + p_n b_n, \tag{8.11}$$

where $b_i > 0$, $p_i \ge 0$, $\sum_{i=1}^{n} p_i = 1$.

Theorem 8.2. *Let V be a Volterra operator (see (8.5)). If $p = (p_1, \ldots, p_n) \in P$, then*

$$\varphi(x) = x_1^{p_1} \cdots x_n^{p_n},$$

is a Lyapunov function for V.

Proof. Without loss of generality we may assume $x \in \mathrm{int} S^{n-1}$ (see Propositions 8.2). From (8.5) we have

$$\varphi(V(x)) = \left[x_1 \left(1 + \sum_{i=1}^{n} a_{1i} x_i\right) \right]^{p_1} \cdots \left[x_n \left(1 + \sum_{i=1}^{n} a_{ni} x_i\right) \right]^{p_n} \tag{8.12}$$

$$= \varphi(x) \prod_{k=1}^{n} \left[1 + \sum_{i=1}^{n} a_{ki} x_i \right]^{p_k}.$$

Denote $b_k = 1 + \sum_{i=1}^{n} a_{ki} x_i$. One has $b_k > 0$, since $x \in \mathrm{int} S^{n-1}$. By (8.11) it follows that

$$\prod_{k=1}^{n} \left[1 + \sum_{i=1}^{n} a_{ki} x_i \right]^{p_k} \le \sum_{k=1}^{n} p_k \left[1 + \sum_{i=1}^{n} a_{ki} x_i \right] = 1 + \sum_{k=1}^{n} p_k \sum_{i=1}^{n} a_{ki} x_i.$$

By $a_{ki} = -a_{ik}$ we obtain

$$\prod_{k=1}^{n} \left[1 + \sum_{i=1}^{n} a_{ki} x_i \right]^{p_k} \le 1 - \sum_{i=1}^{n} \left(\sum_{k=1}^{n} a_{ik} p_k \right) x_i.$$

According to $p \in P$, one has $\sum_{k=1}^{n} a_{ik} p_k \ge 0$ for all $i = 1, \ldots, n$. Thus

$$\sum_{i=1}^{n} \left(\sum_{k=1}^{n} a_{ik} p_k \right) x_i \ge 0,$$

which means

$$\prod_{k=1}^{n} \left[1 + \sum_{i=1}^{n} a_{ki} x_i \right]^{p_k} \le 1, \tag{8.13}$$

for any $x \in \mathrm{int} S^{n-1}$. Hence by (8.12) we obtain $\varphi(V(x)) \le \varphi(x)$, which implies that $\varphi(x^{(m+1)}) \le \varphi(x^{(m)}), m = 0, 1, \ldots$. Consequently $\lim_{m \to \infty} \varphi(x^{(m)})$ exists. So $\varphi(x)$ is a Lyapunov function, which decreases along the trajectory of the dynamical system (8.5). $\qquad\square$

---

[2] https://en.wikipedia.org/wiki/Young27s_inequality_for_products.

Now we are going to give some properties of the following functionals

$$\varphi(x) = \varphi_p(x) = x_1^{p_1} \cdots x_n^{p_n}, \quad \psi(x) = \psi_p(x) = \prod_{k=1}^{n} \left[1 + \sum_{i=1}^{n} a_{ki} x_i \right]^{p_k},$$

defined on S^{n-1}. It is known (see [212]) that if $p_i \geq 0$, and $\sum_{i=1}^{n} p_i = 1$, then the functional φ is concave and has a unique maximum point $p = (p_1, ..., p_n)$ on S^{n-1}, i.e.

$$\max_{x \in S^{n-1}} \varphi(x) = p_1^{p_1} \cdots p_n^{p_n},$$

here for $p_i = 0$ we have assumed that $p_i^{p_i} = 1$.

It is obvious that $\psi(x)$ is also concave on S^{n-1}, since it is the composition of the affine map

$$x \to \left(1 + \sum_{i=1}^{n} a_{1i} x_i, \ldots, 1 + \sum_{i=1}^{n} a_{ni} x_i \right)$$

and the concave function φ. For $p \in P$ by (8.13) one gets

$$\psi(x) = \prod_{k=1}^{n} \left[1 + \sum_{i=1}^{n} a_{ki} x_i \right]^{p_k} \leq 1, \tag{8.14}$$

for any $x \in S^{n-1}$. Now if $y \in P$, then $\sum_{i=1}^{n} a_{ki} y_i \geq 0$, $k = 1, \ldots, n$. Therefore,

$$\psi(y) = \prod_{k=1}^{n} \left[1 + \sum_{i=1}^{n} a_{ki} y_i \right]^{p_k} \geq 1. \tag{8.15}$$

which with (8.14) implies that $\psi(y) = 1$ for any $y \in P$. Thus, if $p \in P$ then

$$\max_{x \in S^{n-1}} \psi(x) = \psi(y) = 1, \quad y \in P.$$

Note that the relation $\sum_{i=1}^{n} a_{ki} y_i \geq 0$, $k = 1, \ldots, n$ with

$$\prod_{k=1}^{n} \left[1 + \sum_{i=1}^{n} a_{ki} y_i \right]^{p_k} = 1$$

yields that $p_{k_0} = 0$ for $\sum_{i=1}^{n} a_{k_0 i} y_i > 0$ and $p_{k_0} > 0$ for $\sum_{i=1}^{n} a_{k_0 i} y_i = 0$. So,

$$p_k \sum_{i=1}^{n} a_{ki} y_i = 0, \quad k = 1, \ldots, n$$

which with $a_{ki} = -a_{ik}$, $\sum_{i=1}^{n} a_{ki} p_i \geq 0$, $\sum_{i=1}^{n} a_{ki} y_i \geq 0$, implies that

$$0 = \sum_{k} p_k \sum_{i=1}^{n} a_{ki} y_i = -\sum_{i} \left(\sum_{k} a_{ki} p_k \right) y_i = 0,$$

consequently

$$y_k \sum_{i=1}^{n} a_{ki} p_i = 0, \ k = 1, \ldots, n.$$

Hence $\lambda_k = \sum_{i=1}^{n} a_{ki} p_i$ is Lagrange multiplier of the extremal problem:

$$p \in P, \ \psi(x) \to \sup, \ x_i \geq 0, \sum_{i=1}^{n} x_i = 1.$$

Therefore (see [212]) if $p \in P$, then

$$1 = \max_{x \in S^{n-1}} \psi(x) = \psi(y) \ \Leftrightarrow \ y \in P. \tag{8.16}$$

Lemma 8.2. *If $\psi(x) \leq 1$ for any $x \in S^{n-1}$ then $p \in P$.*

Proof. First of all we prove that p is a fixed point of V. Denote $u = V^{-1}p$. Then from $\varphi(p) = \varphi(Vu) = \varphi(u)\psi(u) \leq \varphi(u)$ follows that $u = p$. Since function φ has a unique maximum on S^{n-1} at p we have $p = V^{-1}p$ i.e. p is a fixed point.

Now we prove that $p \in P$. Assume this is not a case i.e. $\sum_{i=1}^{n} a_{k'i} p_i < 0$ for some k'. We have $p_{k'} = 0$, since p is a fixed point. Choose k'' such that $p_{k''} > 0$. Then $\sum_{i=1}^{n} a_{k''i} p_i = 0$. Take $x = (x_1, \ldots, x_n)$ by

$$x_i = \begin{cases} \varepsilon & \text{if } i = k' \\ p_{k''} - \varepsilon & \text{if } i = k'' \\ p_i & \text{otherwise.} \end{cases} \tag{8.17}$$

If $0 \leq \varepsilon \leq p_{k''}$ then $x \in S^{n-1}$. Since $\sum_{i=1}^{n} a_{ki} p_i = 0$ for $p_k > 0$, from (8.17) we have $\sum_{i=1}^{n} a_{ki} x_i = \varepsilon \cdot (a_{kk'} - a_{kk''})$. Thus, for all $k = 1, \ldots, n$ we get

$$\left(1 + \sum_{i=1}^{n} a_{ki} x_i\right)^{p_k} = 1 + \varepsilon \cdot p_k \cdot (a_{kk'} - a_{kk''}) + o(\varepsilon).$$

Consequently

$$\psi(x) = 1 + \varepsilon \sum_{k=1}^{n} p_k (a_{kk'} - a_{kk''}) + o(\varepsilon).$$

Since

$$\sum_{k=1}^{n} p_k a_{kk'} = -\sum_{i=1}^{n} a_{k''i} p_i = 0,$$

one has

$$\psi(x) = 1 + \varepsilon \sum_{k=1}^{n} p_k a_{kk'} + o(\varepsilon) = 1 - \varepsilon \sum_{i=1}^{n} a_{k'i} p_i + o(\varepsilon) > 1,$$

for sufficiently small $\varepsilon > 0$. This contradicts to the condition of Lemma. $\square$

Example 8.3. Consider a QSO $V : S^2 \to S^2$ given by

$$\begin{cases} x_1' = x_1(1 - ax_2 - bx_3) \\ x_2' = x_2(1 + ax_1 - cx_3) \\ x_3' = x_3(1 + bx_1 + cx_2) \end{cases}$$

where $0 < a, b, c \le 1$. Note that only solution of

$$\begin{cases} -ax_2 - bx_3 \ge 0 \\ ax_1 - cx_3 \ge 0 \\ bx_1 + cx_2 \ge 0 \end{cases}$$

on S^2 is $p = (1, 0, 0)$. So, $P = \{(1, 0, 0)\}$, and similarly $Q = \{(0, 0, 1)\}$. Consequently, $\varphi(x) = x_1$, $\psi(x) = (1 - ax_2 - bx_3)$. Obviously $\varphi(x) \le 1, \forall x \in S^2$ and $\max_{x \in S^2} \psi(x) = \psi(1, 0, 0) = 1$. From $\psi(y) = 1$, $y \in S^2$ it follows $y = (1, 0, 0)$. If

$$(1 - ax_2 - bx_3)^{p_1}(1 + ax_1 - cx_3)^{p_2}(1 + bx_1 + cx_2)^{p_3} \le 1, \ \forall x \in S^2$$

and $(p_1, p_2, p_3) \in S^2$ then $p_1 = 1$, $p_2 = p_3 = 0$.

Example 8.4. Consider a QSO $V : S^2 \to S^2$ given by

$$\begin{cases} x_1' = x_1(1 - ax_2 + bx_3) \\ x_2' = x_2(1 + ax_1 - cx_3) \\ x_3' = x_3(1 - bx_1 + cx_2) \end{cases}$$

where $0 < a, b, c \le 1$. In this case we have

$$P = Q = \left\{ \left(\frac{c}{a + b + c}, \frac{b}{a + b + c}, \frac{a}{a + b + c} \right) \right\}$$

and

$$\varphi(x) = (x_1^c x_2^b x_3^a)^{\frac{1}{a+b+c}},$$

$$\psi(x) = \left[(1 - ax_2 + bx_3)^c(1 + ax_1 - cx_3)^b(1 - bx_1 + cx_2)^a \right]^{\frac{1}{a+b+c}}.$$

One can see that

1) $\psi(x) \le 1, \forall x \in S^2$.

2) $\max_{x \in S^2} \psi(x) = \psi\left(\frac{c}{a+b+c}, \frac{b}{a+b+c}, \frac{a}{a+b+c} \right) = 1$,

3) $\psi(y) = 1, \ \Rightarrow \ y = \left(\frac{c}{a+b+c}, \frac{b}{a+b+c}, \frac{a}{a+b+c} \right)$.

4) If $(1 - ax_2 + bx_3)^{p_1}(1 + ax_1 - cx_3)^{p_2}(1 - bx_1 + cx_2)^{p_3} \le 1$, for all $x \in S^2$, and $(p_1, p_2, p_3) \in S^2$, then

$$p_1 = \frac{c}{a + b + c}, \ p_2 = \frac{b}{a + b + c}, \ p_3 = \frac{a}{a + b + c}.$$

8.1.3 *The set of limit points of trajectory*

Theorem 8.3. *Let V be a Volterra operator. If $x^0 \in \mathrm{int}\, S^{n-1}$ and $x^0 \neq V(x^0)$ then*

$$\lim_{m \to \infty} \varphi(x^{(m)}) = 0, \quad \sum_{m=0}^{\infty} \varphi(x^{(m)}) < \infty$$

and

$$\omega(x^0) \subset \partial S^{n-1}.$$

Proof. Let us check that $P \cap \omega(x^0) = 0$. If $\bar{p} = (\bar{p}_1, \ldots, \bar{p}_n) \in P$ then we have known that $\overline{\varphi}(x) = x_1^{\bar{p}_1} \cdots x_n^{\bar{p}_n}$ is a Lyapunov function of the dynamical system (8.5) and it reaches its maximum only at the point $\bar{p}$, moreover $V(\bar{p}) = \bar{p}$. Therefore, from $x^0 \neq V(x^0)$ one gets $\varphi(x^0) < \varphi(\bar{p})$. Due to $\varphi(x^{(m)}) \geq \varphi(x^{(m+1)})$, $m = 0, 1, \ldots$ we obtain $\varphi(y) < \varphi(\bar{p})$ for any $y \in \omega(x^0)$. Hence, $\bar{p} \notin \omega(x^0)$. Arbitrariness of $\bar{p} \in P$ implies $P \cap \omega(x^0) = \emptyset$.

Now let us show that $\sum_{m=0}^{\infty} \varphi(x^{(m)}) < \infty$ for any $\varphi(x) = x_1^{p_1} \ldots x_n^{p_n}$, where $p = (p_1, \ldots, p_n) \in P$. From above made arguments we know that the compacts P and $\omega(x^0)$ do not intersect. Note that the functional $\psi(x) = \prod_{k=1}^{n}(1 + \sum_{i=1}^{n} a_{ki} x_i)^{p_k}$ has its maximum 1 at point p. So, for some neighborhood U of $\omega(x^0)$ we have

$$\psi(x) \leq c < 1, \quad \forall x \in U.$$

The trajectory $\{x^{(m)}\}$ belongs to U for all $m \geq m_0$ where m_0 is some number. Hence, $\psi(x^{(m)}) \leq c < 1$, $m \geq m_0$. The equality $\varphi(x^{(m+1)}) = \varphi(x^{(m)}) \cdot \psi(x^{(m)})$ implies

$$\frac{\varphi(x^{(m+1)})}{\varphi(x^{(m)})} \leq c < 1, \quad \forall m \geq m_0,$$

which means $\sum_{m=0}^{\infty} \varphi(x^{(m)}) < \infty$.

From the definition of φ one can see that $\varphi(x) > 0$ if $x \in \mathrm{int}\, S^{n-1}$ and $\varphi(x) = 0$ only if $x \in \partial S^{n-1}$. Consequently, $\omega(x^0) \subset \partial S^{n-1}$. $\qquad\square$

From the proof of Theorem 8.3 we conclude that

$$\omega(x^0) \subset \bigcap_{p \in P} \{x \in S^{n-1} : x_1^{p_1} \cdots x_n^{p_n} = 0\}.$$

Theorem 8.4. *The set $\omega(x^0)$ contains either a single point or infinite.*

Proof. Assume that $\omega(x^0) = \{y^1, \ldots, y^r\}$, where $1 < r < \infty$. Then a collection $\{y^1, y^2, \ldots, y^r\}$ is a cycle, i.e.

$$V(y^1) = y^2, \ \ldots \ , V(y^{r-1}) = y^r, \ V(y^r) = y^1. \tag{8.18}$$

Since $r > 1$, one has $V(x^0) \neq x^0$. So, by Theorem 8.3 we have $\omega(x^0) \subset \partial S^{n-1}$. We know that any face of the simplex is invariant with respect to V (see Proposition 8.2), so from (8.18) we conclude that all $y^1, y^2, \ldots, y^r$ elements belongs to some $n-2$-dimensional face of S^{n-1}. Denote that face by Γ_1. Let V_1 be the restriction of V to Γ_1. By Proposition 8.1 $V_1 : \Gamma_1 \to \Gamma_1$, is Volterra QSO as well. Note that her Γ_1 is a $n-2$-dimensional simplex. From (8.18) we have $V_1(\omega(x^0)) = \omega(x^0)$, and then $\omega(y) = \omega(x^0)$ for any $y \in \omega(x^0)$. Repeating the made argument (using Theorem 8.3) we obtain $\omega(x^0) \subset \Gamma_{n-1}$, which contradicts to $1 < r < \infty$, since Γ_{n-1} contains just one point (a vertex of S^{n-1}). $\qquad\square$

Corollary 8.3. *There is not any periodical orbit of the dynamical system* (8.5).

Theorem 8.5. *If $x^0 \in \mathrm{int} S^{n-1}$, $V(x^0) \neq x^0$ and the trajectory $\{x^{(m)}\}$ converges. Then the limit point of the trajectory belongs to Q.*

Proof. Assume that $x^{(m)} \to \overline{x} = (q_1, \ldots, q_n)$. By Theorem 8.3 we have $\overline{x} \in \partial S^{n-1}$. Denote

$$I = \{1, 2, \ldots, n\}, \ I_0 = \{i : q_i = 0\}, \ I_+ = \{i : q_i > 0\}.$$

Obviously $V(\overline{x}) = \overline{x}$ i.e.

$$q_k = q_k \left(1 + \sum_{i=1}^{n} a_{ki} q_i \right), \quad k \in I. \tag{8.19}$$

If $k \in I_+$, then from (8.19) it follows that

$$\sum_{i=1}^{n} a_{ki} q_i = 0. \tag{8.20}$$

Now assume that for some $k_0 \in I_0$ that $\sum_{i=1}^{n} a_{k_0 i} q_i > 0$. The convergence $x_k^{(m)} \to q_k$ implies the existence a number $m_0 \in \mathbb{N}$ such that

$$\sum_{i=1}^{n} a_{k_0 i} x_i^{(m)} > 0, \quad \forall m > m_0. \tag{8.21}$$

Keeping in mind the invariance of $\mathrm{int} S^{n-1}$ with respect to V, for $x^0 \in \mathrm{int} S^{n-1}$ we have $x_k^{(m)} > 0$ for all $k \in I$ and $m = 0, 1, \ldots$. Now (8.21) with

$$x_{k_0}^{(m+1)} = x_{k_0}^{(m)} \left(1 + \sum_{i=1}^{n} a_{k_0 i} x_i^{(m)} \right), \quad m > m_0$$

implies that $x_{k_0}^{(m+1)} > x_{k_0}^{(m)}$ for any $m > m_0$. This is impossible since $x_{k_0}^{(m)} \to q_{k_0} = 0$. Hence, for any $k \in I_0$ it must hold

$$\sum_{i=1}^{n} a_{ki} q_i \leq 0, \quad k \in I_0.$$

By (8.20) we obtain

$$\sum_{i=1}^{n} a_{ki} q_i \leq 0, \quad \forall k \in I.$$

Consequently $\overline{x} = (q_1, \ldots, q_n) \in Q$. $\qquad\qquad\square$

Theorem 8.6. *If V has an internal isolated fixed point $x^* = (p_1, \ldots, p_n)$ then for any $x^0 \in \mathrm{int}\, S^{n-1}$ with $V(x^0) \neq x^0$ the trajectory $\{x^{(m)}\}$ does not converge.*

Proof. Since $p_k > 0$, $k \in I$ then by

$$p_k = p_k \left(1 + \sum_{i=1}^{n} a_{ki} p_i \right),$$

we have $\sum_{i=1}^{n} a_{ki} p_i = 0$, $k \in I$ and $x^* \in Q$. Assuming that the trajectory $\{x^{(m)}\}$ does converge, by Theorem 8.5 we obtain

$$\overline{x} = \lim_{m \to \infty} x^{(m)} \in Q.$$

Convexity of Q with $x^*, \overline{x} \in Q$ implies that $[x^*, \overline{x}] \subset Q$. Note that $x^* \neq \overline{x}$, since $x^* \in \mathrm{int} S^{n-1}$ and $\overline{x} \in \partial S^{n-1}$. This contradicts to the condition that x^* is isolated. $\qquad\qquad\square$

Corollary 8.4. *If conditions of Theorem 8.6 are satisfied then $\omega(x^0)$ is infinite.*

8.1.4 *The backward trajectories*

The following theorem very useful property of Volterra QSO.

Theorem 8.7. *The Volterra QSO $V : S^{n-1} \to S^{n-1}$ is a homeomorphism.*

Proof. We shall use an induction over n. For $n = 1$ the statement is trivial. Assuming that the statement is true for $1, 2, \ldots, n-1$ we prove it for n. Since any face of S^{n-1} is invariant with respect to V and reduction of V on any face is also a Volterra QSO, by assumption of the induction the operator maps the boundary of S^{n-1} to itself. Therefore, it is sufficient to show that at each interior point the mapping V is a local homeomorphism (see [198]).

A Jacobian of V at point $x = (x_1, \ldots, x_n) \in \mathrm{int}S^{n-1}$ is

$$
J_n = \begin{vmatrix}
1 + \sum_{i=1}^{n} a_{1i}x_i & a_{12}x_1 & \cdots & a_{1n}x_1 \\
a_{21}x_2 & 1 + \sum_{i=1}^{n} a_{2i}x_i & \cdots & a_{2n}x_2 \\
\cdot & \cdot & \cdots & \cdot \\
a_{n1}x_n & a_{n2}x_n & \cdots & 1 + \sum_{i=1}^{n} a_{ni}x_i
\end{vmatrix}.
$$

We prove that $J_n > 0$. For $y_1, \ldots, y_n > 0$ and $a_{ki} = -a_{ik}$ one has

$$
D_n = \begin{vmatrix}
y_1 & a_{12} & \cdots & a_{1n} \\
a_{21} & y_2 & \cdots & a_{2n} \\
\cdot & \cdot & \cdots & \cdot \\
a_{n1} & a_{n2} & \cdots & y_n
\end{vmatrix} \geq y_1 \cdot y_2 \cdots y_n. \tag{8.22}
$$

Indeed, one has

$$
D_1 = y_1 > 0,
$$

$$
D_2 = y_1 y_2 - a_{12}a_{21} = y_1 y_2 + a_{12}^2 \geq y_1 y_2.
$$

Assume that $D_k \geq y_1 \cdots y_k$ for all $k = 1, \ldots, n-1$. Expanding D_n simultaneously by the last row and the last column we obtain

$$
D_n = y_n D_{n-1} - \sum_{i=1}^{n-1} a_{ni} a_{in} D_{n-2}^{(i)}, \tag{8.23}
$$

where D_{n-1} and $D_{n-2}^{(r)}$, $r = 1, \ldots, n-1$ are the main minors of order $n-1$ and $n-2$, respectively. They have the same form like D_n. The equality $a_{ki} = -a_{ik}$ with (8.23) implies that

$$
D_n = y_n D_{n-1} + \sum_{r=1}^{n-1} a_{nr}^2 D_{n-2}^{(r)}.
$$

By assumption $D_{n-2}^{(r)} \geq 0$, one gets

$$
D_n \geq y_n D_{n-1} \geq y_1 y_2 \cdots y_n.
$$

Thus if $y_1, \ldots y_n > 0$, then $D_n > 0$. Returning to the Jacobian J_n we find

$$J_n = x_1 x_2 \cdots x_n \begin{vmatrix} \dfrac{1+\sum_{i=1}^n a_{1i}x_i}{x_1} & a_{12} & \cdots & a_{1n} \\[2mm] a_{21} & \dfrac{1+\sum_{i=1}^n a_{2i}x_i}{x_2} & \cdots & a_{2n} \\[2mm] \cdot & \cdots & \cdots & \cdot \\[2mm] a_{n1} & a_{n2} & \cdots & \dfrac{1+\sum_{i=1}^n a_{ni}x_i}{x_n} \end{vmatrix}$$

$$\geq x_1 \cdots x_n \frac{1+\sum_{i=1}^n a_{1i}x_i}{x_1} \cdots \frac{1+\sum_{i=1}^n a_{ni}x_i}{x_n}$$

$$= \prod_{k=1}^n \left(1 + \sum_{i=1}^n a_{ki}x_i\right).$$

From $x \in \mathrm{int}\,S^{n-1}$ one has $x_k > 0$, and hence $J_n > 0$. So, $J_n \neq 0$ this means V is a local homeomorphism at each point $x \in \mathrm{int}\,S^{n-1}$. $\square$

Corollary 8.5. *For any initial point $x^0 \in S^{n-1}$ the "backward" trajectory*

$$\{x^0, V^{-1}(x^0), V^{-2}(x^0), \ldots\}$$

exists.

Denote by $\alpha(x^0)$ the set of limit points of the backward trajectory started at $x^0 \in S^{n-1}$. It is clear that $\alpha(x^0) \neq \emptyset$ and it is a closed set.

Theorem 8.8. *If $x^0 \in \mathrm{int}\,S^{n-1}$, then $\alpha(x^0) \subset P$.*

Proof. For $p = (p_1, \ldots, p_n) \in P$ the Lyapunov function $\varphi(x) = x_1^{p_1} \cdots x_n^{p_n}$ is monotonic decreasing along any trajectory. Therefore,

$$\varphi(x^0) \leq \varphi(x^{(-1)}) \leq \varphi(x^{(-2)}) \leq \cdots \tag{8.24}$$

where $x^{(-n)} = V^{-n}(x^0)$.

Due to $x^0 \in \mathrm{int}\,S^{n-1}$ one has $\varphi(x^0) > 0$. Hence, from (8.24) we find

$$\lim_{m \to \infty} \varphi(x^{(-m)}) = C > 0. \tag{8.25}$$

By $x^{(-m)} = V(x^{(-m-1)})$ we have

$$\varphi(x^{(-m)}) = \psi(x^{(-m-1)}) \cdot \varphi(x^{(-m-1)}), \tag{8.26}$$

where $\psi(x) = \prod_{k=1}^n (1 + \sum_{i=1}^n a_{ki}x_i)^{p_k}$.

By means of (8.25) with (8.26) one gets

$$\lim_{m \to \infty} \psi(x^{(-m)}) = 1,$$

this means that for any $y \in \alpha(x^0)$ one has $\psi(y) = 1$. Recall that $\psi(y) = 1$ only if $y \in P$. Hence, $\alpha(x^0) \subset P$. $\square$

Theorem 8.9. *The backward trajectories always converge.*

Proof. Without loss of generality we may assume $x^0 \in \mathrm{ri}S^{n-1}$. Indeed, if $x^0 \in \partial S^{n-1}$ then one can consider the corresponding face of the simplex and the reduction of the operator to that face. By Theorem 8.8 we have $\alpha(x^0) \subset P$. Hence, $\alpha(x^0)$ consists of only fixed points of V. Take $p = (p_1, \ldots, p_n) \in \alpha(x^0)$ and consider Lyapunov function $\varphi(x) = x_1^{p_1} \cdots x_n^{p_n}$. According to $p \in \alpha(x^0)$ there is a backward trajectory $\{x^{(-m_i)}\}$ converging to p. Therefore,

$$\lim_{i \to \infty} \varphi(x^{(-m_i)}) = \varphi(p) = p_1^{p_1} \cdots p_n^{p_n}.$$

Then by (8.23) we obtain

$$\lim_{m \to \infty} \varphi(x^{(-m)}) = \varphi(p).$$

Due to fact that $\varphi(p)$ is the maximal value of $\varphi(x)$ which reaches only at p, implies that $\alpha(x^0) = \{p\}$. $\qquad\square$

Example 8.5. Consider a QSO $V : S^3 \to S^3$ given by

$$\begin{cases} x_1' = x_1(1 - x_3 + x_4) \\ x_2' = x_2(1 - x_3 + x_4) \\ x_3' = x_3(1 + x_1 + x_2 - x_4) \\ x_4' = x_4(1 - x_1 - x_2 + x_3). \end{cases} \tag{8.27}$$

Denote $z_1 = (\frac{1}{3}, 0, \frac{1}{3}, \frac{1}{3})$, $z_2 = (0, \frac{1}{3}, \frac{1}{3}, \frac{1}{3})$. Then

$$[z_1, z_2] - \{\lambda z_1 + (1 - \lambda)z_2 : 0 \leq \lambda \leq 1\} - \left\{ \left(\frac{\lambda}{3}, \frac{1-\lambda}{3}, \frac{1}{3}, \frac{1}{3} \right), 0 \leq \lambda \leq 1 \right\}.$$

Solving the corresponding inequality on gets

$$P = Q = [z_1, z_2].$$

Hence, $\varphi(x) = (x_1^\lambda \cdot x_2^{1-\lambda} \cdot x_3 \cdot x_4)^{\frac{1}{3}}$ is a Lyapunov function for the dynamical system (8.27) for every $\lambda \in [0, 1]$. Consequently if $x^0 \in (\mathrm{int}S^3) \setminus P$ then

$$\omega(x^0) \subset \bigcap_{0 \leq \lambda \leq 1} \{x : x_1^\lambda \cdot x_2^{1-\lambda} \cdot x_3 \cdot x_4 = 0\}$$

$$= \{x : x_3 = 0\} \cup \{x : x_4 = 0\} \cup \{x : x_1 = x_2 = 0\} \subset \partial S^3$$

and $\alpha(x^0) \subseteq P$.

For this example, these general estimates of $\omega(x^0)$ and $\alpha(x^0)$ can be improved. In order to do this it is enough to see that (8.27) has the following property

$$\frac{x_1'}{x_2'} = \frac{x_1^0}{x_2^0}.$$

Thus for any $m \in \{0, \pm 1, \pm 2, ...\}$ we have (the "conservation law")

$$\frac{x_1^{(m)}}{x_2^{(m)}} = \frac{x_1^0}{x_2^0}.$$

Consequently, $\alpha(x^0)$ and all points of $\omega(x^0)$ must satisfy the "conservation law". In particular, for $\alpha(x^0) \in P = \{(\frac{\lambda}{3}, \frac{1-\lambda}{3}, \frac{1}{3}, \frac{1}{3}), 0 \leq \lambda \leq 1\}$ we have

$$\frac{\lambda}{1-\lambda} = \frac{x_1^0}{x_2^0}. \tag{8.28}$$

From (8.28) one gets $\lambda = \frac{x_1^0}{x_1^0 + x_2^0}$. Hence

$$\alpha(x^0) = \left(\frac{x_1^0}{3(x_1^0 + x_2^0)}, \frac{x_2^0}{3(x_1^0 + x_2^0)}, \frac{1}{3}, \frac{1}{3} \right).$$

Similarly

$$\omega(x^0) \subset \partial \left\{ \overline{x : \frac{x_1}{x_2} = \frac{x_1^0}{x_2^0}} \right\},$$

where ∂M is relative boundary of M.

8.1.5 *Biological interpretations*

In biology V has the meaning of a population evolution operator. We consider a population consisting of n species (criteria). Let $x^0 = (x_1^0, ..., x_n^0)$ be the probability distribution of species in the initial generations, and $p_{ij,k}$ the probability that individuals in the ith and jth species interbreed to produce an individual of the kth species. In random interbreeding (panmixia) the probability distribution of the species in the first generation are found by $x' = V(x)$ where V is a QSO with coefficients $p_{ij,k}$. Consequently, $x^{(m)} = (x_1^{(m)}, ..., x_n^{(m)})$ is the probability distribution in the mth generation, where $x^{(m)} = V(x^{(m-1)})$. The fixed points of the operator $V : S^{n-1} \to S^{n-1}$ are called *equilibrium states* of the population.

Obviously, the condition $p_{ij,k} = 0$ for $k \neq \{i, j\}$ means that each individual can inherit only the species of the parents. For such biological systems the results of this section admit the following interpretations:

1) The biological system has at least n equilibrium states.

2) The "past" of such biological systems can be uniquely reproduced (Theorem 8.7).

3) The evolution begins in a neighborhood of one of the equilibrium states of the population (Theorem 8.8).

4) As a rule, the population does not tend to an equilibrium state with the passage of time (Theorem 8.6).

5) The "future" of the population is unstable, since certain species turn out to be on the verge of extinction with the passage of time (Theorem 8.3).

8.2 Non-Volterra QSO generated by a product measure

8.2.1 *A non-Volterra QSO*

In this subsection, following [225] we describe a wide class of non-Volterra quadratic stochastic operators using N. Ganikhadjaev's construction of quadratic stochastic operators [64]. By the construction these operators depend on a probability measure μ being defined on the set of all cells which are given on a graph G. We describe a condition on measure μ under which the QSO V_μ generated by μ can be studied using the theory of Volterra QSO.

Let $G = (\mathcal{L}, L)$ be a finite graph and $\{\mathcal{L}_i, i = 1, ..., m\}$ the set of all maximal connected subgraphs of G. Denote by $\Omega_i = \Phi^{\mathcal{L}_i}$ the set of all configurations defined on $\mathcal{L}_i$, $i = 1, ..., m$. Let μ_i be a probability measure defined on Ω_i, such that $\mu_i(\sigma) > 0$ for any $\sigma \in \Omega_i$, $i = 1, ..., m$.

Consider probability measure μ on $\Omega = \Omega_1 \times \cdots \times \Omega_m$ defined as

$$\mu(\sigma) = \prod_{i=1}^{m} \mu_i(\sigma_i), \tag{8.29}$$

where $\sigma = (\sigma_1, ..., \sigma_m)$, with $\sigma_i \in \Omega_i, i = 1, ..., m$.

Note that (see [64]) if $m = 1$ then QSO constructed on G is Volterra QSO.

Theorem 8.10. *The QSO constructed by measure* (8.29) *is reducible to m separate Volterra QSOs.*

Proof. Take $\varphi = (\varphi_1, ..., \varphi_m), \psi = (\psi_1, ..., \psi_m) \in \Omega$. By construction

$$\Omega(G, \varphi, \psi) = \{\sigma = (\sigma_1, .., \sigma_m) \in \Omega : \sigma_i \in \{\varphi_i, \psi_i\}, i = 1, ..., m\}$$

and

$$P_{\varphi\psi,\sigma} = \begin{cases} \prod_{i=1}^{m} \frac{\mu_i(\sigma_i)}{\mu_i(\varphi_i)+\mu_i(\psi_i)}, & \text{if } \sigma \in \Omega(G, \varphi, \psi), \\ 0 & \text{otherwise}, \end{cases} \tag{8.30}$$

where we used the following equality

$$\mu(\Omega(G,\varphi,\psi)) = \sum_{\substack{\sigma_1,\dots,\sigma_m:\\ \sigma_i\in\{\varphi_i,\psi_i\},i=1,\dots,m}} \prod_{i=1}^{m}\mu_i(\sigma_i) = \prod_{i=1}^{m}\big(\mu_i(\varphi_i)+\mu_i(\psi_i)\big).$$

Thus QSO generated by measure (8.29) can be written as

$$\lambda'(\sigma) = \lambda'(\sigma_1,\dots,\sigma_m) = \sum_{\substack{\varphi=(\varphi_1,\dots,\varphi_m):\varphi_i\in\Omega_i\\ \psi=(\psi_1,\dots,\psi_m):\psi_i\in\Omega_i}} \prod_{i=1}^{m} \frac{\mu_i(\sigma_i)\mathbf{1}_{(\sigma_i\in\{\varphi_i,\psi_i\})}}{\mu_i(\varphi_i)+\mu_i(\psi_i)}\lambda(\varphi)\lambda(\psi).$$

$$(8.31)$$

Denote

$$X_{i,\omega} = \sum_{\substack{\sigma\in\Omega:\\ \sigma_i=\omega}} \lambda(\sigma) = \sum_{\substack{\sigma_1,\dots,\sigma_{i-1},\sigma_{i+1},\dots,\sigma_m\\ \sigma_k\in\Omega_k,k\neq i}} \lambda(\sigma_1,\dots,\sigma_{i-1},\omega,\sigma_{i+1},\dots,\sigma_m).$$

$$(8.32)$$

From (8.31) we have

$$X'_{i,\omega} = \sum_{\substack{\sigma\in\Omega:\\ \sigma_i=\omega}} \lambda'(\sigma) = \sum_{\substack{\sigma\in\Omega:\\ \sigma_i=\omega}} \left[\sum_{\substack{\varphi_1,\dots,\varphi_{i-1},\varphi_{i+1},\dots,\varphi_m\\ \psi_1,\dots,\psi_m}} \frac{\mu_i(\omega)}{\mu_i(\omega)+\mu_i(\psi_i)} \right.$$

$$\times \prod_{\substack{j=1\\ j\neq i}}^{m} \frac{\mu_j(\sigma_j)\mathbf{1}_{(\sigma_j\in\{\varphi_j,\psi_j\})}}{\mu_j(\varphi_j)+\mu_j(\psi_j)} \lambda(\varphi_1,\dots,\varphi_{i-1},\omega,\varphi_{i+1},\dots,\varphi_m)\lambda(\psi_1,\dots,\psi_m)$$

$$+ \sum_{\substack{\varphi_1,\dots,\varphi_m\\ \psi_1,\dots,\psi_{i-1},\psi_{i+1},\dots,\psi_m}} \frac{\mu_i(\omega)}{\mu_i(\varphi_i)+\mu_i(\omega)}$$

$$\left. \times \prod_{\substack{j=1\\ j\neq i}}^{m} \frac{\mu_j(\sigma_j)\mathbf{1}_{(\sigma_j\in\{\varphi_j,\psi_j\})}}{\mu_j(\varphi_j)+\mu_j(\psi_j)} \lambda(\varphi_1,\dots,\varphi_m)\lambda(\psi_1,\dots,\psi_{i-1},\omega,\psi_{i+1},\dots,\psi_m) \right]$$

$$= 2 \sum_{\substack{\varphi_1,\dots,\varphi_{i-1},\varphi_{i+1},\dots,\varphi_m\\ \psi_1,\dots,\psi_m}} \frac{\mu_i(\omega)}{\mu_i(\omega)+\mu_i(\psi_i)}$$

$$\times \sum_{\substack{\sigma_1,\dots,\sigma_{i-1},\\ \sigma_{i+1},\dots,\sigma_m}} \prod_{\substack{j=1\\ j\neq i}}^{m} \frac{\mu_j(\sigma_j)\mathbf{1}_{(\sigma_j\in\{\varphi_j,\psi_j\})}}{\mu_j(\varphi_j)+\mu_j(\psi_j)} \lambda(\varphi_1,\dots,\varphi_{i-1},\omega,\varphi_{i+1},\dots,\varphi_m)$$

$$\times \lambda(\psi_1,\dots,\psi_m).$$

$$(8.33)$$

Note that

$$\sum_{\substack{\sigma_1,\ldots,\sigma_{i-1},\sigma_{i+1},\ldots,\sigma_m}} \prod_{\substack{j=1 \\ j\neq i}}^{m} \frac{\mu_j(\sigma_j)\mathbf{1}_{(\sigma_j\in\{\varphi_j,\psi_j\})}}{\mu_j(\varphi_j)+\mu_j(\psi_j)} = 1.$$

Thus from (8.33) we have

$$\text{RHS of (8.33)}$$

$$= \sum_{\substack{\varphi_1,\ldots,\varphi_{i-1}, \\ \varphi_{i+1},\ldots,\varphi_m, \\ \psi_1,\ldots,\psi_m}} \frac{2\mu_i(\omega)}{\mu_i(\omega)+\mu_i(\psi_i)}\lambda(\varphi_1,\ldots,\varphi_{i-1},\omega,\varphi_{i+1},\ldots,\varphi_m)\lambda(\psi_1,\ldots,\psi_m)$$

$$= \sum_{\substack{\varphi_1,\ldots,\varphi_{i-1}, \\ \varphi_{i+1},\ldots,\varphi_m, \\ \psi_1,\ldots,\psi_{i-1}, \\ \psi_{i+1},\ldots,\psi_m}} \lambda(\varphi_1,\ldots,\varphi_{i-1},\omega,\varphi_{i+1},\ldots,\varphi_m)\lambda(\psi_1,\ldots,\psi_{i-1},\omega,\psi_{i+1},\ldots,\psi_m)$$

$$+ \sum_{\psi_i\in\Omega_i\backslash\omega} \frac{2\mu_i(\omega)}{\mu_i(\omega)+\mu_i(\psi_i)}$$

$$\times \sum_{\substack{\varphi_1,\ldots,\varphi_{i-1}, \\ \varphi_{i+1},\ldots,\varphi_m, \\ \psi_1,\ldots,\psi_{i-1}, \\ \psi_{i+1},\ldots,\psi_m}} \lambda(\varphi_1,\ldots,\varphi_{i-1},\omega,\varphi_{i+1},\ldots,\varphi_m)\lambda(\psi_1,\ldots,\psi_m)$$

$$= X_{i,\omega}^2 + \sum_{\psi\in\Omega_i\backslash\omega} \frac{2\mu_i(\omega)}{\mu_i(\omega)+\mu_i(\psi)}X_{i,\omega}X_{i,\psi}.$$

Thus operator (8.31) can be rewritten as

$$X'_{i,\omega} = X_{i,\omega}\left(X_{i,\omega} + \sum_{\psi\in\Omega_i\backslash\omega} \frac{2\mu_i(\omega)}{\mu_i(\omega)+\mu_i(\psi)}X_{i,\psi}\right), \qquad (8.34)$$

where $X_{i,\omega}$ is defined by (8.32), $\omega\in\Omega_i, i=1,\ldots,m$.

Note that $\sum_{\omega\in\Omega_i} X_{i,\omega} = 1$ for any $i=1,\ldots,m$. Using this equality from (8.34) we obtain

$$X'_{i,\omega} = X_{i,\omega}\left(1 + \sum_{\psi\in\Omega_i} \frac{\mu_i(\omega)-\mu_i(\psi)}{\mu_i(\omega)+\mu_i(\psi)}X_{i,\psi}\right). \qquad (8.35)$$

One can see that for each fixed i ($i=1,\ldots,m$) the operator (8.35) is Volterra operator $V^{(i)} : S^{|\Omega_i|-1} \to S^{|\Omega_i|-1}$. $\qquad\square$

8.2.2   *The behavior of the trajectories*

In this subsection using Volterra QSOs (8.35) we shall describe the behavior of trajectories of non-Volterra QSO (8.31).

Denote $a^{(i)}_{\varphi,\psi} = \frac{\mu_i(\varphi)-\mu_i(\psi)}{\mu_i(\varphi)+\mu_i(\psi)}$. One can see that for each fixed $i \in \{1,...,m\}$ the coefficients $a^{(i)}_{\varphi,\psi}$ satisfy the following properties

$$a^{(i)}_{\varphi,\psi} = -a^{(i)}_{\psi,\varphi}, \quad |a^{(i)}_{\varphi,\psi}| \leq 1. \tag{8.36}$$

Thus QSO (8.35) has the same form with a Volterra QSO.

If for any $i \in \{1,...,m\}$ the asymptotical behavior of trajectories of QSO $V^{(i)}$ (i.e. (8.35)) is known, say $X^{(l)}_{i,\omega} \to X^*_{i,\omega}$, $l \to \infty$, then asymptotical behavior of V (i.e. (8.31)), say $\lambda^{(l)}(\sigma) \to \lambda^*(\sigma)$, $l \to \infty$, can be found from system of linear equations

$$\sum_{\sigma\in\Omega:\sigma_i=\omega} \lambda^*(\sigma) = X^*_{i,\omega}, \quad \omega \in \Omega_i, i = 1,...,m. \tag{8.37}$$

The following theorem gives some results from previous sections related to Volterra QSO.

Theorem 8.11.

1. *If $x^{(0)} \in \text{int}\,S^{n-1}$ is not a fixed point (i.e. $Vx^{(0)} \neq x^{(0)}$), then $\omega(x^{(0)}) \subset \partial S^{n-1}$.*
2. *The set $\omega(x^{(0)})$ either consists of a single point or is infinite.*
3. *If a Volterra QSO has an isolated fixed point $x^* \in \text{int}\,S^{n-1}$, then for any initial point $x^{(0)} \neq x^*$, the trajectory $\{x^{(l)}\}$ does not converge.*

Corollary 8.6. *The set of limit points $\nu(\lambda^0)$ of the QSO (8.31) has properties 1–3 mentioned in Theorem 8.11.*

Proof.   1. By Theorem 8.11 there is at least one $\tilde{\omega} \in \Omega_i$ with $X^*_{i,\tilde{\omega}} = 0$. Since $\lambda(\sigma) \geq 0$ from (8.37) we get that $\lambda^*(\sigma) = 0$ for all σ such that $\sigma_i = \tilde{\omega}$. This completes the proof of the property 1. Properties 2, 3 also follow from (8.37). $\qquad\square$

Example 8.6. Consider graph $G = (\mathcal{L}, L)$ with $\mathcal{L} = \{1,2\}$ and $L = \emptyset$. Take $\Phi = \{A, a\}$. Then non-Volterra QSO (8.31) has the form

$$x'_1 = x_1^2 + 2\beta_1 x_1 x_2 + 2\alpha_1 x_1 x_3 + 2\alpha_1\beta_1 x_1 x_4 + 2\alpha_1\beta_1 x_2 x_3$$

$$x'_2 = x_2^2 + 2\beta_2 x_1 x_2 + 2\alpha_1\beta_2 x_2 x_3 + 2\alpha_1 x_2 x_4 + 2\alpha_1\beta_2 x_1 x_4$$

$$x'_3 = x_3^2 + 2\alpha_2 x_1 x_3 + 2\alpha_2\beta_1 x_2 x_3 + 2\beta_1 x_3 x_4 + 2\alpha_2\beta_1 x_1 x_4 \tag{8.38}$$

$$x'_4 = x_4^2 + 2\alpha_2\beta_2 x_1 x_4 + 2\alpha_2 x_2 x_4 + 2\beta_2 x_3 x_4 + 2\alpha_2\beta_2 x_2 x_3$$

where

$$\mu_1 = (\alpha_1, \alpha_2), \ \alpha_j \geq 0, \alpha_1 + \alpha_2 = 1; \quad \mu_2 = (\beta_1, \beta_2), \ \beta_j \geq 0, \beta_1 + \beta_2 = 1.$$

Putting $x_1 + x_2 = X_{1,1}$, $x_3 + x_4 = X_{1,2}$ and $x_1 + x_3 = X_{2,1}$, $x_2 + x_4 = X_{2,2}$ we get the Volterra operators like (8.35):

$$
\begin{aligned}
X'_{1,1} &= X_{1,1} \left(1 + (2\alpha_1 - 1)X_{1,2}\right) \\
X'_{1,2} &= X_{1,2} \left(1 + (2\alpha_2 - 1)X_{1,1}\right)
\end{aligned}
\tag{8.39}
$$

and

$$
\begin{aligned}
X'_{2,1} &= X_{2,1} \left(1 + (2\beta_1 - 1)X_{2,2}\right) \\
X'_{2,2} &= X_{2,2} \left(1 + (2\beta_2 - 1)X_{2,1}\right).
\end{aligned}
\tag{8.40}
$$

For this example from Corollary 8.6 we obtain

Corollary 8.7.

1. Any trajectory of non-Volterra QSO (8.38) has following limit

$$
\lim_{l \to \infty} x^{(l)} =
\begin{cases}
(1,0,0,0), & \text{if } 2\alpha_1 > 1, 2\beta_1 > 1, \\
(0,1,0,0), & \text{if } 2\alpha_1 > 1, 2\beta_1 < 1, \\
(0,0,1,0), & \text{if } 2\alpha_1 < 1, 2\beta_1 > 1, \\
(0,0,0,1), & \text{if } 2\alpha_1 < 1, 2\beta_1 < 1.
\end{cases}
$$

2. If $2\beta_1 = 1$ then $S_1 = \{x : x_3 = x_4 = 0\}$ and $S_2 = \{x : x_1 = x_2 = 0\}$ are the sets of fixed points for (8.38) and for any $x^{(0)} \notin S_1 \cup S_2$

$$
\lim_{l \to \infty} x^{(l)} \in
\begin{cases}
S_1, \text{ if } 2\alpha_1 > 1, \\
S_2 \text{ if } 2\alpha_1 < 1.
\end{cases}
$$

3. If $2\alpha_1 = 1$ then $S_3 = \{x : x_2 = x_4 = 0\}$ and $S_4 = \{x : x_1 = x_3 = 0\}$ are the sets of fixed points for (8.38) and for any $x^{(0)} \notin S_3 \cup S_4$

$$
\lim_{l \to \infty} x^{(l)} \in
\begin{cases}
S_3, \text{ if } 2\beta_1 > 1, \\
S_4 \text{ if } 2\beta_1 < 1.
\end{cases}
$$

4. If $2\alpha_1 = 2b_1 = 1$ then $S_5 = \{x : x_1 = x_2, x_3 = x_4\}$ and $S_6 = \{x : x_2 = x_4, x_1 = x_3\}$ are the sets of fixed points for (8.38).

8.3　Separable quadratic stochastic operators

In this section following [224] we consider QSO (8.1), (8.2) with additional condition

$$P_{ij,k} = a_{ik}b_{jk}, \quad \text{for all } i,j,k \in E \tag{8.41}$$

where $a_{ik}, b_{jk} \in \mathbb{R}$ entries of matrices $A = (a_{ik})$ and $B = (b_{jk})$ such that the conditions (8.1), (8.2) are satisfied for the coefficients (8.41).

Then the QSO V corresponding to the coefficients (8.41) has the form

$$x'_k = (V(x))_k = (A(x))_k \cdot (B(x))_k, \tag{8.42}$$

where $(A(x))_k = \sum_{i=1}^{m} a_{ik}x_i$, $(B(x))_k = \sum_{j=1}^{m} b_{jk}x_j$.

Definition 8.3. The QSO (8.42) is called separable quadratic stochastic operator (SQSO).

Remark 8.3. 1. If A (or B) is the identity matrix then the operator (8.42) becomes a linear Volterra QSO.

2. The following example shows that the condition (8.41) is sufficient for a QSO to be product of two linear operators, but the condition is not necessary: consider matrices

$$A = \begin{pmatrix} 0 & 1 & 0 \\ 1 & 0 & 0 \\ 0 & 0 & 1 \end{pmatrix}, \quad B = \begin{pmatrix} 0 & 1 & 0 \\ 1 & 2 & 2 \\ 0 & 2 & 1 \end{pmatrix}.$$

Then corresponding QSO is

$$\begin{cases} x'_1 = x_2^2 \\ x'_2 = x_1(x_1 + 2x_2 + 2x_3) \\ x'_3 = x_3(2x_2 + x_3). \end{cases}$$

For this operator we can take $P_{13,2} = P_{31,2} = 1$ but from matrices A and B we have $a_{32}b_{12} = 0 \neq P_{31,2}$. But one can easily check that a QSO with coefficients $P_{ij,k}$ can be written as the product of two linear operators $A = (a_{ik})$ and $B = (b_{jk})$ if and only if

$$P_{ij,k} + P_{ji,k} = a_{ik}b_{jk} + a_{jk}b_{ik} \quad \text{for any } i,j,k \in E. \tag{8.43}$$

Thus the condition (8.41) is a particular case of (8.43). In this section for simplicity we assume that (8.41) holds.

8.3.1 *Classification of SQSOs*

From the conditions $P_{ij,k} \geq 0$ and $\sum_{k=1}^{m} P_{ij,k} = 1$ for all i, j it follows the condition on matrices A and B that $a_{ik}b_{jk} \geq 0$, $AB^T = \mathbf{1}$, where B^T is the transpose of B and $\mathbf{1}$ is the matrix with all entries 1's. If $a^{(i)} = (a_{i1}, ..., a_{im})$ is the i-th row of the matrix A and $b^{(j)} = (b_{j1}, ..., b_{jm})$ is the j-th row of the matrix B, then from $AB^T = \mathbf{1}$ we get

$$a^{(i)}b^{(j)} = 1, \quad \text{for all } i, j = 1, ..., m. \tag{8.44}$$

For a fixed j, the above condition implies that

$$A(b^{(j)})^T = (1, 1, ..., 1). \tag{8.45}$$

If $\det(A) \neq 0$, then (8.45) gives $b^{(j)} = b^{(k)}$ for all $k, j = 1, ..., m$, i.e., all rows of B are the same, therefore $\det(B) = 0$. Similarly, if $\det(B) \neq 0$, then all the rows of A must be the same, so $\det(A) = 0$.

But in spite of both A and B have zero determinant we may have the matrices A and B are without all identical rows.

Remark 8.4. If $m = 2$ and $\det(A) = 0$ satisfying the condition (8.44) then one can check that A has the identical rows. But for $m \geq 3$, from $\det(A) = 0$ and the condition (8.44), we may have A and B having not all the same rows. For example, consider

$$A = \begin{pmatrix} b\, y_1\, 1 - y_1 \\ b\, y_2\, 1 - y_2 \\ b\, y_3\, 1 - y_3 \end{pmatrix}, \quad B = \begin{pmatrix} 0 & 1 & 1 \\ 0 & 1 & 1 \\ \frac{1}{2b} & \frac{1}{2} & \frac{1}{2} \end{pmatrix},$$

where $b > 0$, $y_i \in [0, 1]$ for all $i = 1, 2, 3$. One can see that these matrices satisfy the condition (8.44) and $\det(A) = \det(B) = 0$.

So we have three cases for the SQSO (8.42):

Case 1: If $\det(A) = \det(B) = 0$ and both of them have the identical rows, then the SQSO becomes constant, i.e.

$$x'_k = a_{1k}b_{1k}, \quad \text{for} \quad k = 1, ..., m. \tag{8.46}$$

In this case the dynamical system is trivial: independently on initial point $x^{(0)} = x^0 \in S^{m-1}$ all trajectories coincide with $\{x^{(n)}\}$ such that $x_k^{(n)} = a_{1k}b_{1k}$, for $k = 1, ..., m$, $n = 1, 2, ...$

Case 2: If $\det(A) \neq 0$, then B has the same rows and SQSO becomes

$$x'_k = b_{1k} \sum_{i=1}^{m} a_{ik} x_i, \quad \text{for} \quad k = 1, ..., m \tag{8.47}$$

which is a linear stochastic operator.

Remark 8.5. 1. Since B can be uniquely determined by a given A with $\det(A) \neq 0$, the operator (8.47) depends on A only. Moreover, the matrix $\mathbf{P} = (b_{1k} a_{ik})_{i,k=1}^{m}$ is a quadratic stochastic matrix.

2. It is known (see e.g. [248]) that the properties of homogeneous Markov chains with the phase space $E = \{1, ..., m\}$ can by completely determined by the initial distribution $x \in S^{m-1}$ and the stochastic matrix $\mathbf{P}$ i.e. the dynamical system generated by the operator (8.47). The theory of such dynamical systems is known (see for example [248]). Also for the dynamical behavior of general linear operators $\mathbb{R}^m \to \mathbb{R}^m$ see, for example, [40], pages 159–181.

Case 3: If $\det(A) = \det(B) = 0$ but both of them don't have all the identical rows, then the SQSO is

$$x'_k = \left(\sum_{i=1}^{m} a_{ik} x_i \right) \left(\sum_{j=1}^{m} b_{jk} x_j \right), \quad \text{for} \quad k = 1, ..., m. \tag{8.48}$$

Remark 8.6. 1. Since B can *not* be uniquely determined by a given A with $\det(A) = 0$, the operator (8.48) depends on both matrices A and B.

2. By the above mentioned reasons only SQSO (8.48) is interesting to study. In this case, we can have a rich theory of such operators: to find Lyapunov functions; to study fixed points; to determine concepts of tournaments and so on.

8.3.2   *Lyapunov functions of SQSO*

Denote

$$\mathcal{A} = \{(A, B) \; : \; \det(A) = \det(B) = 0, \; AB^T = \mathbf{1}, \text{ and}$$

$$\text{both matrices don't have all the identical rows}\}.$$

Theorem 8.12. *For the dynamical system (8.48), the function* ψ_c : $S^{m-1} \to \mathbb{R}$ *defined by*

$$\psi_c(x) = \sum_{k=1}^{m} c_k x_k \tag{8.49}$$

is a Lyapunov function if $c = (c_1, ..., c_m)^T$ satisfies $c_i \geq 0$ for all $1 \leq i \leq m$ and either $Ac \leq Ic$ or $Bc \leq Ic$ where $A = (a_{ij})$, $B = (b_{ij})$, with $0 \leq a_{ij}, b_{ij} \leq 1$ for all $1 \leq i, j \leq m$, $(A, B) \in \mathcal{A}$ and I is the identity matrix of order m.

Proof. Suppose $Ac \leq Ic$, then we have

$$
\psi_c(x') = \sum_{k=1}^{m} c_k x'_k = \sum_{k=1}^{m} c_k \sum_{i,j=1}^{m} a_{ik} b_{jk} x_i x_j
$$

$$
\leq \sum_{k=1}^{m} c_k \sum_{i,j=1}^{m} a_{ik} x_i x_j = \sum_{k=1}^{m} c_k \sum_{i=1}^{m} a_{ik} x_i
$$

$$
= \sum_{i=1}^{m} \left(\sum_{k=1}^{m} c_k a_{ik} \right) x_i \leq \sum_{i=1}^{m} c_i x_i = \psi_c(x)
$$

as $0 \leq b_{ij} \leq 1$.

Thus, for any n, we have $\psi_c(x^{(n)}) \leq \psi_c(x^{(n-1)})$ and $\underline{c} \leq \psi_c(x^{(n)}) \leq \overline{c}$, with $\underline{c} = \min_i c_i$, $\overline{c} = \max_i c_i$. Consequently, the sequence $\{\psi_c(x^{(n)})\}_{n=0}^{\infty}$ is convergent. Therefore, $\psi_c(x) = \sum_{k=1}^{m} c_k x_k$ is a Lyapunov function for the dynamical system (8.48). $\qquad \square$

Corollary 8.8. *The function defined by*

$$
\phi(x) = \prod_{k=1}^{m} \left(\sum_{i=1}^{m} c_{ik} x_i \right)^{p_k} \tag{8.50}
$$

is a Lyapunov function for the dynamical system (8.48) for any $p_k \in \mathbb{R}^+$ if $c^{(k)} = (c_{k1}, c_{k2}, ..., c_{km})$ satisfies $c_{ij} \geq 0$, for all $i, j = 1, ..., m$ and either $Ac^{(k)} \leq c^{(k)}$ or $Bc^{(k)} \leq c^{(k)}$ for all $k = 1, ..., m$, where $A = (a_{ij})$ and $B = (b_{ij})$, $(A, B) \in \mathcal{A}$ with $0 \leq a_{ij}, b_{ij} \leq 1$ for all $1 \leq i, j \leq m$.

Remark 8.7. To use Theorem 8.12 one has to find non-zero solution (if exists) of the system of inequalities $Ac \leq Ic$ (or $Bc \leq Ic$). It is known (see [135], page 42) that the set $\mathcal{C}$ of all solutions of the system of inequalities is a polyhedral convex cone: if c is in $\mathcal{C}$, then the vector tc for all $t \geq 0$ (the ray or halfline generated by c) are also in $\mathcal{C}$. Such a polyhedral convex cone $\{c . (A - I)c \leq 0\}$ can also be expressed as the convex-cone hull $Q^<$ of a finite set $Q = \{Q_1, ..., Q_q\}$:

$$
Q^< = \{c : c = v_1 Q_1 + ... + v_q Q_q, \ v_i \geq 0\} = \{c : c = Qv, v \geq 0\}.
$$

Concretely, Q is the m by q matrix with $Q_1, ..., Q_q$ as columns. For a method of solving of the system of linear inequalities see [200].

Now we shall give some solutions of the above mentioned system of inequalities.

Proposition 8.3. *If*

$$a_i = \sum_{k=1}^{m} a_{ik} \leq 1, \quad \text{for all} \ \ i = 1, ..., m, \tag{8.51}$$

then the system of inequalities $Ac \leq Ic$ *has nonzero solutions* $c(t) = (tc_1, tc_2, ..., tc_m)$ *with* $c_i = a_i$ *for all* $t > 0$.

Proof. By Remark 8.7 it is enough to prove for $c(1)$, i.e. $t = 1$:

$$\sum_{k=1}^{m} a_{ik} c_k = \sum_{k=1}^{m} a_{ik} a_k \leq \sum_{k=1}^{m} a_{ik} = a_i = c_i. \qquad \square$$

The following example shows that the above condition (8.51) is not necessary:

Example. Consider

$$A = \begin{pmatrix} 1 & 0 & 0 \\ \frac{1}{3} & \frac{1}{2} & \frac{1}{4} \\ \frac{2}{3} & \frac{1}{4} & \frac{1}{8} \end{pmatrix}, \quad B = \begin{pmatrix} 1 & \frac{8-3b_1}{6} & b_1 \\ 1 & \frac{8-3b_2}{6} & b_2 \\ 1 & \frac{8-3b_3}{6} & b_3 \end{pmatrix},$$

where $\frac{2}{3} \leq b_j \leq 1$ for all $j = 1, 2, 3$. One can show that A and B satisfy the condition (8.44). But the condition (8.51) is not satisfied for A. Moreover, the system of inequalities $Ac \leq Ic$ has many non-zero solutions. More precisely the complete set of solutions is

$$C = \left\{ c = (c_1, c_2, c_3) : \ c_1 \geq 0, \ c_2 \geq \frac{11}{9}c_1, \ \frac{16}{21}c_1 + \frac{2}{7}c_2 \leq c_3 \leq 2c_2 - \frac{4}{3}c_1 \right\}.$$

It also can be observed that there is no nonzero solution for the system of inequalities $Bc \leq Ic$.

8.3.3 ω-limit set of SQSO

For (8.46) we have $\omega(x^0) = \{(a_{11}b_{11}, ..., a_{1m}b_{1m})\}$, $\forall x^0 \in S^{m-1}$. But for (8.47) the set $\omega(x^0)$ depends on x^0 and on the properties of the matrix A. The set $\omega(x^0)$ can contain a single point (ergodic case, see [248], page 118), it can be a finite set (non-ergodic or periodic case, see page 121 of [248]).

In this section using the Lyapunov functions described in the previous section we shall give upper estimation of $\omega(x^0)$ for SQSO (8.48). Denote

$$C = \{c \in \mathbb{R}^m : c_i \geq 0, \ c_1 + ... + c_m > 0, \ Ac \leq Ic \ \text{ or } \ Bc \leq Ic\}.$$

Then by Theorem 8.12 we have that ψ_c is a Lyapunov function for any $c \in \mathcal{C}$. That is for any initial point $x^0 \in S^{m-1}$ we have

$$\lim_{n \to \infty} \psi_c(x^{(n)}) = \lambda_c(x^0), \quad c \in \mathcal{C}. \tag{8.52}$$

Thus $\omega(x^0) \subset \{x \in S^{m-1} : \psi_c(x) = \lambda_c(x^0)\}$ for any $c \in \mathcal{C}$ which implies

$$\omega(x^0) \subset \bigcap_{c \in \mathcal{C}} \{x \in S^{m-1} : \psi_c(x) = \lambda_c(x^0)\}. \tag{8.53}$$

The estimation (8.53) is very useful: assume that there are m distinct vectors $c^{(1)}, ..., c^{(m)} \in \mathcal{C}$ such that $\det(C) \neq 0$ where C is the $m \times m$ matrix with rows $c^{(i)}$, $i = 1, ..., m$. Then system of equations $\psi_{c^{(i)}}(x) = \lambda_{c^{(i)}}(x^0)$, $i = 1, ..., m$ has unique solution $x = x^*$ which by (8.53) gives that $\omega(x^0) = \{x^*\}$. If there is no any collection $c^{(i)}$, $i = 1, ..., m$ with $\det(C) \neq 0$, then RHS of (8.53) is a uncountable set. Note that RHS of (8.53) can not be empty set since $\omega(x^0) \neq \emptyset$, because $\{x^{(n)}\}_{n=0}^{\infty} \subset S^{m-1}$ and S^{m-1} is a compact set.

8.4 Quasi-strictly non-Volterra QSO

In [232] a notion of strictly non-Volterra QSOs was introduced, and it was proved that an arbitrary strictly non-Volterra quadratic stochastic operator on the two-dimensional simplex has a unique fixed point, which is not attracting. Moreover under some conditions on parameters of this operator it was constructed some periodic trajectories.

In this section following [90] we consider a non-Volterra QSO, called quasi-strictly non-Volterra defined on the two-dimensional simplex which has the form

$$V : \begin{cases} x_1' = \alpha x_2^2 + c x_3^2 + 2 x_2 x_3, \\ x_2' = a x_1^2 + d x_3^2 + 2 x_1 x_3, \\ x_3' = b x_1^2 + \beta x_2^2 + e x_3^2 + 2 x_1 x_2, \end{cases} \tag{8.54}$$

where $\alpha, \beta, a, b, c, d, e \geq 0$ and

$$a + b = 1, \quad \alpha + \beta = 1, \quad c + d + e = 1. \tag{8.55}$$

8.4.1 *Fixed point of the operator*

A fixed point is a solution of the following system

$$\begin{cases} x_1 = \alpha x_2^2 + c x_3^2 + 2x_2 x_3, \\[4pt] x_2 = a x_1^2 + d x_3^2 + 2x_1 x_3, \\[4pt] x_3 = b x_1^2 + \beta x_2^2 + e x_3^2 + 2x_1 x_2. \end{cases} \tag{8.56}$$

Theorem 8.13. *The non-Volterra QSO (8.54) has a unique fixed point* $x^* = (x_1^*, x_2^*, x_3^*) \in S^2$ *in all cases, except when* $e = 1$. *In the case* $e = 1$, *there are two fixed points of the system, one of which is* $(0, 0, 1)$.

Proof.   We shall consider all possible cases on a and α.

1) Let $\alpha \neq 0, a \neq 0$. Substituting $x_1 = 1 - x_2 - x_3$ into the first equation of (8.56) gives

$$x_2 = \frac{-2x_3 - 1 + \sqrt{4(1 - \alpha c)x_3^2 + 4(1 - \alpha)x_3 + 1 + 4\alpha}}{2\alpha} \geq 0, \tag{8.57}$$

where $x_3 \in [0, \frac{\sqrt{1+4c}-1}{2c}]$ if $c \neq 0$ and $x_3 \in [0, 1]$ if $c = 0$. Similarly, the second equation in (8.56) gives

$$x_1 = \frac{-2x_3 - 1 + \sqrt{4(1 - ad)x_3^2 + 4(1 - a)x_3 + 1 + 4a}}{2a} \geq 0, \tag{8.58}$$

where $x_3 \in [0, \frac{\sqrt{1+4d}-1}{2d}]$ if $d \neq 0$ and $x_3 \in [0, 1]$ if $d = 0$.

Now we may substitute (8.57) and (8.58) into $1 = x_1 + x_2 + x_3$, allowing $x = x_3$ and $f(x) = x$. This gives the function

$$f(x) = \frac{\alpha\sqrt{4(1 - ad)x^2 + 4(1 - a)x + 1 + 4a}}{2(\alpha + a - \alpha a)}$$
$$+ \frac{a\sqrt{4(1 - \alpha c)x^2 + 4(1 - \alpha)x + 1 + 4\alpha} - \alpha - a - 2\alpha a}{2(\alpha + a - \alpha a)}.$$

Now define

$$g(x) = 4(1 - ad)x^2 + 4(1 - a)x + 1 + 4a > 0,$$
$$h(x) = 4(1 - \alpha c)x^2 + 4(1 - \alpha)x + 1 + 4\alpha > 0.$$

Thus, we have

$$g'(x) = 8(1 - ad)x + 4(1 - a) \geq 0, \quad h'(x) = 8(1 - \alpha c)x + 4(1 - \alpha) \geq 0,$$
$$g''(x) = 8(1 - ad) \geq 0, \qquad\qquad h''(x) = 8(1 - \alpha c)x \geq 0.$$

Then,

$$f(x) = \frac{\alpha\sqrt{g(x)}}{2(\alpha + a - \alpha a)} + \frac{a\sqrt{h(x)} - \alpha - a - 2\alpha a}{2(\alpha + a - \alpha a)}.$$

Differentiating $f(x)$ gives

$$f'(x) = \frac{\alpha g'(x)}{4(\alpha + a - \alpha a)\sqrt{g(x)}} + \frac{ah'(x)}{4(\alpha + a - \alpha a)\sqrt{h(x)}} \geq 0,$$

$$f''(x) = \frac{\alpha}{4(\alpha + a - \alpha a)} \frac{g(x)g''(x) - \frac{1}{2}g'(x)^2}{\sqrt{g(x)^3}}$$

$$+ \frac{a}{4(\alpha + a - \alpha a)} \frac{h(x)h''(x) - \frac{1}{2}h'(x)^2}{\sqrt{h(x)^3}} \geq 0.$$

These inequalities follow from (8.55) as well as the fact that $\alpha + a - \alpha a > 0$. Moreover, substituting the values for $g(x)$, $h(x)$, and their derivatives into the inequalities

$$g(x)g''(x) - \tfrac{1}{2}g'(x)^2 \geq 0, \; h(x)h''(x) - \tfrac{1}{2}h'(x)^2 \geq 0,$$

gives

$$(4(1 - ad)x^2 + 4(1 - a)x + 1 + 4a)(1 - ad) - (2(1 - ad)x + (1 - a))^2 \geq 0,$$
$$(4(1 - \alpha c)x^2 + 4(1 - \alpha)x + 1 + 4\alpha)(1 - \alpha c) - (2(1 - \alpha c)x + (1 - \alpha))^2 \geq 0.$$

This can be reduced to

$$\alpha(-6 + a + d + 4ad) \leq 0, \; a(-6 + \alpha + c + 4\alpha c) \leq 0.$$

Thus the inequalities $g(x)g''(x) - \tfrac{1}{2}g'(x)^2 \geq 0$ and $h(x)h''(x) - \tfrac{1}{2}h'(x)^2 \geq 0$ can be demonstrated to be true.

Moreover,

$$f(0) = \frac{\alpha(\sqrt{1 + 4a} - 1 - a) + a(\sqrt{1 + 4\alpha} - 1 - \alpha)}{2(\alpha + a - \alpha a)} > 0,$$

$$f(1) = \frac{a\sqrt{9 - ad} + a\sqrt{9 - \alpha c} - \alpha - a - 2\alpha a}{2(\alpha + a - \alpha a)} \leq 1,$$

which follows from the above inequalities in addition to the fact that

$$\sqrt{9 - ad} \leq 3, \; \sqrt{9 - \alpha c} \leq 3.$$

Moreover, both $\sqrt{9 - ad}$ and $\sqrt{9 - \alpha c}$ can only be simultaneously equal to 3 when $c = d = 0$ (and thus when $e - 1$). This means that $f(1) = 1$ when $e = 1$ and $f(1) < 1$ otherwise.

Thus the function is increasing and convex in $[0, 1]$. Therefore, for $e \in [0, 1)$ the system has a unique fixed point, since $f(x)$ will only intersect the line x at one point in the domain $[0, 1]$. When $e = 1$ the system has a fixed point $(0, 0, 1)$. It can also be directly shown that when $e = 1$, $f(\frac{9}{10}) < \frac{9}{10}$.

This demonstrates that the function $f(x)$ must cross the line x prior to reaching the value $f(1) = 1$. This proves that there are two fixed points when $e = 1$.

2) Let $\alpha \neq 0, a = 0$ can be handled similarly. Substituting $x_1 = 1 - x_2 - x_3$ into the first equation of (8.56) gives

$$x_2 = \frac{-2x_3 - 1 + \sqrt{4(1 - \alpha c)x_3^2 + 4(1 - \alpha)x_3 + 1 + 4\alpha}}{2\alpha} \geq 0, \qquad (8.59)$$

where $x_3 \in [0, \frac{\sqrt{1+4c}-1}{2c}]$ if $c \neq 0$ and $x_3 \in [0, 1]$ if $c = 0$. The second equation in (8.56) gives

$$x_1 = \frac{1 - x_3 - dx_3^2}{2x_3 + 1} \geq 0, \qquad (8.60)$$

where $x_3 \in [0, \frac{\sqrt{1+4d}-1}{2d}]$ if $d \neq 0$ and $x_3 \in [0, 1]$ if $d = 0$. This restriction ensures that x_1 is positive.

Directly substituting (8.59) and (8.60) into $1 = x_1 + x_2 + x_3$, allowing $x = x_3$, gives

$$1 = x + \frac{1 - x - dx^2}{2x + 1} + \frac{-2x - 1 + \sqrt{4(1 - \alpha c)x^2 + 4(1 - \alpha)x + 1 + 4\alpha}}{2\alpha}.$$

Solving for the x in the first two terms of the above equation (but not any of the x in the (8.59) term), and allowing $F_\pm(x) = x$, gives the functions

$$F_\pm(x) = \frac{1 + 2\alpha + 2x - \sqrt{h(x)} \pm \sqrt{q(x)}}{2\alpha(2 - d)},$$

where

$$h(x) = 4(1 - \alpha c)x^2 + 4(1 - \alpha)x + 1 + 4\alpha \geq 0,$$

$$q(x) = 2\alpha(2 - d)\left(1 - \sqrt{h(x)} + 2x\right) + \left(1 - \sqrt{h(x)} + 2x + 2\alpha\right)^2.$$

Lemma 8.3. $F_\pm(x) = x$ *at a unique point when* $e \in [0, 1)$ *and at two points (one of which is* $F_+(1) = 1$ *when* $e = 1$*).*

Proof. Note that when $q(x) = 0$,

$$\begin{aligned}
x_q = \frac{1}{2c}\Big[&-5 + d + \sqrt{12 - 8d + d^2} \\
&+ \Big(37 + 2d^2 - 10\sqrt{12 - 8d + d^2} + 2d\left(-9 + \sqrt{12 - 8d + d^2}\right) \\
&- 2c\Big(2 - d - \sqrt{12 - 8d + d^2} + \alpha\Big(14 + d^2 - 4\sqrt{12 - 8d + d^2} \\
&+ d\left(-8 + \sqrt{12 - 8d + d^2}\right)\Big)\Big)\Big)^{\frac{1}{2}}\Big].
\end{aligned}$$

The curve $F_\pm$ is imaginary when $x < x_q$. When $x = x_q$, $F_+(x) = F_-(x)$, and when $x > x_q$, $F_+(x) > F_-(x)$.

Moreover, at $x = 1$ we have

$$F_+(1) = \frac{3 + 2\alpha - \sqrt{h(1)} + \sqrt{q(1)}}{2\alpha(2 - d)} \geq 1,$$

where

$$h(1) = 9 - 4\alpha c \geq 0,$$
$$q(1) = 2\alpha(2 - d)\left(3 - \sqrt{h(1)}\right) + \left(3 - \sqrt{h(1)} + 2\alpha\right)^2.$$

It can be readily shown that in the case that $e = 1$, $F_+(1) = 1$. When $e \in [0, 1)$, we would like to show that $F_+(1) > 1$. This can be reduced to showing that $\sqrt{q(1)} + 2\alpha d + 3 > \sqrt{h(1)} + 2\alpha$. It can be easily shown that $\sqrt{q(1)} > 2\alpha$. Additionally, when $d = 0$, $3 > \sqrt{h(1)}$. When $d \neq 0$, $3 \geq \sqrt{h(1)}$, but $2\alpha d > 0$. The culmination of the above facts proves the inequality $\sqrt{q(1)} + 2\alpha d + 3 > \sqrt{h(1)} + 2\alpha$ to be always true.

Thus, $F_+(1) > 1$ when $e < 1$. It can be similarly shown that $F_-(1) < 1$. Analysis of the derivations of $F_\pm$ shows that the following inequalities are true.

$$F'_+(x) = \frac{1}{2\alpha(2 - d)}\left(2 - \frac{h'(x)}{2\sqrt{h(x)}} + \frac{q'(x)}{2\sqrt{q(x)}}\right) \geq 0,$$

$$F'_-(x) = \frac{1}{2\alpha(2 - d)}\left(2 - \frac{h'(x)}{2\sqrt{h(x)}} - \frac{q'(x)}{2\sqrt{q(x)}}\right) \leq 0,$$

$$F''_+(x) = \frac{-1}{2\alpha(2 - d)}\left(\frac{h(x)h''(x) - \frac{1}{2}(h'(x))^2}{\sqrt{(h(x))^3}} - \frac{q(x)q''(x) - \frac{1}{2}(q'(x))^2}{\sqrt{(q(x))^3}}\right) \leq 0,$$

for all $x > x_q$.

The above demonstrates that $F_\pm(x)$ must intersect the line x at least once on $[0, 1)$. Because $F_+(x)$ is increasing and concave, and $F_-(x)$ is decreasing, $F_\pm(x)$ and x will only intersect once on $[0, 1)$. Moreover, $F_+(1) = 1$ only when $e = 1$. Thus the lemma is proven. $\qquad\square$

Therefore, a direct extension of the above lemma shows that when $e = 1$, the system (8.54) has two fixed points—one of which is $(0, 0, 1)$. And when $e < 1$, it has a unique fixed point.

3) The case $\alpha = 0, a \neq 0$ can be handled analogously to the previous case.

4) Let $\alpha = a = 0$. The system is therefore

$$\begin{cases} x_1 = cx_3^2 + 2x_2x_3 \\[2mm] x_2 = dx_3^2 + 2x_1x_3 \\[2mm] x_3 = (x_1 + x_2)^2 + ex_3^2 = (1 - x_3)^2 + ex_3^2. \end{cases} \tag{8.61}$$

The proof of the fourth case will follow directly from the following lemma.

Lemma 8.4. *If $e \in [0,1)$, the system has a unique fixed point, (x_1^*, x_2^*, x_3^*), where*

$$x_1^* = \frac{7 + 24d - 2e - 19de - 5e^2 + 2e^2d + (3 - 11d + 4de + 3d^2)\sqrt{5 - 4e}}{2(1 + e)(-13 + 6\sqrt{5 - 4e} + 6e + e^2)},$$

$$x_2^* = 1 - \frac{7 + 24d - 2e - 19de - 5e^2 + 2e^2d + (3 - 11d + 4de + 3d^2)\sqrt{5 - 4e}}{2(1 + e)(-13 + 6\sqrt{5 - 4e} + 6e + e^2)}$$

$$\quad - \frac{3 - \sqrt{5 - 4e}}{2(1 + e)},$$

$$x_3^* = \frac{3 - \sqrt{5 - 4e}}{2(1 + e)}.$$

However, when $e = 1$ the system has two fixed points: $(0, 0, 1)$ and $(\frac{1}{4}, \frac{1}{4}, \frac{1}{2})$.

Proof. Examine each possible case.

- Let $e \in [0, 1)$. Solving the third equation of (8.61) for x_3 gives $x_3^* = \frac{3 \pm \sqrt{T}}{2(1+e)} > 0$ where $T = 5 - 4e > 0$.

  Assume for the purpose of contradiction that $\frac{3+\sqrt{T}}{2(1+e)} \leq 1$. This can be reduced to $\sqrt{T} \leq 2e - 1$. If $e \leq \frac{1}{2}$, then $2e - 1 \leq 0$ and $\sqrt{T} \geq 0$, so $\frac{3+\sqrt{T}}{2(1+e)} \nleq 1$. Moreover, if $e > \frac{1}{2}$ then $0 < \sqrt{T} \leq 2e - 1$ which can be reduced to $1 \leq e$ which is not true for $e \in [0, 1)$. Thus $\frac{3+\sqrt{T}}{2(1+e)} \nleq 1$ for any $e \in [0, 1)$ and is not a fixed point of $x_3' = (1 - x_3)^2 + ex_3^2$.

  It can be proved that $\frac{3-\sqrt{T}}{2(1+e)} \leq 1$ because it can be reduced to $\sqrt{T} \geq 1 - 2e$. As it was shown previously that $\sqrt{T} \leq 1 - 2e$ is false for all $e \in [0, 1)$, it must be that $\sqrt{T} > 1 - 2e$. Therefore,

$$x_3^* = \frac{3 - \sqrt{5 - 4e}}{2(1 + e)},$$

  is a unique fixed point of the system.

Substituting x_3^* into the first two equations of (8.61) gives $x_1 = cx_3^{*2} + 2x_2 x_3^*$ and $x_2 = dx_3^{*2} + 2x_1 x_3^*$. Substituting this value of x_2 into x_1 and reducing gives

$$x_1^* = \frac{(3x_3^* - 1)(c + 2dx_3^*)}{5 + e - 12x_3^*},$$

which can be written as

$$x_1^* = \frac{7 + 25d - 2e - 19de - 5e^2 + 2e^2d + (3 - 11d + 4de + 3d^2)\sqrt{5 - 4e}}{2(1 + e)(-13 + 6\sqrt{5 - 4e} + 6e + e^2)}.$$

Moreover, we know that $x_2^* = 1 - x_1^* - x_3^*$ which yields

$$x_2^* = 1 - \frac{3 - \sqrt{5 - 4e}}{2(1 + e)}$$

$$-\frac{7 + 25d - 2e - 19de - 5e^2 + 2e^2d + (3 - 11d + 4de + 3d^2)\sqrt{5 - 4e}}{2(1 + e)(-13 + 6\sqrt{5 - 4e} + 6e + e^2)}.$$

Thus, (x_1^*, x_2^*, x_3^*) is a unique fixed point of the system.

- Let $e = 1$. The system is therefore

$$\begin{cases} x_1 = 2x_2 x_3, \\ x_2 = 2x_1 x_3, \\ x_3 = (1 - x_3)^2 + x_3^2. \end{cases} \tag{8.62}$$

Moreover, we know that $x_3^* = \frac{3 \pm \sqrt{(5 - 4e)}}{2(1 + e)}$; therefore, $x_3^* = \frac{1}{2}$ or 1. When $x_3^* = \frac{1}{2}$, it follows from (8.62) that $x_1 = x_2$, and it follows from $x_1 + x_2 + x_3 = 1$ that $x_1 + x_2 = \frac{1}{2}$. Thus, $x_1^* = x_2^* = \frac{1}{4}$ and the point $(\frac{1}{4}, \frac{1}{4}, \frac{1}{2}) \in S^2$ is a fixed point of the system. When $x_3^* = 1$, it follows from $x_1 + x_2 + x_3 = 1$ that $x_1^* = x_2^* = 0$. Thus, the point $(0, 0, 1) \in S^2$ is a second fixed point of the system.

$\square$

By the above cases, all possible values for the system are considered and the theorem is proved.

$\square$

8.4.2 *The type of the fixed point*

To find the type of a fixed point we use $x_3 = 1 - x_1 - x_2$ to rewrite QSO (8.54) as follows:

$$V : \begin{cases} x_1' = c - 2cx_1 + cx_1^2 + 2(c - 1)x_1 x_2 + 2(1 - c)x_2 + (\alpha + c - 2)x_2^2, \\ x_2' = d - 2dx_2 + dx_2^2 + 2(d - 1)x_1 x_2 + 2(1 - d)x_1 + (a + d - 2)x_1^2, \end{cases}$$

where $(x_1, x_2) \in \{(x, y) : x, y \geq 0, \ 0 \leq x + y \leq 1\}$ and x_1, x_2, are the first two coordinates of a point lying in the two-dimensional simplex.

The Jacobian, $J(x^*)$, has the representation

$$\begin{pmatrix} -2c(1 - x_1^* - x_2^*) - 2x_2^* & 2(\alpha - 1)x_2^* + 2(1 - c)(1 - x_1^* - x_2^*) \\ 2(a - 1)x_1^* + 2(1 - d)(1 - x_1^* - x_2^*) & -2d(1 - x_1^* - x_2^*) - 2x_1^* \end{pmatrix}. \tag{8.63}$$

The Jacobian (8.63) has the eigenvalues $\lambda_{1,2} = ex_3^* - 1 \pm \sqrt{D(a, \alpha, c, e)}$, where

$$D \equiv D(a, \alpha, c, e) = (ex_3^* - 1)^2 + 4ex_3^{*2}$$
$$+ 4[(b\beta - 1)x_1^* x_2^* + (a(1 - c) - 1)x_1^* x_3^* + (\alpha(1 - d) - 1)x_2^* x_3^*]. \tag{8.64}$$

The classification of these eigenvalues is as follows:

$$\begin{cases} \text{If } D < -1 + (1 - ex_3^*)^2, & \text{the fixed point is repelling;} \\[2ex] \text{If } D = -1 + (1 - ex_3^*)^2, & \text{the fixed point is nonhyperbolic;} \\[2ex] \text{If } -1 + (1 - ex_3^*)^2 < D < 0, & \text{the fixed point is attracting;} \\[2ex] \text{If } D = 0 \text{ and } ex_3^* = 0, & \text{the fixed point is nonhyperbolic;} \\[2ex] \text{If } D = 0 \text{ and } ex_3^* > 0, & \text{the fixed point is attracting;} \\[2ex] \text{If } 0 < D < e^2 x_3^{*2}, & \text{the fixed point is attracting;} \\[2ex] \text{If } D = e^2 x_3^{*2}, & \text{the fixed point is nonhyperbolic;} \\[2ex] \text{If } e^2 x_3^{*2} < D < (2 - ex_3^*)^2, & \text{the fixed point is a saddle point;} \\[2ex] \text{If } D = (2 - ex_3^*)^2, & \text{the fixed point is nonhyperbolic;} \\[2ex] \text{If } (2 - ex_3^*)^2 < D, & \text{the fixed point is repelling.} \end{cases} \tag{8.65}$$

In [232] it was proven that strictly non-Volterra QSOs with $m = 3$ have a unique fixed point and that the type of the hyperbolic fixed point can never be attracting. However, in the system (8.54), the introduction of the parameter e has caused an attracting fixed point to become possible, as evidenced by the following example.

Example. When $c = \beta = \frac{5}{8}$, $d = b = 0$, $\alpha = e = \frac{3}{8}$, and $a = 1$ the system (8.54) can be written

$$V : \begin{cases} x_1' = \frac{3}{8}x_2^2 + \frac{5}{8}x_3^2 + 2x_2x_3, \\[2mm] x_2' = x_1^2 + 2x_1x_3, \\[2mm] x_3' = \frac{5}{8}x_2^2 + \frac{3}{8}x_3^2 + 2x_1x_2. \end{cases} \qquad (8.66)$$

The fixed point of this system is $(\frac{1}{3}, \frac{1}{3}, \frac{1}{3})$. Substituting these values into the eigenvalues of the Jacobian gives $\lambda_{1,2} = -\frac{7}{8} \pm \frac{1}{8}\sqrt{\frac{13}{3}}\, i$. Therefore, $|\lambda_{1,2}| = \sqrt{\frac{5}{6}} < 1$, and $(\frac{1}{3}, \frac{1}{3}, \frac{1}{3})$ is attracting.

In the case where $\alpha = a = 0$ (i.e. system (8.61)), the eigenvalues of the Jacobian can be written

$$\lambda_{1,2} = ex_3^* - 1 \pm \sqrt{1 + 2(e-2)x_3^* + (4+e^2)x_3^{*2}},$$

where $x_3^* = \frac{3 - \sqrt{5-4e}}{2(1+3)}$. It can be proven that $|\lambda_1| < 1$ and $|\lambda_2| > 1$ for all e. The inequality $|\lambda_1| < 1$ can be proven from the facts that

$$0 \leq 1 + 2(e-2)x_3^* + (4+e^2)x_3^{*2},$$

$$0 < ex_3^* + \sqrt{1 + 2(e-2)x_3^* + (4+e^2)x_3^{*2}},$$

$$x_3^{*2} + ex_3^* < \frac{3}{4} + x_3^*.$$

The last inequality, $x_3^{*2} + ex_3^* < \frac{3}{4} + x_3^*$, follows from the fact that $x_3^{*2} \leq x_3^*$ and from substituting the value of x_3^* into the inequality $ex_3^* < \frac{3}{4}$, which gives $3e < 3 + 2e\sqrt{5-4e}$. Additionally, $|\lambda_2| > 1$ can be proven from the fact that $ex_3^* - 1 - \sqrt{1 + 2(e-2)x_3^* + (4+e^2)x_3^{*2}} < -1$. This can be reduced to the quadratic $1 + (2e-4)x_3^* + 4x_3^{*2}$, which is always positive. This means that the fixed point x_3^* is a saddle point for all $e \in [0, 1)$. In the case that $a = \alpha = 0$ and $e = 1$, the fixed point $(\frac{1}{4}, \frac{1}{4}, \frac{1}{2})$ is a saddle point and $(0, 0, 1)$ is a repeller.

Additionally, for all cases where $e = 1$ and $(0, 0, 1)$ is a second fixed point of the system, it can be seen that $\lambda_{1,2} = \pm 2$. Thus the second fixed point that occurs when $e = 1$ is always a repeller.

Remark 8.8. A non-Volterra QSO with $m = 3$ (8.54) also has a unique fixed point in all cases except $e = 1$ for which there are two fixed points, the point $(0, 0, 1)$ is always a repeller. Moreover, the fixed point of a non-Volterra QSO may be attracting.

8.4.3   *The ω-limit set*

In this section we shall describe the ω-limit set of trajectories under certain parameter restrictions. Let $x^0 = (x_1^0, x_2^0, x_3^0) \in S^2$ be the initial point and let $\{x^{(n)}, n = 0, 1, 2, \dots\}$ be the trajectory of x^0 under the action of the operator (8.54); that is,

$$x^{(n)} = (x_1^{(n)}, x_2^{(n)}, x_3^{(n)}) = V(x^{(n-1)}), \qquad n = 1, 2, \dots \quad x^{(0)} = x^0.$$

For simplicity we shall examine the case in which $a = \alpha = 0$; therefore the operator can be written as (8.61), i.e.,

$$\begin{cases} x_1' = cx_3^2 + 2x_2x_3, \\ x_2' = dx_3^2 + 2x_1x_3, \\ x_3' = (1 - x_3)^2 + ex_3^2, \end{cases} \tag{8.67}$$

which demonstrates that the trajectory of the third coordinate $\{x_3^{(n)}\}$ is defined by the dynamical system of

$$\varphi(x) = (1 - x)^2 + ex^2.$$

8.4.4   *Case $e = 1$*

In this case operator has the form (denoted by V_1)

$$V_1 : \quad \begin{cases} x_1' = 2x_2x_3, \\ x_2' = 2x_1x_3, \\ x_3' = (1 - x_3)^2 + x_3^2. \end{cases} \tag{8.68}$$

This operator has been studied in [129]: One can see that

$$\mathrm{Fix}(V_1) = \{(1/4, 1/4, 1/2), (0, 0, 1)\}.$$

The following lemma describes all periodic points of the operator.

Lemma 8.5. *For the operator (8.68) the following hold*

a) $\mathrm{Per}_2(V_1) = \{(x, y, 1/2) \in S^2 : x + y = 1/2\}.$
b) $\mathrm{Per}_n(V_1) = \{\lambda \in S^2 : V_1^n(\lambda) = \lambda\} = \emptyset, \quad n \geq 3.$

Proof.   a) 2-periodic points are solutions $\lambda = (x, y, z)$ to $V_1^2(\lambda) = \lambda$, where

$$V_1^2 : \quad \begin{cases} x^{(2)} = 2^2 xzf(z), \\ y^{(2)} = 2^2 yzf(z), \\ z^{(2)} = f^2(z), \end{cases} \tag{8.69}$$

here $f(z) = 2z^2 - 2z + 1$. Thus for $z \in [0,1]$ we have equation $f^2(z) = z$. It is clear that fixed points of f are solutions to this equation. Therefore we consider the following

$$\frac{f^2(z) - z}{f(z) - z} = \frac{8z^4 - 16z^3 + 12z^2 - 5z + 1}{2z^3 - 3z + 1} = 4z^2 - 2z + 1 = 0.$$

But this equation does not have solution in $[0,1]$. By Sharkovskii's theorem [40] we have that $f^n(z) = z$ does not have solution (except fixed points) for all $n \geq 2$. Moreover $f^2(z) = z$ has solutions $z = 1/2$ and $z = 1$. In case $z = 1$ from $V^2(\lambda) = \lambda$ we obtain $\lambda = (0,0,1) \in \mathrm{Fix}(V_1)$. For $z = 1/2$ from $V^2(\lambda) = \lambda$ we get $x = x$, $y = y$, i.e. for any $(x, y, 1/2)$ with $x + y = 1/2$, we have

$$V_1^2(x, y, 1/2) = V_1(y, x, 1/2) = (x, y, 1/2).$$

b) We have

$$V_1^3 : \begin{cases} x^{(3)} = 2^3 yz f(z) f^2(z), \\ y^{(3)} = 2^3 xz f(z) f^2(z), \\ z^{(3)} = f^3(z). \end{cases} \tag{8.70}$$

By mathematical induction we get

$$V_1^n : \begin{cases} x^{(n)} = \left[2^n \prod_{i=0}^{n-1} f^i(z)\right] \left[\left(\frac{1+(-1)^n}{2}\right) x + \left(\frac{1-(-1)^n}{2}\right) y\right], \\ y^{(n)} = \left[2^n \prod_{i=0}^{n-1} f^i(z)\right] \left[\left(\frac{1+(-1)^n}{2}\right) y + \left(\frac{1-(-1)^n}{2}\right) x\right], \\ z^{(n)} = f^n(z). \end{cases} \tag{8.71}$$

By the proof of part a) we know that $f^n(z) = z$ has two solutions $z = 1/2$ and $z = 1$. Let $z = 1/2$ then

$$2^n \prod_{i=0}^{n-1} f^i\left(\frac{1}{2}\right) = 1$$

and from $V^n(\lambda) = \lambda$ we get

$$\begin{cases} x = \left(\frac{1+(-1)^n}{2}\right) x + \left(\frac{1-(-1)^n}{2}\right) y, \\ y = \left(\frac{1+(-1)^n}{2}\right) y + \left(\frac{1-(-1)^n}{2}\right) x. \end{cases} \tag{8.72}$$

From this system for even n we get arbitrary x, y (with $x + y = 1/2$) is solution, that is $(x, y, 1/2) \in \mathrm{Per}_2(V_1)$. If n is odd then $x = y$ and by $x + y = 1/2$ we get $x = y = 1/4$, i.e. $(1/4, 1/4, 1/2) \in \mathrm{Fix}(V_1)$. This completes the proof. $\qquad \square$

One can see that $V_1(x, y, 0) = (0, 0, 1)$, for any $(x, y, 0) \in S^2$.

For $\theta \in [0, \infty)$, denote

$$M_\theta = \begin{cases} \{(x, y, z) \in S^2 : xy = 0\}, & \text{if } \theta = 0, \\ \{(x, y, z) \in S^2 : x = \theta y \ \text{ or } \ x = \frac{1}{\theta}y\}, & \text{if } \theta \in (0, \infty). \end{cases}$$

Note that M_θ is invariant with respect to V_1 for any $\theta \in [0, \infty)$. Moreover

$$S^2 = \bigcup_{\theta \in [0,\infty)} M_\theta.$$

Theorem 8.14. *If $e = 1$, then*

1. *For any initial point (x_1^0, x_2^0, x_3^0), with $x_3^0 = 0$ or $x_3^0 = 1$ we have*

$$\lim_{n \to \infty} V_1^n(x_1^0, x_2^0, x_3^0) = (0, 0, 1).$$

2. *For any initial point $x^0 = (x_1^0, x_2^0, x_3^0)$, with $x_3^0 \neq 0$ and $x_3^0 \neq 1$, there exists $\theta \in [0, +\infty)$, such that $x^0 \in M_\theta$. Moreover,*

$$\lim_{n \to \infty} V_1^n(x_1^0, x_2^0, x_3^0) = \begin{cases} \left(\frac{\theta}{2(\theta+1)}, \frac{1}{2(\theta+1)}, \frac{1}{2} \right), n = 2k, \\ \left(\frac{1}{2(\theta+1)}, \frac{\theta}{2(\theta+1)}, \frac{1}{2} \right), n = 2k+1, \end{cases} \quad k = 1, 2, 3, \ldots$$

see Fig. 8.1.

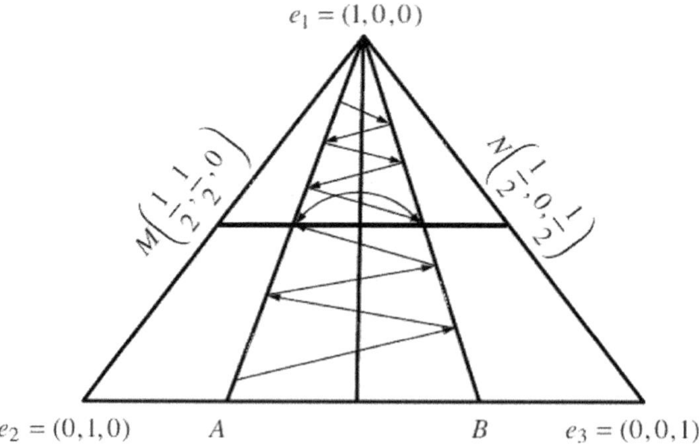

Fig. 8.1 The figure presenting some limits mentioned in Theorem 8.14. Here $M_\theta = Ae_1 \cup e_1 B$, $M_0 = e_2 e_1 \cup e_1 e_3$.

Proof. 1. If $x_3^0 = 0$ then $x_3^{(1)} = 1$ thus it suffices to prove the assertion for $x_3^0 = 1$. In this case one gets $x_3^{(n)} = 1$ for all $n \geq 1$.

2. Take a $\theta \in (0, \infty)$ and $\lambda = (x, y, z) \in M_\theta$ then we have $(\theta+1)y+z = 1$ or $(\frac{1}{\theta} + 1)y + z = 1$. For definiteness assume $(\theta + 1)y + z = 1$ then the operator V_1 on this line has the form

$$y' = 2\theta yz, \quad z' = f(z).$$

It is clear that $\lim_{n \to \infty} f^n(z) = 1/2$ and from

$$(\theta + 1)y^{(2k)} + z^{(2k)} = 1, \quad \left(\frac{1}{\theta} + 1\right) y^{(2k+1)} + z^{(2k+1)} = 1$$

we get

$$\lim_{n \to \infty} y^{(n)} = \begin{cases} \frac{1}{2(\theta+1)}, & \text{if } n = 2k \\ \frac{\theta}{2(\theta+1)}, & \text{if } n = 2k + 1, \end{cases} \quad k = 0, 1, 2, \ldots$$

This completes the proof. $\square$

8.4.5 *Case $e \in [0, 1)$*

Now we consider the operator (8.67) for all $e \neq 1$. By the above given results we know that this operator has a unique fixed point:

$$\text{Fix}(V) = \{x^* = (x_1^*, x_2^*, x_3^*)\},$$

which is never attractive.

Let us describe periodic points of the operator. By (8.67) the sequence $x^{(n+1)} = V^n(x^0)$ has the form

$$\begin{cases} x_1^{(n+1)} = c(x_3^{(n)})^2 + 2x_2^{(n)} x_3^{(n)}, \\ x_2^{(n+1)} = d(x_3^{(n)})^2 + 2x_1^{(n)} x_3^{(n)}, \\ x_3^{(n+1)} = \varphi(x_3^{(n)}). \end{cases} \quad (8.73)$$

Lemma 8.6. *If $1/4 \leq e < 1$ then the operator (8.67) does not have any n-periodic point, $n \geq 2$, different from the fixed point x^*.*

Proof. First we give analysis of the equation $\varphi(\varphi(x)) = x$, (for existence of 2-periodic points) which has solutions x_3^*, $\overline{x}_3 = \frac{1+\sqrt{1-4e}}{2(1+e)}$, and $\overline{\overline{x}}_3 = \frac{1-\sqrt{1-4e}}{2(1+e)}$, where $\varphi(\overline{x}_3) = \overline{\overline{x}}_3$ and vice-versa. These numbers exist if and only iff $e < 1/4$, and when $e = 1/4$ the three numbers coincide. Thus for $1/4 \leq e < 1$ there are no 2-periodic points of φ. By Sharkovskii's theorem

[40] we have that $\varphi^n(x) = x$ does not have solution $x \neq x_3^*$ for all $n \geq 2$. Thus, $x_3^{(n)} = x_3$ has unique solution $x_3 = x_3^*$ for any $n \geq 1$. Using this fact we reduce the equation $V^n(x) = x$ (with $x = (x_1, x_2, x_3)$) of n-periodic points to the equation $L^n(\hat{x}) = \hat{x}$ (with $\hat{x} = (x_1, x_2)$), where L is the linear operator given by

$$L: \quad \begin{cases} x' = c(x_3^*)^2 + 2x_3^* y, \\[2mm] y' = d(x_3^*)^2 + 2x_3^* x. \end{cases}$$

It is then easy to see that this linear operator has a unique fixed point (x_1^*, x_2^*) (the first two coordinates of the fixed point x^*). This fixed point is attractive and therefore by the known theorem of liner dynamical systems (see Chapter 3 of [56]), we see that all trajectories of the linear operator tend to the fixed point. Therefore this linear operator has no periodic points except x^*. $\qquad\square$

Lemma 8.7. *If $0 \leq e < 1/4$, then the operator (8.67) has 2-periodic points, $(\overline{x}_1, \overline{x}_3, \overline{x}_3)$ and $(\overline{\overline{x}}_1, \overline{\overline{x}}_2, \overline{\overline{x}}_3)$ — different from the fixed point x^* — which are described explicitly below. Moreover the operator does not have any n-periodic point for all $n \geq 3$.*

Proof. It can be seen from $|\varphi'(x_3^*)| = |1 - \sqrt{5 - 4e}|$ that if $e \in [0, \tfrac{1}{4})$, then x_3^* is a repelling fixed point of $\varphi(x)$. As mentioned in the proof of the previous lemma, if $e < \tfrac{1}{4}$ then the function $\varphi(\varphi(x))$ has fixed points x_3^*, $\overline{x}_3 = \frac{1+\sqrt{1-4e}}{2(1+e)}$, and $\overline{\overline{x}}_3 = \frac{1-\sqrt{1-4e}}{2(1+e)}$, i.e., $\varphi^2(\overline{x}_3) = \overline{x}_3$, and $\varphi^2(\overline{\overline{x}}_3) = \overline{\overline{x}}_3$. By substituting $x_3 = \overline{x}_3, \overline{\overline{x}}_3$ in the first and second equations of the system $V^2(x) = x$ and solving it with respect to x_1 and x_2, we get

$$\overline{x}_1 = \frac{2de + (1-e-2e^2)c + (2de - c(1+e))\sqrt{1-4e}}{2(e-1)^2(e+1)},$$

$$\overline{\overline{x}}_1 = \frac{2d + (1-e-2e^2)c + (-2d + c(e+1))\sqrt{1-4e}}{2(e-1)^2(e+1)},$$

$$\overline{x}_2 = \frac{2ce + (1-e-2e^2)d + (2ce - d(1+e))\sqrt{1-4e}}{2(e-1)^2(e+1)}, \tag{8.74}$$

$$\overline{\overline{x}}_2 = \frac{2c + (1-e-2e^2)d + (-2c + d(e+1))\sqrt{1-4e}}{2(e-1)^2(e+1)}.$$

Now we show that the operator has no n-periodic point if $n \geq 3$. One can see that for each solution $x = \tilde{x}_3$ of $\varphi^n(x) = x$, one gets a unique $(\tilde{x}_1, \tilde{x}_2)$ from $V^n(x) = x$. Therefore the number of periodic points of V is equal to the number of periodic points of φ. Now we show that φ does not have n-periodic points for any $n \geq 3$. Taking $h(x) = ax + b$ one can see that the function φ is topologically conjugate to the logistic map $\xi(x) = \mu x(1 - x)$

with $\mu = 1 + \sqrt{5 - 4e}$. For $e \in [0, \frac{1}{4})$ we have $\mu \in (3, 1 + \sqrt{5}]$. For the logistic map the following is known (see [247]):

If μ between 3 and $1 + \sqrt{5} \approx 3.236$, then ξ has one 2-periodic orbit and all trajectories (except when started at the fixed point) will approach this 2-periodic orbit.

From this fact, by the conjugacy argument, it follows that φ, and thus V, do not have n-periodic point for any $n \geq 3$. $\qquad\square$

Remark 8.9. The conjugacy argument mentioned in the proof of Lemma 8.7 can be also used to give an alternative proof of Lemma 8.6. In this case $\mu = 1 + \sqrt{5 - 4e} \in (2, 3]$ and (see [247]) the function ξ has no periodic points (except fixed points). All trajectories will converge to the non-zero fixed point.

Theorem 8.15. *Let $e \in [0, 1)$.*

1. *If $e < \frac{1}{4}$ then there exists an open set $\mathcal{U} \subset S^2$ such that $\overline{x}, \overline{\overline{x}} \in \mathcal{U}$ and for any $x^0 = (x_1^0, x_2^0, x_3^0) \in \mathcal{U}$ we have*

$$\lim_{n \to \infty} V^n(x^0) = \begin{cases} \overline{x}, & \text{if } x_3^0 \neq x_3^* \text{ and } n = 2k, \\ x^*, & \text{if } x_3^0 = x_3^*, \\ \overline{\overline{x}}, & \text{if } x_3^0 \neq x_3^* \text{ and } n = 2k + 1, \end{cases} \tag{8.75}$$

   where $x^* = (x_1^*, x_2^*, x_3^*)$ is fixed point and $\overline{x} = (\overline{x}_1, \overline{x}_3, \overline{x}_3)$, $\overline{\overline{x}} = (\overline{\overline{x}}_1, \overline{\overline{x}}_2, \overline{\overline{x}}_3)$ are periodic points described above.

2. *If $e \geq \frac{1}{4}$ then there exists an open set $U \subset S^2$ such that $x^* \in U$ and for any $x^0 = (x_1^0, x_2^0, x_3^0) \in U$ we have*

$$\lim_{n \to \infty} V^n(x_1^0, x_2^0, x_3^0) = (x_1^*, x_2^*, x_3^*).$$

Proof. 1) For $e \in [0, \frac{1}{4})$, it can be seen from $|\varphi'(x_3^*)| = |1 - \sqrt{5 - 4e}|$ that x_3^* is a repelling fixed point of $\varphi(x)$. Additionally, when $e < \frac{1}{4}$ the fixed points $\overline{x}_3$ and $\overline{\overline{x}}_3$ of function $g(x) = \varphi(\varphi(x))$ are attracting, which follow from $|g'(\overline{x}_3)| < 1$ and $|g'(\overline{\overline{x}}_3)| < 1$. Define the operator $W : [0, 1]^2 \to [0, 1]^2$ by the first and the last coordinate of the operator V:

$$W : \begin{cases} x_1' = cx_3^2 + 2x_3(1 - x_1 - x_3), \\ x_3' = \varphi(x_3). \end{cases} \tag{8.76}$$

Now using the Jacobian of the operator $W(W(x))$ one can see that the 2-periodic orbit $\{\overline{x}, \overline{\overline{x}}\}$ is a unique attracting orbit, and the fixed point x^*

is a saddle point of V. The operator V has the following invariant sets:

$$\gamma = \{(x_1, x_2, x_3) \in S^2 : x_3 = x_3^*\},$$
$$\Gamma = \{(x_1, x_2, x_3) \in S^2 : x_3 = \overline{x}_3 \text{ or } \overline{\overline{x}}_3\}.$$

Note that if $x^0 \in \gamma$ then $\lim_{n \to \infty} x_3^{(n)} = x_3^*$. If $x^0 \in \Gamma$ then $\lim_{n \to \infty} x_3^{(n)} = \overline{x}_3$ when n is even and $\lim_{n \to \infty} x_3^{(n)} = \overline{\overline{x}}_3$ when n is odd, in these cases the trajectory for x_1 can be written as

$$x_1^{(n+2)} = c\overline{x}_2^2 + 2(d\overline{x}_3^2 + 2x_1^{(n)}\overline{x}_3)\overline{\overline{x}}_3 \quad \text{for even } n,$$

and

$$x_1^{(n+2)} = c\overline{x}_2^2 + 2(d\overline{\overline{x}}_3^2 + 2x_1^{(n)}\overline{\overline{x}}_3)\overline{x}_3 \quad \text{for odd } n.$$

The existence of the limit (8.75) follows from general theory of dynamical systems (see [40]) and the uniqueness of the attracting 2-periodic points.

2) Next we shall consider when $e = \frac{1}{4}$. It can be seen from $|\varphi'(x_3^*)| = |1 - \sqrt{5 - 4e}|$ that when $e = \frac{1}{4}$, x_3^* is nonhyperbolic. It can be shown that the quadratic function $\varphi(x) - \frac{2}{5}$ has roots at $\frac{2}{5}$ and $\frac{6}{5}$. Additionally, $\varphi(x) - \frac{2}{5}$ is concave for all x. Therefore,

$$\begin{cases} \varphi(x) - \frac{2}{5} > 0 \Rightarrow \varphi(x) > \frac{2}{5}, & x < \frac{2}{5}, \\ \\ \varphi(x) - \frac{2}{5} < 0 \Rightarrow \varphi(x) < \frac{2}{5}, & x > \frac{2}{5} \end{cases}$$

which demonstrates that $x_3^{(n)}$ oscillates between $[0, \frac{2}{5})$ and $(\frac{2}{5}, 1]$. Moreover, it can be demonstrated that

$$\begin{cases} x < \varphi(\varphi(x)), & x < \frac{2}{5}, \\ \\ x > \varphi(\varphi(x)), & x > \frac{2}{5}. \end{cases}$$

Thus, $\lim x_3^{(n)} = \frac{2}{5}$.

It can be seen from $|\varphi'(x_3^*)| = |1 - \sqrt{5 - 4e}|$ that when $e \in (\frac{1}{4}, 1)$, x_3^* is an attracting fixed point of $\varphi(x)$.

Therefore, $x_3^{(n)}$ will converge to x_3^* when $e \geq \frac{1}{4}$. On the invariant line γ a trajectory of this operator is as follows:

$$x^{(n)} = (x_1^{(n)}, 1 - x_1^{(n)} - x_3^*, x_3^*),$$

where $x_1^{(n)}$ satisfies the equality

$$x_1^{(n+1)} = x_3^*(2 - (2 - c)x_3^* - 2x_1^{(n)}). \tag{8.77}$$

It follows from (8.77) that $\lim_{n \to \infty} x_1^{(n)} = x_1^*$. Therefore, when $\lim_{n \to \infty} x_3^{(n)} = x_3^*$, then $\lim_{n \to \infty} V(x^0) = (x_1^*, x_2^*, x_3^*)$. $\qquad \square$

A complete understanding of non-Volterra QSOs would not only be a significant advance in the field of mathematical genetics and dynamical systems, but it would also answer questions about the modeling of populations that have complex genetic structures for certain traits.

Bibliographical notes. In [120] the notion of Volterra QSO was introduced. In this chapter we only gave some results of [69]–[72]. For complete theory of Volterra QSO see [68], [69]–[72], [155]. But non-Volterra QSOs were not in completely studied. Because there is no any general theory which can be applied for investigation of non-Volterra operators. In this chapter we mainly used [40], [90], [129], [224] and [225]. To the best of our knowledge, there are few papers devoted to non-Volterra operators [62], [71], [73]–[75], [219]–[222], [224], [231]–[234].

Chapter 9

Dynamics of sex-linked population

This chapter is devoted to two-sex populations, the discrete-time dynamical systems of such populations have not been fully studied yet. We present some results related to Volterra QSO of two-sex populations and a class of non-Volterra QSOs with a preferences on a type of females and males. The last section presents a predator-prey model with the assumption that birth and death rates for male and female preys are different. Conditions for the global stability of the boundary equilibrium and for the permanence of the prey and predators are obtained. It is shown that the predation will not alter the sex ratio of the prey eventually when it is not sex-biased.

9.1　Bisexual population: Volterra operators

Recall (see Section 1.4) the evolution operator of bisexual population: we suppose that the set of females can be partitioned into types indexed by $\{1, 2, \dots, n\}$ and, similarly, that the male types are indexed by $\{1, 2, \dots, \nu\}$.

The population is described by its state vector (x, y) in $S = S^{n-1} \times S^{\nu-1}$.

Let $P_{ik,j}^{(f)}$ and $P_{ik,l}^{(m)}$ be inheritance coefficients defined as the probability that a female offspring is type j and, respectively, that a male offspring is of type l, when the parental pair is ik $(i, j = 1, \dots, n;$ and $k, l = 1, \dots, \nu)$.

Let $z' = (x', y')$ be the state of the offspring population at the birth stage. This is obtained from inheritance coefficients as

$$x'_j = \sum_{i,k=1}^{n,\nu} P_{ik,j}^{(f)} x_i y_k$$

$$y'_l = \sum_{i,k=1}^{n,\nu} P_{ik,l}^{(m)} x_i y_k \tag{9.1}$$

337

where

$$P_{ik,j}^{(f)} \geq 0, \quad \sum_{j=1}^{n} P_{ik,j}^{(f)} = 1$$

$$P_{ik,l}^{(m)} \geq 0, \quad \sum_{l=1}^{\nu} P_{ik,l}^{(m)} = 1. \tag{9.2}$$

The dynamics of this operator has not been completely studied yet. In [230] some results on dynamical system generated by the operator (9.1) are obtained. Here we shall give some results (without proofs) from this paper.

The evolution operator (9.1) is called a Volterra quadratic stochastic operator (VQSO) of a bisexual population if the hereditary coefficients (9.2) satisfy the conditions

$$p_{ik,j}^{(f)} = 0, \quad j \notin \{i,k\} \ \ (1 \leq i,j \leq n, 1 \leq k \leq \nu),$$

$$p_{ik,l}^{(m)} = 0, \quad l \notin \{i,k\} \ \ (1 \leq i \leq n, 1 \leq k,l \leq \nu). \tag{9.3}$$

For definiteness, we set $n \leq \nu$. Then an arbitrary VQSO of a bisexual population has the form

$$W : \begin{cases} x_j' = x_j \left(1 + \sum_{\substack{k=1 \\ k \neq j}}^{\nu} (p_{jk,j}^{(f)} - 1)y_k \right) + y_j \left(\sum_{\substack{k=1 \\ k \neq j}}^{n} p_{kj,j}^{(f)} x_k \right), & (1 \leq j \leq n) \\[2em] y_l' = y_l \left(1 + \sum_{\substack{k=1 \\ k \neq l}}^{n} (p_{kl,l}^{(m)} - 1)x_k \right) + x_l \left(\sum_{\substack{k=1 \\ k \neq l}}^{\nu} p_{lk,l}^{(m)} y_k \right), & (1 \leq l \leq n) \\[2em] y_l' = y_l \left(1 + \sum_{\substack{k=1 \\ k \neq l}}^{n} (p_{kl,l}^{(m)} - 1)x_k \right), & (n < l \leq \nu). \end{cases}$$

$$\tag{9.4}$$

For any initial state $(x^{(0)}, y^{(0)})$ the mapping (9.4) uniquely defines the trajectory

$$\{(x^{(t)}, y^{(t)})\}_{t=0}^{\infty} :$$

$$(x^{(t+1)}, y^{(t+1)}) = W((x^{(t)}, y^{(t)})) = W^{(t+1)}((x^{(0)}, y^{(0)})), t = 0, 1, 2, \ldots$$

The set of the limit points of the trajectory that starts from the point $(x^{(0)}, y^{(0)})$ is called its limit set and is denoted by $\omega(x^{(0)}, y^{(0)})$.

As usual our main problem is to study $\omega(x^{(0)}, y^{(0)})$ for each given $(x^{(0)}, y^{(0)}) \in S := S^{n-1} \times S^{\nu-1}$. This problem can be (particularly) solved by using Lyapunov functions:

A continuous functional $\varphi : S \to \mathbb{R}$ is called a Lyapunov function if the limit

$$\lim_{t \to \infty} \varphi(x^{(t)}, y^{(t)})$$

exists for any initial point $(x^{(0)}, y^{(0)}) \in S$.

One can see that if

$$\lim_{t\to\infty} \varphi(x^{(t)}, y^{(t)}) = C$$

then

$$\omega(x^0, y^0) \subset \varphi^{-1}(C).$$

Therefore the Lyapunov functions are very useful to give an upper estimation of the set of limit points.

Theorem 9.1. *If $n < \nu$, then*

i) $\varphi(x, y) = \prod\limits_{j=n+1}^{\nu} y_j$ *is a Lyapunov function.*

ii) *If $p_{ij,k}^{(m)} \neq 1$ for all $j = 1, ..., r$, $k = r+1, ..., n$, then $\varphi(x, y) = \prod\limits_{j=n+1}^{\nu} y_j^{b_j}$*
is a Lyapunov function, where $(b_{n+1}, ..., b_\nu) \in S^{\nu-n-1}$.
Moreover

$$\sum_{t=0}^{\infty} \varphi(x^{(t)}, y^{(t)}) < +\infty.$$

iii) *If $p_{ij,k}^{(m)} \neq 1$ for all $j = 1, ..., r$, $k = r+1, ..., n$, then $\omega(x^0, y^0) \subset S^{n-1} \times S^{n-1}$.*

Theorem 9.2. *If $n = \nu$ and $p_{jk,j}^{(f)} + p_{jk,j}^{(m)} < 1, p_{kj,j}^{(m)} + p_{kj,j}^{(f)} < 1$, for all $j = 1, r$, $k = r+1, n$, then for any $r < n$ the function $\varphi(x, y) = \sum\limits_{j=1}^{r}(x_j + y_j)$ is a Lyapunov function. Moreover,*

$$\sum_{t=0}^{\infty} \varphi(x^{(t)}, y^{(t)}) < +\infty.$$

From this theorem we get

$$\lim_{t\to\infty} x_j^{(t)} = \lim_{t\to\infty} y_j^{(t)} = 0, \quad \forall j = 1, ..., r.$$

Let $\mathcal{V}$ denote the set of all VQSOs of a bisexual population defined on $S^{n-1} \times S^{\nu-1}$. The matrix $\mathbb{P} = \left((p_{ik,j}^{(f)}), (p_{ik,l}^{(m)})\right)$ of a VQSO can be considered as a point in the space $\mathbb{R}^{2\nu n-(n+\nu)}$.

Theorem 9.3. *The following assertions are true:*

1. $\mathcal{V}$ is a convex compact subset in $\mathbb{R}^{2\nu n-(n+\nu)}$.

2. The set $\mathrm{Extr}\mathcal{V}$ *of extreme points of the set* $\mathcal{V}$ *has the form*

$$\mathrm{Extr}\mathcal{V} = \{W \in \mathcal{V} : \text{ the matrix} \mathbb{P} \text{ of} W \text{ consists only} 0 \text{ and} 1\}.$$

3. $|\mathrm{Extr}\mathcal{V}| = 2^{2\nu n - (n+\nu)}$.

Example 9.1. Let $n = \nu = 2$. In this case, by the theorem, there exist 16 extreme VQSOs. Here are some of them:

$$W_1(x_1, x_2; y_1, y_2) = (x_1 + x_2 y_1, x_2 y_2; x_1 + x_2 y_1, x_2 y_2);$$
$$W_2(x_1, x_2; y_1, y_2) = (x_1 + x_2 y_1, x_2 y_2; x_1, x_2);$$
$$W_3(x_1, x_2; y_1, y_2) = (x_1, x_2; x_1 + x_2 y_1, x_2 y_2);$$
$$W_4(x_1, x_2; y_1, y_2) = (x_1, x_2; x_1, x_2);$$
$$W_5(x_1, x_2; y_1, y_2) = (x_1, x_2; y_1, y_2).$$

Denote $e_i = (\delta_{1i}, \delta_{2i}, ..., \delta_{ni}) \in S^{n-1}$, $i = 1, ..., n$ and $\tilde{e}_j = (\delta_{1j}, \delta_{2j}, ..., \delta_{\nu j}) \in S^{\nu-1}$, $j = 1, ..., \nu$ vertices of simplex, where δ_{ij} is the Kronecker's delta.

Consider $(n + \nu)$-dimensional vector $\eta_{ij} = (e_i, \tilde{e}_j)$.

Theorem 9.4.

1) For any $i = 1, 2, ..., \min\{n, \nu\}$, *the vector* η_{ii} *is a fixed point of any VQSO* $W \in \mathcal{V}$;

2) For any $i \neq j$, $i, j = 1, 2, ..., \min\{n, \nu\}$, *the exists a VQSO* $W_{(ij)} \in \mathcal{V}$, *for which* $\{\eta_{ij}, \eta_{ji}\}$ *is a two-periodic trajectory.*

Let us give an example of a VQSO of a bisexual population that has a 2-periodic orbit.

Example 9.2. Let $n = \nu = 2$, $\eta_{12} = (1, 0, 0, 1)$ and $\eta_{21} = (0, 1, 1, 0)$. Define an operator $W : S^1 \times S^1 \to S^1 \times S^1$ as follows:

$$W(x_1, x_2; y_1, y_2) = (y_1, y_2; x_1, x_2).$$

We have that $W(W(\eta_{12}) = W(\eta_{21}) = \eta_{12}$, i.e., $\{\eta_{12}, \eta_{21}\}$ is a periodic orbit.

Remark 9.1. Comparing the properties of the VQSO of free population with the properties of VQSO of bisexual population we readily see that trajectories of these operators have substantially different properties. For example, in the bisexual population case (as distinct from the free population) there are periodic trajectories.

9.2 Dynamical systems with a preference of a type of females and males

Consider the special case (given by (4.20)) of heredity coefficients:

$$
P_{ik,j}^{(f)} = \begin{cases} a_{ij} & \text{if } k = 1 \\ 1 & \text{if } k \neq 1, j = 1 \\ 0 & \text{if } k \neq 1, j \neq 1 \end{cases}, \quad
P_{ik,l}^{(m)} = \begin{cases} b_{kl} & \text{if } i = 1 \\ 1 & \text{if } i \neq 1, l = 1 \\ 0 & \text{if } i \neq 1, l \neq 1. \end{cases} \quad (9.5)
$$

The matrices $A = (a_{ij})$ and $B = (b_{kl})$ satisfy the following conditions

$$
a_{ij} \geq 0, \quad \sum_{j=1}^{n} a_{ij} = 1, \quad i = 1,\ldots,n; \quad b_{kl} \geq 0, \quad \sum_{l=1}^{\nu} b_{kl} = 1, \quad k = 1,\ldots,\nu,
$$

$$(9.6)$$

i.e. both matrices are stochastic.

For these coefficients (9.5) the operator defined by (9.1) has the following form

$$
W : \quad \begin{aligned}
x_1' &= \sum_{i=1}^{n} \left(a_{i1} y_1 + \sum_{k=2}^{\nu} y_k \right) x_i \\
x_j' &= y_1 \sum_{i=1}^{n} a_{ij} x_i, \quad j \neq 1 \\
y_1' &= \sum_{k=1}^{\nu} \left(b_{k1} x_1 + \sum_{i=2}^{n} x_i \right) y_k \\
y_l' &= x_1 \sum_{k=1}^{\nu} b_{kl} y_k, \quad l \neq 1.
\end{aligned}
\qquad (9.7)
$$

Note that (9.7) maps S to itself.

In this section we shall study the behavior of the trajectory $z^{(m)} = W^m(z)$, $m \geq 1$ for each given initial point $z \in S$. But this problem is a rather difficult in general, because the dynamical system generated by operator (9.7) is a complicated non linear system. Therefore, following [43], we shall consider two particular cases of the system.

9.2.1 *A simple case: a hard constraint*

Let us start with a particular case of (9.7) and completely study the corresponding dynamical system.

Consider the case

$$
a_{ij} = \begin{cases} 1 & \text{if } i = j \\ 0 & \text{if } i \neq j \end{cases}, \quad
b_{kl} = \begin{cases} 1 & \text{if } k = l \\ 0 & \text{if } k \neq l, \end{cases} \qquad (9.8)
$$

then the corresponding operator has the following form

$$
H : \begin{cases}
x_1' = 1 - y_1(1 - x_1) \\
x_j' = y_1 x_j, \quad j = 2, \ldots, n \\
y_1' = 1 - x_1(1 - y_1) \\
y_l' = x_1 y_l, \quad l = 2, \ldots, \nu.
\end{cases}
\tag{9.9}
$$

Let $\mathrm{Fix}(H)$ be the set of fixed points of H, i.e., $H(z) = z$, $z = (x, y) \in S$. By Proposition 4.5 we get

Proposition 9.1. *The set of fixed points is*

$$
\mathrm{Fix}(H) = \left\{ ((1, 0, \ldots, 0), (y_1, \ldots, y_\nu)) : \sum_{k=1}^{\nu} y_k = 1 \right\} \cup
$$
$$
\left\{ ((x_1, \ldots, x_n), (1, 0, \ldots, 0)) : \sum_{i=1}^{n} x_i = 1 \right\}.
$$

To find the type of a fixed point of the operator (9.9) we write the Jacobi matrix, replacing x_1 by $1 - \sum_{i=2}^{n} x_i$ and y_1 by $1 - \sum_{j=2}^{\nu} y_j$, which is $(n-1) \times (\nu - 1)$-matrix of the form:

$$
J(s) = J_H(s) = \begin{pmatrix}
y_1 & 0 & \cdots & 0 & -x_2 & -x_2 & \cdots & -x_2 \\
0 & y_1 & \cdots & 0 & -x_3 & -x_3 & \cdots & -x_3 \\
\vdots & \vdots & \cdots & \vdots & \vdots & \vdots & \cdots & \vdots \\
0 & 0 & \cdots & y_1 & -x_n & -x_n & \cdots & -x_n \\
-y_2 & -y_2 & \cdots & -y_2 & x_1 & 0 & \cdots & 0 \\
-y_3 & -y_3 & \cdots & -y_3 & 0 & x_1 & \cdots & 0 \\
\vdots & \vdots & \cdots & \vdots & \vdots & \vdots & \cdots & \vdots \\
-y_\nu & -y_\nu & \cdots & -y_\nu & 0 & 0 & \cdots & x_1
\end{pmatrix}.
$$

One can see that $J(((1, 0, \ldots, 0), (y_1, \ldots, y_\nu)))$ has eigenvalues equal to 1, y_1, and $J(((x_1, \ldots, x_n), (1, 0, \ldots, 0)))$ has eigenvalues 1 and x_1 therefore all fixed points are not hyperbolic.

The following theorem completely describes the limit points of trajectories for operator H.

Theorem 9.5. *Let* $z^{(0)} = ((x_1^{(0)}, \ldots, x_n^{(0)}), (y_1^{(0)}, \ldots, y_\nu^{(0)})) \in S$ *be an initial point*

(i) If $x_1^{(0)} \neq 0$ *and* $y_1^{(0)} \neq 0$ *then*

$$\lim_{m \to +\infty} z^{(m)}$$

$$= \begin{cases} ((1,0,\ldots,0),(1,0,\ldots,0)), & \text{if } x_1^{(0)} = y_1^{(0)} \\ ((1,0,\ldots,0),(1 - x_1^{(0)} + y_1^{(0)}, y_2^*, \ldots, y_\nu^*)), & \text{if } x_1^{(0)} > y_1^{(0)} \\ ((1 + x_1^{(0)} - y_1^{(0)}, x_2^*, \ldots, x_n^*),(1,0,\ldots,0)), & \text{if } x_1^{(0)} < y_1^{(0)}, \end{cases}$$

where

$$x_j^* = \frac{(y_1^{(0)} - x_1^{(0)})x_j^{(0)}}{1 - x_1^{(0)}}, \quad j = 2, \ldots, n;$$

$$y_l^* = \frac{(x_1^{(0)} - y_1^{(0)})y_l^{(0)}}{1 - y_1^{(0)}}, \quad l = 2, \ldots, \nu.$$

(ii) If $x_1^{(0)} = 0$ *then*

$$z^{(m)} = H(z^{(0)}) = \left((1 - y_1^{(0)}, y_1^{(0)} x_2^{(0)}, \ldots, y_1^{(0)} x_n^{(0)}), (1,0,\ldots,0) \right),$$

for all $m \geq 1$.

(iii) If $y_1^{(0)} = 0$ *then*

$$z^{(m)} = H(z^{(0)}) = \left((1,0,\ldots,0), (1 - x_1^{(0)}, x_1^{(0)} y_2^{(0)}, \ldots, x_1^{(0)} y_\nu^{(0)}) \right),$$

for all $m \geq 1$.

Proof.   From (9.9) we get

$$H^{m+1}(z^{(0)}) : \begin{cases} x_1^{(m+1)} = 1 - y_1^{(m)}(1 - x_1^{(m)}) \\ x_j^{(m+1)} = y_1^{(m)} x_j^{(m)}, \quad j = 2, \ldots, n \\ y_1^{(m+1)} = 1 - x_1^{(m)}(1 - y_1^{(m)}) \\ y_l^{(m+1)} = x_1^{(m)} y_l^{(m)}, \quad l = 2, \ldots, \nu. \end{cases} \tag{9.10}$$

Since $y_1^{(m)} \in [0,1]$ we have

$$x_1^{(m+1)} = 1 - y_1^{(m)}(1 - x_1^{(m)}) \geq 1 - (1 - x_1^{(m)}) = x_1^{(m)},$$

$$x_j^{(m+1)} \leq x_j^{(m)}, \quad j = 2, \ldots, n.$$

Thus $x_1^{(m)}$ is a non-decreasing sequence, which bounded from above by 1 and for each $j = 2, \ldots, n$ the sequence $x_j^{(m)}$ is non-increasing and with lower bound 0. Consequently, each $x_j^{(m)}$ has a limit say α_j, $j = 1, \ldots, n$. Similarly one shows that $y_l^{(m)}$ also has a limit, say β_l, $l = 1, \ldots, \nu$. Hence, the limit $\lim_{m \to \infty} z^{(m)} = z_*$ exists. It is clear that $z_* \in \mathrm{Fix}(H)$, and $z_* = z_*(z^{(0)})$, i.e., it depends on the initial point $z^{(0)}$. Now we shall find $z_*(z^{(0)})$. Subtracting from the first equality of (9.10) the third one we get

$$x_1^{(m+1)} - y_1^{(m+1)} = x_1^{(m)} - y_1^{(m)}, \quad m = 0, 1, 2, \ldots \tag{9.11}$$

Iterating these equalities we obtain

$$x_1^{(m)} = y_1^{(m)} + (x_1^{(0)} - y_1^{(0)}), \quad m = 1, 2, \ldots \tag{9.12}$$

Proof of (i). From the first and third equations of (9.10), and (9.12) for limit values α_1 and β_1 we get the following equations

$$\begin{cases} \alpha_1 = 1 - \beta_1(1 - \alpha_1) \\ \beta_1 = 1 - \alpha_1(1 - \beta_1) \\ \alpha_1 = \beta_1 + (x_1^{(0)} - y_1^{(0)}). \end{cases} \tag{9.13}$$

One can see that the system (9.13) has following solution:

$$(\alpha_1, \beta_1) = \begin{cases} (1, 1), & \text{if } x_1^{(0)} = y_1^{(0)} \\ (1, 1 - x_1^{(0)} + y_1^{(0)}), & \text{if } x_1^{(0)} > y_1^{(0)} \\ (1 + x_1^{(0)} - y_1^{(0)}, 1), & \text{if } x_1^{(0)} < y_1^{(0)}. \end{cases} \tag{9.14}$$

Thus if $x_1^{(0)} = y_1^{(0)}$ then $z_* = z_*(z^{(0)}) = ((1, 0, \ldots, 0), (1, 0, \ldots, 0))$. Consider now two cases:

Case $x_1^{(0)} > y_1^{(0)}$: In this case $y_1^{(0)} < 1$. By (9.14) we have

$$\lim_{m \to \infty} x_j^{(m)} = 0, \quad j = 2, \ldots, n; \quad \lim_{m \to \infty} y_1^{(m)} = 1 - x_1^{(0)} + y_1^{(0)}. \tag{9.15}$$

Now we shall calculate $\lim_{m \to \infty} y_l^{(m)}$, $l = 2, \ldots, \nu$. From the third and last equalities of (9.10) we get

$$1 - y_1^{(m+1)} = x_1^{(m)}(1 - y_1^{(m)})$$

$$= x_1^{(m)} x_1^{(m-1)}(1 - y_1^{(m-1)}) = \cdots = (1 - y_1^{(0)}) \prod_{k=0}^{m} x_1^{(k)}. \tag{9.16}$$

$$y_l^{(m+1)} = x_1^{(m)} y_l^{(m)} = x_1^{(m)} x_1^{(m-1)} y_l^{(m-1)}$$

$$= \cdots = y_l^{(0)} \prod_{k=0}^{m} x_1^{(k)}, \quad l = 2, \ldots, \nu. \tag{9.17}$$

Taking limit from both side of (9.16) and using (9.15) we obtain

$$\lim_{m \to \infty} \prod_{k=0}^{m} x_1^{(k)} = \frac{x_1^{(0)} - y_1^{(0)}}{1 - y_1^{(0)}}.$$

Using this equality from (9.17) we get

$$\lim_{m \to \infty} y_l^{(m)} = \frac{(x_1^{(0)} - y_1^{(0)}) y_l^{(0)}}{1 - y_1^{(0)}}, \quad l = 2, \ldots, \nu.$$

Case $x_1^{(0)} < y_1^{(0)}$: This case is similar to the previous case and one obtains

$$\lim_{m \to \infty} x_j^{(m)} = \frac{(y_1^{(0)} - x_1^{(0)}) x_j^{(0)}}{1 - x_1^{(0)}}, \quad j = 2, \ldots, n.$$

These equalities complete the proof of part (i).

Proof of (ii). Let $x_1^{(0)} = 0$, $y_1^{(0)} \neq 0$. In this case by (9.10) we get

$$H(z^{(0)}) : \begin{cases} x_1^{(1)} = 1 - y_1^{(0)} \\ x_j^{(1)} - y_1^{(0)} x_j^{(0)}, \quad j = 2, \ldots, n \\ y_1^{(1)} = 1 \\ y_l^{(1)} = 0, \quad l = 2, \ldots, \nu. \end{cases} \tag{9.18}$$

Thus by Proposition 9.1 we have $H(z^{(0)}) \in \text{Fix}(H)$. This completes the proof of (ii).

Proof of (iii) is similar to the case (ii). $\qquad \square$

9.2.2 *All possible constraints for* $n = \nu = 2$

Consider the case $n = \nu = 2$. In this case since

$$x_1 + x_2 = y_1 + y_2 = a_{11} + a_{12} = a_{21} + a_{22} = b_{11} + b_{12} = b_{21} + b_{22} = 1$$

denoting

$$a = a_{12}, \quad b = a_{22}, \quad c = b_{12}, \quad d = b_{22}, \quad x = x_2, \quad y = y_2 \tag{9.19}$$

the operator (9.7) can be reduced to the following operator

$$T: \begin{cases} x' = (1-y)(a+(b-a)x) \\ y' = (1-x)(c+(d-c)y), \end{cases} \tag{9.20}$$

where $a, b, c, d, x, y \in [0,1]$. Thus we get a quadratic operator with four independent parameters. In this subsection we shall study the dynamical system generated by this operator.

9.2.2.1 *Fixed points*

To find fixed points of (9.20) we should solve the following

$$x = (1-y)(a+(b-a)x)$$
$$y = (1-x)(c+(d-c)y). \tag{9.21}$$

Denote by $\mathrm{Fix}(T)$ the set of all fixed points of the operator T given by (9.20).

From the first equation of the system we find

$$(1-(b-a)(1-y))x = a(1-y). \tag{9.22}$$

Case: $(1-y)(b-a) = 1$. In this case from (11.16) we get $a = 0$ or $y = 1$. If $a = 0$ then $(1-y)b = 1$, i.e., $y = 1 - 1/b$. This gives non-negative y only if $b = 1$, i.e. in this case $y = 0$. Under these assumptions from the second equation of the system (9.21) we get $c = 0$ or $x = 1$. Assume $c = 0$ then it is easy to see that $(x,0)$ is a solution of the system for any $x \in [0,1]$. In case $c \neq 0$ we get solution $(1,0)$.

Case: $(1-y)(b-a) \neq 1$. From (11.16) we obtain

$$x = \frac{a(1-y)}{a(1-y)+1-b(1-y)}. \tag{9.23}$$

From this equality it is clear that for each $y \in [0,1]$ the corresponding x also will be in $[0,1]$. Substituting (9.23) in the second equation of (9.21) we get

$$[b(d-c)+a-b]y^2 + [2bc-bd-a+b-c+d-1]y+c(1-b) = 0. \tag{9.24}$$

Subcase: $b(d-c) = b-a$ and $c(b-1) = 1-d$. Then $c(1-b) = 0$. Under these conditions we get $a = c = 0$, $d = 1$ and $b \neq 1$ which gives fixed points $(0,y)$ for any $y \in [0,1]$. Moreover if $a = c \neq 0$, $b = d = 1$, then the following points are fixed

$$P(y) = \left(\frac{a(1-y)}{a(1-y)+y}, y \right), \quad \forall y \in [0,1].$$

Subcase: $b(d-c) = b-a$ and $c(b-1) \neq 1-d$. Then the following point is a fixed point

$$P_0 = \left(\frac{a(1-d)}{a(1-d) + c(1-b) + (1-b)(1-d)}, \frac{c(1-b)}{c(1-b) + (1-d)} \right).$$

Subcase: $b(d-c) < b-a$. In this case we have

$$D = [2bc - bd - a + b - c + d - 1]^2 - 4[b(d-c) - (b-a)]c(1-b) > 0.$$

Thus equation (9.24) has two solutions $y_1 < y_2$. By Vieta's theorem one can see that $y_1 < 0$. For y_2 we have

$$y_2 = \frac{(2bc - bd - a + b - c + d - 1) + \sqrt{D}}{2(b(c-d) + b - a)}.$$

Subcase: $b(d-c) > b-a$. In this case we assume

$$D = [2bc - bd - a + b - c + d - 1]^2 - 4[b(d-c) - (b-a)]c(1-b) > 0.$$

Thus equation (9.24) has two solutions $0 < y_1 < y_2$, with

$$y_{1,2} = \frac{-(2bc - bd - a + b - c + d - 1) \pm \sqrt{D}}{2(b(d-c) + a - b)}.$$

Choose parameters a, b, c, d such that y_1 or/and y_2 belong in $[0,1]$ then corresponding value of x is defined by (9.23). Denote this fixed point by $Q_i = (x_i, y_i)$, $i = 1, 2$.

Summarizing we get the following

Proposition 9.2. *The set* $\mathrm{Fix}(T)$ *has the following form*

$$\mathrm{Fix}(T) = \begin{cases} \{(0,0)\}, & if\ a = c = 0, b \neq 1, d \neq 1, \\ \{(x,0),\ x \in [0,1]\}, & if\ a = c = 0, b = 1, \\ \{(0,y),\ y \in [0,1]\}, & if\ a = c = 0, d = 1, \\ \{(1,0)\}, & if\ a = 0, b = 1, c \neq 0, \\ \{P(y),\ y \in [0,1]\}, & if\ a = c \neq 0, b = d = 1, \\ \{P_0\}, & if\ b(d-c) = b - a, c(b-1) \neq 1 - d, \\ \{Q_2\}, & if\ b(d-c) < b - a, \\ \{Q_1, Q_2\}, & otherwise. \end{cases}$$

Remark 9.2. Note that the case $a = c = 0$, $b = d = 1$ is a particular case of operator (9.9). Therefore in the sequel of this section we consider the case when $a + c \neq 0$ or $b + d \neq 2$. Thus we will consider the cases where the fixed points $P(y)$, P_0, Q_i, $i = 1, 2$ exist.

9.2.2.2 *Type of fixed points*

To study type of fixed points consider Jacobi matrix of the operator (9.20) at a fixed point $z = (x, y)$:

$$J(z) = J_T(z) = \begin{pmatrix} (b-a)(1-y) & -(a+(b-a)x) \\ -(c+(d-c)y) & (d-c)(1-x) \end{pmatrix}.$$

For $P(y)$ we have $a = c \neq 0$ and $b = d = 1$. In this case one can find two eigenvalues: $\lambda_1 = 1$ and

$$\lambda_2 = \lambda_2(a, y) = \frac{(1-a)y - [a + (1-a)y]^2}{(1-a)y + a}.$$

Now we shall prove that $|\lambda_2| < 1$ for any $a \in (0,1)$ and $y \in [0,1]$. Denote $t = a + (1-a)y$. We have $t \in [a, 1]$. The function λ_2 can be written as

$$\lambda_2 = \lambda_2(t) = 1 - t - \frac{a}{t}.$$

One can see that

$$-a = \lambda_2(a) = \lambda_2(1) = \min_{t \in [a,1]} \lambda_2(t) \leq \lambda_2(t) \leq \max_{t \in [a,1]} \lambda_2(t) = \lambda_2(\sqrt{a}) = 1 - 2\sqrt{a}.$$

From this inequalities we get

$$|\lambda_2| \leq \max\{a, |1 - 2\sqrt{a}|\} < 1, \quad \forall a \in (0,1).$$

Thus the fixed point $P(y)$ is not hyperbolic.

At the fixed point P_0 the eigenvalues of $J(P_0)$ has very long expression. Using Maple and varying the parameters a, b, c, d (taking into account the conditions $b(d-c) = b - a$ and $c(b-1) \neq 1 - d$) one can check that P_0 is attractive. For example,

- if $a = 0.18$, $b = 0.2$, $c = 0.3$, $d = 0.4$ then eigenvalues are approximately -0.197 and 0.298.
- if $a = 0$, $b = 0$, $c = 0.9$, $d = 0.1$ then eigenvalues are -0.8 and 0.
- if $a = 0.81$, $b = 0.9$, $c = 0.7$, $d = 0.8$ then eigenvalues are -0.743, 0.845.

Moreover, giving values of three parameters and plot the eigenvalues as functions of the remaining parameter, one can see that the absolute value of the functions are less than 1. Thus based on numerical computations we conjecture that P_0 is attractive for any parameters a, b, c, d with conditions $b(d-c) = b - a$ and $c(b-1) \neq 1 - d$.

Checking the type of Q_1 and Q_2 is also very difficult, because corresponding eigenvalues have very long expression depending on four parameters. Using Maple one can get the following results for Q_2 in case $b(d-c) < b - a$:

- if $a = 0.1$, $b = 0.9$, $c = 0.2$, $d = 0.3$ then eigenvalues are approximately -0.0205, 0.7662.
- if $a = 0.7$, $b = 0.9$, $c = 0.2$, $d = 0.3$ then eigenvalues are approximately -0.3245, 0.5340.
- if $a = 0.7$, $b = 0.71$, $c = 0.2$, $d = 0$ then eigenvalues are -0.3949, 0.3367.

Thus we also have conjecture that Q_2 is an attracting point for any a, b, c, d with $b(d - c) < b - a$. Similar computations can be done for Q_1.

9.2.2.3 *Dynamics for $a = c$, $b = d$*

In this case the operator (9.20) has the following form

$$T : \begin{cases} x' = (1 - y)(a + (b - a)x) \\ y' = (1 - x)(a + (b - a)y). \end{cases} \tag{9.25}$$

The following lemma gives full description of fixed points.

Lemma 9.1. *The set of fixed points $\mathrm{Fix}(T)$ of the operator (9.25) has the following form*

(a) If $b = 1$ then

$$\mathrm{Fix}(T) = \begin{cases} \{(1, 0)\} \cup \{(0, y), \forall y \in [0, 1]\}, & \text{if } a = 0 \\ \left\{ \left(x, \frac{a(1-x)}{a(1-x)+a} \right), \forall x \in [0, 1] \right\}, & \text{if } a \neq 0. \end{cases}$$

(b) If $b \neq 1$ then there is a unique fixed point, i.e.,

$$\mathrm{Fix}(T) = \begin{cases} \{(\frac{a}{1+a}, \frac{a}{1+a})\}, & \text{if } a = b \\ \{(x_*, x_*)\}, & \text{if } a \neq b \end{cases}$$

where

$$x_* = \frac{1 - b + 2a - \sqrt{(1 - b)^2 + 4a}}{2(a - b)}.$$

Proof. Follows from detailed (but simple) analysis of the system

$$\begin{cases} x = (1 - y)(a + (b - a)x) \\ y = (1 - x)(a + (b - a)y). \end{cases} \tag{9.26}$$

For example, subtracting from the first equation of the system the second one we get $x - y = b(x - y)$, i.e., $x = y$ if $b \neq 1$. Consequently in the case $b \neq 1$ the solutions are given by solution of $x = (1 - x)(a - (b - a)x)$. $\qquad \square$

Lemma 9.2. *Type of fixed points mentioned in Lemma 9.1 as the following*

(a) If $b = 1$ the fixed points mentioned in part (a) of Lemma 9.1 are not hyperbolic.

(b) If $b \in [0,1)$ then for fixed points mentioned in part (b) we have

$$\left(\frac{a}{1+a}, \frac{a}{1+a} \right) \quad \text{is attractive}$$

and

$$(x_*, x_*) = \begin{cases} \text{attractive,} & \text{if } 0 \le a < 1 - \frac{(1-b)^2}{4} \\[2mm] \text{nonhyperbolic,} & \text{if } a = 1 - \frac{(1-b)^2}{4} \\[2mm] \text{saddle,} & \text{if } 1 - \frac{(1-b)^2}{4} < a \le 1. \end{cases}$$

Proof.   (a) Simple computations show that for each fixed point mentioned in part (a) of Lemma 9.1 one of eigenvalues of the corresponding Jacobian is equal to 1 and the second eigenvalue is less than 1.

     (b) For the case $a = b < 1$ the eigenvalues are $-a$ and a, therefore $(\frac{a}{1+a}, \frac{a}{1+a})$ is attractive. For the case $a \ne b$ we have the eigenvalues $\lambda_1 = b < 1$ and $\lambda_2 = 1 - \sqrt{(1-b)^2 + 4a}$. One can see that

$$|\lambda_2| = \begin{cases} < 1, & \text{if } 0 \le a < 1 - \frac{(1-b)^2}{4} \\[2mm] 1, & \text{if } a = 1 - \frac{(1-b)^2}{4} \\[2mm] > 1, & \text{if } 1 - \frac{(1-b)^2}{4} < a \le 1. \end{cases}$$

This completes the proof. $\square$

     Limit points for $a = b$. One very simple but interesting case is $a = b$. In this case we have

$$x^{(n+1)} = a(1 - y^{(n)}) = a - a^2 + a^2 x^{(n-1)} = a - a^2 + a^3 - a^3 y^{(n-2)} = \cdots =$$

$$\frac{a(1 - (-a)^{n+1})}{1+a} + (-a)^{n+1} x^{(0)} \quad (\text{or } y^{(0)} \text{ depending on parity of } n).$$

Similarly for $y^{(n+1)}$ we obtain

$$y^{(n+1)} = \frac{a(1 - (-a)^{n+1})}{1+a} + (-a)^{n+1} y^{(0)} \quad (\text{or } x^{(0)} \text{ depending on parity of } n).$$

Consequently, we have the following

$$\lim_{n\to\infty} (x^{(n)}, y^{(n)}) = \left(\frac{a}{1+a}, \frac{a}{1+a} \right), \quad \text{if } 0 \le a = b < 1.$$

Moreover if $a = b = 1$ then for any initial point $(x^{(0)}, y^{(0)})$ its trajectory $(x^{(n)}, y^{(n)})$ is a 2-periodic sequence:

$$(x^{(n)}, y^{(n)}) = \begin{cases} (x^{(0)}, y^{(0)}), & \text{if } n \text{ is even} \\ (1 - y^{(0)}, 1 - x^{(0)}), & \text{if } n \text{ is odd.} \end{cases}$$

Limit points for $a \ne b$. For $(x^{(0)}, y^{(0)}) \in [0,1]^2$, let $(x^{(n)}, y^{(n)})$ be the trajectory generated by actions of operator (9.25).

The following lemma is useful.

Lemma 9.3. *The following equalities hold*

(1) $x^{(n+1)} - y^{(n+1)} = b(x^{(n)} - y^{(n)})$ *for any* $n = 0, 1, 2, \ldots$.
(2) *If* $b = 1$ *then* $x^{(n)} - y^{(n)} = x^{(0)} - y^{(0)}$ *for any* $n = 1, 2, \ldots$.
(3) *If* $b \ne 1$ *then*

$$\lim_{n\to\infty} \left(x^{(n)} - y^{(n)} \right) = 0.$$

Proof. From (9.25) we have

$$\begin{cases} x^{(n+1)} = (1 - y^{(n)})(a + (b - a)x^{(n)}) \\ y^{(n+1)} = (1 - x^{(n)})(a + (b - a)y^{(n)}). \end{cases} \tag{9.27}$$

Subtracting from the first equation the second one we get (1). The assertions (2) and (3) follow from (1). $\qquad\square$

One can see that the set $I = \{(x, y) \in [0,1]^2 : x = y\}$ is invariant with respect to operator (9.25), i.e., $T(I) \subset I$. Let us study the dynamical system on I.

Reducing the operator T on the set I we get

$$x' = f(x) = (1 - x)(a + (b - a)x).$$

This function has the following properties: $f(0) = a$, $f(1) = 0$ and if $2a \ge b$ then $f(x)$ is a decreasing function, so maximal value of f is a. If $2a < b$ then $f(x)$ is increasing for $x \in [0, \frac{b-2a}{2(b-a)}]$ and decreasing for $x \in [\frac{b-2a}{2(b-a)}, 1]$, then maximal value of f is $\frac{b^2}{4(b-a)}$, which is less than 1 for any $b > 2a$. Thus $f : [0,1] \to [0,1]$.

For any $a \neq b$ the function $f(x)$ has a unique fixed point $x_* \in [0,1]$ which is defined in Lemma 9.1. It is known that $x_* \in [0,1]$ is attractive if $|f'(x_*)| < 1$, saddle if $|f'(x_*)| = 1$ and repeller if $|f'(x_*)| > 1$. Solving these inequalities we get

$$x_* = \begin{cases} \text{attractive, if } 0 \leq a < 1 - \frac{(1-b)^2}{4} \\[2mm] \text{saddle,} \quad \text{if } a = 1 - \frac{(1-b)^2}{4} \\[2mm] \text{repeller,} \quad \text{if } 1 - \frac{(1-b)^2}{4} < a \leq 1, \end{cases}$$

where $b \in [0,1)$.

We will apply the following lemma (see page 70 of [126]).

Lemma 9.4. *Let $f : [0,1] \to [0,1]$ be a continuous function with a fixed point $x_* \in (0,1)$. Assume that f is differentiable at x_* and that $f'(x_*) < -1$. Then there exist p_1, p_2, $0 \leq p_1 < x_* < p_2 \leq 1$, such that $p_1 = f(p_2)$ and $p_2 = f(p_1)$.*

Note that the condition $f'(x_*) < -1$ of Lemma 9.4 is satisfied iff $1 - \frac{(1-b)^2}{4} < a \leq 1$, i.e., when the fixed point is repeller, then by Lemma 9.4 it follows that there are two 2-periodic points, denoted by p_1 and p_2, which are solutions of the system $p_1 = f(p_2)$, $p_2 = f(p_1)$. Thus p_1 and p_2 are solutions of $f^2(x) = x$ which are different from the fixed point x_*. Hence to find 2-periodic points one has to solve the following equation

$$\frac{f^2(x) - x}{f(x) - x} = 0.$$

This equation has solutions

$$p_1 = \frac{2a - b - 1 - \sqrt{(1-b)^2 - 4(1-a)}}{2(a-b)},$$

$$p_2 = \frac{2a - b - 1 + \sqrt{(1-b)^2 - 4(1-a)}}{2(a-b)}.$$

Recall that a periodic point p with of period n is called an attracting periodic point if $|(f^n)'(p)| < 1$.

Lemma 9.5. *Both 2-periodic points p_1 and p_2 are attracting.*

Proof. Since $(f^2)'(x) = f'(f(x))f'(x)$ we have

$$(f^2)'(p_1) = f'(f(p_1))f'(p_1) = f'(p_2)f'(p_1) = (f^2)'(p_2) = 4 - 4a + 2b - b^2.$$

Using condition $1 - \frac{(1-b)^2}{4} < a \leq 1$, $b \neq 1$ one can check that

$$|4 - 4a + 2b - b^2| < 1. \qquad \square$$

Summarizing we get

Theorem 9.6. *For an initial point* $(x^{(0)}, y^{(0)}) \in [0,1]^2$ *its trajectory* $(x^{(n)}, y^{(n)})$ *generated by the operator (9.25) has the following properties*

1.
$$\lim_{n\to\infty} (x^{(n)}, y^{(n)}) = \left(\frac{a}{1+a}, \frac{a}{1+a}\right), \quad \text{if } 0 \le a = b < 1.$$

2. *If* $a = b = 1$ *then*
$$(x^{(n)}, y^{(n)}) = \begin{cases} (x^{(0)}, y^{(0)}), & \text{if } n \text{ is even} \\ (1 - y^{(0)}, 1 - x^{(0)}), & \text{if } n \text{ is odd}. \end{cases}$$

3. *Let* $b \neq 1$ *and* $0 \le a < 1 - \frac{(1-b)^2}{4}$. *Then there is an open neighborhood* U *of* (x_*, x_*) *such that if* $(x^{(0)}, y^{(0)}) \in U$, *then*
$$\lim_{n\to\infty} (x^{(n)}, y^{(n)}) = (x_*, x_*).$$

4. *Let* $b \neq 1$ *and* $1 - \frac{(1-b)^2}{4} < a \le 1$. *Then there is an open neighborhood* V *of* (x_*, x_*) *such that, if* $(x^{(0)}, y^{(0)}) \in V \setminus \{(x_*, x_*)\}$, *with* $x^{(0)} = y^{(0)}$ *then there exists* $k > 0$ *such that* $(x^{(k)}, y^{(k)}) \notin V$. *Moreover, there is a curve* $\gamma \subset [0,1]^2$ *through* (x_*, x_*) *which is invariant with respect to* T *and for any initial point* $(x^{(0)}, y^{(0)}) \in \gamma$ *all trajectories tend to* (x_*, x_*).

5. *If* $1 - \frac{(1-b)^2}{4} < a \le 1$ *and* $x^{(0)} = y^{(0)} < x_*$ *(resp.* $x^{(0)} = y^{(0)} > x_*$*) then there is an open neighborhood* $W \subset [0,1]$ *of the 2-periodic orbit* $\{p_1, p_2\} \subset [0,1]$ *such that if* $x^{(0)} \in W$, *then*
$$\lim_{n\to\infty} (x^{(n)}, y^{(n)}) = \begin{cases} (p_1, p_1), & \text{if } n = 2m, \ m \to \infty \\ (p_2, p_2), & \text{if } n = 2m+1, \ m \to \infty \end{cases}$$
$$\left(\text{resp. } \lim_{n\to\infty} (x^{(n)}, y^{(n)}) = \begin{cases} (p_2, p_2), & \text{if } n = 2m, \ m \to \infty \\ (p_1, p_1), & \text{if } n = 2m+1, \ m \to \infty \end{cases} \right).$$

Proof. The proof follows from above mentioned results by applying Theorem 6.3 and Theorem 6.5 of page 216 in [40]. $\qquad\square$

Remark 9.3. In part 4 of Theorem 9.6 the set I (resp γ) is the local unstable (resp. stable) manifold at (x_*, x_*) (see [40]). Note that the part 2 of this theorem gives a continuum set of 2-periodic orbits: any point of $[0,1]$ generates a 2-periodic orbit. But in part 5 we have only one 2-periodic orbit $\{(p_1, p_1), (p_2, p_2)\}$ of the operator (9.25). This periodic orbit attracts other non-periodic trajectories.

9.2.3 *Biological interpretation*

In biology, a population genetics is the study of the distributions and changes of allele (type) frequency in a population.

The results formulated in previous sections have the following biological interpretations:

Let $z^{(0)} = (x^{(0)}, y^{(0)}) \in \mathcal{S}$ be an initial state, i.e., $x^{(0)}$ (resp. $y^{(0)}$) is the probability distribution on the set $\{1, \ldots, n\}$ (resp. $\{1, \ldots, \nu\}$) of genotypes. Assume the trajectory $z^{(m)}$ of this point has a limit $z = ((x_1, \ldots, x_n), (y_1, \ldots, y_\nu))$ this means that the future of the population is stable: female genotypes i survives with probability x_i and male genotype j survives with probability y_j. A genotype will disappear if its probability is zero.

The case of hard constraint:

(a) The population has a continuum set of equilibrium states (Proposition 9.1), it stays in a neighborhood of one of the equilibrium states of the population (stable fixed point).

(b) If initially the type 1 of females *and* males have a positive *equal* probability, then in the future of the evolution the type 1 of both sexes will survive with probability 1 and consequently all other types will disappear. If initially the type 1 of females (resp. males) have positive probability which is strictly large than initial probability of the type 1 of males (resp. females) then type 1 of females (resp. males) survives with probability 1, but all types of males (resp. females) survive with probabilities *less* than 1 (interpretation of the part (i) of Theorem 9.5).

(c) If initially the type 1 of females is not present (i.e., has probability zero) but the type 1 of males has a positive probability, say $y_1^{(0)}$, then in the future of the evolution the type 1 of females survives with probability $1 - y_1^{(0)}$, moreover another types, say i, of females survives if and only if it was present initially (i.e. $x_i^{(0)} > 0$) the probability of its survive is $y_1^{(0)} x_i^{(0)}$; for males only type 1 survives with probability 1 ((ii) of Theorem 9.5). Moreover, the population goes to its equilibrium starting from the first generation.

Similar interpretation also can be made for the part (iii).

Note that from the parts (ii) and (iii) of Theorem 9.5 it follows that if initially the type 1 of both sexes are not present then still this type 1 will survive with probability 1 (as in the case (b)). Moreover, this initial state, say $((0, x_2^{(0)}, \ldots, x_n^{(0)}), (0, y_2^{(0)}, \ldots, y_\nu^{(0)}))$, goes to the equilibrium state $((1, 0, \ldots, 0), (1, 0, \ldots, 0))$ in the next (first) generation. This is a consequence of the hard constraint which is assumed for the population.

The case of $n = \nu = 2$:

(d) Depending on the parameters the population my have one, two or infinitely many (continuum) equilibrium states (Proposition 9.2). The population stays in a neighborhood of one of the equilibrium states of the population (stable fixed point).

(e) In case $0 \le a = b = c = d < 1$ the types of the population will survive. The future of the population is stable which has equilibrium $\left(\left(\frac{1}{1+a}, \frac{a}{1+a}\right), \left(\frac{1}{1+a}, \frac{a}{1+a}\right)\right)$ (part 1 of Theorem 9.6).

(f) In case $a = b = c = d = 1$ the population is periodic, i.e., the probability of a type will 2-periodically increase and decrease. For example, if the probability of the type 1, (i.e., $1 - x^{(0)}$) is less then $\frac{1}{2}$ the in the next generation the probability will be large than $\frac{1}{2}$, after that it will be less than $\frac{1}{2}$ and so on. Moreover, the population goes to this periodic state starting at the first generation (part 2 of Theorem 9.6).

(g) Under conditions of the part 3 of Theorem 9.6 all types of the population will survive if the initial state of the population is sufficiently close to the equilibrium state $((1 - x^*, x^*), (1 - x^*, x^*))$ and in the future the state of population will be in the neighborhood of this equilibrium state.

(h) Under conditions of the part 4 of Theorem 9.6 there are some initial states of the population for which in the future it will stay in a neighborhood of the equilibrium state $((1 - x^*, x^*), (1 - x^*, x^*))$. While there are other initial states for which the trajectory of the states will go out from a neighborhood of the equilibrium.

(i) Under conditions of part 5 of Theorem 9.6, depending on the initial state the population may go to the 2-periodic state.

Remark 9.4. 1) Note that in biology there are some species where population have reasonably predictable patterns of change although the full reasons for population periodicity is one of the major unsolved ecological problems. There are a number of factors which influence population change such as availability of food, predators, diseases and climate. The periodicity mentioned in items (f) and (i) can be examples of such biological phenomenon.

2) Recall that a quadratic stochastic operator V is a Bernstein mapping if $V^2 = V$. This property is just the stationarity principle. Note that the equality $V^2 = V$ means that for any $z \in S$ the point $V(z)$ is a fixed point for V. As a corollary of Theorems 9.5 and 9.6 we can say that the operators (9.9) and (9.20) are not Bernstein mapping. But for the operator (9.9) the condition $H^2(z) = H(z)$ is satisfied for any $z = (x, y) \in S$ with $x_1 y_1 = 0$.

9.3 A predator-prey system

In this section following [259] we consider a predator-prey model where the prey population is separated into female and male populations.

In this model, random mating, i.e., without the formation of permanent male-female couples, is the prevailing and only arrangement for propagation of the prey population.

It is interesting to know how the predator population influences the sex ratio of the Prey population, i.e., the proportion of the female prey to the male prey.

Consider the following two-sex model [150]:

$$\begin{cases} \dot{x} = b_1 y - d_1 x - \frac{1}{k}(x+y)x \\ \dot{y} = (b_2 - d_2)y - \frac{1}{k}(x+y)y, \end{cases} \tag{9.28}$$

where $x(t)$ and $y(t)$ denote the numbers of males and females, respectively, in the population at time t; d_1 and d_2 are per-capita death rates for males and females, b_1 and b_2 per-capita death rates for males and females, and k carrying capacity of the population.

Liu [150] shows that the unique positive equilibrium is globally asymptotically stable as long as it exits.

It is known that many populations can be separated into female and male populations. Due to some reasons, male (female) populations are predated more easily than female (male) populations (see [121]). Therefore, it is important to consider how predation influence the dynamics of prey population, especially how to influence the sex ratio of prey population. Moreover it is known (see [259]) that the ratios in the models with vital processes independent of age always tend to a constant. Therefore, for a simplicity in mathematics, we consider a predator-prey model with the prey population based on the two-sex model ([259]).

Let $x(t)$ be the number of the male prey, $y(t)$ the number of the female prey and $z(t)$ the number of predators at time t.

Let us give some assumptions:

1. There is the sex difference in the prey population but not in the predator population;
2. the prey is limited only by predators, and in its absence would grow according to the logistic law.
3. In the absence of the prey, predators die off and it is depensatory.
4. The functional response of predators is of Holling type II[1] and the predation is not sex biased. Let $\phi_1(x,y)$, $\phi_2(x,y)$ be the functional responses of predator to male and female Prey separately, then

$$\phi_1(x,y) = \frac{ax}{1 + b(x+y)}, \quad \phi_2(x,y) = \frac{ay}{1 + b(x+y)}.$$

Then a predator-prey system is given by the following equations

$$\begin{cases} \dot{x} = b_1 y - d_1 x - \frac{1}{k}(x+y)x - \phi_1(x,y)z, \\ \dot{y} = ry - \frac{1}{k}(x+y)y - \phi_2(x,y)z \\ \dot{z} = z[-d - cz + e(\phi_1(x,y) + \phi_2(x,y))], \end{cases} \tag{9.29}$$

where b_1, d_1 and k are as in the model (9.28), e is the conversion factor of prey to predator, d the death rate of predator and c the density-restricting factor. All above parameters are positive.

If $\frac{ae-bd}{d} \le \frac{1}{rk}$, then system (9.29) has only two boundary equilibria

$$E_0(0,0,0), \quad E_1\left(\frac{b_1 rk}{A}, \frac{(d_1+r)kr}{A}, 0\right),$$

where $A = b_1 + d_1 + r$. If $\frac{ae-bd}{d} > \frac{1}{rk}$, then system (9.29) may have one, two or three positive equilibria besides boundary equilibria E_0 and E_1. Denote

$$\Pi = \left\{ (x,y,z) \in \mathbb{R}^3 : \frac{x}{b_1} = \frac{y}{d_1+r} \right\}.$$

Theorem 9.7. *For any positive solution $(x(t), y(t), z(t))$ of system (9.29) we have*

$$\lim_{t \to +\infty} \frac{x}{y} = \frac{b_1}{d_1+r},$$

that is, the ω limit sets of all positive solutions belong to plane Π.

---

[1] https://en.wikipedia.org/wiki/Functional_response: A functional response in ecology is the intake rate of a consumer as a function of food density (i.e. the amount of food available in a given ecotope). By C. S. Holling, functional responses are classified into three types, which are called Holling's type I, II, and III. Type II functional response is characterized by a decelerating intake rate, which follows from the assumption that the consumer is limited by its capacity to process food. Often one assumes that processing of food and searching for food are mutually exclusive behaviors.

Proof. Denote

$$L(x,y,z) = \frac{1}{2}[(d_1+r)x - b_1 y]^2.$$

Calculating the derivative of V along the solutions of system (9.29), we obtain

$$\frac{dL}{dt} = [(d_1+r)x - b_1 y][(d_1+r)\dot{x} - b_1\dot{y}]$$

$$= -[d_1 + \frac{x+y}{k} + \frac{az}{1+b(x+y)}][(d_1+r)x - b_1 y]^2 \leq 0.$$

Consequently, from this inequality we get

$$\frac{dL}{dt} = 0 \quad \Leftrightarrow \quad (x(t), y(t), z(t)) \in \Pi.$$

Therefore, $\lim_{t\to+\infty} \frac{x}{y} = \frac{b_1}{d_1+r}$. $\qquad\qquad \square$.

Remark 9.5. Theorem 9.7 indicates that when the predation is not sex-biased the eventual proportion of male prey to female prey is a constat $\frac{d_1+r}{b_1}$ which dose not depend on the parameters of predator, i.e., the predator's predation of prey will not change the sexual ratio in the end.

Substituting $x = \frac{b_1 y}{d_1+r}$ into the second and third equations of system (9.29) we get

$$\begin{cases} \dot{y} = y\left[r - \frac{A}{k(d_1+r)}y - \frac{a(d_1+r)z}{d_1+r+bAy}\right] \\ \dot{z} = z\left[-d - cz + \frac{aeAy}{d_1+r+bAy}\right]. \end{cases} \tag{9.30}$$

Lemma 9.6. *There exits a positive constant M such that for any positive solution $(y(t), z(t))$ of system (9.30) we have*

$$\lim_{t\to+\infty} \sup \; y(t) < M, \quad \lim_{t\to+\infty} \sup \; z(t) < M.$$

Consequently, system (9.30) is point dissipative.

Proof. Follows from the density-dependentness of the prey and predators. $\qquad\qquad \square$

Theorem 9.8. *If* $\frac{ae-bd}{d} < \frac{1}{rk}$, *then the boundary equilibrium* $E_1\left(\frac{b_1 rk}{A}, \frac{(d_1+r)kr}{A}, 0\right)$ *is globally asymptotically stable.*

Proof. If $\frac{ae-bd}{d} < \frac{1}{rk}$, then it is easy to see that the system (9.30) has only two boundary equilibria $E_0'(0,0)$ and $E_1'\left(\frac{(d_1+r)kr}{A},0\right)$ and has no positive equilibrium. To complete the proof it is sufficient to show that E_1' is a globally asymptotically stable equilibrium of system (9.30). We have that the Jacobian matrixes of system (9.30) at equilibria E_0' and E_1' are

$$J_0 = \begin{pmatrix} r & 0 \\ 0 & -d \end{pmatrix}, \quad J_1 = \begin{pmatrix} -r & * \\ 0 & \frac{kr(ae-bd)-d}{bkr+1} \end{pmatrix}$$

respectively, where $*$ denotes a constant only depending on the parameters. Therefore, E_1' is locally asymptotically stable and E_0' is a saddle whose unstable manifold belongs to y axis and whose stable manifold is z axis. Hence it is impossible that E_0' is in the ω-limit set of positive solutions of (9.30). Moreover, periodic solution does not exit since there is not positive equilibrium. Since system (9.30) is point dissipative. Recall Poincaré–Bendixson theorem [253] which says that given a differentiable real dynamical system defined on an open subset of the plane, then every non-empty compact ω-limit set of an orbit, which contains only finitely many fixed points, is either a fixed point, a periodic orbit, or a connected set composed of a finite number of fixed points together with homoclinic and heteroclinic orbits connecting these. Moreover, there is at most one orbit connecting different fixed points in the same direction. Therefore the Poincaré–Bendixson theorem yields that the ω limit set of positive solution of (9.30) must be equilibrium E_1'. Therefore, E_1' is globally attractive. $\sqcup$

Let us present the persistence theory for infinite dimensional systems from [89], [163], [267].

Let S be a collection of sets and X be a complete metric space. A bounded set $B \subset X$ dissipates S-sets under a continuous map $T : X \to X$ if for any $C \in S$ there is an integer $n_0(C)$ such that $T^n(C) \subset B$ for $n \geq n_0(C)$. If $S = \{\{x\} : x \in X\}$ one says T is point dissipative. If $S = \{J \subset X : J \text{ is compact}\}$ one says T is compact dissipative.

Suppose that $X^0 \subset X$, $X_0 \subset X$, $X^0 \cap X_0 = 0$. Assume that $T(t)$ is a semigroup on X satisfying

$$T(t) : X^0 \to X^0, \quad T(t) : X_0 \to X_0. \tag{9.31}$$

Let $T_b(t)$ be the restriction of $T(t)$ on X_0 and let A_b be the global attractor for $T_b(t)$.

Theorem 9.9. [89] *Suppose that $T(t)$ satisfies (9.31) and*

(i) there is a $t_0 \geq 0$ such that $T(t)$ is compact for $t > t_0$;

(ii) $T(t)$ is point dissipative in X;

(iii) $\tilde{A}_b = \cup_{x \in A_b} \omega(x)$ is isolated and has an acyclic covering $\tilde{M}$, where

$$\tilde{M} = \{M_1, M_2, \ldots, M_n\};$$

(iv) $W^s(M_i) \cap X^0 = $ for $i = 1, \ldots, n$, (where W^s denotes a stable manifold [40])

Then X_0 is a uniform repellor with respect to X^0, i.e., there is an $\epsilon > 0$ such that for any $x \in X^0$,

$$\lim_{t \to +\infty} \inf \, \mathrm{dist}(T(t)x, X_0) \geq \epsilon.$$

Theorem 9.10. *The male population x and female population y in system (9.29) are permanent.*

Proof. It is sufficient to prove that the female population y in system (9.30) is permanent. Denote

$$X_0 = \{(y, z) : y \geq 0, z = 0\}, \quad X^0 = \{(y, z) : y > 0, z > 0\}.$$

By Lemma 9.6 and the proof of Theorem 9.8, since the conditions of Theorem 9.9 are satisfied, one completes the proof. $\quad\square$

Theorem 9.11. *If $\frac{ae-bd}{d} > \frac{1}{rk}$, then system (9.29) is permanent.*

Proof. The proof is similar to that of Theorem 9.10 by finding that the equilibrium E_1' is a saddle if $\frac{ae-bd}{d} > \frac{1}{rk}$. $\quad\square$

Bibliographical notes. The chapter is based on [40], [43], [89], [121], [150], [163], [230], [253], [259], [267]. For further reading the theory of two-sex populations dynamics see[2] [33], [37], [53], [58], [59], [66], [86], [87], [109]–[114], [149], [162], [176], [178], [180], [181], [182], [187], [199], [207] and the references therein.

---

[2] https://en.wikipedia.org/wiki/Population_dynamics.

Chapter 10

Dynamical systems generated by a gonosomal evolution operator

In this chapter we consider dynamical systems generated by gonosomal evolution operators of sex linked inheritance. The main biological system of this chapter is a hemophilia, which is a group of hereditary genetic disorders that impair the body's ability to control blood clotting or coagulation, which is used to stop bleeding when a blood vessel is broken. The evolution of such system is studied by a quadratic gonosomal operator. This operator is considered as a mapping from $\mathbb{R}^n$, $n \geq 2$ to itself. In particular, for a gonosomal operator at $n = 4$ we explicitly give all (two) fixed points. Then limit points of the trajectories of the corresponding dynamical system are studied. In the case $n = 4$, for the normalized gonosomal operator we show uniqueness of fixed point and prove that this point is globally attractive. Biologically the main result of this chapter means that the hemophilia illness asymptotically disappears.

10.1 Bisexual population: gonosomal evolution operator

In many cases, the sex determination is genetic, in particular, it is controlled by two chromosomes called gonosomes. Gonosomal inheritance is a mode of inheritance that is observed for traits related to a gene encoded on the sex chromosomes.

Following [228] we consider an example of sex-linked inheritance. Haemophilia is a lethal recessive X-linked disorder: a female carrying two alleles for hemophilia die. Therefore if we denote by X^h the gonosome X carrying the hemophilia, there are only two female genotypes: XX and XX^h (X^hX^h is lethal) and two male genotypes: XY and X^hY. We have four types of crosses (see[1] Fig. 10.1):

$$XX \times XY \rightarrowtail \tfrac{1}{2}XX, \;\; \tfrac{1}{2}XY;$$

$$XX \times X^hY \rightarrowtail \tfrac{1}{2}XX^h, \;\; \tfrac{1}{2}XY;$$

$$XX^h \times XY \rightarrowtail \tfrac{1}{4}XX, \tfrac{1}{4}XX^h, \tfrac{1}{4}XY, \tfrac{1}{4}X^hY;$$

$$XX^h \times X^hY \rightarrowtail \tfrac{1}{3}XX^h, \tfrac{1}{3}XY, \tfrac{1}{3}X^hY.$$

$$(10.1)$$

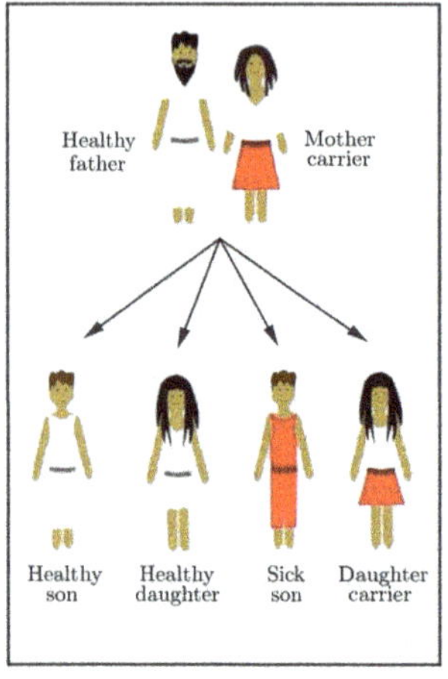

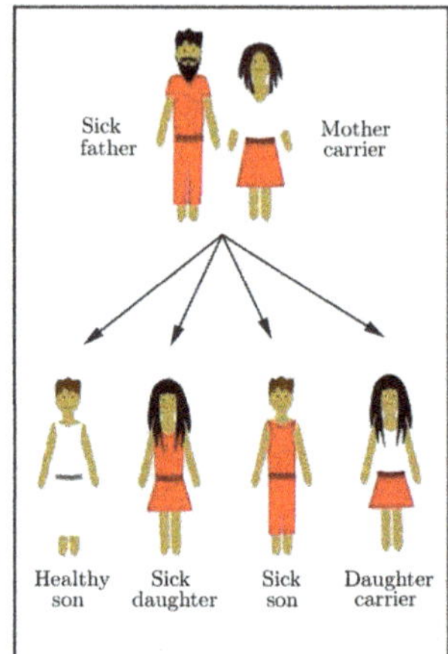

Fig. 10.1   Schematic three types (at least one of them contains X^h) of crosses.

Let $F = \{XX, XX^h\}$ and $M = \{XY, X^hY\}$ be sets of genotypes. Assume state of the set F is given by a real vector (x, y) and state of M by a real vector (u, v). Then a state of $F \cup M$ is given by the vector $s = (x, y, u, v) \in \mathbb{R}^4$. If $s' = (x', y', u', v')$ is a state of the system $F \cup M$ in the next generation then by the rule (10.1) we get the evolution operator $W : \mathbb{R}^4 \to \mathbb{R}^4$ defined by

$$W : \begin{cases} x' = \tfrac{1}{2}xu + \tfrac{1}{4}yu \\[4pt] y' = \tfrac{1}{2}xv + \tfrac{1}{4}yu + \tfrac{1}{3}yv \\[4pt] u' = \tfrac{1}{2}xu + \tfrac{1}{2}xv + \tfrac{1}{4}yu + \tfrac{1}{3}yv \\[4pt] v' = \tfrac{1}{4}yu + \tfrac{1}{3}yv. \end{cases} \qquad (10.2)$$

This example can be generalized: suppose that the set of female types is $F = \{1, 2, \ldots, n\}$ and the set of male types is $M = \{1, 2, \ldots, \nu\}$. Let

---

[1]https://es.123rf.com/photo_57627441_el-mecanismo-de-la-herencia-de-la-hemofilia-infografia-ilustracion-del-vector-en-el-fondo-aislado-.html.

$x = (x_1, \ldots, x_n) \in \mathbb{R}^n$ be a state of F and $y = (y_1, \ldots, y_\nu) \in \mathbb{R}^\nu$ be a state of M.

Consider $\gamma_{ik,j}^{(f)}$ and $\gamma_{ik,l}^{(m)}$ as some inheritance real coefficients (not necessary probabilities) with

$$\sum_{j=1}^{n} \gamma_{ik,j}^{(f)} + \sum_{l=1}^{\nu} \gamma_{ik,l}^{(m)} = 1. \tag{10.3}$$

Consider an evolution operator $W : \mathbb{R}^{n+\nu} \to \mathbb{R}^{n+\nu}$ defined as

$$W : \begin{cases} x'_j = \sum_{i,k=1}^{n,\nu} \gamma_{ik,j}^{(f)} x_i y_k, & j = 1, \ldots, n \\[2mm] y'_l = \sum_{i,k=1}^{n,\nu} \gamma_{ik,l}^{(m)} x_i y_k, & l = 1, \ldots, \nu. \end{cases} \tag{10.4}$$

This operator is called gonosomal evolution operator. This means that the association $s = (x, y) \in \mathbb{R}^{n+\nu} \to s' = (x', y') \in \mathbb{R}^{n+\nu}$ defines a map W. The population evolves by starting from an arbitrary state s, then passing to the state $s' = W(s)$ (in the next 'generation'), then to the state $s'' = W(W(s))$, and so on. Thus, states of the population described by the following discrete-time dynamical system

$$s^{(0)}, \quad s^{(1)} = W(s^{(0)}), \quad s^{(2)} = W^2(s^{(0)}), \quad s^{(3)} = W^3(s^{(0)}), \ldots \tag{10.5}$$

where $s^{(0)} \in \mathbb{R}^{n+\nu}$ is a given initial point and $W^n(s) = \underbrace{W(W(...W(s))...)}_{n}$

denotes the n times iteration of W to s.

The main problem is to describe the limit points of the trajectory $\{s^{(n)}\}_{n=0}^{\infty}$ for arbitrary given $s^{(0)}$.

10.2 Dynamical system generated by the operator (10.2)

Note that operator (10.4) describes evolution of a hemophilia. The dynamical systems generated by gonosomal operator (10.4) is complicated. In this section we study the dynamical system generated by gonosomal operator (10.2), which is a particular case of (10.4), obtained by $n = \nu = 2$ and the following coefficients:

$$\gamma_{11,1}^{(f)} = \tfrac{1}{2} \quad \gamma_{11,2}^{(f)} = 0 \quad \gamma_{11,1}^{(m)} = \tfrac{1}{2} \quad \gamma_{11,2}^{(m)} = 0$$

$$\gamma_{12,1}^{(f)} = 0 \quad \gamma_{12,2}^{(f)} = \tfrac{1}{2} \quad \gamma_{12,1}^{(m)} = \tfrac{1}{2} \quad \gamma_{12,2}^{(m)} = 0$$

$$\gamma_{21,1}^{(f)} = \tfrac{1}{4} \quad \gamma_{21,2}^{(f)} = \tfrac{1}{4} \quad \gamma_{21,1}^{(m)} = \tfrac{1}{4} \quad \gamma_{21,2}^{(m)} = \tfrac{1}{4}$$

$$\gamma_{22,1}^{(f)} = 0 \quad \gamma_{22,2}^{(f)} = \tfrac{1}{3} \quad \gamma_{22,1}^{(m)} = \tfrac{1}{3} \quad \gamma_{22,2}^{(m)} = \tfrac{1}{3}.$$

10.2.1   *Fixed points*

Let us find all fixed points of W given by (10.2), i.e. we solve the following system of equations

$$
W : \begin{cases}
x = \frac{1}{2}xu + \frac{1}{4}yu \\[2mm]
y = \frac{1}{2}xv + \frac{1}{4}yu + \frac{1}{3}yv \\[2mm]
u = \frac{1}{2}xu + \frac{1}{2}xv + \frac{1}{4}yu + \frac{1}{3}yv \\[2mm]
v = \frac{1}{4}yu + \frac{1}{3}yv.
\end{cases}
\tag{10.6}
$$

One can see that $s_0 = (0,0,0,0)$ is a solution to system (10.6).

To find another solution from the first equation of this system we get $(4 - 2u)x = yu$. Assuming $u = 2$ from this equation we get $y = 0$ then the last equation of (10.6) gives $v = 0$ consequently from the third equation of the system we get $x = 2$. Thus we obtained the solution $s_2 = (2,0,2,0)$.

From the last equation we get $(12 - 4y)v = 3yu$, assume first that $y = 3$ then this equation gives $u = 0$, consequently the first equation of the system (10.6) gives $x = 0$. Then from the second equation we get $v = 3$. But these values do not satisfy the third equation of the system. Assume now $u \neq 2$ and $y \neq 3$ then we have

$$
x = \frac{yu}{4 - 2u}, \quad v = \frac{3yu}{12 - 4y}.
\tag{10.7}
$$

Using (10.7) from the second and third equations of the system (10.6) we obtain

$$
\begin{cases}
16(3 - y)(2 - u) = 3u(8 - 4u + yu) \\[2mm]
16(3 - y)(2 - u) = y(24 - yu).
\end{cases}
$$

From the first equation of the last system we find

$$
y = \frac{12(u^2 - 6u + 8)}{3u^2 - 16u + 32}.
$$

Substituting this to the second equation of the last system we obtain the following equation

$$
(u - 2)^2 (u - 8) \left(3u^2 - 14u + 24\right) = 0.
$$

This equation gives $u = 2$, $u = 8$ and $3u^2 - 14u + 24 = 0$. For case $u = 2$ we have solution s_2 mentioned above. The case $u = 8$ gives $y = 3$ which does not give solution of the system (10.6) as was discussed above. Thus only remains $3u^2 - 14u + 24 = 0$ which does not have real solutions.

We proved the following

Proposition 10.1. *The gonosomal operator (10.2) has exactly two fixed points:* $s_0 = (0,0,0,0)$ *and* $s_2 = (2,0,2,0)$.

10.2.2 Dynamics on invariant sets

To find the type of a fixed point of the operator (10.2) we write the Jacobi matrix:

$$J(s) = J_W = \begin{pmatrix} \frac{1}{2}u & \frac{1}{4}u & \frac{1}{2}x + \frac{1}{4}y & 0 \\ \frac{1}{2}v & \frac{1}{4}u + \frac{1}{3}v & \frac{1}{4}y & \frac{1}{2}x + \frac{1}{3}y \\ \frac{1}{2}u + \frac{1}{2}v & \frac{1}{4}u + \frac{1}{3}v & \frac{1}{2}x + \frac{1}{4}y & \frac{1}{2}x + \frac{1}{3}y \\ 0 & \frac{1}{4}u + \frac{1}{3}v & \frac{1}{4}y & \frac{1}{3}y \end{pmatrix}.$$

One can see that $J(s_0)$ has all eigenvalues equal to 0, therefore s_0 is an attracting point.

The Jacobian $J(s_2)$ has eigenvalues $-\frac{1}{2}, 0, 1, 2$, therefore the fixed point is not hyperbolic.

A set A is called invariant with respect to W if $W(A) \subset A$.

Denote

$$O = \{(0, 0, u, v) \in \mathbb{R}^4 : u, v \in \mathbb{R}\} \cup \{(x, y, 0, 0) \in \mathbb{R}^4 : x, y \in \mathbb{R}\},$$
$$I = \{s = (x, y, u, v) \in \mathbb{R}^4 : y = v = 0\},$$
$$J = \{s \in I : x = u\},$$
$$P = \{s = (x, y, u, v) \in \mathbb{R}^4 : x \geq 0, y \geq 0, u \geq 0, v \geq 0\},$$
$$Q_a = \{s = (x, y, u, v) \in P : x + y + u + v \leq a\}, \qquad a \in [0, 4],$$
$$\mathcal{N} = \{s = (x, y, u, v) \in \mathbb{R}^4 : x \leq 0, y \leq 0, u \leq 0, v \leq 0\},$$
$$\mathcal{N}_0 = \{s = (x, y, u, v) \in \mathbb{R}^4 : x \leq 0, y \leq 0, u \geq 0, v \geq 0\},$$
$$\mathcal{N}_1 = \{s = (x, y, u, v) \in \mathbb{R}^4 : x \geq 0, y \geq 0, u \leq 0, v \leq 0\}.$$

Lemma 10.1. *1) The sets I, J, P and Q_a ($a \in [0, 4]$) are invariant with respect to W.*
2) $W(O) = \{(0, 0, 0, 0)\}$.
3) $W(Q_a) \subset Q_{a^2/4}$.
4) $W(\mathcal{N}) \subset P$.
5) $W(\mathcal{N}_0) \subset \mathcal{N}$, $W(\mathcal{N}_1) \subset \mathcal{N}$.

Proof.   We give the proof for Q_a, for other sets it simply follows from (10.2). Take any $s = (x, y, u, v) \in Q_a$ then we have $0 \leq x + y \leq a$ and $0 \leq u + v \leq a$. From (10.2) we get $x' \geq 0$, $y' \geq 0$, $u' \geq 0$, $v' \geq 0$ and

$$x' + y' + u' + v' = (x + y)(u + v) \leq \left(\frac{x + y + u + v}{2} \right)^2 \leq \frac{a^2}{4}.$$

Thus $s' = (x', y', u', v') \in Q_{\frac{a^2}{4}} \subset Q_a$. $\qquad\qquad \square$

Reduce W on J then we get the mapping $x' = f(x) = \frac{1}{2}x^2$. This function has two fixed points $x = 0$ and $x = 2$. Moreover, 0 is attractive ($f'(0) = 0 < 1$) and 2 is repeller ($f'(2) = 2 > 1$). Take an initial point $x_0 \in J$ and iterate the function f, then we get

$$x_n = f^n(x_0) = 2^{-(1+2+2^2\ldots+2^{n-1})}x_0^{2^n} = 2^{-2^n+1}x_0^{2^n} = 2\left(\frac{x_0}{2}\right)^{2^n}.$$

Hence we have

$$\lim_{n\to\infty} x_n = \begin{cases} 0, & \text{if } |x_0| < 2 \\ 2, & \text{if } |x_0| = 2 \\ +\infty, & \text{if } |x_0| > 2. \end{cases}$$

Now reduce the operator W on I:

$$V : \begin{cases} x' = \frac{1}{2}xu \\ u' = \frac{1}{2}xu. \end{cases}$$

Thus for any $t_0 = (x_0, 0, u_0, 0) \in I$ we have $W(x_0, 0, u_0, 0) \in J$. Consequently, we have full characterization of the dynamical system on the invariant set J, i.e., we proved the following

Proposition 10.2. *For any initial point $t_0 = (x_0, 0, u_0, 0) \in I$ we have*

$$\lim_{n\to\infty} W^n(t_0) = \begin{cases} (0,0,0,0) & \text{if } |x_0 u_0| < 4 \\ (2,0,2,0) & \text{if } |x_0 u_0| = 4 \\ +\infty, & \text{if } |x_0 u_0| > 4. \end{cases}$$

Let us now consider the dynamical system on the other sets (which may intersect with J).

Lemma 10.2. *Let $a \in [0,4)$. Then for any initial point $s = (x, y, u, v) \in Q_a$ we have*

$$\lim_{n\to\infty} W^n(s) = (0,0,0,0). \tag{10.8}$$

Proof. Let $f(a) = a^2/4$. By Lemma 10.1 we have

$$W^n(Q_a) \subset W^{n-1}(Q_{f(a)}) \subset W^{n-2}(Q_{f^2(a)}) \subset \cdots \subset Q_{f^n(a)}.$$

One can see that $f(x)$ has two fixed points 0 and 4. Moreover, 0 is attracting point and 4 is repelling point. For any $a \in [0,4)$ we have $\lim_{n\to\infty} f^n(a) = 0$. Consequently, we get

$$\lim_{n\to\infty} W^n(Q_a) \subset Q_0 = \{(0,0,0,0)\}. \qquad \square$$

Lemma 10.3. *For an initial point $s = (x, y, u, v) \in Q_4$ the following hold*

(i) if there is $k \geq 0$ such that $y^{(k)} v^{(k)} \neq 0$ then (10.8) is satisfied,
(ii) if $y^{(k)} v^{(k)} = 0$ for any $k \geq 0$ then

$$\lim_{n \to \infty} W^n(s) = (2, 0, 2, 0), \tag{10.9}$$

where $y^{(k)}$ and $v^{(k)}$ are second and fourth coordinates of the vector $W^k(s)$.

Proof. In the case $a = 4$ we have $0 \leq x + y + u + v \leq 4$. From this inequality it follows that

$$0 \leq x + y \leq 2 \quad \text{or} \quad 0 \leq u + v \leq 2,$$

if both $x + y$ and $u + v$ large than 2 then their sum is large than 4. Without loss of generality we assume that $t = x + y \leq 2$ then $u + v \leq 4 - t$. Hence we have

$$x' + y' + u' + v' = (x + y)(u + v) \leq t(4 - t) = \begin{cases} 4, & \text{if } t = 2 \\ < 4, & \text{if } t < 2 \end{cases}$$

hence $W(s) \in Q_{t(4-t)}$, i.e., for $t < 2$ the case is reduced to the case of $a < 4$. Consider now the case $t = 2$. Then if $u + v < 2$ we can reduce the case to the case $a < 4$. But if

$$t = x + y = 2 \quad \text{and} \quad u + v = 2 \tag{10.10}$$

then from (10.2) we get

$$x' + y' + u' + v' = 4, \quad x' + y' = 2 - \frac{yv}{6} \quad \text{and} \quad u' + v' = 2 + \frac{yv}{6}. \tag{10.11}$$

Consequently, if $yv \neq 0$ then $W^2(s) \in Q_{4-(\frac{yv}{6})^2}$. By (10.10) we have $0 \leq y \leq 2$ and $0 \leq v \leq 2$, hence $0 < 4 - (\frac{yv}{6})^2 < 4$. Thus condition (10.10) together with $yv \neq 0$, by Lemma 10.2 gives (10.8).

Let now $yv = 0$ then (10.11) is reduced to the case (10.10). Repeating above argument we see that if $y'v' \neq 0$ then $W^3(s) \in Q_{4-(\frac{y'v'}{6})^2}$, otherwise we iterate the argument again. By this way one can show that if (10.10) is satisfied and there exists $k \geq 0$ such that $y^{(k)} v^{(k)} \neq 0$ then we have (10.8).

Suppose now (10.10) is satisfied and

$$y^{(k)} v^{(k)} = 0 \quad \text{for any } k \geq 0 \tag{10.12}$$

then similarly to (10.11) we get

$$x^{(n)} + y^{(n)} + u^{(n)} + v^{(n)} = 4,$$

$$x^{(n)} + y^{(n)} = 2, \tag{10.13}$$

$$u^{(n)} + v^{(n)} = 2, \quad \text{for any } n \geq 0.$$

To complete the proof we need the following

Lemma 10.4. *If conditions (10.10) and (10.12) are satisfied then*

$$y^{(k)} = v^{(k)} = 0 \quad \text{for any } k \geq 0.$$

Proof. From (10.2) we get

$$y^{(k+1)} = \tfrac{1}{2}x^{(k)}v^{(k)} + v^{(k+1)}$$
$$v^{(k+1)} = \left(\tfrac{1}{4}u^{(k)} + \tfrac{1}{3}v^{(k)}\right)y^{(k)}.$$

$$(10.14)$$

Now using (10.13) and (10.12) from (10.14) we get

$$y^{(k+1)} = \tfrac{1}{2}(2 - y^{(k)})v^{(k)} + v^{(k+1)} = v^{(k)} + v^{(k+1)}$$
$$v^{(k+1)} = \left(\tfrac{1}{4}(2 - v^{(k)}) + \tfrac{1}{3}v^{(k)}\right)y^{(k)} = \tfrac{1}{2}y^{(k)}.$$

$$(10.15)$$

If $v^{(0)} = 0$ then from the first equation of (10.15) we get $y^{(1)} = v^{(1)} = 0$. Consequently the second equation gives $v^{(2)} = 0$. Then using the first equation we get $y^{(2)} = 0$ and so on, we get $y^{(k)} = v^{(k)} = 0$ for any $k \geq 0$.

If $y^{(0)} = 0$ then the second equation gives $v^{(1)} = 0$. Assume $v^{(0)} = v \neq 0$ then from the first equation we get $y^{(1)} = v + v^{(1)} = v$. Then $v^{(2)} = \tfrac{1}{2}v$. Consequently, $y^{(2)} = \tfrac{1}{2}v$. Now condition $y^{(2)}v^{(2)} = 0$ gives $v = 0$. This completes the proof. $\square$

Now by Lemma 10.4 and property (10.13) we get

$$x^{(n)} = 2 \quad \text{and} \quad u^{(n)} = 2, \quad \text{for any } n \geq 0.$$

This completes the proof Lemma 10.3. $\square$

Lemma 10.5. *If $s = (x, y, u, v) \in P$ is an initial point with $x + y + u + v > 4$, for which*

(a) *if there exists $k \geq 0$ such that $(x^{(k)} + y^{(k)})(u^{(k)} + v^{(k)}) < 4$ then (10.8) is satisfied.*

(b) *if $\max\{\tfrac{xu}{4}, \tfrac{yu}{16}, \tfrac{yv}{9}\} > 1$ then*

$$\lim_{n \to \infty} W^n(s) = \infty, \quad \text{i.e. at least one coordinate of } W^n(s) \text{ goes to } \infty.$$

Proof. (a) This simply follows from the equality

$$x^{(k+1)} + y^{(k+1)} + u^{(k+1)} + v^{(k+1)} = (x^{(k)} + y^{(k)})(u^{(k)} + v^{(k)}).$$

Indeed, from this equality it follows that $W(s^{(k+1)}) \in Q_4$. Since Q_4 is invariant the part (a) follows from Lemma 10.2.

(b) Let us prove it for the case $\max\{\frac{xu}{4}, \frac{yu}{16}, \frac{yv}{9}\} = \frac{xu}{4} > 1$. For other cases the proof is similar. From (10.2) for any $s = (x, y, u, v) \in P$ we get

$$x^{(k+1)} \geq \frac{1}{2} x^{(k)} u^{(k)}, \quad u^{(k+1)} \geq \frac{1}{2} x^{(k)} u^{(k)}, \quad k \geq 0. \tag{10.16}$$

Iterating these inequalities we obtain

$$x^{(k+1)} \geq \frac{1}{2} x^{(k)} u^{(k)} \geq 2^{-(1+2+2^2+\cdots+2^k)} (xu)^{2^k} = 2 \left(\frac{xu}{4}\right)^{2^k}, \quad k \geq 0. \tag{10.17}$$

Similarly

$$u^{(k+1)} \geq 2 \left(\frac{xu}{4}\right)^{2^k}, \quad k \geq 0.$$

This completes the proof. $\qquad\square$

Denote

$$P_0 = \{s = (x, y, u, v) \in P : (x + y)(u + v) < 4\},$$
$$F = \{s = (x, y, u, v) \in P : x + y + u + v > 4, \ \max\{\tfrac{xu}{4}, \tfrac{yu}{16}, \tfrac{yv}{9}\} > 1\}.$$

Summarizing above-mentioned results we get the following

Theorem 10.1. *If $s = (x, y, u, v) \in \mathbb{R}^4$ is such that*

(i) one of the following conditions is satisfied

> *1) $s \in P_0$;*
> *2) $s \in Q_4$ and the condition of part (i) of Lemma 10.3 is hold;*
> *3) $s \in \mathcal{N}, \ W(s) \in P_0$;*
> *4) $s \in \mathcal{N}_0, \ W^2(s) \in P_0$;*
> *5) $s \in \mathcal{N}_1, \ W^2(s) \in P_0$*

> *then*

$$\lim_{n \to \infty} W^n(s) = (0, 0, 0, 0).$$

(ii) one of the following conditions is satisfied

> *a) $s \in F$;*
> *b) $s \in \mathcal{N}, \ W(s) \in F$;*
> *c) $s \in \mathcal{N}_0, \ W^2(s) \in F$;*
> *d) $s \in \mathcal{N}_1, \ W^2(s) \in F$*

> *then*

$$\lim_{n \to \infty} W^n(s) = +\infty.$$

Proof. (i) The case 1) follows from Lemma 10.2 and Lemma 10.5. The case 2) is result of Lemma 10.3. By Lemma 10.1 we have $W(\mathcal{N}) \subset P$, $W^2(\mathcal{N}_0) \subset P$ and $W^2(\mathcal{N}_1) \subset P$. Consequently, parts 3)–5) follow from 1).

Part (ii) is a result of Lemma 10.5 and Lemma 10.1. $\qquad\square$

10.3   A normalized gonosomal operator

Note that the gonosomal operator (10.4) does not map the simplex

$$S^{n+\nu-1} = \left\{ s = (x_1,\ldots,x_n,y_1,\ldots,y_\nu) \in \mathbb{R}^{n+\nu} : \right.$$

$$\left. x_i \geq 0, y_j \geq 0, \sum_{i=1}^{n} x_i + \sum_{j=1}^{\nu} y_j = 1 \right\}$$

to itself, since

$$\sum_{i=1}^{n} x_i' + \sum_{j=1}^{\nu} y_j' = \left(\sum_{i=1}^{n} x_i\right)\left(\sum_{j=1}^{\nu} y_j\right) \tag{10.18}$$

is not equal to 1 in general.

We denote

$$\mathcal{O} = \left\{ s \in S^{n+\nu-1} : (x_1,\ldots,x_n) = (0,\ldots,0) \text{ or } (y_1,\ldots,y_\nu) = (0,\ldots,0) \right\}.$$

$$\mathcal{S}^{n,\nu} = S^{n+\nu-1} \setminus \mathcal{O}.$$

One can see that $W(\mathcal{O}) = \{(0,\ldots,0)\}$. So the points from $\mathcal{O}$ do not give any contribution to the dynamical system generated by W.

Therefore we introduce the normalized gonasomal operator as the following. Consider the coefficients of the operator (10.4) with the following properties

$$\gamma_{ik,j}^{(f)} \geq 0, \quad \gamma_{ik,l}^{(m)} \geq 0,$$

$$\sum_{j=1}^{n} \gamma_{ik,j}^{(f)} + \sum_{l=1}^{\nu} \gamma_{ik,l}^{(m)} = 1, \quad \text{for all } i,k,j,l. \tag{10.19}$$

An normalized evolution operator V, with coefficients (10.19) is defined as

$$V : \begin{cases} x_j' = \dfrac{\sum_{i,k=1}^{n,\nu} \gamma_{ik,j}^{(f)} x_i y_k}{\left(\sum_{i=1}^{n} x_i\right)\left(\sum_{j=1}^{\nu} y_j\right)}, \, j = 1,\ldots,n \\[2ex] y_l' = \dfrac{\sum_{i,k=1}^{n,\nu} \gamma_{ik,l}^{(m)} x_i y_k}{\left(\sum_{i=1}^{n} x_i\right)\left(\sum_{j=1}^{\nu} y_j\right)}, \, l = 1,\ldots,\nu. \end{cases} \tag{10.20}$$

Proposition 10.3. *The operator V defined by (10.20) with coefficients (10.19) maps $\mathcal{S}^{n,\nu}$ to itself if and only if the following condition*

$$\left(\gamma_{ik,1}^{(f)},\ldots,\gamma_{ik,n}^{(f)},\gamma_{ik,1}^{(m)},\ldots,\gamma_{ik,\nu}^{(m)}\right) \in \mathcal{S}^{n,\nu}, \quad \text{for all } i,k. \tag{10.21}$$

is satisfied.

Proof. *Necessity.* Suppose for any $s \in \mathcal{S}^{n,\nu}$ we have $s' = V(s) \in \mathcal{S}^{n,\nu}$ then we shall show that (10.21) is satisfied. Assume that (10.21) is not true, then there is $i_0 \in \{1,\dots,n\}$ and $k_0 \in \{1,\dots,\nu\}$ such that

$$\left(\gamma^{(f)}_{i_0 k_0,1},\dots,\gamma^{(f)}_{i_0 k_0,n}\right) = (0,\dots,0), \tag{10.22}$$

or

$$\left(\gamma^{(m)}_{i_0 k_0,1},\dots,\gamma^{(m)}_{i_0 k_0,\nu}\right) = (0,\dots,0). \tag{10.23}$$

Consider the case (10.22) (the case (10.23) is similar). Take now some $s \in \mathcal{S}^{n,\nu}$ such that $x_{i_0} \neq 0$, $x_i = 0$ for $i \neq i_0$ and $y_{k_0} \neq 0$, $y_k = 0$ for $k \neq k_0$. Then for this s we have

$$x'_j = \frac{\sum_{i,k=1}^{n,\nu} \gamma^{(f)}_{ik,j} x_i y_k}{\left(\sum_{i=1}^{n} x_i\right)\left(\sum_{j=1}^{\nu} y_j\right)} = \frac{\gamma^{(f)}_{i_0 k_0,j} x_{i_0} y_{k_0}}{\left(\sum_{i=1}^{n} x_i\right)\left(\sum_{j=1}^{\nu} y_j\right)} = 0,$$

for all $j = 1,\dots,n$, i.e., $s' \in \mathcal{O}$. This is contradiction to the assumption that $s' \in \mathcal{S}^{n,\nu}$.

Sufficiency. Assume the conditions (10.19) and (10.21) are satisfied, we want to show that if $s \in \mathcal{S}^{n,\nu}$ then $s' = V(s) \in \mathcal{S}^{n,\nu}$. By the construction of the operator (10.20) it is easy to see that $s' \in S^{n+\nu+1}$ so it remains to show that $s' \notin \mathcal{O}$. Assume that $s' \in \mathcal{O}$, i.e., $(x'_1,\dots,x'_n) = (0,\dots,0)$ (the case $(y'_1,\dots,y'_\nu) = (0,\dots,0)$ is similar). Then by (10.20) we should have

$$\sum_{i,k=1}^{n,\nu} \gamma^{(f)}_{ik,j} x_i y_k = 0, \quad \text{for each } j = 1,\dots,n. \tag{10.24}$$

Since $s \in \mathcal{S}^{n,\nu}$ there is $i_0 \in \{1,\dots,n\}$ and $k_0 \in \{1,\dots,\nu\}$ such that $x_{i_0} > 0$ and $y_{k_0} > 0$. From our conditions it follows that $\gamma^{(f)}_{ik,j} x_i y_k \geq 0$, for all i, j, k. Hence from (10.24) we get

$$\gamma^{(f)}_{i_0 k_0,j} x_{i_0} y_{k_0} = 0, \quad \text{for each } j = 1,\dots,n,$$

consequently,

$$\gamma^{(f)}_{i_0 k_0,j} = 0 \text{ for each } j = 1,\dots,n.$$

This is contradiction to the condition (10.21). $\qquad\square$

A fixed point $s = (x_1,\dots,x_n,y_1,\dots,y_\nu)$ of the gonosomal operator (10.4) is called non-negative and normalizeable if all coordinates of this point are non-negative and $\sum_{i=1}^{n} x_i + \sum_{k=1}^{\nu} y_k > 0$.

Proposition 10.4. *There is one-to-one correspondence between non-negative and normalizeable fixed points of (10.4) and all fixed points of (10.20).*

Proof. Let $s = (x_1, \ldots, x_n, y_1, \ldots, y_\nu)$ be a non-negative and normalizeable fixed point of (10.4). Denote $Z = \sum_{i=1}^{n} x_i + \sum_{k=1}^{\nu} y_k$, and consider the point

$$\tilde{s} = (x_1/Z, \ldots, x_n/Z, y_1/Z, \ldots, y_\nu/Z).$$

By (10.18) for the fixed point we have

$$Z = \left(\sum_{i=1}^{n} x_i \right) \left(\sum_{k=1}^{\nu} y_k \right). \tag{10.25}$$

Using formula (10.25) one can see that $\tilde{s}$ is a fixed point of (10.20).

Now let $\tilde{s} = (\tilde{x}_1, \ldots, \tilde{x}_n, \tilde{y}_1, \ldots, \tilde{y}_\nu)$ be a fixed point of (10.20), i.e. it satisfies the following system

$$\begin{cases} \tilde{x}_j = \dfrac{\sum_{i,k=1}^{n,\nu} \gamma_{ik,j}^{(f)} \tilde{x}_i \tilde{y}_k}{\left(\sum_{i=1}^{n} \tilde{x}_i \right) \left(\sum_{j=1}^{\nu} \tilde{y}_j \right)}, \; j = 1, \ldots, n \\[3ex] \tilde{y}_l = \dfrac{\sum_{i,k=1}^{n,\nu} \gamma_{ik,l}^{(m)} \tilde{x}_i \tilde{y}_k}{\left(\sum_{i=1}^{n} \tilde{x}_i \right) \left(\sum_{j=1}^{\nu} \tilde{y}_j \right)}, \; l = 1, \ldots, \nu. \end{cases} \tag{10.26}$$

Denote

$$\tilde{Z} = \left(\sum_{i=1}^{n} \tilde{x}_i \right) \left(\sum_{k=1}^{\nu} \tilde{y}_k \right). \tag{10.27}$$

Dividing both side of (10.26) to $\tilde{Z}$ one can see that the following point is a fixed point of (10.4):

$$s = (\tilde{x}_1/\tilde{Z}, \ldots, \tilde{x}_n/\tilde{Z}, \tilde{y}_1/\tilde{Z}, \ldots, \tilde{y}_\nu/\tilde{Z}). \qquad \square$$

In this section we consider the normalized version of the evolution operator (10.2), i.e.,

$$V : \begin{cases} x' = \dfrac{2xu + yu}{4(x + y)(u + v)} \\[2ex] y' = \dfrac{6xv + 3yu + 4yv}{12(x + y)(u + v)} \\[2ex] u' = \dfrac{6xu + 6xv + 3yu + 4yv}{12(x + y)(u + v)} \\[2ex] v' = \dfrac{3yu + 4yv}{12(x + y)(u + v)}. \end{cases} \tag{10.28}$$

One can see that the operator (10.28) satisfies the conditions of Proposition 10.3, hence $V : \mathcal{S}^{2,2} \to \mathcal{S}^{2,2}$.

Theorem 10.2. *The operator* $V : S^{2,2} \to S^{2,2}$ *given by* (10.28) *has unique nonhyperbolic fixed point* $s_0 = (\frac{1}{2}, 0, \frac{1}{2}, 0)$ *and for any initial point* $s \in S^{2,2}$ *we have*

$$\lim_{m \to \infty} V^m(s) = s_0. \tag{10.29}$$

Proof. This proof is based on [4]. Consider a trajectory

$$s^{(m)} = (x^{(m)}, y^{(m)}, u^{(m)}, v^{(m)}) = V^m(s), \quad m = 0, 1, 2, \dots$$

$\square$

To prove Theorem 10.2 we use the following lemmas.

Lemma 10.6. *Let* $s^{(0)} = (x^{(0)}, y^{(0)}, u^{(0)}, v^{(0)}) \in S^{2,2}$ *be any initial point. Then, for all non-negative integers* m, *it holds that*
(i) $x^{(m+1)} \le u^{(m+1)}$ *and* $v^{(m+1)} \le y^{(m+1)} \le u^{(m+1)}$
(ii) and that

$$\frac{1}{8} \le \frac{u^{(m+1)}}{4(u^{(m+1)} + v^{(m+1)})} \le x^{(m+2)} \le \frac{u^{(m+1)}}{2(u^{(m+1)} + v^{(m+1)})} \le \frac{1}{2},$$

$$0 \le \frac{v^{(m+1)}}{3(u^{(m+1)} + v^{(m+1)})} \le y^{(m+2)} \le \frac{u^{(m+1)} + 2v^{(m+1)}}{4(u^{(m+1)} + v^{(m+1)})} \le \frac{1}{2},$$

$$\frac{1}{4} \le \frac{2x^{(m+1)} + y^{(m+1)}}{4(x^{(m+1)} + y^{(m+1)})} \le u^{(m+2)} \le \frac{3x^{(m+1)} + 2y^{(m+1)}}{6(x^{(m+1)} + y^{(m+1)})} \le \frac{1}{2},$$

$$0 \le \frac{y^{(m+1)}}{4(x^{(m+1)} + y^{(m+1)})} \le v^{(m+2)} \le \frac{y^{(m+1)}}{3(x^{(m+1)} + y^{(m+1)})} \le \frac{1}{3}.$$

Proof. (i) By (10.28), we have

$$\begin{cases} x^{(m+1)} = \dfrac{2x^{(m)}u^{(m)} + y^{(m)}u^{(m)}}{4(x^{(m)} + y^{(m)})(u^{(m)} + v^{(m)})}, \\[2ex] y^{(m+1)} = \dfrac{6x^{(m)}v^{(m)} + 3y^{(m)}u^{(m)} + 4y^{(m)}v^{(m)}}{12(x^{(m)} + y^{(m)})(u^{(m)} + v^{(m)})}, \\[2ex] u^{(m+1)} = \dfrac{6x^{(m)}u^{(m)} + 6x^{(m)}v^{(m)} + 3y^{(m)}u^{(m)} + 4y^{(m)}v^{(m)}}{12(x^{(m)} + y^{(m)})(u^{(m)} + v^{(m)})}, \\[2ex] v^{(m+1)} = \dfrac{3y^{(m)}u^{(m)} + 4y^{(m)}v^{(m)}}{12(x^{(m)} + y^{(m)})(u^{(m)} + v^{(m)})}. \end{cases} \tag{10.30}$$

Consequently, we obtain the following inequalities

$$u^{(m+1)} - x^{(m+1)} = \frac{v^{(m)}(3x^{(m)} + 2y^{(m)})}{6(x^{(m)} + y^{(m)})(u^{(m)} + v^{(m)})} \geq 0,$$

$$u^{(m+1)} - y^{(m+1)} = \frac{x^{(m)}u^{(m)}}{2(x^{(m)} + y^{(m)})(u^{(m)} + v^{(m)})} \geq 0,$$

$$y^{(m+1)} - v^{(m+1)} = \frac{x^{(m)}v^{(m)}}{2(x^{(m)} + y^{(m)})(u^{(m)} + v^{(m)})} \geq 0,$$

which complete the proof of the first part.

(ii) Follows from the first part. $\qquad\square$

Now we make the notations

$$\alpha^{(m)} := \frac{y^{(m+2)}}{x^{(m+2)}}, \quad \beta^{(m)} := \frac{v^{(m+2)}}{u^{(m+2)}}, \quad m = 0, 1, \ldots \qquad (10.31)$$

Lemma 10.7. *For any initial point* $s^{(0)} = (x^{(0)}, y^{(0)}, u^{(0)}, v^{(0)}) \in S^{2,2}$ *and for any nonnegative integer* m *the following hold:*

$$0 \leq \alpha^{(m)} \leq 4, \quad 0 \leq \beta^{(m)} \leq 1 \qquad (10.32)$$

Proof. From the first and second inequalities of the part (ii) of Lemma 10.6, for any initial point $s^{(0)} \in S^{2,2}$ and for any nonnegative integer m we obtain

$$0 = \frac{\min\left\{y^{(m+2)}\right\}}{\max\left\{x^{(m+2)}\right\}} \leq \alpha^{(m)} \leq \frac{\max\left\{y^{(m+2)}\right\}}{\min\left\{x^{(m+2)}\right\}} = 4.$$

By the part (i) of Lemma 10.6 we have the inequality $v^{(m+2)} \leq u^{(m+2)}$ then

$$\beta^{(m)} = \frac{v^{(m+2)}}{u^{(m+2)}} \leq 1.$$

Moreover, from the third and fourth inequalities of the part (ii) of Lemma 10.6 we get the lower bound for $\beta^{(m)}$:

$$\beta^{(m)} \geq \frac{\min\left\{v^{(m+2)}\right\}}{\max\left\{u^{(m+2)}\right\}} = 0.$$

This completes the proof. $\qquad\square$

Next, using the system of equations (10.30), we obtain

$$\alpha^{(m+1)} = \frac{6\beta^{(m)} + 3\alpha^{(m)} + 4\alpha^{(m)}\beta^{(m)}}{6 + 3\alpha^{(m)}},$$

$$\beta^{(m+1)} = \frac{3\alpha^{(m)} + 4\alpha^{(m)}\beta^{(m)}}{6 + 6\beta^{(m)} + 3\alpha^{(m)} + 4\alpha^{(m)}\beta^{(m)}} \qquad (10.33)$$

which yields the nonlinear dynamical system

$$F : \begin{cases} \alpha' = \dfrac{6\beta + 3\alpha + 4\alpha\beta}{6 + 3\alpha}, \\[2mm] \beta' = \dfrac{3\alpha + 4\alpha\beta}{6 + 6\beta + 3\alpha + 4\alpha\beta} \end{cases} \tag{10.34}$$

with the initial point $(\alpha^{(0)}, \beta^{(0)}) \in \Delta$, where

$$\Delta := \{(\alpha, \beta) \in \mathbb{R}^2 : 0 \leq \alpha \leq 4, \ 0 \leq \beta \leq 1\}. \tag{10.35}$$

Note that $(0,0)$ is the unique non-hyperbolic fixed point of F with the eigenvalues $\lambda_1 = 1$, $\lambda_2 = -\frac{1}{2}$. Recall that a fixed point of the operator F is called hyperbolic if its Jacobian at the fixed point has no eigenvalues on the unit circle.

Fixed point $(\alpha, \beta) = (0,0)$ of the dynamical system (10.34) corresponds to the fixed point $s_0 = (\frac{1}{2}, 0, \frac{1}{2}, 0)$ of the dynamical system (10.28).

Lemma 10.8. *For any initial point $(\alpha, \beta) \in \Delta$, we have*
(i) $F(\alpha, \beta) \in \Omega \subset \Delta$, where

$$\Omega = \{(\alpha, \beta) \in \mathbb{R}^2 : 0 \leq \alpha \leq 2, \ 0 \leq \beta \leq 1\};$$

(ii) $\beta' \leq \alpha'$ and $\alpha^{(2)} + \beta^{(2)} \leq \alpha' + \beta'$, where α', β' are defined by (10.34).

Proof. For the proof of the first part it suffices to observe that

$$0 \leq \alpha' = \frac{6\beta + 3\alpha + 4\alpha\beta}{6 + 3\alpha} = 2 - \frac{6(1 - \beta) + 4\alpha(1 - \beta) + (6 - \alpha)}{6 + 3\alpha} \leq 2$$

and that

$$0 \leq \beta' = \frac{3\alpha + 4\alpha\beta}{6 + 6\beta + 3\alpha + 4\alpha\beta} = 1 - \frac{6 + 6\beta}{6 + 6\beta + 3\alpha + 4\alpha\beta} \leq 1,$$

while the first claim of the second part follows by observing that

$$\begin{aligned} \alpha' - \beta' &= \frac{6\beta + 3\alpha + 4\alpha\beta}{6 + 3\alpha} - \frac{3\alpha + 4\alpha\beta}{6 + 6\beta + 3\alpha + 4\alpha\beta} \\[2mm] &= \frac{4(9\beta + 9\beta^2 + 9\alpha\beta + 12\alpha\beta^2 + 3\alpha^2\beta + 4\alpha^2\beta^2)}{3(2 + \alpha)(6 + 6\beta + 3\alpha + 4\alpha\beta)} \geq 0. \end{aligned}$$

The last claim follows from the relations

$$\begin{aligned} &\alpha' + \beta' - (\alpha^{(2)} + \beta^{(2)}) \\[2mm] &= \alpha' + \beta' - \left(\frac{6\beta' + 3\alpha' + 4\alpha'\beta'}{6 + 3\alpha'} + \frac{3\alpha' + 4\alpha'\beta'}{6 + 6\beta' + 3\alpha' + 4\alpha'\beta'} \right) \\[2mm] &\geq \frac{12(\alpha' - \beta') + 6(\alpha'^2 - \beta'^2) + 4\alpha'\beta'(\alpha' - \beta')}{3(2 + \alpha')(6 + 6\beta' + 3\alpha' + 4\alpha'\beta')} \end{aligned}$$

and the fact that $\alpha' \geq \beta'$. $\qquad\square$

Corollary 10.1. *For any initial point $(\alpha, \beta) \in \Delta$, it holds that*

$$0 \leq \beta^{(m)} \leq \alpha^{(m)}, \quad m = 1, 2, \dots \tag{10.36}$$

and that

$$\alpha^{(m+1)} + \beta^{(m+1)} \leq \alpha^{(m)} + \beta^{(m)}, \quad m = 1, 2, \dots \tag{10.37}$$

In particular, the sequence $\{\alpha^{(m)} + \beta^{(m)}\}_{m \geq 1}$ is convergent.

Lemma 10.9. *For any initial point $(\alpha, \beta) \in \Delta$ the following hold*

$$\lim_{m \to \infty} \alpha^{(m)} = \lim_{m \to \infty} \beta^{(m)} = 0. \tag{10.38}$$

Proof. These sequences are bounded by (10.32) therefore Bolzano–Weierstrass theorem[2] ensures the existence of real numbers $a \in [0, 4]$, $b \in [0, 1]$ and subsequences $\{\alpha^{(m_k)}\}_{k \geq}$ and $\{\beta^{(m_k)}\}_{k \geq 1}$ such that

$$\lim_{k \to \infty} \alpha^{(m_k)} = a, \quad \lim_{k \to \infty} \beta^{(m_k)} = b. \tag{10.39}$$

Since the sequence $\{\alpha^{(m)} + \beta^{(m)}\}_{m \geq 1}$ is convergent (see Corollary 10.1), we deduce from (10.39) that

$$\lim_{m \to \infty} \left(\alpha^{(m)} + \beta^{(m)}\right) = \lim_{k \to \infty} \left(\alpha^{(m_k+1)} + \beta^{(m_k+1)}\right) = \lim_{k \to \infty} \left(\alpha^{(m_k)} + \beta^{(m_k)}\right)$$

$$= \lim_{k \to \infty} \alpha^{(m_k)} + \lim_{k \to \infty} \beta^{(m_k)} = a + b. \tag{10.40}$$

Furthermore, (10.36) and (10.39) imply that

$$0 \leq b \leq a. \tag{10.41}$$

Next, in view of (10.33), we can write

$$\alpha^{(m_k+1)} + \beta^{(m_k+1)} = \frac{6\beta^{(m_k)} + 3\alpha^{(m_k)} + 4\alpha^{(m_k)}\beta^{(m_k)}}{6 + 3\alpha^{(m_k)}}$$
$$+ \frac{3\alpha^{(m_k)} + 4\alpha^{(m_k)}\beta^{(m_k)}}{6 + 6\beta^{(m_k)} + 3\alpha^{(m_k)} + 4\alpha^{(m_k)}\beta^{(m_k)}}.$$

Letting $k \to \infty$ and using (10.39), (10.40), we get the equation

$$a + b = \frac{6b + 3a + 4ab}{6 + 3a} + \frac{3a + 4ab}{6 + 6b + 3a + 4ab} \tag{10.42}$$

which can be written, equivalently, as

$$a\big(12(a - b) + 6(a^2 - b^2) + 4ab(a - b) + 6a + 3a^2 + 15ab + 8a^2 b\big) = 0.$$

---

[2]https://en.wikipedia.org/wiki/Bolzano-Weierstrass_theorem.

By the constraint (10.41), it follows that the latter equation has a unique solution

$$(a, b) = (0, 0).$$

In particular, we have

$$\lim_{m \to \infty} \left(\alpha^{(m)} + \beta^{(m)} \right) = 0. \tag{10.43}$$

However, this observation together with the inequality

$$0 \le \beta^{(m)} \le \alpha^{(m)} \le \alpha^{(m)} + \beta^{(m)}, \quad m = 1, 2, \dots$$

imply that both of the sequence $\{\alpha^{(m)}\}_{m \ge 1}$ and $\{\beta^{(m)}\}_{m \ge 1}$ converge to zero as $m \to \infty$. This completes the proof. $\square$

Corollary 10.2. *The result* (10.38) *gives*

$$\lim_{m \to \infty} y^{(m)} = \lim_{m \to \infty} v^{(m)} = 0. \tag{10.44}$$

Proof. From Lemma 10.6(ii) and (10.31) we get

$$0 \le y^{(m+2)} = \alpha^{(m)} x^{(m+2)} \le \frac{1}{2} \alpha^{(m)}, \quad m = 0, 1, 2, \dots$$

$$0 \le v^{(m+2)} = \beta^{(m)} u^{(m+2)} < \frac{1}{2} \beta^{(m)}, \quad m = 0, 1, 2, \dots$$

Hence, (10.44) holds.

On the other hand, in view of (10.30), we have

$$x^{(m+3)} = \frac{2 + \alpha^{(m)}}{4(1 + \alpha^{(m)})(1 + \beta^{(m)})},$$

$$u^{(m+3)} = \frac{6 + 6\beta^{(m)} + 3\alpha^{(m)} + 4\alpha^{(m)}\beta^{(m)}}{12(1 + \alpha^{(m)})(1 + \beta^{(m)})},$$

implying the convergence of the sequences $\{x^{(m)}\}_{m \ge 1}$ and $\{u^{(m)}\}_{m \ge 1}$ with

$$\lim_{m \to \infty} x^{(m)} = \lim_{m \to \infty} u^{(m)} = \frac{1}{2}. \tag{10.45}$$

Now (10.29) follows from (10.44) and (10.45). This completes proof of Theorem 10.2. $\square$

Biological interpretations: Let $s = (x, y, u, v) \in \mathcal{S}^{2,2}$ be an initial state (the probability distribution on the set $\{XX, XX^h; XY, X^hY\}$ of genotypes). Theorem 10.2 says that, as a rule, the population tends to the equilibrium state $s_0 = (1/2, 0, 1/2, 0)$ with the passage of time, i.e. the future of the population is stable: genotypes XX and XY are survived always, but the genotypes XX^h and X^hY (therefore hemophilia) will disappear in the future. It follows that hemophilia is maintained in a population only if it occurs mutations on the genes coding for the coagulation factors.

Bibliographical notes. The chapter is based on [4], [40], [68], [228] and many internet sources. Some generalizations of the hemophilia model can be read in [5] and [263], see also [36], [146], [250], [265].

Chapter 11

Dynamical system and evolution algebra of mosquito population

This chapter is devoted to a discrete-time dynamical system, generated by an evolution operator of a mosquito population. Under some conditions on the parameters of the system it has two saddle fixed points. We construct an evolution algebra, taking its matrix of structural constants equal to the Jacobian of the evolution operator at a fixed point. Idempotent and absolute nilpotent elements, simplicity properties, and some limit points of the evolution operator corresponding to the evolution algebra are studied.

11.1   The model of mosquito population

There are many papers devoted to several kind of mathematical models of mosquito population dynamics (see for example [35], [52], [148], [245] and the references therein).

In this chapter we consider a discrete-time dynamical system, generated by an evolution operator of a mosquito population. This is a six-dimensional quadratic dynamical system.

Following [154] we give a mathematical model of mosquito dispersal. It is known[1] that all mosquito species life cycle has four distinct stages:

(i) Egg (denote by E the variable) — hatches when exposed to water.

(ii) Larva (L) — the stage after eggs hatch, lives in water.

(iii) Pupa (P) — "tumbler" does not feed; stage just before emerging as adult.

(iv) Adult (A) — flies short time after emerging and after its body parts have hardened.

---

[1]Source http://www.mosquito.org/page/lifecycle.

379

The stages (i)–(iii) occur in water, but the adult is an active flying insect. Only the female mosquito bites and feeds on human or animal blood.

The female mosquito obtains a blood meal, and after that lays the eggs directly on or near water.

The eggs can survive dry conditions for a few months. The eggs hatch in water and a mosquito larva or "wriggler" emerges. The length of time to hatch depends on water temperature, food and type of mosquito.

The larva lives in the water, feeds and develops into the third stage of the life cycle called, a pupa or "tumbler." The pupa also lives in the water but no longer feeds. It takes from two days to a week for a mosquito to emerge from a pupa.

The life cycle usually takes up two weeks, but depending on conditions, it can range from four days to as long as a month.

The adult mosquito emerges onto the water's surface and flies away, ready to begin its life cycle. The adults are classified as host seeking adults (A_h), resting adults (A_r), and oviposition or breeding site seeking adults (A_o).

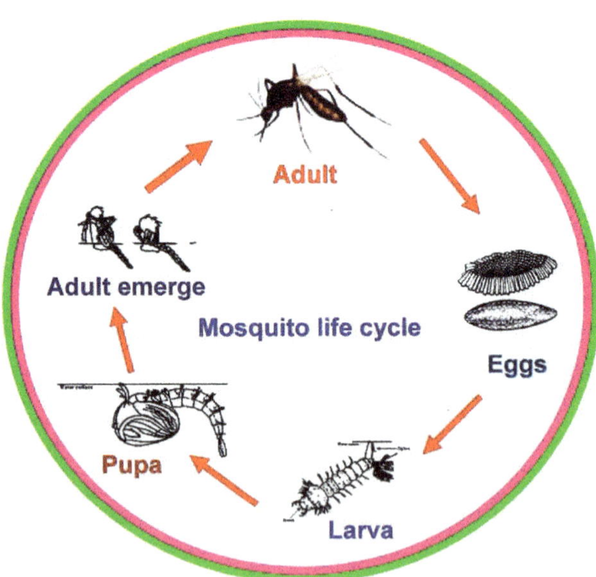

Fig. 11.1   The life cycle of mosquito. Source: internet.

Thus at time moment $t \geq 0$ the state of the population is given by the 6-dimensional non-negative density vector $(E(t),\ L(t),\ P(t),\ A_h(t),\ A_r(t),\ A_o(t))$. The following system of differential equations describes mosquito dynamics (see [154]):

$$
\begin{aligned}
\frac{dE}{dt} &= b\rho_{A_o}A_o - (\mu_E + \rho_E)E, \\[4pt]
\frac{dL}{dt} &= \rho_E E - (\mu_{1L} + \mu_{2L}L + \rho_L)L, \\[4pt]
\frac{dP}{dt} &= \rho_L L - (\mu_P + \rho_P)P, \\[4pt]
\frac{dA_h}{dt} &= \rho_P P + \rho_{A_o}A_o - (\mu_{A_h} + \rho_{A_h})A_h, \\[4pt]
\frac{dA_r}{dt} &= \rho_{A_h}A_h - (\mu_{A_r} + \rho_{A_r})A_r, \\[4pt]
\frac{dA_o}{dt} &= \rho_{A_r}A_r - (\mu_{A_o} + \rho_{A_o})A_o,
\end{aligned}
\tag{11.1}
$$

with initial conditions $E(0), L(0), P(0), A_h(0), A_r(0), A_o(0)$, the description and values of parameters of this system of equations are given in the following table (see [154] and references therein).

In [154] the following results were proved

Theorem 11.1. *The continuous-time dynamical system (11.1) has the following properties*

- *If $(E(0), L(0), P(0), A_h(0), A_r(0), A_o(0)) \in \mathbb{R}_+^6$ then the system (11.1) has a unique solution $(E(t), L(t), P(t), A_h(t), A_r(t), A_o(t)) \in \mathbb{R}_+^6$, for and $t \geq 0$.*
- *There are two equilibrium points $P_0 = (0,0,0,0,0,0)$ (called mosquito-free) and $P_e = (E^*, L^*, P^*, A_h^*, A_r^*, A_0^*)$ (called persistent).*
- *The mosquito-free (resp. persistent) equilibrium is locally stable when $R_0 < 1$ (resp. $R_0 > 1$) and unstable if $R_0 > 1$ (resp. $R_0 < 1$), where R_0 is population reproduction number (which depends on all parameters of the model).*

In this chapter following [229] we study discrete-time dynamical systems and evolution algebras corresponding to the system of differential equations (11.1).

Mosquito control manages the population of mosquitoes to make sure that there is a limit value for the population. This is a particular case of a limited population model (see e.g. pages 5–6 [40]). It is given by the logistic map (a quadratic mapping), often cited as an archetypal example of how complex, chaotic behavior can arise from very simple non-linear discrete-time dynamical system. However, the continuous-time analog of

Table 11.1   The description and values of parameters. Denote by $\mathbb{R}_+^6$ the set of 6-dimensional vectors with non-negative coordinates.

Parameter	Description	Units	Range
b	number of female eggs laid per oviposition	-	50–300
ρ_E	egg hatching rate into larvae	day^{-1}	0.33–1.0
ρ_L	rate at which larvae develop into pupae	day^{-1}	0.08–0.17
ρ_P	rate at which pupae develop into adult/emergence rate	day^{-1}	0.33–1.0
μ_E	egg mortality rate	day^{-1}	0.32–0.80
μ_{1L}	density-independent larvae mortality rate	day^{-1}	0.30–0.58
μ_{2L}	density-dependent larvae mortality rate	day^{-1}, mosq.$^{-1}$	0.0–1.0
μ_P	pupae mortality rate	day^{-1}	0.22–0.52
ρ_{A_h}	rate at which host seeking mosquitoes enter the resting state	day^{-1}	0.322–0.598
ρ_{A_r}	rate at which resting mosquitoes enter oviposition site searching state	day^{-1}	0.30–0.56
ρ_{A_o}	oviposition rate	day^{-1}	3.0–4.0
μ_{A_h}	mortality rate of mosquitoes of searching for hosts	day^{-1}	0.125–0.233
μ_{A_r}	mortality rate of resting mosquitoes	day^{-1}	0.0034–0.01
μ_{A_o}	mortality rate of mosquitoes searching for oviposition sites	day^{-1}	0.41–0.56

this quadratic dynamical system has no cyclic behavior or other fluctuations in the population. This example, showing that continuous and discrete time dynamics of the same system may be essentially different, motivates us to consider discrete-time version of the system (11.1).

11.2   Discrete-time dynamics of mosquito populations

To simplify notations, we denote

$$
\begin{aligned}
&H = A_h,\ R = A_r,\ O = A_o,\ \theta = \rho_O,\ e = \rho_E,\\
&a = \rho_L,\ p = \rho_P,\ h = \rho_H,\ r = \rho_R,\\
&\hat{e} = \mu_E + \rho_E,\ \hat{l}_1 = \mu_{1L} + \rho_L,\ \hat{l}_2 = \mu_{2L},\ \hat{p} = \mu_P + \rho_P,\\
&\hat{h} = \mu_H + \rho_H,\ \hat{r} = \mu_R + \rho_R,\ \hat{\theta} = \mu_O + \rho_O.
\end{aligned}
\tag{11.2}
$$

Then discrete-time version of (11.1) has the following form

$$E_{n+1} = b\theta O_n + (1 - \hat{e})E_n,$$

$$L_{n+1} = eE_n + \left(1 - \hat{l}_1 - \hat{l}_2 L_n\right) L_n,$$

$$P_{n+1} = aL_n + (1 - \hat{p})P_n,$$

$$H_{n+1} = pP_n + \theta O_n + (1 - \hat{h})H_n, \tag{11.3}$$

$$R_{n+1} = hH_n + (1 - \hat{r})R_n,$$

$$O_{n+1} = rR_n + (1 - \hat{\theta})O_n,$$

where n is a non-negative integer number and

$$(E_n, L_n, P_n, H_n, R_n, O_n) = (E(n), L(n), P(n), H(n), R(n), O(n)).$$

Consider operator $M : \mathbb{R}_+^6 \to \mathbb{R}_6^+$, $v = (E, L, P, H, R, O) \to v' = M(v) = (E', L', P', H', R', O')$, defined by

$$M : \begin{cases} E' = b\theta O + (1 - \hat{e})E, \\[2mm] L' = eE + \left(1 - \hat{l}_1\right) L - \hat{l}_2 L^2, \\[2mm] P' = aL + (1 - \hat{p})P, \\[2mm] H' = pP + \theta O + (1 - \hat{h})H, \\[2mm] R' = hH + (1 - \hat{r})R, \\[2mm] O' = rR + (1 - \hat{\theta})O. \end{cases} \tag{11.4}$$

By Table 11.1 and (11.2) we can see that some parameters of the operator (11.4) are strongly positive, but others may be negative and non-negative too.

Then the dynamical system can be written as $v_{n+1} = M(v_n) = \underbrace{M(M(\dots M(v_0))\dots)}_{n \text{ times}}$, $n \geq 0$. We are interested in investigating the limit $\lim_{n \to \infty} v_n$, for any initial condition $v_0 \in \mathbb{R}_+^6$.

11.2.1 *Fixed points of M*

A fixed point v is solution to $v = M(v)$, i.e. the solution of the following system

$$
\begin{cases}
E = b\theta O + (1 - \hat{e})E, \\
L = eE + \left(1 - \hat{l}_1\right)L - \hat{l}_2 L^2, \\
P = aL + (1 - \hat{p})P, \\
H = pP + \theta O + (1 - \hat{h})H, \\
R = hH + (1 - \hat{r})R, \\
O = rR + (1 - \hat{\theta})O,
\end{cases}
\Rightarrow
\begin{cases}
b\theta O - \hat{e}E = 0, \\
eE - \hat{l}_1 L - \hat{l}_2 L^2 = 0, \\
aL - \hat{p}P = 0, \\
pP + \theta O - \hat{h}H = 0, \\
hH - \hat{r}R = 0, \\
rR - \hat{\theta}O = 0.
\end{cases}
\tag{11.5}
$$

The following proposition describes all fixed points.

Proposition 11.1. *The operator M has two fixed points:*

$$
P_0 = (0,0,0,0,0,0), \quad P_1 = (E^*, L^*, P^*, H^*, R^*, O^*),
$$

where

$$
E^* = \frac{b\theta}{\hat{e}}O^*, \quad L^* = CO^*, \quad P^* = \frac{a}{\hat{p}}CO^*, \quad H^* = \frac{\hat{r}}{r}\frac{\hat{\theta}}{h}O^*, \quad R^* = \frac{\hat{\theta}}{r}O^*,
$$

$$
C = \frac{\theta}{a}\frac{\hat{p}}{p}\left(\frac{\hat{h}}{h}\frac{\hat{r}}{r}\frac{\hat{\theta}}{\theta} - 1\right), \quad O^* = \frac{eb\theta - \hat{e}\hat{l}_1 C}{\hat{l}_1 \hat{l}_2 C^2}.
$$

Proof. Since equations of the system (11.5) are linear (except the second one), the proof consists simple computations. $\qquad\square$

To find the type of a fixed point of the operator (11.4) we write the Jacobi matrix:

$$
J(v) = J_M(v) =
\begin{pmatrix}
1 - \hat{e} & 0 & 0 & 0 & 0 & b\theta \\
e & 1 - \hat{l}_1 - 2\hat{l}_2 L & 0 & 0 & 0 & 0 \\
0 & a & 1 - \hat{p} & 0 & 0 & 0 \\
0 & 0 & p & 1 - \hat{h} & 0 & \theta \\
0 & 0 & 0 & h & 1 - \hat{r} & 0 \\
0 & 0 & 0 & 0 & r & 1 - \hat{\theta}
\end{pmatrix}.
\tag{11.6}
$$

The equation for the eigenvalues is $\det(J(v) - \lambda I) = 0$ which is[2]

$$\det(J(v) - \lambda I) = (1 - \hat{e} - \lambda)(1 - \hat{l}_1 - 2\hat{l}_2 L - \lambda)(1 - \hat{p} - \lambda)$$

$$\times \left[(1 - \hat{h} - \lambda)(1 - \hat{r} - \lambda)(1 - \hat{\theta} - \lambda) + hr\theta\right] - abehpr\theta = 0. \qquad (11.7)$$

This is a polynomial equation of order 6, and can not be solved, in general. But one can have numerical solutions by using Maple, for concrete given parameters $a, b, e, \hat{e}, h, \hat{h}, p, \hat{p}, r, \hat{r}, \theta, \hat{\theta}, \hat{l}_1, \hat{l}_2, \epsilon$, therefore we do not list them here. After having some values of eigenvalues one can give condition on absolute values of them to satisfy Definition 4.16.

For example, we take 'baseline' values of parameters mentioned in [154]:

$$a = 0.14, \quad b = 100, \quad e = 0.5, \quad \hat{e} = 1.06, \quad h = 0.46, \quad \hat{h} = 0.64, \quad p = 0.5,$$

$$\hat{p} = 0.87, r = 0.43, \quad \hat{r} = 0.4343, \quad \theta = 3, \quad \hat{\theta} = 3.41, \quad \hat{l}_1 = 0.58, \quad \hat{l}_2 = 0.05. \qquad (11.8)$$

Then for the case $L = 0$ the equation (11.7) has the form:

$$(0.06 + \lambda)(0.42 - \lambda)(0.13 - \lambda)$$

$$\times \left[(0.36 - \lambda)(0.5657 - \lambda)(2.41 + \lambda) - 0.5934\right] - 2.0769 = 0.$$

Maple gives the following six solutions of the last equation:

$$\lambda_1 = 1.256222386, \quad \lambda_2 = -2.35164464,$$

$$\lambda_3 = 0.611373493 + 0.7919408816i, \quad \lambda_4 = 0.611373493 - 0.7919408816i,$$

$$\lambda_5 = -0.5608123659 + 0.6228748264i, \quad \lambda_6 = -0.5608123659 - 0.6228748264i.$$

We have

$$|\lambda_1| = 1.256222386, \quad |\lambda_2| = 2.35164464, \quad |\lambda_3| = |\lambda_4| = 1.000947908,$$

$$|\lambda_5| = |\lambda_6| = 0.7024835591.$$

Therefore for the baseline parameters the fixed point P_0 is a saddle point.

The case of $L = L^*$: (see Proposition 11.1), for baseline parameters we have $L^* = 209.2580821$. Putting this value in the equation (11.7) (in case of baseline parameters) we obtain the following equation

$$(0.06 + \lambda)(20.50580821 + \lambda)(0.13 - \lambda)$$

$$\times \left[(0.36 - \lambda)(0.5657 - \lambda)(2.41 + \lambda) - 0.5934\right] + 2.0769 = 0.$$

---

[2]This equation is given in page 201 of [154] too, but there the non-zero term $-abehpr\theta$ is missed.

Using Maple we get the following six solutions:
$$\lambda_1 = 0.936104284, \quad \lambda_2 = -0.3491645453,$$
$$\lambda_3 = -2.331267091, \quad \lambda_4 = -20.50580883,$$
$$\lambda_5 = 0.1650139855 - 0.3234822044i, \quad \lambda_6 = 0.1650139855 + 0.3234822044i.$$
Hence $|\lambda_1| < 1$, $|\lambda_2| < 1$, $|\lambda_5| = |\lambda_6| = 0.131870352 < 1$, but $|\lambda_3| > 1$, $|\lambda_4| > 1$. Therefore for the baseline parameters the fixed point P_1 is a saddle point.

From the known theorem about stable and unstable manifolds (see [40] and [76]) we get the following result

Proposition 11.2. *If parameters of the operator M given by (11.4) are as in (11.8) then*

a. *There is a two-dimensional (resp. four-dimensional) smooth manifold, denoted by $W^s(P_0)$ (resp. $W^s(P_1)$) such that for any initial vector $v \in W^s(P_i)$ one has $M(v) \in W^s(P_i)$ (invariance) and*
$$\lim_{n\to\infty} M^n(v) = P_i, \, i = 0, 1.$$

b. *There is a four-dimensional (resp. two-dimensional) smooth manifold, denoted by $W^u(P_0)$ (resp. $W^u(P_1)$) and a neighborhood $V(P_i)$ of P_i such that for any initial vector $v \in W^u(P_i) \cap V(P_i)$, there exists $k = k(v) \in N$ that $M^k(v) \notin V(P_i)$, $i = 0, 1$.*

The set $W^s(P_i)$ is known as a stable manifold and $W^u(P_i)$ is an unstable manifold (see [56] for notations of stable manifold, stable eigenspace etc.)

Using linearity of five coordinates of the operator (11.4) one can solve the equation $M(M(v)) = v$, which gives 2-periodic points of the dynamical system. Since there are 15 parameters the solutions have very long formulas, but computer analysis will be helpful to see that for some parameters there are two-periodic points different from the fixed points.

11.3 Evolution algebras of mosquito population

11.3.1 *Definitions*

Recall the definition of an evolution algebra: Let $(\mathcal{E}, \cdot)$ be an algebra over a field $\mathbb{K}$ ($\mathbb{R}$ or $\mathbb{C}$). If $\mathcal{E}$ admits a basis $\{e_1, e_2, \dots\}$, such that
$$e_i \cdot e_j = \begin{cases} 0, & \text{if } i \neq j; \\ \sum_k a_{ik} e_k, & \text{if } i = j, \end{cases}$$

then this algebra is called an *evolution algebra* [254], [26]. The basis is called a natural basis. Denote by $\mathbf{A} = (a_{ij})$ the matrix of the structural constants of the evolution algebra $\mathcal{E}$.

For an evolution algebra $\mathcal{E}$ and $k \geq 1$ we introduce the following sequence

$$\mathcal{E}^k = \sum_{i=1}^{k-1} \mathcal{E}^i \mathcal{E}^{k-i}, \tag{11.9}$$

where $\mathcal{E}^2 := \{a \cdot b : a, b \in \mathcal{E}\}$.

Since $\mathcal{E}$ is a commutative algebra we obtain

$$\mathcal{E}^k = \sum_{i=1}^{\lfloor k/2 \rfloor} \mathcal{E}^i \mathcal{E}^{k-i},$$

where $\lfloor x \rfloor$ denotes the integer part of x.

Definition 11.1. An evolution algebra $\mathcal{E}$ is called nilpotent if there exists some $n \in \mathbb{N}$ such that $\mathcal{E}^n = 0$. The smallest n such that $\mathcal{E}^n = 0$ is called the index of nilpotency.

Define an algebra structure on the vector space $\mathbb{R}^6$ which is closely related to the map defined by (11.4). Let $\mathbb{E}_\epsilon \equiv \mathbb{E}_{\epsilon,M}$ be a 6-dimensional evolution algebra over the set of real numbers, with the natural basis $\{e_1, ..., e_6\}$ and multiplication table $e_i e_j = 0$ if $i \neq j$,

$$e_1^2 = (1 - \hat{e})e_1 + b\theta e_6,$$

$$e_2^2 = ee_1 + (1 - \hat{l}_1 - 2\hat{l}_2\epsilon)e_2,$$

$$e_3^2 = ae_2 + (1 - \hat{p})e_3,$$

$$e_4^2 = pe_3 + (1 - \hat{h})e_4 + \theta e_6, \tag{11.10}$$

$$e_5^2 = he_4 + (1 - \hat{r})e_5,$$

$$e_6^2 = re_5 + (1 - \hat{\theta})e_6,$$

where all parameters coincide with parameters of the operator (11.4) and $\epsilon \in \{0, L^*\}$, with L^* is defined in Proposition 11.1. Thus the matrix of structural constants $\mathbf{A}_\epsilon - (a_{ij})$ of this algebra $\mathbb{E}_\epsilon$ coincides with Jacobi matrix defined in (11.6) calculated at fixed points (at P_0 for $\epsilon = 0$, at P_1 for $\epsilon = L^*$). Namely, we consider two evolution algebras $\mathbb{E}_\epsilon$, where $\epsilon = 0, L^*$ respectively, corresponding to matrices $\mathbf{A}_\epsilon$ of structural constants given by

$$\mathbf{A}_0 = J_M(P_0), \quad \mathbf{A}_{L^*} = J_M(P_1), \tag{11.11}$$

where P_i is fixed point given in Proposition 11.1.

By formula (11.10), for any two vectors $x = (x_1, \ldots, x_6)$, $y = (y_1, \ldots, y_6) \in \mathbb{E}_\epsilon$ we get the following multiplication

$$xy = \Big((1-\hat{e})x_1y_1 + ex_2y_2,\ (1-\hat{l}_1 - 2\hat{l}_2\epsilon)x_2y_2 + ax_3y_3,\ (1-\hat{p})x_3y_3 + px_4y_4,$$

$$(1-\hat{h})x_4y_4 + hx_5y_5,\ (1-\hat{r})x_5y_5 + rx_6y_6,\ (1-\hat{\theta})x_6y_6 + b\theta x_1y_1 + \theta x_4y_4 \Big).$$

$$(11.12)$$

11.3.2 *Idempotent and absolute nilpotent elements*

Note that x^2, by (11.12) has the form

$$x^2 = \Big((1-\hat{e})x_1^2 + ex_2^2,\ (1-\hat{l}_1 - 2\hat{l}_2\epsilon)x_2^2 + ax_3^2,\ (1-\hat{p})x_3^2 + px_4^2,$$

$$(1-\hat{h})x_4^2 + hx_5^2,\ (1-\hat{r})x_5^2 + rx_6^2,\ (1-\hat{\theta})x_6^2 + b\theta x_1^2 + \theta x_4^2 \Big). \qquad (11.13)$$

Proposition 11.3. *The algebra $\mathbb{E}_\epsilon$ has unique absolute nilpotent element* $x = (0, \ldots, 0)$.

Proof. We have

$$\det(\mathbf{A}_\epsilon) = (1-\hat{e})(1-\hat{l}_1 - 2\hat{l}_2\epsilon)(1-\hat{p})[(1-\hat{h})(1-\hat{r})(1-\hat{\theta}) + hr\theta] - abehpr\theta.$$

$$(11.14)$$

One can see that if $\det(\mathbf{A}_\epsilon) \neq 0$ then the absolute nilpotent element is unique for the algebra $\mathbb{E}_\epsilon$, i.e. $x^2 = 0$ has unique solution $x = (0, \ldots, 0)$. In case $\det(\mathbf{A}_\epsilon) = 0$ we use Table 11.1 (see also (11.2)), from which we have that $h > 0$ and $\hat{h} < 1$, therefore from the 4-th equation of $x^2 = 0$, i.e., $(1-\hat{h})x_4^2 + hx_5^2 = 0$ we get $x_4 = x_5 = 0$, then consequently, by the direction $5 \to 6 \to 1 \to 2 \to 3$ we get $x_6 = x_1 = x_2 = x_3 = 0$. $\qquad \square$

Lemma 11.1. *Let $\mathbf{A}_\epsilon$ be the structure matrix of the evolution algebra $\mathbb{E}_\epsilon$*

- *If $(1-\hat{e})(1-\hat{l}_1 - 2\hat{l}_2\epsilon)(1-\hat{p}) = 0$ then $\det(\mathbf{A}_\epsilon) \neq 0$.*
- $\mathrm{rank}(\mathbf{A}_\epsilon) \in \{5, 6\}$.

Proof. The first part follows from the formula (11.14) knowing that $abehpr\theta \neq 0$. For second part, since

$$\mathbf{A}_\epsilon = \begin{pmatrix} 1-\hat{e} & 0 & 0 & 0 & 0 & b\theta \\ e & 1-\hat{l}_1-2\hat{l}_2 L & 0 & 0 & 0 & 0 \\ 0 & a & 1-\hat{p} & 0 & 0 & 0 \\ 0 & 0 & p & 1-\hat{h} & 0 & \theta \\ 0 & 0 & 0 & h & 1-\hat{r} & 0 \\ 0 & 0 & 0 & 0 & r & 1-\hat{\theta} \end{pmatrix}$$

and the parameters e, a, o, h, r are non-zero, it follows that the corresponding columns are linearly independents and hence $\text{rank}(\mathbf{A}_\epsilon) \geq 5$. This completes the proof. $\square$

For the idempotent elements we should solve the following system

$$\begin{aligned} x_1 &= (1-\hat{e})x_1^2 + ex_2^2, \\ x_2 &= (1-\hat{l}_1-2\hat{l}_2\epsilon)x_2^2 + ax_3^2, \\ x_3 &= (1-\hat{p})x_3^2 + px_4^2, \\ x_4 &= (1-\hat{h})x_4^2 + hx_5^2, \\ x_5 &= (1-\hat{r})x_5^2 + rx_6^2, \\ x_6 &= (1-\hat{\theta})x_6^2 + b\theta x_1^2 + \theta x_4^2. \end{aligned} \tag{11.15}$$

By Table 11.1 and (11.2) we have that

$$1-\hat{h} > 0, \quad 1-\hat{r} > 0, \quad 1-\hat{\theta} < 0.$$

Lemma 11.2. *For the system given in (11.15), the following assertions hold*

(1) If $x_i = 0$ for some $i = 1, 2, \ldots, 6$ in (11.15) then $x_1 = \cdots = x_6 = 0$.
(2) Let $1-\hat{e} > 0$, $1-\hat{p} > 0$, $1-\hat{l}_1-2\hat{l}^2\epsilon > 0$. If $(x_1, \ldots, x_6)$ is a solution to (11.15) then

$$x_1 \in \left(0, \frac{1}{1-\hat{e}}\right], \quad x_2 \in \left(0, \frac{1}{1-\hat{l}_1-2\hat{l}_2\epsilon}\right], \quad x_3 \in \left(0, \frac{1}{1-\hat{p}}\right],$$

$$x_4 \in \left(0, \frac{1}{1-\hat{h}}\right], \quad x_5 \in \left(0, \frac{1}{1-\hat{r}}\right], \quad x_6 \in \left(-\infty, \frac{1}{1-\hat{\theta}}\right] \cup (0, +\infty).$$

(3) If $1 - \hat{e} < 0$ or $1 - \hat{p} < 0$ or $1 - \hat{l}_1 - 2\hat{l}^2\epsilon < 0$ then the corresponding x_i may be positive and negative (as x_6).

Proof. 1. Straightforward.

2. From the first equation of (11.15) we have

$$x_1 - (1 - \hat{e})x_1^2 = ex_2^2 \geq 0, \quad \text{i.e.} \quad x_1 \in \left[0, \frac{1}{1 - \hat{e}}\right].$$

Other parts of the lemma can be obtained similarly, taking into account the sign of the parameters. $\square$

For t such that $t - at^2 \geq 0$ we introduce the following function

$$f_{a,b}(t) = \left(\frac{t - at^2}{b}\right)^{1/2}, \quad a \in \mathbb{R}, \quad b > 0.$$

The following lemma reduces the nonlinear system (11.15) of six unknown to equations with only one unknown.

Lemma 11.3. *If $(x_1, \ldots, x_6)$ is a solution to (11.15) then x_1 satisfies one of the following equations:*

$$\pm g(x_1) = (1 - \hat{\theta})(g(x_1))^2 + b\theta x_1^2 + \theta(\varphi(x_1))^2, \tag{11.16}$$

where

$$g(x) = f_{1-\hat{r},r}(f_{1-\hat{h},h}(\varphi(x))),$$

and[3]

$$\varphi(x) = f_{1-\hat{p},p}(\pm f_{1-\hat{l}_1-2\hat{l}_2\epsilon,a}(\pm f_{1-\hat{e},e}(x))).$$

Proof. From the first equation of (11.15) we have

$$x_2 = \pm f_{1-\hat{e},e}(x_1) \tag{11.17}$$

and from the second equation we have

$$x_3 = \pm f_{1-\hat{l}_1-2\hat{l}_2\epsilon,a}(x_2). \tag{11.18}$$

Note that the sign $\pm$ in (11.17) depends on the values of $1 - \hat{l}_1 - 2\hat{l}_2\epsilon$: if the last number is positive then $x_2 \in [0, \frac{1}{1-\hat{l}_1-2\hat{l}_2\epsilon}]$; if $1 - \hat{l}_1 - 2\hat{l}_2\epsilon < 0$ then $x_2 \in (-\infty, \frac{1}{1-\hat{l}_1-2\hat{l}_2\epsilon}] \cup [0, +\infty)$; but if $1 - \hat{l}_1 - 2\hat{l}_2\epsilon = 0$ then $x_2 \geq 0$. The sign $\pm$ in (11.18) depends on value of $1 - \hat{p}$ similarly as in (11.17).

---

[3]Depending on sign of parameters $1 - \hat{p}$, $1 - \hat{l}_1 - 2\hat{l}^2\epsilon$ the function φ has 4 forms. For example, in case of part (2) of Lemma 11.2 the function is $\varphi(x) = f_{1-\hat{p},p}(f_{1-\hat{l}_1-2\hat{l}_2\epsilon,a}(f_{1-\hat{e},e}(x)))$, and if $1 - \hat{p} < 0$, $1 - \hat{l}_1 - 2\hat{l}^2\epsilon > 0$ then $\varphi(x) = f_{1-\hat{p},p}(-f_{1-\hat{l}_1-2\hat{l}_2\epsilon,a}(f_{1-\hat{e},e}(x)))$.

From the third equation we get

$$x_4 = f_{1-\hat{p},p}(x_3). \tag{11.19}$$

In this equation do not consider the case '$-$' because $1 - \hat{h} > 0$ and value of x_4 by the fourth equation of (11.15) can be $x_4 \in [0, \frac{1}{1-\hat{h}}]$. Similarly, from the fourth equation for x_5 we have (positive value):

$$x_5 = f_{1-\hat{h},h}(x_4). \tag{11.20}$$

Consequently, for x_6 we have (because $1 - \hat{\theta} < 0$):

$$x_6 = \pm f_{1-\hat{r},r}(x_5). \tag{11.21}$$

Now using (11.17)–(11.21), we can write each x_i, $i = 2, \ldots, 6$ as a function of x_1. After that, from the sixth equation of (11.15) we get (11.16). $\qquad\square$

From proof of Lemma 11.3 we get

Corollary 11.1. *If $(x_1, \ldots, x_6)$ is a solution to (11.15) then $x_4 \geq 0$, $x_5 \geq 0$.*

For each given concrete values of parameters, one can solve equation (11.16) using a computer (Maple or Mathematica).

Here we illustrate this for the case of parameters (11.8) and algebra $\mathbb{E}_0$, i.e. $\epsilon = 0$. In this case we have $1 - \hat{e} = -0.06 < 0$, $1 - \hat{p} = 0.13 > 0$, and $1 - \dot{l}_1 = 0.42 > 0$. We should have a solution, x_1, to (11.16), such that $x_1 \in (-\infty, -\frac{1}{0.06}) \cup (0, +\infty)$. This x_1 should satisfy one of the following two equations (the parameters as in (11.8))

$$g(x_1) = -2.41(g(x_1))^2 + 300x_1^2 + 3(\varphi(x_1))^2, \tag{11.22}$$

$$-g(x_1) = -2.41(g(x_1))^2 + 300x_1^2 + 3(\varphi(x_1))^2, \tag{11.23}$$

where

$$g(x) = f_{0.5657,0.43}(f_{0.36,0.46}(\varphi(x))),$$

$$\varphi(x) = f_{0.13,0.5}(f_{0.42,0.14}(f_{-0.06,0.5}(x))).$$

Case: (11.22). As Maple analysis shows the equation (11.22) has a solution x_1, which by (11.17)–(11.21) generates all x_i:

$$\begin{aligned}
&x_1 = 0.0003369672, \quad x_2 = 0.02596050896, \quad x_3 = 0.4282643609, \\
&x_4 = 0.8993564519, \quad x_5 = 1.149832995, \quad x_6 = 0.966788043.
\end{aligned} \tag{11.24}$$

Case: (11.23). The equation (11.23) has a solution x_1, which by (11.17)–(11.21) generates all x_i:

$$x_1 = 0.000260712, \quad x_2 = 0.02283488902, \quad x_3 = 0.4019229449,$$
$$x_4 = 0.8728373021, \quad x_5 = 1.140721661, \quad x_6 = -0.9700237961. \tag{11.25}$$

Summarizing we have

Proposition 11.4. *The evolution algebra* $\mathbb{E}_\epsilon$ *corresponding to baseline parameters (11.8) has at least three idempotent elements:* $(0,\ldots,0)$ *and the elements given by the coordinates (11.24) and (11.25).*

One can see the following equalities (see page 27 of [254]):

$$e_i^m = a_{ii}^{m-2}e_i^2, \quad m \geq 2.$$

By Table 11.1 we have that $a_{ii} \in (0,1)$ for $i = 2,3,4,5$ and $a_{ii} > 1$ for $i = 1,6$. Therefore we have

$$\lim_{m \to \infty} e_i^m = \begin{cases} 0, & \text{if } i = 2,3,4,5 \\ \infty, & \text{if } i = 1,6. \end{cases}$$

Comparing with Theorem 3.10 one can see that the algebra $\mathbb{E}_\epsilon$ is not nilpotent, because for example e_i^n which is not zero for any $n \geq 1$.

11.3.3 *Simplicity*

We recall that an algebra is simple whenever it has non-zero product and has no non-zero proper ideals.

Definition 11.2. As Definitions 3.1 of [26], let $B = \{e_i : i \in \Lambda\}$ be a natural basis of an evolution algebra E and let $i_0 \in \Lambda$. The first-generation descendents of i_0 are the elements of the subset $D^1(i_0)$ given by:

$$D^1(i_0) := \left\{ k \in \Lambda \mid e_{i_0}^2 = \sum_k a_{i_0 k}e_k \text{ with } a_{i_0 k} \neq 0 \right\}.$$

In an abbreviated form, $D^1(i_0) := \{j \in \Lambda \mid a_{i_0 j} \neq 0\}$.

Similarly, j is a *second-generation descendent* of i_0 whenever $j \in D^1(k)$ for some $k \in D^1(i_0)$. Therefore,

$$D^2(i_0) = \bigcup_{k \in D^1(i_0)} D^1(k).$$

By recurrence, define the set of *m-th-generation descendants* of i_0 as

$$D^m(i_0) = \bigcup_{k \in D^{m-1}(i_0)} D^1(k).$$

Finally, the set of *descendants* of i_0 is defined as the subset of Λ given by

$$D(i_0) = \bigcup_{m \in \mathbb{N}} D^m(i_0).$$

Simple finite-dimensional evolution algebras were characterized in Corollary 4.10 of [26] as those evolution algebras E with a natural basis $B = \{e_i : i \in \Lambda\}$ such that the determinant of the corresponding matrix of structural constants is non-zero and $\Lambda = D(i)$ for every $i \in \Lambda$.

Theorem 11.2. *The algebra $\mathbb{E}_\epsilon$ is simple if and only if* $\det(\mathbf{A}_\epsilon) \neq 0$.

Proof. We use the above-mentioned Corollary 4.10 [26]. One can see that the graph associated with matrix $\mathbf{A}_\epsilon$ is cyclic, in the sense that given two vertices there is always a path from one to the other one (see Remark 4.11 [26]). Consequently, for the algebra of mosquito the condition "$\Lambda = D(i)$ for every $i \in \Lambda$" is always satisfied. Thus from condition $\det(\mathbf{A}_\epsilon) \neq 0$ it follows that the algebra is simple.

Conversely, assume that $\mathbb{E}_\epsilon$ is simple, then from Corollary 4.10 [26] it follows that $\det(\mathbf{A}_\epsilon) \neq 0$. $\qquad\square$

Next we determine how are the non-zero proper ideals of $\mathbb{E}_\epsilon$ whenever this algebra is not simple.

According to Table 11.1 and equations (11.1), (11.2) we recall ranges of parameters:

$$1 - \hat{e} \in [-0.8, 0.33], \qquad \theta \in [3, 4], \qquad 1 - \hat{l}_1 - 2\hat{l}_2\epsilon \in [0.25, 0.62 - 2\epsilon],$$

$$e \in [0.33, 1], \qquad 1 - \hat{r} \in [0.43, 0.697], \qquad 1 - \hat{p} \in [-0.52, 0.45],$$

$$l \in [0.08, 0.17], \qquad 1 - \hat{\theta} \in [-3.56, -2.41], \qquad 1 - \hat{h} \in [0.169, 0.553],$$

$$p \in [0.33, 1], \qquad b\theta \in [150, 1200], \qquad h \in [0.322, 0.598],$$

$$r \in [0.3, 0.56], \qquad a \in [0.08, 0.17].$$

$$\tag{11.26}$$

Theorem 11.3. *Let I be a non-zero proper ideal of $\mathbb{E}_\epsilon$ (i.e. the case* $\det(\mathbf{A}_\epsilon) = 0$). *Then*

a) $\dim I = 5$ and $I = lin\{c_1^2, ..., e_5^2\} = lin\{e_1^2, ..., e_6^2\}$.
b) $e_k \notin I$ for every $i = 1, 2, ..., 6$.

Proof. a) Since $\det(\mathbf{A}_\epsilon) = 0$ (otherwise $\mathbb{E}_\epsilon$ is simple by the above theorem) one can check that e_6^2 is a linear combination of the linearly independent vectors $e_1^2, e_2^2, e_3^2, e_4^2, e_5^2$ (see Lemma 1 in Section 3.2) so that

$$lin\left\{e_1^2, e_2^2, e_3^2, e_4^2, e_5^2\right\} = lin\left\{e_1^2, e_2^2, e_3^2, e_4^2, e_5^2, e_6^2\right\}. \tag{11.27}$$

Let I be a non-zero proper ideal of $\mathbb{E}_\epsilon$ and let

$$k_0 := \min\{k \in \{1, 2, 3, 4, 5, 6\} : \pi_k(I) \neq 0\},$$

where π_k denotes the projection over $\mathbb{K}e_k$. For this k_0 there exists $x \in I$ and $c \neq 0$ such that $x = ce_{k_0} + \dots$. Multiplying x to e_{k_0} we get $xe_{k_0} = ce_{k_0}^2 \in I$ therefore $e_{k_0}^2 \in I$. Now using (11.10) we get that $e_k^2 \in I$ for *any* $k = 1, 2, ..., 6$, so that

$$I = lin\{e_1^2, e_2^2, e_3^2, e_4^2, e_5^2\} = lin\{e_1^2, e_2^2, e_3^2, e_4^2, e_5^2, e_6^2\}$$

and $\dim I = 5$.

b) To prove part b) we show that if $e_k \in I$ for some $k = 1, 2, \dots, 6$ then $e_i \in I$ for any $i = 1, \dots, 6$, i.e., $I = \mathbb{E}_\epsilon$, which is non-proper. Start from $e_1 \in I$:

1) if $e_1 \in I$ then by (11.10) and (11.26) there are c_6, d_6 such that $e_6 = c_6 e_1^2 - d_6 e_1 \in I$;
2) Now from $e_6 \in I$ by the last equality of (11.10) we see that there are c_5 and d_5 such that $e_5 = c_5 e_6^2 - d_5 e_6 \in I$;
3) From $e_5 \in I$ by (11.10) it follows that there are c_4 and d_4 such that $e_4 = c_4 e_5^2 - d_4 e_5 \in I$;
4) Using 1)–3), i.e., $e_4^2, e_4, e_6 \in I$, by (11.10) there are c_3, d_3 and $\tilde{d}_3$ such that $e_3 = c_3 e_4^2 - d_3 e_4 - \tilde{d}_3 e_6 \in I$;
5) Since $e_3 \in I$ we have c_2 and d_2 such that $e_2 = c_2 e_3^2 - d_2 e_3 \in I$;
6) Since $e_2 \in I$ we have c_1 and d_1 such that $e_1 = c_1 e_2^2 - d_1 e_2 \in I$.

Thus we have the following cycle:

$$e_1 \in I \Rightarrow e_6 \in I \Rightarrow e_5 \in I \Rightarrow e_4 \in I \Rightarrow e_3 \in I \Rightarrow e_2 \in I \Rightarrow e_1 \in I. \quad (11.28)$$

Note that in the cycle (11.28) if we start from any $e_k \in I$, $k = 1, 2, 3, 5, 6$ (except $k = 4$) then the cycle runs recurrently depending only on previous 'state'. But if we start from $k = 4$ then as part 4) above shows that to run the cycle we need $e_6 \in I$. To avoid this we construct the following new cycle:

i) start from $e_4 \in I$ then, since $e_5^2 \in I$, by (11.10) and (11.26) there are α_5, β_5 such that $e_5 = \alpha_5 e_5^2 - \beta_5 e_4 \in I$;
ii) using $e_5 \in I$, $e_6^2 \in I$ by the last equality of (11.10) we see that there are α_6 and β_6 such that $e_6 = \alpha_6 e_6^2 - \beta_6 e_5 \in I$;
iii) using $e_4, e_4^2, e_6 \in I$ from 4-th equation of (11.10) we get $e_3 = \alpha_3 e_4^2 - \beta_3 e_4 - \gamma_3 e_6 \in I$;

iv) using $e_3 \in I$, $e_3^2 \in I$ by the third equation of (11.10) it follows that there are α_2 and β_2 such that $e_2 = \alpha_2 e_3^2 - \beta_2 e_3 \in I$.

v) Finally, from the second equation of (11.10) we get $e_1 = \alpha_1 e_2^2 - \beta_1 e_2 \in I$. Thus we constructed the following cycle

$$e_4 \in I \Rightarrow e_5 \in I \Rightarrow e_6 \in I \Rightarrow e_3 \in I \Rightarrow e_2 \in I$$
$$\Rightarrow e_1 \in I \Rightarrow \text{continued by (11.28)}. \tag{11.29}$$

These two cycles show that if $e_k \in I$ for *some* $k = 1, \ldots 6$ then $e_i \in I$ for *all* $i = 1, \ldots, 6$. Then it follows that $I = \mathbb{E}_\epsilon$, i.e., I is not proper. $\qquad\square$

From the proof of the theorem we get

Corollary 11.2. *Let I be a non-zero ideal of $\mathbb{E}_\epsilon$ then it is proper if and only if $e_k \notin I$ for every $i = 1, 2, ..., 6$.*

The above result shows that $\mathbb{E}_\epsilon$ is not decomposable as a direct sum of ideals.

The Jacobson radical of a commutative algebra without a unit (as $\mathbb{E}_\epsilon$) is defined as the intersection of all its maximal modular ideals. We note that if the algebra has a unit then maximal modular ideals are nothing but maximal ideals. But this is not the case $\mathbb{E}_\epsilon$ because as showed in page 22 of [254] if an evolution algebra has a unit then it is a non-zero trivial algebra.

We recall that an algebra is said to be semisimple if its Jacobson radical is zero and it said to be a radical algebra whenever the algebra has no ideals of this type.

In Corollary 3.12 of [264] maximal modular ideals of an evolution algebra were characterized in it terms of modular indexes (those index i such that $\pi_i(e_i^2) \neq 0$ and $\pi_i(e_k^2) = 0$ if $k \neq i$). Since the structure matrix $\mathbf{A}_\epsilon$ has not a modular index unless $1 - \hat{e} \neq 0$, the following result follows:

Proposition 11.5. *The algebra $\mathbb{E}_\epsilon$ is a radical algebra.*

The above proposition shows that $\mathbb{E}_\epsilon$ has a 'bad' behavior. In fact the algebras that enjoy nice properties (as the automatic continuity of every surjective homomorphisms from a Banach algebra onto them) are the semisimple ones (that is those whose Jacobson radical is zero). Thus $\mathbb{E}_\epsilon$ is a chaotic algebra.

By (11.7) it follows that $\lambda = 1$ is an eigenvector of $\mathbf{A}_\epsilon$ iff

$$\hat{e}\hat{p}(\hat{l}_1 + 2\hat{l}_2\epsilon)[\hat{h}\hat{r}\hat{\theta} - hr\theta] = abehpr\theta. \tag{11.30}$$

In the next subsection, depending on the fact that $\lambda = 1$ be or not an eigenvalue of $\mathbf{A}_\epsilon$ we will describe limit points of the dynamical system (generated by a linear operator), as well as initial points to reach them.

11.3.4   *A subset of limit points of the evolution operator*

Following [254] define an evolution operator as a linear map $\mathcal{L} \equiv \mathcal{L}_\epsilon$ to be $\mathcal{L} : \mathbb{E}_\epsilon \to \mathbb{E}_\epsilon$ as $\mathcal{L}(x) = \mathbf{A}_\epsilon x$, i.e., $\mathbf{A}_\epsilon$ is the matrix of the linear map. This also can be written as $\mathcal{L}(x) = \ell x$, with $\ell = e_1 + e_2 + \cdots + e_6$.

Denote by $\sigma(\mathbf{A}_\epsilon)$ the set of all eigenvalues λ_i, $i = 1, \ldots, 6$ (spectrum) of $\mathbf{A}_\epsilon$.

Let $\mathcal{L}^n$ be n-th iteration of $\mathcal{L}$.

Proposition 11.6. *Let $\lambda \in \sigma(\mathbf{A}_\epsilon)$ and $1 \notin \sigma(\mathbf{A}_\epsilon)$. For an eigenvector c corresponding to λ (i.e. $\mathcal{L}(c) = \lambda c$) define $b_c = (\mathcal{L} - I)^{-1} c$ then $\mathcal{L}$ has unique fixed point 0 and*

$$
\lim_{n \to \infty} \mathcal{L}^n(b_c) = \begin{cases} 0, & \text{if } |\lambda| < 1 \\ \infty, & \text{if } |\lambda| > 1, \\ b_c, & \text{if } n = 2k, \lambda = -1, \\ b_c + c, & \text{if } n = 2k + 1, \lambda = -1. \end{cases}
$$

Proof.   Using $\mathcal{L}(x) = \ell x$ and $\ell c = \lambda c$ we get

$$
c = (\mathcal{L} - I)b_c \Rightarrow \ell b_c = b_c + c.
$$

$$
\mathcal{L}(b_c) = \ell b_c = b_c + c,
$$

$$
\mathcal{L}^2(b_c) = \ell(\ell b_c) = \ell(b_c + c) = \ell b_c + \ell c = b_c + c + \lambda c = b_c + (\lambda + 1)c.
$$

$$
\mathcal{L}^3(b_c) = \ell(b_c + (\lambda+1)c) = \ell b_c + (\lambda+1)\ell c = b_c + c + (\lambda+1)\lambda c = b_c + (\lambda^2 + \lambda + 1)c.
$$

Using induction on n we show that

$$
\mathcal{L}^n(b_c) = b_c + \left(\sum_{i=0}^{n-1} \lambda^i \right) c.
$$

For $n = 1, 2, 3$ we have already checked this formula, now assuming that it is true for n, we show it for $n + 1$:

$$
\mathcal{L}^{n+1}(b_c) = \ell \left(b_c + \left(\sum_{i=0}^{n-1} \lambda^i \right) c \right) = b_c + c + \left(\sum_{i=0}^{n-1} \lambda^i \right) \lambda c = b_c + \left(\sum_{i=0}^{n} \lambda^i \right) c.
$$

Case: $|\lambda| < 1$. Then

$$
\lim_{n \to \infty} \sum_{i=0}^{n-1} \lambda^i = \frac{1}{1 - \lambda}.
$$

$$\lim_{n \to \infty} \mathcal{L}^n(b_c) = b_c + \frac{c}{1 - \lambda}.$$

We show that $(\mathcal{L} - I)^{-1}c + \frac{c}{\lambda}$ is a fixed point:

$$\mathcal{L}\left((\mathcal{L} - I)^{-1}c + \frac{c}{1 - \lambda} \right) = \ell\left(b_c + \frac{c}{1 - \lambda} \right) = \ell b_c + \frac{\ell c}{1 - \lambda}$$

$$= b_c + c + \frac{\lambda c}{1 - \lambda} = b_c + \frac{c}{1 - \lambda} = (\mathcal{L} - I)^{-1}c + \frac{c}{1 - \lambda}.$$

Now we prove that if $1 \notin \sigma(\mathbf{A}_\epsilon)$ then only 0 is a fixed point of $\mathcal{L}$: assume it has some non-zero fixed point, say x^*, i.e. $\mathcal{L}(x^*) = x^*$. Then any element of the form $x = \alpha x^*$ satisfies the equation $\mathcal{L}(x) = x$, this is a contradiction to the condition that $1 \notin \sigma(\mathbf{A}_\epsilon)$. Thus $(\mathcal{L} - I)^{-1}c + \frac{c}{1-\lambda} = 0$.

Case: $|\lambda| > 1$. In this case

$$\lim_{n \to \infty} \sum_{i=0}^{n-1} \lambda^i = \lim_{n \to \infty} \frac{\lambda^n - 1}{\lambda - 1} = \begin{cases} +\infty, & \text{if } \lambda > 1, \\ \pm\infty, & \text{depending on parity of } n, \text{ if } \lambda < -1 \end{cases}$$

Case: $\lambda = -1$

$$\lim_{n \to \infty} \sum_{i=0}^{n-1} \lambda^i = \lim_{n \to \infty} \frac{(-1)^n - 1}{-2} = \begin{cases} 0, & \text{if } n = 2k, \\ 1, & \text{if } n = 2k + 1 \end{cases}$$

and the proof is completed. $\qquad\square$

Let us give some remarks.

Remark 2. Concerning Proposition 11.6 note that

1. By assumption $1 \notin \sigma(\mathbf{A}_\epsilon)$ we have $\lambda \neq 1$.
2. The infinite value of the limit means that $|\mathcal{L}^n| \to +\infty$ as $n \to \infty$.
3. The case $\lambda = -1$ means that $\mathcal{L}(c) = -c$, for any eigenvector c. Then it follows that any such c is a two-periodic point, i.e. $\mathcal{L}^2(c) = \mathcal{L}(-c) = -\mathcal{L}(c) = c$.
4. From conditions of Proposition 11.6 remains the case $|\lambda| = 1$ and λ is a complex number. In this case one can show that the limit does not exist, because the sequence will depend on $\lambda^n = \cos(n\varphi) + i\sin(n\varphi)$. Depending on $\varphi \in [0, 2\pi]$ the set of limit points can be a finite or an infinite set.
5. Note that for the baseline parameters (mentioned in the previous section), from Proposition 11.6 it follows that the corresponding operator has limit 0 or ∞.

Denote

$$\mathcal{M} = \lin\{c \in \mathrm{Ker}(\mathbf{A}_\epsilon - \lambda I) : \lambda \in \sigma(\mathbf{A}_\epsilon), |\lambda| < 1\},$$

where $\lin(S)$ denotes the set of all finite linear combinations of elements of S.

Proposition 11.7. *The following assertions hold*

(i) *If* $1 \in \sigma(\mathbf{A}_\epsilon)$ *then for any* $v \in \mathcal{M}$ *and any* $b \in \mathrm{Ker}(\mathbf{A}_\epsilon - I)$ *(i.e. b is a fixed point of $\mathcal{L}$) the following holds*

$$\lim_{n\to\infty} \mathcal{L}^n(v + b) = b.$$

(ii) $1 \notin \sigma(\mathbf{A}_\epsilon)$ *then for any* $v \in \mathcal{M}$ *the following holds*

$$\lim_{n\to\infty} \mathcal{L}^n(v) = 0.$$

(iii) *Let* $\rho(\mathbf{A}_\epsilon)$ *be the spectral radius of* $\mathbf{A}_\epsilon$, *i.e. the largest absolute value of its eigenvalues. Then* $\lim_{n\to\infty} \mathcal{L}^n$ *exists if and only if* $\rho(\mathbf{A}_\epsilon) < 1$ *or* $\rho(\mathbf{A}_\epsilon) = 1$, *where* $\lambda = 1$ *is the only eigenvalue on the unit circle, and* $\lambda = 1$ *is semisimple. Moreover, when the limit exists,*

$$\lim_{n\to\infty} \mathcal{L}^n = \text{the projector onto the set of}$$

all eigenvectors associated with $\lambda = 1$ *(eigenspace)*

along the range of a matrix $I - \mathbf{A}_\epsilon$ *(image space).*

Proof. (i) Any vector $v \in \mathcal{M}$ has the form

$$v = \sum_i \alpha_i c_i, \quad c_i \in \mathrm{Ker}(\mathbf{A}_\epsilon - \lambda_i I), |\lambda_i| < 1.$$

Consequently,

$$\mathcal{L}(v + b) = \mathcal{L}(v) + \mathcal{L}(b) = \sum_i \alpha_i \lambda_i c_i + b.$$

By induction one can see that

$$\mathcal{L}^n(v + b) = \sum_i \alpha_i \lambda_i^n c_i + b. \tag{11.31}$$

Taking limit from the both sides of this equality completes the proof.

(ii) This follows from Proposition 11.6.

(iii) This is a known theorem (see page 630 of [165]). $\qquad\square$

Remark 3. Concerning to Proposition 11.7 note that

1. In case $1 \in \sigma(\mathbf{A}_\epsilon)$ (resp. $1 \notin \sigma(\mathbf{A}_\epsilon)$) we have $\det(\mathbf{A}_\epsilon - I) = 0$ (resp. $\neq$ 0) therefore $(\mathbf{A}_\epsilon - I)b = 0$ has infinitely many (resp. unique) solutions, having the form αb (resp. 0), which are fixed points of the operator $\mathcal{L}$.
2. In case when limit of $\mathcal{L}^n(x)$ does not exist, for some $x \in \mathbb{E}_\epsilon$, but the set of limit points is finite, then the sequence $\{\mathcal{L}^n(x)\}$ is asymptotically period, say with a period p. For such a sequence, one can use the part (iii) of Proposition 11.7 for the linear function $\mathcal{L}^p$, to investigate the limit $\lim_{k\to\infty} \mathcal{L}^{pk+i}(x)$, $i = 0, 1, \ldots, p - 1$.

11.4 Biological interpretations

A population biologist is interested in the long-term behavior of the population of certain species or collection of species. Namely, the population biologist is interested in what happens to an initial population of members. Does the population vanishes as time goes on, leading to extinction of the species? Does the population become arbitrarily large? Here we give some answers to these questions related to the mosquito population.

Each point (vector) $v = (v_1, \ldots, v_6) \in \mathbb{R}_+^6$ can be considered as a state of the mosquito population, which is a measure on the set $\{E, L, P, A_h, A_r, A_o\}$. If, for example, the value of v_2 is close to zero, biologically, this means that the contribution of the larva stage L will be small in future of the population.

Each fixed point is an equilibrium state, a stable and unstable manifold biologically means that if an initial point (state) of the population is from the stable (resp. unstable) manifold then in future the state of the population goes close (resp. far) to (resp. from) the state described by the fixed point.

As it was mentioned before, for each state $x \in \mathbb{E}_\epsilon$ of the mosquito population, the state of its next generation is given by the evolution map $V(x) = x^2$ (see [155]) with respect to the multiplication (11.13). Therefore, an absolute nilpotent element is a state of the population which dies in next generation. Proposition 11.3 says that the mosquito population has no any such (non-zero) state. An idempotent element is a state of the population which does not change in the next generation, Proposition 11.4 says that the population with baseline parameters (11.8) has at least three such states.

Ideals on evolution algebras have biological meaning that when the system reach an ideal then all individuals in future generations will stay in it, i.e. an ideal is closed subpopulation.

Under condition of Theorem 11.2 the mosquito population has not a closed subpopulation (distinct from the full population itself). But in case of Theorem 11.3 there is a closed subpopulation which has 5 generators.

Results of Proposition 11.6 and Proposition 11.7 show some sets of initial states the population started from them tend to zero as time goes on, leading to extinction of the population and some other initial states which lead the population to become stable or arbitrarily large.

Bibliographical notes. As was noted above there are several kind of mathematical models of mosquito population dynamics (see for example [35], [52], [148], [245] and the references therein). In this chapter we considered a discrete-time dynamical system, generated by an evolution operator of a mosquito population, its continuous time version was considered in [154] (see also the references therein). This chapter is based on recent paper [229] and many internet sources.

For further reading the theory of mosquito populations dynamics see [7], [34], [35], [52], [108], [152], [261], [266], [272] and the references therein.

Chapter 12

On ocean ecosystem discrete time dynamics generated by ℓ-Volterra operators

This chapter is devoted to a discrete-time dynamical system generated by a nonlinear operator of ocean ecosystem. Under some conditions on the parameters of the system its evolution operator reduced to a ℓ-Volterra quadratic stochastic operator mapping two-dimensional simplex to itself. It is show that (under some conditions on parameters) this ℓ-Volterra operator may have up to three or a countable set of fixed points. These fixed points may be attracting, repelling or saddle points. The limit behaviors of trajectories of the dynamical system are studied. It is shown that independently on values of parameters and on initial point all trajectories converge. Thus the quadratic operator is regular. Some biological interpretations of these results are given.

12.1 The discrete time ecosystem

12.1.1 *What is an ecosystem?*

An ecosystem is a community made up of living organisms and nonliving components such as air, water and mineral soil.[1]

In ecosystems the energy flows are somewhat difficult to measure and to model. The most successful ecosystem models have concentrated on the balance of essential elements such as *carbon*, *nitrogen* and *phosphorus* rather than explicitly on energy [23], [186].

Following [23], [186] we consider an example of plankton in ocean limited by the essential element nitrogen, and assume that the system is closed to nitrogen. Plankton are of two kinds, *phytoplankton*, or plant plankton, which photosynthesis and require essential elements, and *zooplankton*, or

---

[1]https://en.wikipedia.org/wiki/Ecosystem.

animal plankton, which feed on phytoplankton. It is assumed that all the action takes place in a well-mixed surface layer of the ocean.

Let N be the concentration of nitrogen available for uptake, measured as mass per unit surface area of the ocean, P the concentration of phytoplankton, and Z the concentration of zooplankton, both measured in the same currency, i.e. as mass of nitrogen incorporated in the plankton per unit surface area of the ocean.

Nitrogen (as dissolved gas or compounds) is taken up from the ocean and incorporated into phytoplankton. It is incorporated into zooplankton through consumption of phytoplankton. It is recycled from the plankton of the ocean through death and excretion.

In [1] the authors developed moment closure approximations to represent micro-scale spatial variability in the concentrations of nutrients, phytoplankton and zooplankton in an NPZ model. For the NPZ closure model the following are showed: the stability domains increases with micro-scale variability, increases the biomass of zooplankton, and the coefficient of variation of phytoplankton increases with micro-scale variability.

At time moment $t \geq 0$ the state of the ecosystem is given by the vector $(N(t), P(t), Z(t))$.

In [23] the following model of ocean ecosystem processes is given:

$$\begin{cases} \frac{dN}{dt} = aP + bZ - cNP \\ \frac{dP}{dt} = cNP - dPZ - aP \\ \frac{dZ}{dt} = dPZ - bZ \end{cases} \tag{12.1}$$

where $a, b, c, d \in R$, it follows from the system that

$$\frac{d}{dt}(N + P + Z) = 0, \quad \text{so that} \quad N + P + Z = \text{const},$$

i.e., this is law of conservation of mass for nitrogen and this constant represents the total concentration of nitrogen, both available for uptake (free) and incorporated in the plankton (bound).

In this chapter following [226] we consider the discrete time version of (12.1). This system is a dynamical system generated by a 2-Volterra quadratic stochastic operator (QSO) (see Section 8.1 for definitions).

Note that, each QSO (denoted by V) can be uniquely defined by a cubic matrix $\mathbb{P} = (P_{ij,k})_{i,j,k=1}^m$ with conditions

$$P_{ij,k} \geq 0, \quad P_{ij,k} = P_{ji,k}, \quad \sum_{k=1}^m P_{ij,k} = 1, \quad (i, j, k = 1, ..., m). \tag{12.2}$$

For a given $\lambda^{(0)} \in S^{m-1}$ the *trajectory* (orbit)

$$\{\lambda^{(n)}\}, \quad n = 0, 1, 2, ... \text{of } \lambda^{(0)}$$

under the action of QSO V is defined by

$$\lambda^{(n+1)} = V(\lambda^{(n)}), \quad \text{where} \quad n = 0, 1, 2, ...$$

Recall definition of ℓ-Volterra QSO. Fix $\ell \in \{1, ..., m\}$ and assume that elements $P_{ij,k}$ of the matrix $\mathbb{P}$ satisfy

$$P_{ij,k} = 0 \quad \text{if} \quad k \notin \{i, j\} \quad \text{for any} \quad k \in \{1, ..., \ell\}, i, j \in E;$$

$$P_{ij,k} > 0 \quad \text{for at least one pair} \quad (i, j), i \neq k, \quad j \neq k, \quad \forall k \in \{\ell + 1, ..., m\}. \tag{12.3}$$

Definition 12.1. For any fixed $\ell \in \{1, ..., m\}$, the QSO defined by coefficients (12.2) and (12.3) is called ℓ-Volterra QSO.

Definition 12.2. A QSO V is called regular if for any initial point $\lambda^{(0)} \in S^{m-1}$, the limit

$$\lim_{n \to \infty} V^n(\lambda^{(0)})$$

exists.

Consider a discrete time process of ocean ecosystem (12.1), which has the following form

$$V : \begin{cases} x^{(1)} = x(1 - b + dy) \\ y^{(1)} = y(1 - a - dx + cz) \\ z^{(1)} = z(1 - cy) + ay + bx \end{cases} \tag{12.4}$$

where $x = Z, y = P, z = N, a, b, c, d \in R$.

As usual the **main problem** of investigation is to study the set of limit points of trajectories of the operator (12.4).

A fixed point p for a mapping $F : R^m \to R^m$ is a solution to the equation $F(p) = p$. By the continuity of the operator V (12.4) its limit points are fixed points for the operator V.

12.1.2 *Evolution operator as a 2-Volterra QSO*

Note that the operator (12.4) has a form of 2-Volterra QSO, but the parameters of this operator are not related to $P_{ij,k}$. Here to make some relations with $P_{ij,k}$ we find conditions on parameters of (12.4) rewriting it in the form of QSO. One can see that $x^{(1)} + y^{(1)} + z^{(1)} = x + y + z$, to embed a vector (x, y, z) in the set S^2 we assume $x + y + z = 1$ and $x, y, z \geq 0$.

Using $x + y + z = 1$, the system (12.4) can be written as the following:

$$\begin{cases} x^{(1)} = x[(1 - b)x + (1 - b + d)y + (1 - b)z] \\ y^{(1)} = y[(1 - a - d)x + (1 - a)y + (1 - a + c)z] \\ z^{(1)} = z[x + (1 - c)y + z] + ay + bx. \end{cases} \qquad (12.5)$$

Third equation of the system (12.5) can be written as the following:

$$z^{(1)} = z[x + (1 - c)y + z] + ay(x + y + z) + bx(x + y + z)$$

$$= z[(1 + b)x + (1 + a - c)y + z] + bx^2 + (a + b)xy + ay^2. \qquad (12.6)$$

We consider a QSO for the case $m = 3$:

$$\begin{cases} x^{(1)} = P_{11,1}x^2 + 2P_{12,1}xy + 2P_{13,1}xz + 2P_{23,1}yz + P_{22,1}y^2 + P_{33,1}z^2 \\ y^{(1)} = P_{11,2}x^2 + 2P_{12,2}xy + 2P_{13,2}xz + 2P_{23,2}yz + P_{22,2}y^2 + P_{33,2}z^2 \\ z^{(1)} = P_{11,3}x^2 + 2P_{12,3}xy + 2P_{13,3}xz + 2P_{23,3}yz + P_{22,3}y^2 + P_{33,3}z^2. \end{cases}$$
$$(12.7)$$

From equation (12.6) and by the systems (12.5) and (12.7) we have the following relations:

$$P_{11,1} = 1 - b, \quad 2P_{12,1} = 1 - b + d, \quad 2P_{13,1} = 1 - b,$$

$$P_{23,1} = P_{22,1} = P_{33,1} = 0, \quad P_{22,2} = 1 - a, \quad 2P_{12,2} = 1 - a - d,$$

$$2P_{23,2} = 1 - a + c, \quad P_{11,2} = P_{13,2} = P_{33,2} = 0, \quad P_{11,3} = b, \qquad (12.8)$$

$$2P_{12,3} = a + b, \quad 2P_{13,3} = 1 + b, \quad 2P_{23,3} = 1 + a - c,$$

$$P_{22,3} = a, \quad P_{33,3} = 1.$$

Proposition 12.1. *The operator (12.4) maps S^2 to itself if and only if*

$$0 \leq a \leq 1, \quad 0 \leq b \leq 1, \quad -(1-a) \leq c \leq 1+a, \quad -(1-b) \leq d \leq 1-a. \quad (12.9)$$

Moreover, under condition (12.9) and $a + b \neq 0$ the operator is a 2-Volterra QSO.

Proof. Follows from solving of inequalities which are obtained from conditions (12.2) and (12.3) for $P_{ij,k}$ given by equalities (12.8). $\qquad \square$

Remark 12.1. If $a = b$, $c = 0, d = 0$ then the operator (12.4) coincides with operator (5.1) in [221]. By the condition (5.3) in the paper [221] and by (12.8) we get the following:

$$P_{11,1} = P_{22,2} \Rightarrow a = b,$$

$$P_{12,1} = P_{12,2} \Rightarrow d = 0,$$

$$P_{13,1} = P_{23,2} \Rightarrow c = 0.$$

Remark 12.2. In the sequel of this chapter we consider operator (12.4) with four parameters a, b, c, d which satisfy condition (12.9). This operator maps S^2 to itself and we are interested to study the behavior of the trajectory of any initial point $(x, y, z) \in S^2$ under iterations of the operator (12.4).

12.2   Case $cd(c + d) = 0$

Case $c = d = 0$. In this case the operator (12.4) becomes a linear operator:

$$V : \begin{cases} x^{(1)} = x(1 - b) \\ y^{(1)} = y(1 - a) \\ z^{(1)} = z + ay + bx. \end{cases} \tag{12.10}$$

Case: $a = b = 0$. In this case the operator is *id* map, i.e., $V(x, y, z) = (x, y, z)$.

Case: $a \neq 0, b = 0$. We have

$$\lim_{n \to \infty} V^n(x, y, z) = \lim_{n \to \infty} (x^{(n)}, y^{(n)}, z^{(n)})$$

$$= \lim_{n \to \infty} (x, (1 - a)^n y, z^{(n)}) = (x, 0, 1 - x), \tag{12.11}$$

thus in this case there are infinitely many fixed points $(x, 0, 1 - x)$ and the trajectory started at any initial point (x, y, z) has the limit point $(x, 0, 1-x)$.

The case $a = 0$, $b \neq 0$ is similar, and the limit point is $(0, y, 1 - y)$.

Case: $ab \neq 0$. In this case the operator has a unique fixed point $(0, 0, 1)$. Moreover, for any initial point $(x, y, z) \in S^2$ we have

$$\lim_{n \to \infty} V^n(x, y, z) = \lim_{n \to \infty} (x^{(n)}, y^{(n)}, z^{(n)}) = (0, 0, 1), \tag{12.12}$$

with $x^{(n)} = (1 - b)^n x, y^{(n)} = (1 - a)^n y.$

Case: $c = 0, d \neq 0$. In this case the third coordinate of the operator has a linear form:

$$\begin{cases} x^{(1)} = x(1 - b + dy) \\ y^{(1)} = y(1 - a - dx) \\ z^{(1)} = z + ay + bx. \end{cases} \qquad (12.13)$$

Case: $a = b = 0$. To find fixed points we solve $V(x, y, z) = (x, y, z)$, i.e.

$$x = x(1 + dy), \quad y = y(1 - dx), \quad z = z.$$

This system has infinitely many solutions, i.e., the following are the set of fixed points:

$$F_1 = \{(x, y, z) \in S^2 : y = 0\}, \quad F_2 = \{(x, y, z) \in S^2 : x = 0\}.$$

For the trajectory $(x^{(n)}, y^{(n)}, z^{(n)})$ we have

$$\begin{cases} x^{(n+1)} = x^{(n)}(1 + dy^{(n)}) \\ y^{(n+1)} = y^{(n)}(1 - dx^{(n)}) \\ z^{(n+1)} = z^{(n)} = z, \end{cases} \qquad (12.14)$$

i.e. the third coordinate does not depend on n, i.e., $z^{(n)} = z$. Now using $x^{(n)} + y^{(n)} + z^{(n)} = 1$ from the first equality of (12.14) we get $x^{(n+1)} = x^{(n)}(1 + d(1 - x^{(n)} - z))$, thus the behavior of $x^{(n)}$ is given by the one-dimensional dynamical system generated by the function

$$f(x) = x(1 + d(1 - x - z)) = -dx^2 + [1 + d(1 - z)]x, \quad x \in [0, 1]$$

here d and z are considered as parameters.

Note that $f(x)$ has two fixed points $x = 0$ and $x = 1 - z$.

Assume $d > 0$ then $f'(0) = 1 + d(1 - z) > 1, z \neq 1$ (note that $z = 1$ gives the fixed point $(0, 0, 1)$), and $f'(1 - z) = 1 - d(1 - z) < 1$. Hence 0 is repelling and $1 - z$ is attracting fixed point for f. To find two-periodic points of $f(x)$ one has to solve $f(f(x)) = x$. Note that the fixed points of f also solutions to $f(f(x)) = x$. To find the two-periodic points different from the fixed points one has to solve

$$\frac{f(f(x)) - x}{f(x) - x} = 0.$$

This equation is a quadratic equation whose discriminant is $D = d^2(z - 1)^2 - 4 < 0$, for any $d \in [-1, 1]$ and $z \in [0, 1]$. Therefore, the function f has no any two periodic point (different from fixed points) then by Sarkovskii's theorem (see page 62 of [40]) f has no any periodic (except fixed) point.

For initial point (x, y, z) we have $x = 1 - y - z \le 1 - z$, moreover, for $d > 0$ from (12.14) one can see that $x^{(n)}$ is increasing, therefore it has a limit. Thus $1 - z$ is globally attracting, i.e., for any initial point $x \in (0, 1]$ we have $\lim_{n \to \infty} x^{(n)} = 1 - z$.

Similarly in case $d < 0$ we have 0 is attracting and $1 - z$ is repeller fixed point for f. Moreover, for any initial point $x \in [0, 1)$ we have

$$\lim_{n \to \infty} x^{(n)} = 0.$$

Thus, if $d > 0$ (resp. $d < 0$) then for any initial point $(x, y, z) \in S^2$ we have the following

$$\lim_{n \to \infty} V^n(x, y, z) = \lim_{n \to \infty} (x^{(n)}, y^{(n)}, z^{(n)}) = (1 - z, 0, z) \quad (\text{resp. } (0, 1 - z, z)),$$
$$(12.15)$$

because in this case there is no fixed points in $\mathrm{int} S^2 = \{(x, y, z) \in S^2 : xyz > 0\}$.

Case: $a = 0, b \ne 0$. Then the operator (12.4) has the following form

$$\begin{cases} x^{(1)} = x(1 - b + dy) \\ y^{(1)} = y(1 - dx) \\ z^{(1)} = z + bx. \end{cases} \qquad (12.16)$$

One can see that the operator (12.16) has infinitely many fixed points given by the following set

$$F_3 = \{(x, y, z) \in S^2 : x = 0\}.$$

Moreover, It does not have a fixed point outside of F_3.

We have $z^{(1)} = z + bx \ge z$, consequently $z^{(n+1)} \ge z^{(n)}$, i.e. $z^{(n)}$ increasing. Since $z^{(n)} \le 1$ it has a limit.

In addition, if $d > 0$ (resp. $d < 0$) then $y^{(1)} = y(1 - dx) \le y$ (resp. $x^{(1)} = x(1 - b + dy) \le x$), consequently, $y^{(n)}$ (resp. $x^{(n)}$) is decreasing sequence and it has a limit point. Hence, for any initial point $(x, y, z) \in S^2$ there exists the limit

$$\lim_{n \to \infty} V^n(x, y, z) = \lim_{n \to \infty} (x^{(n)}, y^{(n)}, z^{(n)}) = (0, \bar{y}, 1 - \bar{y}), \qquad (12.17)$$

because all limit points are fixed points belong to the set F_3, where $\lim_{n \to \infty} y^{(n)} = \bar{y}$.

Case: $a \ne 0, b = 0$. This case is similar to the previous case the operator (12.4) has the form

$$\begin{cases} x^{(1)} = x(1 + dy) \\ y^{(1)} = y(1 - a - dx) \\ z^{(1)} = z + ay. \end{cases} \qquad (12.18)$$

The operator (12.18) also has the following infinite set of fixed points

$$F_4 = \{(x, y, z) \in S^2 : y = 0\}.$$

Here also $z^{(1)} = z + ay \geq z$ so $z^{(n)}$ increasing, if $d > 0$ then $x^{(1)} = x(1 + dy) \geq x$, $y^{(1)} = y(1 - a - dx) \leq y$ and $x^{(n)}$ increasing, $y^{(n)}$ decreasing. If $d < 0$ then $x^{(1)} = x(1 + dy) \leq x$ so $x^{(n)}$ decreasing sequence. Hence, for any initial point $(x, y, z) \in S^2$ we have

$$\lim_{n \to \infty} V^n(x, y, z) = \lim_{n \to \infty} (x^{(n)}, y^{(n)}, z^{(n)}) = (\bar{x}, 0, 1 - \bar{x}), \qquad (12.19)$$

because all fixed points belong to the set F_4, where $\lim_{n \to \infty} x^{(n)} = \bar{x}$.

Remark 12.3. In limits (12.17) and (12.19) we only know existence of $\bar{y}$ and $\bar{x}$. Of course, these values depend on the initial point (x, y, z), i.e. $\bar{y} = \bar{y}(x, y, z)$ and $\bar{x} = \bar{x}(x, y, z)$. But finding an explicit form of these values seems difficult problem. This problem can be reduced to following non-linear recursive equations:

For finding of $z^{(n)}$ in (12.17):

$$z^{(n+2)} = -\frac{d}{b}(z^{(n+1)})^2 + \frac{(2 - b)d}{b}z^{(n+1)}z^{(n)}$$

$$-\frac{(1 - b)d}{b}(z^{(n)})^2 + (2 - b + d)z^{(n+1)} - (1 - b + d)z^{(n)}, \qquad (12.20)$$

where $z^{(0)} = z$ and $z^{(1)} = z + bx$.

For finding of $z^{(n)}$ in (12.19):

$$z^{(n+2)} = \frac{d}{a}(z^{(n+1)})^2 - \frac{(2 - a)d}{a}z^{(n+1)}z^{(n)}$$

$$+\frac{(1 - a)d}{a}(z^{(n)})^2 + (2 - a - d)z^{(n+1)} - (1 - a - d)z^{(n)}, \qquad (12.21)$$

where $z^{(0)} = z$ and $z^{(1)} = z + ay$. There is no any general method to solve these recursive equations (with given initial values). But formulas (12.20) and (12.21) are useful for using of a computer program.

Case: $c \neq 0, d = 0$. In this case the system (12.4) has the form

$$\begin{cases} x^{(1)} = x(1 - b) \\ y^{(1)} = y(1 - a + cz) \\ z^{(1)} = z(1 - cy) + ay + bx. \end{cases} \qquad (12.22)$$

For the operator (12.22) fixed points are

$$
F_5 = \begin{cases}
(0,0,1) & \text{if } b \neq 0, c < 0 \\[2mm]
\{(0,0,1),(0,1-\frac{a}{c},\frac{a}{c})\} & \text{if } b \neq 0, c > 0 \\[2mm]
\{(x,0,1-x),(x,1-x-\frac{a}{c},\frac{a}{c})\} & \text{if } b = 0, c > 0 \\[2mm]
\{(0,0,1),(x,0,1-x)\} & \text{if } b = 0, c < 0.
\end{cases}
$$

Case: $b = 0$. Then $x^{(n)} = x$ for any $x \in [0,1]$, consequently

$$
y^{(n+1)} = y^{(n)}(1 - a + c(1 - x - y^{(n)})).
$$

Hence $y^{(n)}$ is given by a dynamical system of

$$
g(y) = y(1 - a + c(1 - x - y)) = -cy^2 + (1 - a + c(1 - x))y,
$$

where a, c, x are parameters. This function has two fixed points: $y = 0$ and $y = y_* = 1 - x - \frac{a}{c}$. If $c > 0$ then 0 is repelling and y_* is attracting fixed point for g. Moreover, for any initial point $y \in (0,1]$ we have

$$
\lim_{n \to \infty} y^{(n)} = y_*.
$$

Similarly in case $c < 0$ we have 0 is attracting and y_* is repeller fixed point for g. Moreover, for any initial point $y \in [0,1)$ we have

$$
\lim_{n \to \infty} y^{(n)} = 0.
$$

Thus, if $c > 0$ (resp. $c < 0$) then for any initial point $(x, y, z) \in S^2$ we have the following

$$
\lim_{n \to \infty} V^n(x, y, z) = \lim_{n \to \infty} (x^{(n)}, y^{(n)}, z^{(n)})
$$
$$
= (x, y_*, 1 - x - y_*) \quad (\text{resp. } (x, 0, 1 - x)). \qquad (12.23)
$$

Case: $b \neq 0$. Then $x^{(n)} = (1 - b)^n x$ and when $c \leq a$, $(0,0,1)$ unique fixed point, $y^{(1)} = y(1 - a + cz) \leq y$, i.e. the sequence $y^{(n)}$ is monotone decreasing so for any initial point $(x, y, z) \in S^2$

$$
\lim_{n \to \infty} (x^{(n)}, y^{(n)}, z^{(n)}) = (0,0,1).
$$

If $c > a$ then the set

$$
H = \left\{ (x, y, z) \in S^2 : z > \frac{a}{c} \right\}
$$

is an invariant. Indeed, if $z > \frac{a}{c}$ then

$$1 - x - y > \frac{a}{c} \Leftrightarrow cy < c - a - cx \Leftrightarrow$$

$$1 - cy > 1 - (c - a) + cx > 0 \Leftrightarrow z^{(1)} = z(1 - cy) + ay > \frac{a}{c}.$$

One can see that if initial point $(x, y, z) \in H$ then the sequence $y^{(n)}$ has limit as an increasing and bounded sequence, if $(x, y, z) \notin H$ then for $y^{(n)}$ we have two possibilities: stays outside of H and decreasingly goes to $1 - \frac{a}{c}$; after finite steps goes inside of H and increasingly converges to the same limit.

Hence, for any initial point $(x, y, z) \in S^2$ we have

$$\lim_{n \to \infty} (x^{(n)}, y^{(n)}, z^{(n)}) = \begin{cases} (0, 1 - \frac{a}{c}, \frac{a}{c}), & \text{if } y > 0 \\ (0, 0, 1), & \text{if } y = 0. \end{cases} \tag{12.24}$$

Case: $c = -d \neq 0$. Here the operator (12.4) has the form

$$\begin{cases} x^{(1)} = x(1 - b - cy) \\ y^{(1)} = y(1 - a + cx + cz) \\ z^{(1)} = z(1 - cy) + ay + bx. \end{cases} \tag{12.25}$$

For the operator (12.25) fixed points are

$$F_6 = \begin{cases} (0, 0, 1) & \text{if } c \leq a, b \neq 0 \\ \{(x, 0, 1 - x)\} & \text{if } c \leq a, b = 0 \\ \{(0, 0, 1), (0, 1 - \frac{a}{c}, \frac{a}{c})\} & \text{if } c > a, b \neq 0 \\ \{(x, 0, 1 - x), (0, 1 - \frac{a}{c}, \frac{a}{c})\} & \text{if } c > a, b = 0. \end{cases}$$

Case: $0 < c \leq a$. Then $x^{(1)} = x(1 - b - cy) \leq x$, $y^{(1)} = y(1 - a + cx + cz) = y(1 - (a - c(x + y))) \leq y$, i.e. the sequences $x^{(n)}$ and $y^{(n)}$ are decreasing.

Case: $c \leq a, c < 0$. Then $y^{(1)} = y(1 - a + cx + cz) \leq y$, $z^{(1)} = z(1 - cy) + ay + bx \geq z$, i.e. the sequences $y^{(n)}$ and $z^{(n)}$ are monotone. Thus, if $b \neq 0$ then for any initial point $(x, y, z) \in S^2$ we have

$$\lim_{n \to \infty} (x^{(n)}, y^{(n)}, z^{(n)}) = (0, 0, 1) \tag{12.26}$$

(since $(0, 0, 1)$ is unique fixed point).

If $b = 0$ then for any initial point $(x, y, z) \in S^2$

$$\lim_{n \to \infty} (x^{(n)}, y^{(n)}, z^{(n)}) = (\tilde{x}, 0, 1 - \tilde{x}), \tag{12.27}$$

where $\lim_{n \to \infty} x^{(n)} = \tilde{x}$.

Case: $c > a$. In the case $c > a$ for any $(x, y, z) \in S^2$ we have (this is a particular case of Theorem 12.2 given in the next section)

$$\lim_{n \to \infty} (x^{(n)}, y^{(n)}, z^{(n)}) = \begin{cases} (0, 0, 1), & \text{if } y = 0, b \neq 0 \\ (x, 0, 1 - x), & \text{if } y = 0, b = 0 \\ (0, 1 - \frac{a}{c}, \frac{a}{c}), & \text{if } y > 0. \end{cases} \tag{12.28}$$

Thus independently on parameters a, b, c, d and initial point (x, y, z) the limit of trajectory exists. Summarizing above mentioned results (i.e., (12.11), (12.12), ..., (12.28)) we get

Theorem 12.1. *If* $cd(c + d) = 0$ *then for any* $(x, y, z) \in S^2$ *we have*

$$\lim_{n \to \infty} V^n(x, y, z) = \begin{cases} (x, y, z), & \text{if } a = b = c = d = 0 \\ (x, 0, 1 - x), & \text{if } a \neq 0, b = c = d = 0 \\ (0, y, 1 - y), & \text{if } a = c = d = 0, b \neq 0 \\ (0, 0, 1), & \text{if } ab \neq 0, \ c = d = 0 \\ (1 - z, 0, z), & \text{if } a = b = c = 0, d > 0 \\ (0, 1 - z, z), & \text{if } a = b = c = 0, d < 0 \\ (0, \bar{y}, 1 - \bar{y}), & \text{if } a = c = 0, bd \neq 0 \\ (\bar{x}, 0, 1 - \bar{x}), & \text{if } b = c = 0, ad \neq 0 \\ (x, 1 - x - \frac{a}{c}, \frac{a}{c}), & \text{if } c > 0, b = d = 0 \\ (x, 0, 1 - x), & \text{if } b = d = 0, c < 0 \\ (0, 0, 1), & \text{if } b \neq 0, \ c \leq a, d = 0, \ \text{ or } \ y = 0 \\ (0, 1 - \frac{a}{c}, \frac{a}{c}) & \text{if } b \neq 0, \ c > a, d = 0, \ y > 0 \\ (0, 0, 1), & \text{if } b \neq 0, \ c = -d \neq 0, 0 < c \leq a, \\ (\tilde{x}, 0, 1 - \tilde{x}), & \text{if } b = 0, \ c = -d \neq 0, c \leq a, \\ (0, 0, 1), & \text{if } b \neq 0, \ c = -d, a < c, y = 0, \\ (x, 0, 1 - x), & \text{if } b = 0, \ c = -d, a < c, y = 0, \\ (0, 1 - \frac{a}{c}, \frac{a}{c}) & \text{if } a < c \ \ y > 0, \end{cases}$$

where $\bar{x}, \bar{y}$ *and* $\tilde{x}$ *are some functions of the initial point* (x, y, z).

12.3   Case $cd(c+d) \neq 0$

12.3.1   *Fixed points of the operator (12.4)*

To find fixed points of operator V given by (12.4) we have to solve

$$\begin{cases} x = x(1 - b + dy) \\ y = y(1 - a - dx + cz) \\ z = z(1 - cy) + ay + bx. \end{cases} \tag{12.29}$$

Lemma 12.1. *The fixed points of the operator (12.4) are*

$$\bar{\lambda}_0 = (0,0,1), \quad \bar{\lambda}_1 = \left(0, 1 - \frac{a}{c}, \frac{a}{c}\right) \quad (c \geq a, c \neq 0),$$

and

$$\bar{\lambda}_2 = \left(\frac{cd - ad - bc}{d(c+d)}, \frac{b}{d}, \frac{a - b + d}{c + d}\right)$$

where there are conditions to the parameters, in other words we consider all coordinates of $\bar{\lambda}_2$ between 0 and 1:

$$d > 0, \quad d \geq b, \quad a - b \leq c, \quad c \geq 0, \quad cd - ad - bc \geq 0. \tag{12.30}$$

Proof.   Consider the following cases
   Case: $x = 0, y = 0$. One can see from (12.29) that $(0,0,1)$ is a fixed point.
   Case: $x = 0, y \neq 0$. From (12.29) we get system of equations

$$x = 0, \quad y(1 - a + cz) = y, \quad z(1 - cy) + ay = z,$$

which has a unique solution $\left(0, 1 - \frac{a}{c}, \frac{a}{c}\right)$ for $c > 0, c \geq a$.
   Case: $x \neq 0, y = 0$. In this case we do not have fixed point, because from the first equation of (12.29) we get $x = x(1 - b)$ which has no non-zero solution for $b \neq 0$.
   Case: $x \neq 0, y \neq 0$. In this case the system (12.29) is reduced to the system

$$1 = 1 - b + dy, \quad 1 = 1 - a - dx + cz, \quad z = z(1 - cy) + ay + bx$$

from which we get the third fixed point $\left(\frac{cd-ad-bc}{d(c+d)}, \frac{b}{d}, \frac{a-b+d}{c+d}\right)$. $\square$

   By using $x + y + z = 1$ in (12.4), we obtain the following mapping:

$$W : \begin{cases} x' = x(1 - b + dy) \\ y' = y(1 - a + c - (c + d)x - cy) \end{cases} \quad \text{where } x + y \leq 1. \tag{12.31}$$

For the operator W there are three fixed points with conditions (12.30):

$$\lambda_0 = (0,0), \ \lambda_1 = \left(0, \frac{c-a}{c}\right) \ (c > 0, c \geq a), \ \text{and} \ \lambda_2 = \left(\frac{cd - ad - bc}{d(c+d)}, \frac{b}{d}\right).$$

Proposition 12.2. *The following relations are true:*

(1)

$$\lambda_0 = \begin{cases} \text{nonhyperbolic, if } \ b = 0 \ \text{ or } \ a = c \\ \text{attractive,} \qquad \text{if } \ a > c \\ \text{saddle,} \qquad \text{if } \ a < c. \end{cases}$$

(2)

$$\lambda_1 = \begin{cases} \text{nonhyperbolic, if } \ a = c \ \text{ or } \ b = d(1 - \frac{a}{c}) \\ \text{attractive,} \qquad \text{if } \ b > d(1 - \frac{a}{c}) \\ \text{saddle,} \qquad \text{if } \ b < d(1 - \frac{a}{c}). \end{cases}$$

(3)

$$\lambda_2 = \begin{cases} \text{nonhyperbolic, if } \ b = 0 \text{ or } c \leq a \text{ or } cd - ad - bc = 0 \text{ or } b = d \\ \text{attractive,} \qquad \text{if } \ c > 0, c > a, d > b > 0, cd - ad - bc > 0. \end{cases}$$

Proof.   (1) Note that the Jacobian of the system (12.31) has the form

$$\mathbf{J} = \begin{bmatrix} 1 - b + dy & dx \\ -(c+d)y & 1 \quad a + c - (c+d)x - 2cy \end{bmatrix}. \tag{12.32}$$

The Jacobian at the fixed point $\lambda_0 = (0,0)$ has the form

$$\mathbf{J}(\lambda_0) = \begin{bmatrix} 1 - b & 0 \\ 0 & 1 - a + c. \end{bmatrix}.$$

Clearly, if $b = 0$ or $a = c$ then the eigenvalues μ_1 or μ_2 of the matrix $\mathbf{J}(\lambda_0)$ are equal to one, i.e., λ_0 is nonhyperbolic and in all other cases λ_0 is hyperbolic.

If $b > 0, a \neq c$ then one eigenvalue $\mu_1 = 1 - b < 1$ is correct for all b. If $a > c$ and $a < c$ then $|\mu_2| < 1$ and $|\mu_2| > 1$ respectively. Hence, λ_0 is an attracting and saddle point when $a > c$ and $a < c$ respectively, repelling case does not occur.

(2) At the point λ_1 the Jacobian is:

$$\mathbf{J}(\lambda_1) = \begin{bmatrix} 1 - b + \frac{(c-a)d}{c} & 0 \\ -\frac{(c+d)(c-a)}{c} & 1 + a - c \end{bmatrix}.$$

One can see that if $a = c$ or $b = d(1 - \frac{a}{c})$ then at least one eigenvalue of the matrix $\mathbf{J}(\lambda_1)$ is equal to one, so λ_1 is nonhyperbolic.

If $a \neq c$ then by the condition $c \geq a$ one eigenvalue $\mu_1 = 1 - c + a$ is always less than one. Moreover, if $b > d(1 - \frac{a}{c})$ then λ_1 is an attractive, and if $b < d(1 - \frac{a}{c})$ then λ_1 is a saddle point. Here also repelling case does not exist.

(3) Next we consider the Jacobian at the fixed point λ_2:

$$\mathbf{J}(\lambda_2) = \begin{bmatrix} 1 & \frac{cd-ad-bc}{c+d} \\ -\frac{(c+d)b}{d} & \frac{d-bc}{d} \end{bmatrix}.$$

$\circ$ If $b = 0$ or $cd - ad - bc = 0$ then λ_2 is nonhyperbolic.
$\circ$ If $c = 0$ then $a = 0$ and $\lambda_2 = (0, \frac{b}{d})$ is nonhyperbolic again.
$\circ$ If $b = d$ then by the condition $cd - ad - bc \geq 0$ we have $a = 0$, it means λ_2 is nonhyperbolic.
$\circ$ If $0 < c \leq a$ then by the condition $cd - ad - bc \geq 0$ we have $c = a$ and $b = 0$ and λ_2 is nonhyperbolic again.
$\circ$ If $d > b > 0, c > a, c > 0$ then the eigenvalues of the matrix $\mathbf{J}(\lambda_2)$ are:

$$\mu_{1,2} = \frac{2d - bc \pm \sqrt{D}}{2d},$$

where

$$D = b(bc^2 - 4d(cd - ad - bc)).$$

We study this eigenvalues:
If $b \geq \frac{4d^2(c-a)}{c(c+4d)}$ then

$$D = b^2c^2 - 4bd(cd - ad - bc) \geq 0$$

and since $D < b^2c^2$ we have $\mu_{1,2} < 1$. Furthermore, we have to check the condition $\mu_{1,2} > -1$:

If $d > \frac{bc(b+2)}{4+bc-ab}$ then λ_2 is a attractive fixed point.

If $d < \frac{bc(b+2)}{4+bc-ab}$ then λ_2 is a saddle fixed point.

If $d = \frac{bc(b+2)}{4+bc-ab}$ then λ_2 is nonhyperbolic fixed point.

But by the conditions (12.9)

$$\frac{bc(b+2)}{4+bc-ab} - d = \frac{b^2c + 2bc - 4d - bcd + abd}{4+bc-ab} < \frac{d^2c + 2dc - 4d - bcd + cbd}{4+bc-ab}$$

$$= \frac{d(cd + 2c - 4)}{4+bc-ab} < \frac{d((1+a)(1-a) + 2(1+a) - 4)}{4+bc-ab} = -\frac{d(a-1)^2}{4+bc-ab} < 0.$$

It follows that

$$d > \frac{bc(b+2)}{4 + bc - ab}.$$

If $b < \frac{4d^2(c-a)}{c(c+4d)}$ then $D < 0$ and we obtain the complex eigenvalues.

If $a < \frac{c(2d^2-b^2c)}{2d^2}$ then $|\mu_1| = |\mu_2| = \frac{\sqrt{(2d-bc)^2+D}}{2d} < 1$ and λ_2 is attracting fixed point.

If $a > \frac{c(2d^2-b^2c)}{2d^2}$ then $|\mu_1| = |\mu_2| > 1$ and λ_2 is repelling fixed point.

If $a = \frac{c(2d^2-b^2c)}{2d^2}$ then $|\mu_1| = |\mu_2| = 1$ and λ_2 is nonhyperbolic fixed point.

By the conditions to the parameters again we have:

$$\frac{c(2d^2 - b^2c)}{2d^2} - a = \frac{2cd^2 - b^2c^2 - 2ad^2}{2d^2} = \frac{2cd^2 - 2ad^2 - 2bcd + 2bcd - b^2c^2}{2d^2}$$

$$= \frac{2d(cd - ad - bc) + bc(2d - bc)}{2d^2} > \frac{2d(cd - ad - bc) + bc(2b - bc)}{2d^2} > 0,$$

because $cd - ad - bc > 0, c \leq 2$. It follows that

$$a < \frac{c(2d^2 - b^2c)}{2d^2}.$$

Consequently, for $cd - ad - bc > 0, c > a, c > 0, d > b > 0$ the fixed point λ_2 is attracting. $\square$

12.3.2   *The limit points*

We consider the operator (12.31) with initial point $\lambda^{(0)} = (x^{(0)}, y^{(0)})$ and will study when some of fixed points λ_0, λ_1 and λ_2 is a limit point.

If $x^{(0)} = y^{(0)} = 0$ then the limit point is λ_0.

The invariant sets with respect to W, (12.31), are

$$M_1 = \{\lambda = (x,y) : x = 0\}, \quad M_2 = \{\lambda = (x,y) : y = 0\},$$

i.e., $W(M_i) \subset M_i, i = 1, 2.$

Consider restriction of the operator (12.31) on invariant sets:

Case M_1. In this case the restriction is

$$y^{(1)} = y(1 - a + c - cy) = \varphi(y), \quad y \in [0, 1].$$

Note that $\varphi(y)$ has a unique fixed point $y = 0$ if $a > c$ and two fixed points $y = 0, y = 1 - \frac{a}{c}$ if $c \geq a$. Since $\varphi'(0) = 1 - a + c$ the fixed point 0 is attracting iff $c < a$, non-hyperbolic if $c = a$ and repelling if $c > a$. Similarly, by $\varphi'(1 - \frac{a}{c}) = 1 + a - c$ we have that $1 - \frac{a}{c}$ is attracting if $c > a$ and non-hyperbolic if $c = a$. Moreover, for $a \geq c$ the sequence $y^{(n)} = \varphi^n(y^{(0)})$ is a decreasing. Thus for any $y^{(0)} \in [0, 1]$ we have

$$\lim_{n \to \infty} y^{(n)} = \begin{cases} 0, & \text{if } a \geq c \\ 1 - \frac{a}{c}, & \text{if } a < c. \end{cases}$$

Case M_2. In this case the restriction is

$$x^{(1)} = (1 - b)x, \quad x \in [0, 1].$$

One can see that

$$\lim_{n \to \infty} x^{(n)} = \lim_{n \to \infty} (1 - b)^n x = \begin{cases} 0, \text{ if } b > 0, \quad or \quad x = 0, \\ x, \text{ if } b = 0. \end{cases}$$

Thus behavior of trajectories are clear if the initial point is taken on invariant sets M_i, $i = 1, 2$.

We assume now that $x^{(0)} \neq 0, y^{(0)} \neq 0$.

Case 1. If $c < a$ then λ_1 and λ_2 do not belong to the simplex (because they have negative coordinates), so λ_0 is a unique attracting fixed point. We show that the fixed point is globally attracting, i.e. $\forall \lambda^{(0)} \in S^2$ one has

$$\lim_{n \to \infty} \lambda^{(n)} = \bar{\lambda}_0.$$

One can see that

$$z^{(1)} = z^{(0)}(1 - cy^{(0)}) + ay^{(0)} + bx^{(0)} = z^{(0)} - y^{(0)}(cz^{(0)} - a) + bx^{(0)}$$

$$\geq z^{(0)} - y^{(0)}(cz^{(0)} - a) \geq z^{(0)} \quad (\text{since } cz^{(0)} - a \leq 0).$$

Consequently, the sequence $z^{(n)}$ is increasing and bounded, i.e., it has a limit:

$$\lim_{n \to \infty} z^{(n)} = \bar{z} \geq 0.$$

Next, we consider two cases:

1.1. If $d < 0$ then $x^{(1)} = x^{(0)}(1 - b + dy^{(0)}) \leq x^{(0)}(1 - b) \leq x^{(0)}$, by iterating we get that $x^{(n+1)} \leq x^{(n)}, \forall n \in N$ and there exists the limit

$$\lim_{n \to \infty} x^{(n)} = \bar{x}.$$

By $x^{(n)} + y^{(n)} + z^{(n)} = 1$ it follows that there exists the limit of sequence $y^{(n)}$,

$$\lim_{n \to \infty} y^{(n)} = \bar{y}.$$

1.2. If $d \geq 0$ then

$$y^{(1)} = y^{(0)}(1-a-dx^{(0)}+cz^{(0)}) \leq y^{(0)}(1-a-dx^{(0)}+c) \leq y^{(0)}(1-dx^{(0)}) \leq y^{(0)}$$

and similarly, in this case there exist limits of sequences $x^{(n)}, y^{(n)}$, i.e.,

$$\lim_{n\to\infty} \lambda^{(n)} = \bar{\lambda} = (\bar{x}, \bar{y}, \bar{z}).$$

Since a limit point must be a fixed point and in this case we have a unique fixed point on the simplex, we get

$$\lim_{n\to\infty} \lambda^{(n)} = \bar{\lambda}_0 = (0,0,1).$$

Thus, λ_0 is a globally attracting fixed point.

Case 2. If $c > a, d < b$ then $x^{(1)} = x^{(0)}(1-b+dy^{(0)}) \leq x^{(0)}(1-b+d) < x^{(0)}$ and we check the monotonicity of the sequence $y^{(n)}$. In this case the following lemma is useful:

Lemma 12.2. *If $c > a, d < b$ for the operator (12.4) we have*
   a) If $c+d \geq 0$ then the set $\mathcal{E} = \{(x,y,z) : z \geq \frac{a+dx}{c}\}$ is an invariant.
   b) If $c+d < 0$ then the set $F = \{(x,y,z) : z < \frac{a+dx}{c}\}$ is an invariant.

Proof.   a) For any point $(x,y,z) \in \mathcal{E}$ we have

$$z \geq \frac{a+dx}{c} \Leftrightarrow 1-x-y \geq \frac{a+dx}{c} \Leftrightarrow cy \leq c-a-(c+d)x.$$

Then $1-cy \geq 1-(c-a-(c+d)x) > 0$. Next we show that $z^{(1)} \geq \frac{a+dx^{(1)}}{c}$, i.e.

$$z(1-cy) + ay + bx \geq \frac{a+dx(1-b+dy)}{c}$$

which is equivalent to

$$z \geq \frac{a+dx-bdx+d^2xy-acy-bcx}{c(1-cy)} \Leftrightarrow \frac{a+dx}{c}$$

$$\geq \frac{a+dx-bdx+d^2xy-acy-bcx}{c(1-cy)}$$

$$\Leftrightarrow (c+d)(b-dy) \geq 0$$

the last inequality is true, since $b-dy > 0, c+d \geq 0$. Hence, $z^{(1)} \geq \frac{a+dx^{(1)}}{c}$;
   b) If $(x,y,z) \in F$ then $z < \frac{a+dx}{c}$ and we have to show that $z^{(1)} < \frac{a+dx^{(1)}}{c}$, i.e.

$$z(1-cy) < \frac{a+dx-bdx+d^2xy-acy-bcx}{c}.$$

i) If $1 - cy > 0$ then

$$z < \frac{a + dx - bdx + d^2 xy - acy - bcx}{c(1 - cy)} \Leftrightarrow$$

$$\frac{a + dx}{c} < \frac{a + dx - bdx + d^2 xy - acy - bcx}{c(1 - cy)}$$

from this we have the inequality $(c + d)(b - dy) < 0$ which is true.

ii) If $1 - cy < 0$ then

$$z > \frac{a + dx - bdx + d^2 xy - acy - bcx}{c(1 - cy)} \Leftrightarrow$$

$$\frac{a + dx}{c} > \frac{a + dx - bdx + d^2 xy - acy - bcx}{c(1 - cy)}$$

$$\Leftrightarrow \frac{dy(c + d) - b(c + d)}{(1 - cy)} < 0.$$

Hence we have again correct inequality $(c + d)(dy - b) > 0$ (since $c + d < 0, dy - b < 0$). Thus, $z^{(1)} < \frac{a + dx^{(1)}}{c}$. $\square$

Now we continue the proof of monotonicity of the sequence $y^{(n)}$. Using by Lemma 12.2 we have that if an initial point $\lambda^0 = (x^0, y^0, z^0)$ taken from the set $\mathcal{E}$ and $c + d \geq 0$ then the sequence $y^{(n)}$ is increasing (since $y^{(n+1)} = y^{(n)}(1 - a - dx^{(n)} + cz^{(n)}) \geq y^{(n)}, \forall n \in N$), it has the limit and we obtain

$$\lim_{n \to \infty} \lambda^{(n)} = \bar{\lambda}_1,$$

because λ_1 is the unique attracting fixed point in $\mathcal{E}$.

If $\lambda^{(0)} \notin \mathcal{E}$ then we consider two cases:

(i) If $\lambda^{(n)} \notin \mathcal{E}, \forall n \in N$ then $y^{(n)}$ is decreasing (since $y^{(n+1)} < y^{(n)}(1 - a - dx + cz^{(n)}) < y^{(n)}, \forall n \in N$);

(ii) If $\lambda^{(k)} \in \mathcal{E}$, for some $k \in N$ then $y^{(n)}$ is increasing for $n > k$.

Similarly, if an initial point $\lambda^{(0)} = (x^{(0)}, y^{(0)}, z^{(0)})$ taken from the set F and $c + d < 0$ then the sequence $y^{(n)}$ is decreasing and for the case $\lambda^{(n)} \notin F$ the sequence $y^{(n)}$ has limit point again. Consequently, if $y^{(0)} = 0$ then

$$\lim_{n \to \infty} \lambda^{(n)} = \lambda_0.$$

Thus, λ_0 is a saddle fixed point and λ_1 is globally attracting fixed point.

Case 3. $c > a, d > b > 0$. In this case numerical analysis show that the coordinates of the vector $\lambda^{(n)}$ are not monotone, so it is not easy to see the limit properties of the trajectory. Therefore we study these limits numerically for concrete values of parameters.

3.1. $a = 1/4, b = 1/2, c = 1, d = 3/4$. Then by the system (12.31) we get

$$\begin{cases} x' = 0.5x + 0.75xy \\ y' = 1.75y - 1.75xy - y^2. \end{cases} \tag{12.33}$$

For this system $\lambda_2 = (\frac{1}{21}; \frac{2}{3}; \frac{2}{7}) \approx (0.0476; 0.6667; 0.2857)$. Next we divide the simplex to the four parts as following (Fig. 12.1):

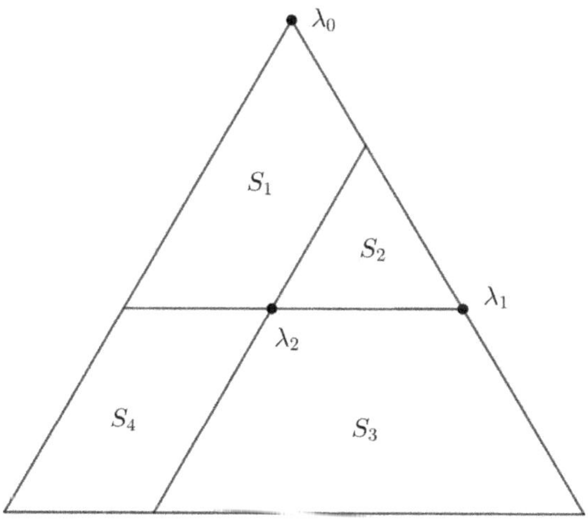

Fig. 12.1

By using Wolfram Mathematica 7.0 one finds limit points of initial points from S_1, S_2, S_3 and S_4 respectively (Fig. 12.2).

i) If $(x^{(0)}, y^{(0)}) = (0.1, 0.6) \in S_1$ then after 100 iterations we get

$$(x^{(100)}, y^{(100)}) = (0.0476346, 0.666636).$$

ii) If $(x^{(0)}, y^{(0)}) = (0.02, 0.68) \in S_2$ then

$$(x^{(100)}, y^{(100)}) = (0.0475566, 0.666789)$$

iii) If $(x^{(0)}, y^{(0)}) = (0.05, 0.68) \in S_3$ then

$$(x^{(100)}, y^{(100)}) = (0.0476215, 0.666662).$$

iv) If $(x^{(0)}, y^{(0)}) = (0.07, 0.66) \in S_4$ then

$$(x^{(100)}, y^{(100)}) = (0.0476302, 0.666645).$$

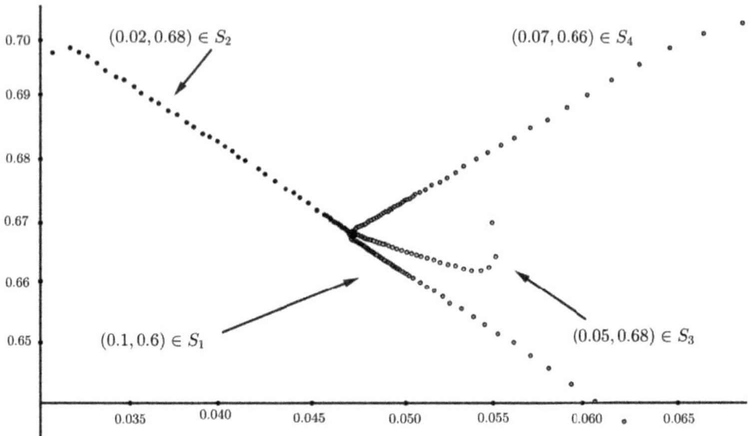

Fig. 12.2 Blue: $(0.1, 0.6) \in S_1$; black: $(0.02, 0.68) \in S_2$; green: $(0.05, 0.68) \in S_3$; red: $(0.07, 0.66) \in S_4$.

3.2. $a = 1/6, b = 1/3, c = 4/3, d = 2/3$. Then by the system (12.31) we have

$$\begin{cases} x' = \frac{2}{3}x + \frac{2}{3}xy \\ y' = \frac{13}{6}y - 2xy - \frac{4}{3}y^2. \end{cases} \tag{12.34}$$

For this system $\lambda_2 = (0.25; 0.5; 0.25)$ and we have (Fig. 12.3)

i) If $(0.26, 0.48) \in S_1$ then

$$(x^{(100)}, y^{(100)}) = (0.25, 0.5).$$

ii) If $(0.22, 0.52) \in S_2$ then

$$(x^{(100)}, y^{(100)}) = (0.25, 0.5).$$

iii) If $(0.24, 0.52) \in S_3$ then

$$(x^{(100)}, y^{(100)}) = (0.25, 0.5).$$

iv) If $(0.29, 0.48) \in S_4$ then

$$(x^{(100)}, y^{(100)}) = (0.25, 0.5).$$

To sum up for the initial point $\lambda^{(0)} = (x^{(0)}, y^{(0)}, z^{(0)}) \in int S^2$ we have

$$\lim_{n \to \infty} \lambda^{(n)} = \bar{\lambda}_2.$$

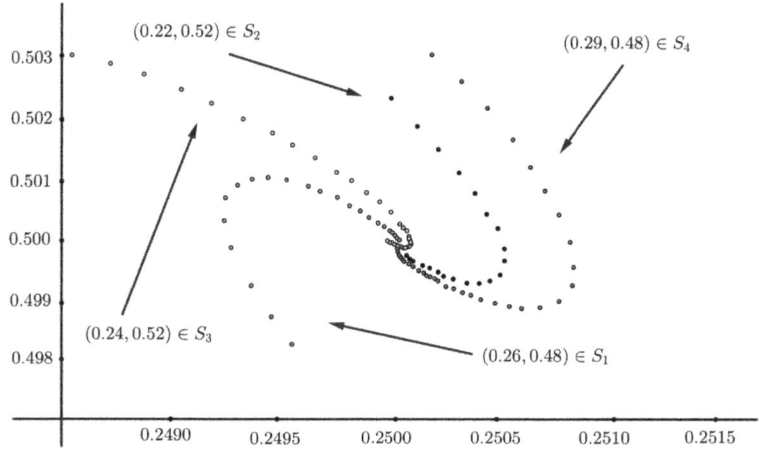

Fig. 12.3   Blue: $(0.26, 0.48) \in S_1$; black: $(0.22, 0.52) \in S_2$; green: $(0.24, 0.52) \in S_3$; red: $(0.29, 0.48) \in S_4$.

We get phase portraits of the trajectories of (12.4) shown in Fig. 12.4, Fig. 12.5 and Fig. 12.6.

Summarizing we obtain the following:

Theorem 12.2. *Let* $cd(c + d) \neq 0$ *and* $\lambda^{(0)} = (x^{(0)}, y^{(0)}, z^{(0)})$ *be an initial point. Then the following three cases hold*

(1) If $c < a$ *then*

$$\lim_{n \to \infty} \lambda^{(n)} = \bar{\lambda}_0 = (0, 0, 1).$$

(2) If $c > a, d < b$ *then*

$$\lim_{n \to \infty} \lambda^{(n)} = \begin{cases} (x^{(0)}, 0, 1 - x^{(0)}) & \text{if } y^{(0)} = 0, b = 0 \\ \bar{\lambda}_0 & \text{if } y^{(0)} = 0, b \neq 0 \\ \bar{\lambda}_1 & \text{if } y^{(0)} > 0. \end{cases} \tag{12.35}$$

(3) If $c > a, d > b > 0$ *then*

$$\lim_{n \to \infty} \lambda^{(n)} = \begin{cases} \bar{\lambda}_2 = \left(\frac{cd - ad - bc}{d(c+d)}, \frac{b}{d}, \frac{a - b + d}{c + d} \right) & \text{if } x^{(0)} \neq 0, y^{(0)} \neq 0 \\ \bar{\lambda}_0 & \text{if } y^{(0)} = 0 \\ \bar{\lambda}_1 & \text{if } x^{(0)} = 0. \end{cases} \tag{12.36}$$

Remark 12.4. We were able to prove the first line of the formula (12.36) only for a small neighborhood of $\bar{\lambda}_2$, because this fixed point is an attracting point (see Proposition 12.2).

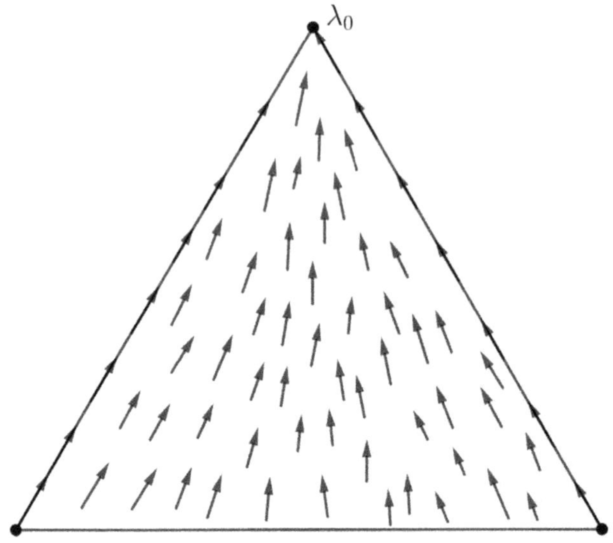

Fig. 12.4   If $c < a$.

Remark 12.5. To give mathematically full analysis of the operator (12.4) we considered the cases $cd(c+d) = 0$ and $cd(c+d) \neq 0$. But if $cd(c+d) = 0$ then some above mentioned results (when $y^{(n)}$ has zero limit) have not a biological meaning. Below biological interpretation of all remaining cases are presented.

12.4   Biological interpretations

In [23] for the continuous-time case the steady (stable) states of the system of equations (12.1) are found. Usually in models of ecosystems species at alternate levels in the food chain benefit from an increase in nutrient supply. It is known that if the supply of the nutrient which limits phytoplankton growth is increased, it is not the phytoplankton but the predatory zooplankton that benefit.

The results formulated in previous sections have the following biological interpretations:

Let $\lambda^{(0)} = (x^{(0)}, y^{(0)}, z^{(0)}) \in S^2$ be an initial state, i.e., $\lambda^{(0)}$ is the probability distribution on the set $\{N, P, Z\}$ of ecosystem.

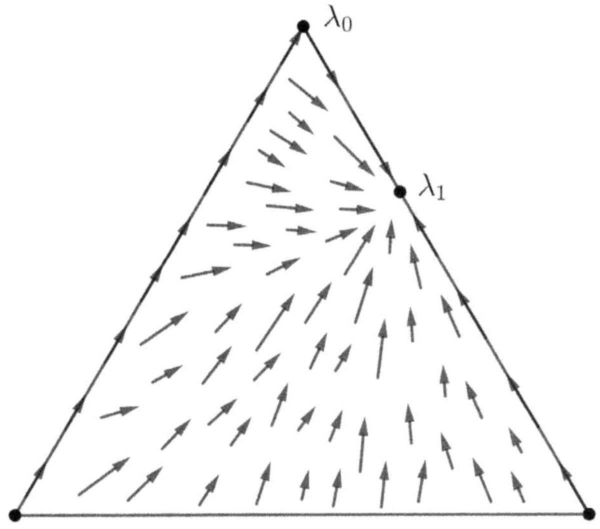

Fig. 12.5   If $c > a, d < b$.

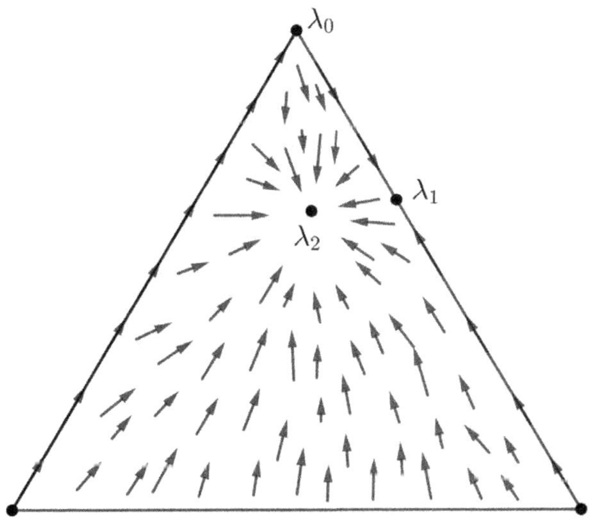

Fig. 12.6   If $c > a, d > b$.

Assume the trajectory $\lambda^{(m)}$ of this point has a limit $\lambda^* = (x^*, y^*, z^*)$ (equilibrium state) this means that the future of the system is stable: each N, P, Z survives with probability x^*, y^*, z^* respectively. For example, the nitrogen, N, of the system will disappear if its probability x^* is zero.

Each fixed point of the operator (12.4) is an equilibrium state and Theorem 12.1 gives that system may have a continuum set (but in under conditions of Theorem 12.2 it has up to three) of equilibrium states, the system stays in a neighborhood of one of the equilibrium states (stable fixed point), which depends on the initial state.

Moreover, as in continuous time ([23]) and in discrete time (Theorem 12.1 and Theorem 12.2) for $c < a$, the only steady state is the trivial one $\bar{\lambda}_0 = (0, 0, 1)$. Plankton levels are very low. As c increases past a, $\bar{\lambda}_0$ loses its stability as an eigenvalue passes through zero, and $\bar{\lambda}_1$ becomes stable. When c increases still further, there is a second critical value at $b < d(1 - \frac{a}{c})$ (Proposition 12.2) where $\bar{\lambda}_1$ loses its stability as an eigenvalue passes through zero, and $\bar{\lambda}_2$ becomes stable.

Bibliographical notes. This chapter is based on [226], [227]. For the continuous time version and further reading about ecosystem see [17], [23], [186], [190], [208], [210], [239], [260], [271], [274] and the references therein.

Bibliography

[1] V. M. Abraham, *A note on train algebras*, Proc. Edinburgh Math. Soc., **20**(1) (1976), 53–58.

[2] V. M. Abraham, *Linearising quadratic transformations in genetic algebras*, Thesis, Univ. London (1975).

[3] G. Abrams, A. Louly, E. Pardo, C. Smith, *Flow invariants in the classification of Leavitt path algebras*, J. Algebra **333** (2011), 202–231.

[4] A. T. Absalamov, *The global attractiveness of the fixed point of a gonosomal evolution operator*, arXiv:1905.13083v2.

[5] A. T. Absalamov, U. A. Rozikov, *The dynamics of gonosomal evolution operators*, arXiv:1809.09357.

[6] A. A. Albert, *On the power-associativity of rings*, Summa Brasil. Math. **2** (1948), 21–32.

[7] D. Aldila, H. Seno, *A Population Dynamics Model of Mosquito-Borne Disease Transmission, Focusing on Mosquitoes' Biased Distribution and Mosquito Repellent Use*, Bull. Math. Biol. **81**(12) (2019), 4977–5008.

[8] T. Ali, N. Gohain, *Some algebraic aspects and evolution of genetic code*, Applied analysis in biological and physical sciences, 201–215, Springer Proc. Math. Stat., 186, Springer, New Delhi, 2016.

[9] R. Almeida, N. R. O. Bastos, M. T. T. Monteiro, *A fractional Malthusian growth model with variable order using an optimization approach*, Stat. Optim. Inf. Comput. **6**(1) (2018), 4–11.

[10] A. Amei, J. Xu, *Inference of genetic forces using a Poisson random field model with non-constant population size*, J. Statist. Plann. Inference **203** (2019), 57–69.

[11] P. Anupam, M. Sandip, S. S. Lan, Y. Hidekatsu, *Micro-scale variability enhances trophic transfer and potentially sustains biodiversity in plankton ecosystems*, J. Theoret. Biol. 412 (2017), 86–93.

[12] S. Asmussen, *Applied probability and queues*, 2nd ed., Applications of Mathematics (New York), 51, Stochastic Modelling and Applied Probability, Springer-Verlag, New York, 2003.

[13] J. C. Avise, *The Hope, Hype, and Reality of Genetic Engineering*, 2014. New York: Oxford Univ. Press.

[14] N. Bacaër, *A short history of mathematical population dynamics*, Springer-Verlag London, Ltd., London, 2011.

[15] R. Bai, H. Liu, M. Zhang, *3-Lie algebras realized by cubic matrices*, Chin. Ann. Math. Ser. B **35**(2) (2014), 261–270.

[16] R. J. Baxter, *Exactly solved models in statistical mechanics*, Reprint of the 1982 original. Academic Press, Inc. [Harcourt Brace Jovanovich, Publishers], London, (1989).

[17] A. Beckmann, C.-E. Schaum, I. Hense, *Phytoplankton adaptation in ecosystem models*, J. Theoret. Biol. **468** (2019), 60–71.

[18] J. Bergman, D. Schrempf, C. V. Kosiol, *Claus Inference in population genetics using forward and backward, discrete and continuous time processes*, J. Theoret. Biol. **439** (2018), 166–180.

[19] S. N. Bernstein, *The solution of a mathematical problem related to the theory of heredity*, Uchn. Zapiski NI Kaf. Ukr. Otd. Mat. (1924) 83–115 (Russian).

[20] Z. S. Boxonov, U. A. Rozikov, *A dynamical system of temperature-dependent sex linked inheritance*, Uzbek Math. J. **2** (2018), 17–31.

[21] V. Bouche, *A new instability phenomenon in the Malthus-Verhulst model*, J. Phys. A: Math. Gen., **15**(6) (1982) 1841–1849.

[22] L. Breuer, D. Baum, *An introduction to queueing theory and matrix-analytic methods*, Springer, Dordrecht, 2005.

[23] N. Britton, *Essential Mathematical Biology*, Springer, London, 2003.

[24] I. Burdujan, *Two classes of algebras arising from genetics*, Annual Symposium on Mathematics Applied in Biology and Biophysics. Sci. Ann. Univ. Agric. Sci. Vet. Med. **45** (2002), 23–32.

[25] C. Y. Cabrera, M. Kanuni, M. M. Siles, *Basic ideals in evolution algebras*, Linear Algebra Appl. **570** (2019), 148–180.

[26] C. Y. Cabrera, M. M. Siles, M. V. Velasco, *Evolution algebras of arbitrary dimension and their decompositions*, Linear Algebra Appl. **495** (2016), 122–162.

[27] L. M. Camacho, A. Kh. Khudoyberdiyev, B. A. Omirov, *On the property of subalgebras of evolution algebras*, Algebr. Represent. Theory. **22**(2) (2019), 281–296.

[28] L. M. Camacho, J. R. Gomez, B. A. Omirov, R. M. Turdibaev, *Some properties of evolution algebras*, Bull. Korean Math. Soc. **50**(5) (2013), 1481–1494.

[29] J. M. Casas, M. Ladra, B. A. Omirov, U. A. Rozikov, *On evolution algebras*, Algebra Colloquium. **21**(2) (2014), 331–342.

[30] J. M. Casas, M. Ladra, B. A. Omirov, U. A. Rozikov, *On nilpotent index and dibaricity of evolution algebras*, Linear Algebra Appl. **439**(1) (2013), 90–105.

[31] J. M. Casas, M. Ladra, U. A. Rozikov, *A chain of evolution algebras*, Linear Algebra Appl. **435**(4) (2011), 852–870.

[32] J. M. Casas, M. Ladra, U. A. Rozikov, *Markov processes of cubic stochastic matrices: Quadratic stochastic processes*, Linear Algebra Appl. **575** (2019), 273–298.

[33] O. Castanos, U. U. Jamilov, U. A. Rozikov, *On Volterra quadratic stochastic operators of a two-sex population on* $S^1 \times S^1$, Uzbek Math. J. **3** (2018) 37–50.

[34] N. Chitnis, D. Hardy, T. Smith, *A periodically-forced mathematical model for the seasonal dynamics of malaria in mosquitoes*, Bull. Math. Biol. **74**(5) (2012), 1098–1124.

[35] N. Chitnis, T. Smith, R. Steketee, *A mathematical model for the dynamics of malaria in mosquitoes feeding on a heterogeneous host population*, Journal of Biological Dynamics, **2**(3) (2008), 259–285.

[36] M.-H. Chung, C. K. Kim, *Non-random mating involving inheritance of social status*, J. Comput. Biol. **17**(5) (2010), 745–754.

[37] M. H. Cortez, S. Patel, *The effects of predator evolution and genetic variation on predator-prey population-level dynamics*, Bull. Math. Biol. **79**(7) (2017), 1510–1538.

[38] R. Costa, *On the derivations of gametic algebras for polyploidy with multiple alleles*, Bol. Soc. Brasil. Mat. **13**(2) (1982), 69–81.

[39] M. A. Couto, F. J. C. Gutiérrez, *Dibaric algebras*, Proyecciones **19**(3) (2000), 249–269.

[40] R. L. Devaney, *An introduction to chaotic dynamical system*, Westview Press, 2003.

[41] F. Diego, K. Jónsdótti, *Associative operations on a three-element set*, TMME **5**(2–3) (2008), 257–268.

[42] A. Dzhumadil'daev, B. A. Omirov, U. A. Rozikov, *On a class of evolution algebras of "chicken" population*, Inter. Jour. Math. **25**(8) (2014), 1450073 (19 pages).

[43] A. Dzhumadil'daev, B. A. Omirov, U. A. Rozikov, *Constrained evolution algebras and dynamical systems of a bisexual population*, Linear Algebra Appl. **496** (2016), 351–380.

[44] A. Elduque, A. Labra, *On nilpotent evolution algebras*, Linear Algebra Appl. **505** (2016), 11–31.

[45] A. Elduque, A. Labra, *Evolution algebras and graphs*, J. Algebra Appl. **14**(7) (2015), 1550103, 10 pp.

[46] L. E. Elgolz, S. B. Norkin, *Introduction to the theory of differential equations with a deviating argument*, 1971, Moscow: Nauka.

[47] C. L. Epstein, C. A. Pop, *Transition probabilities for degenerate diffusions arising in population genetics*, Probab. Theory Related Fields. **173**(1–2) (2019), 537–603.

[48] I. M. H. Etherington, *Genetic algebras*, Proc. Roy. Soc. Edinburgh 59 (1939) 242–258.

[49] I. M. H. Etherington, *Non-associative ulyebra and the symbolism of genetics*, Proc. Roy. Soc. Edinburgh. Sect. B. **61** (1941) 24–42.

[50] O. J. Falcón, R. M. Falcón, J. Núñez, *Algebraic computation of genetic patterns related to three-dimensional evolution algebras*, Appl. Math. Comput. **319** (2018), 510–517.

[51] O. J. Falcón, R. M. Falcón, J. Núñez, *Classification of asexual diploid organisms by means of strongly isotopic evolution algebras defined over any field*, J. Algebra. **472** (2017), 573–593.

[52] J. Z. Farkas, S. A. Gourley, R. Liu, A.-A. Yakubu, *Modelling Wolbachia infection in a sex-structured mosquito population carrying West Nile virus*, J. Math. Biol. **75** (2017), 621–647.

[53] M. W. Feldman, *Dynamical systems from evolutionary population genetics*, Lectures in the sciences of complexity (Santa Fe, NM, 1988), 501–526, Santa Fe Inst. Stud. Sci. Complexity Lectures, I, Addison-Wesley, Redwood City, CA, 1989.

[54] P. Fortini, R. Barakat, *An algorithm for gene frequency changes for linked autosomal loci based on genetic algebras*, J. Math. Anal. Appl. **83** (1981), 135–143.

[55] A. Gallegos, T. Plummer, D. Uminsky, C. Vega, C. Wickman, M. Zawoiski, *A mathematical model of a crocodilian population using delay-differential equations*, J. Math. Biol. **57**(5) (2008), 737–754.

[56] O. Galor, *Discrete dynamical systems*, Springer, Berlin, 2007.

[57] G. Galperin, A. Zemlyakov, *Mathematical billiards*, Nauka, Moscow, 1990 (in Russian).

[58] N. N. Ganikhodjaev, U. U. Jamilov, *Contracting quadratic operators of bisexual population*, Appl. Math. Inf. Sci. **9**(5) (2015), 2645–2650.

[59] N. N. Ganikhodzhaev, U. U. Zhamilov, R. T. Mukhitdinov, *Nonergodic quadratic operators for a two-sex population*, Ukrainian Math. J. **65**(8) (2014), 1282–1291.

[60] N. N. Ganikhodjaev, *On stochastic processes generated by quadric operators*, J. Theoret. Probab. **4**(4) (1991), 639–653.

[61] N. N. Ganikhodjaev, H. Akin, F. M. Mukhamedov, *On the ergodic principle for Markov and quadratic stochastic processes and its relations*, Linear Algebra Appl. **416**(2–3), (2006) 730–741.

[62] N. N. Ganikhodjaev, R. T. Mukhitdinov, *On a class of non-Volterra quadratic operators*, Uzbek Math. (3–4) (2003), 65–69.

[63] N. N. Ganikhodjaev, U. A. Rozikov, *On quadratic stochastic operators generated by Gibbs distributions*, Regular and Chaotic Dynamics. **11**(4) (2006), 467–473.

[64] N. N. Ganikhodjaev, *An application of the theory of Gibbs distributions to mathematical genetics*, Doklady Math. **61** (2000) 321–323.

[65] N. N. Ganikhodjaev, *Algebras with genetic realization and corresponding evolutionary population dynamics*, Topics in functional analysis and algebra, 169–178, Contemp. Math., 672, Amer. Math. Soc., Providence, RI, 2016.

[66] N. N. Ganikhodjaev, R. N. Ganikhodjaev, U. U. Jamilov, *Quadratic stochastic operators and zero-sum game dynamics*, Ergodic Theory Dynam. Systems **35**(5) (2015), 1443–1473.

[67] R. N. Ganikhodzhaev, F. M. Mukhamedov, A. T. Pirnapasov, I. Qaralleh, *Genetic Volterra algebras and their derivations*, Comm. Algebra **46**(3) (2018), 1353–1366.

[68] R. N. Ganikhodzhaev, F. M. Mukhamedov, U. A. Rozikov, *Quadratic stochastic operators and processes: results and open problems*, Inf. Dim. Anal. Quant. Prob. Rel. Fields., **14**(2) (2011), 279–335.

[69] R. N. Ganikhodzhaev, *Quadratic stochastic operators, Lyapunov functions and tournaments*, Acad. Sci. Sb. Math. **76**(2) (1993), 489–506.

[70] R. N. Ganikhodzhaev, *A chart of fixed points and Lyapunov functions for a class of discrete dynamical systems*, Math. Notes. **56**(5–6) (1994), 1125–1131.

[71] R. N. Ganikhodzhaev, D. B. Eshmamatova, *Quadratic automorphisms of a simplex and the asymptotic behavior of their trajectories*, Vladikavkaz. Mat. Zh. **8**(2) (2006), 12–28. (Russian)

[72] R. N. Ganikhodzhaev, *Quadratic stochatic operators and asymptotic behavior of their trajectories*, Thesis Doctor of Sciences, Tashkent. (1995), (Russian)

[73] R. N. Ganikhodzhaev, A. M. Dzhurabaev, *The set of equilibrium states of quadratic stochastic operators of type V_π.*, Uzbek. Mat. Zh. 3 (1998), 23–27.

[74] R. N. Ganikhodzhaev, R. E. Abdirakhmanova, *Description of quadratic automorphisms of a finite-dimensional simplex*, Uzbek. Mat. Zh. 1 (2002) 7–16.

[75] R. N. Ganikhodzhaev, A. I. Eshniyazov, *Bistochastic quadratic operators*, Uzbek. Mat. Zh. 3 (2004) 29–34.

[76] J. Gayon, M. Cobb, *Darwinism's struggle for survival: Heredity and the hypothesis of natural selection*, Cambridge University Press, 1998.

[77] H.-O. Georgii, *Gibbs Measures and Phase Transitions*, 2nd ed. De Gruyter Studies in Mathematics, **9**. De Gruyter, Berlin, 2011.

[78] M. Gerstenhaber, *On the deformation of rings and algebras*, Ann. Math. (2) **79** (1964), 59–103.

[79] H. Gonshor, *Special train algebras arising in genetics*, Proc. Edinburgh Math. Soc. (2) **12** (1960), 41–53.

[80] H. Gonshor, *Special train algebras arising in genetics*, II, Proc. Edinburgh Math. Soc. (2) **14** (1965), 333–358.

[81] H. Gonshor, *Contributions to genetic algebras*, Proc. Edinburgh Math. Soc. **17** (1971), 289–298.

[82] H. Gonshor, *Derivations in genetic algebras*, Comm. Algebra **16**(8) (1988), 1525–1542.

[83] J. Gray, *A history of abstract algebra. From algebraic equations to modern algebra*, Springer Undergraduate Mathematics Series. Springer, Cham, 2018. xxiv+414 pp.

[84] A. N. Griskhov, *On the genetic property of Bernstein algebras*, Soviet Math. Dokl., **35** (1987), 489–492 (Russian).

[85] M. Gromov, *Crystals, proteins, stability and isoperimetry*, Bull. Amer. Math. Soc. (N.S.) **48** (2) (2011), 229–257.

[86] K. P. Hadeler, R. Waldstätter, A. Wörz-Busekros, *Models for pair formation in bisexual populations*, J. Math. Biol. **26** (1989), 635–649.

[87] K. P. Hadeler, *Pair formation models with maturation period*, J. Math, Biol. **32** (1993), 1–15.

[88] P. Hänggi, H. Thomas, *Time evolution, correlations, and linear response of non-Markov processes*, Zeitschrift Phys. B. **26** (1977), 85–92.

[89]   J. K. Hale, P. Waltman, *Persistence in infinite-dimensional systems*, SlAM J. Math. Anal. **20** (1989), 388–396.

[90]   A. J. M. Hardin, U. A. Rozikov, *A quasi-strictly non-Volterra quadratic stochastic operator*, Qualit. Theory Dyn. Syst. **18**(3) (2019), 1013–1029.

[91]   A. S. Hegazi, H. Abdelwahab, *Nilpotent evolution algebras over arbitrary fields*, Linear Algebra Appl. **486** (2015), 345–360.

[92]   I. R. Hentzel, L. A. Peresi, P. Holgate, *On k-th order Bernstein algebras and stability at the $k+1$ generation in polyploids*, IMA J. Math. Appl. in Med. and Biol., **7** (1990), 33–40.

[93]   I. Heuch, *The linear algebra for linked loci with mutation*, Math. Biosci. **16** (1973) 263–271.

[94]   I. Heuch, *Genetic algebras for systems with linked loci*, Math. Biosci. **34**(1-2) (1977), 35–47.

[95]   I. Heuch, *Partial and complete sex linkage in infinite populations*, J. Math. Biol. **1**(4) (1974/75), 331–343.

[96]   I. Heuch, *k loci linked to a sex factor in haploid individuals*, Biometrische Z. **14**(1) (1972), 7–68.

[97]   A. Hillion, Mathematical theories of populations, What Do I Know? (French), 2258. Presses Universitaires de France, Paris, 1986. 128 pp.

[98]   J. Hofbauer, K. Sigmund, *Evolutionary games and population dynamics*, Cambridge Univ. Press, Cambridge, 1998.

[99]   J. Hofbaver, K. Sigmund, *The theory of evolution and dynamical systems*, Cambridge Univ. Press, 1988.

[100]   P. Holgate, *The genetic algebra of k linked loci*, Proc. London Math. Sot. **l8** (1968), 315–327.

[101]   P. Holgate, *Genetic algebras associated with sex linkage*, Proc. Edinburgh Math. Soc. (2) **17** (1970/71), 113–120.

[102]   P. Holgate, *Characterisations of genetic algebras*, J. London Math. Soc. (2) **6** (1972), 169–174.

[103]   P. Holgate, *Genetic algebras satisfying Bernstein's stationarity principle*, J. London Math. Soc. (2), **9** (1975), 613–623.

[104]   P. Holgate, *The interpretation of derivations in genetic algebras*, Linear Algebra Appl. **85** (1987), 75–79.

[105]   P. Holgate, *Biometric and chromosome algebras*, J. Appl. Probab. **29**(2) (1992), 247–254.

[106]   A. J. Homburg, U. U. Jamilov, M. Scheutzow, *Asymptotics for a class of iterated random cubic operators*, Nonlinearity 32 (2019), no. 10, 3646–3660.

[107]   J. E. Hornos, Y. M. M. Hornos, *Perspectives in the algebraic approach to the genetic code*, Resenhas **2**(4) (1996), 363–371.

[108]   X. Huang, H. Deng, F. Chen, *Note on the stability property of the trivial equilibrium point of a stage structured mosquito population model*, Appl. Math. E-Notes **19** (2019), 465–468.

[109]   M. Iannelli, M. Martcheva, *A semigroup approach to the well posedness of the two-sex population model*, Dynam. Systems. Appl. **6** (1997), 353–369.

[110] M. Iannelli, F. Milner, *The basic approach to age-structured population dynamics. Models, methods and numerics*, Lecture Notes on Mathematical Modelling in the Life Sciences. Springer, Dordrecht, 2017.

[111] M. Iannelli, A. Pugliese, *An introduction to mathematical population dynamics*, Along the trail of Volterra and Lotka. Unitext, 79. La Matematica per il 3+2. Springer, Cham, 2014.

[112] A. N. Imomkulov, *Isomorphicity of two dimensional evolution algebras and evolution algebras corresponding to their idempotents*, Uzbek Math. J. **2** 2018, 63–73.

[113] H. Inaba, R. Saito, N. Bacaër, *An age-structured epidemic model for the demographic transition*, J. Math. Biol. **77**(5) (2018), 1299–1339.

[114] H. Inaba, *Age-structured population dynamics in demography and epidemiology*, Springer, Singapore, 2017.

[115] M. Izumi, *The flow of weights and the Cuntz-Pimsner algebras*, Comm. Math. Phys. **357**(1) (2018), 203–229.

[116] N. Jacobson, *Lie algebras*, Interscience Tracts in Pure and Applied Mathematics, No. 10, Interscience Publishers (a division of John Wiley & Sons), New York-London, 1962.

[117] U. U. Jamilov, A. Yu. Khamraev, M. Ladra, *On a Volterra cubic stochastic operator*, Bull. Math. Biol. **80**(2) (2018), 319–334.

[118] U. U. Jamilov, M. Ladra, *On identically distributed non-Volterra cubic stochastic operator*, J. Appl. Nonlinear Dyn. **6**(1) (2017), 79–90.

[119] U. U. Jamilov, M. Scheutzow, M. Wilke-Berenguer, *On the random dynamics of Volterra quadratic operators*, Ergodic Theory Dynam. Systems **37**(1) (2017), 228–243.

[120] R. Jenks, *Quadratic differential systems for interactive population models*, J. Differ. Equations. **5** (1969), 497–514.

[121] D. T. Jennings, M. W. Houseweart, *Sex-biased predation by web-spinning spiders (Araneae) on spruce budworm moths*, J. Arachnol. **17** (1989), 179–194.

[122] G. Jumarie, *New stochastic fractional models for Malthusian growth, the Poissonian birth process and optimal management of populations*, Math. Comput. Modelling **44**(3–4) (2006), 231–254.

[123] S. Karlin, *Mathematical models, problems, and controversies of evolutionary theory*, Bull. Amer. Math. Soc. (N.S.) **10**(2) (1984), 221–274.

[124] S. Karlin, S. Lessard, *On the optimal sex-ratio: a stability analysis based on a characterization for one-locus multiallele viability models*, J. Math. Biol. **20**(1) (1984), 15–38.

[125] Y. Kawamura, *Structure of cubic matrix mechanics*, Progr. Theoret. Phys. **109**(1) (2003), 1–10.

[126] H. Kesten, *Quadratic transformations: A model for population growth*, I, II, Adv. Appl. Probab. **2**(2) (1970) 1–82; 179–228.

[127] A. Yu. Khamraev, *On a Volterra-type cubic operator*, Uzbek. Mat. Zh. 3: (2009), 65–71.

[128] A. Yu. Khamraev, *On cubic operators of Volterra type*, Uzbek. Mat. Zh. 2: (2004), 79–84.

[129] A. Yu. Khamraev, *On the dynamics of a quasistrictly non-Volterra quadratic stochastic operator*, (Russian) Ukrain. Mat. Zh. **71**(8) (2019), 1116–1122.

[130] A. Kh. Khudoyberdiyev, B. A. Omirov, I. Qaralleh, *Few remarks on evolution algebras*, J. Algebra Appl. **14**(4) (2015), 1550053, 16 pp.

[131] G. Kirlinger, *Permanence in Lotka-Volterra equations: linked prey-predator systems*, Math. Biosci. **82**(2) (1986), 165–191.

[132] A. N. Kolmogorov, *On analytic methods in probability theory*, Uspekhi Mat. Nauk. 5, 5–41 (1938) (Russian), English transl., Selected works of A. N. Kolmogorov, Vol. II, Kluwer, Dordrecht 1992, article 9.

[133] V. Kolokoltsov, *Nonlinear Markov processes and kinetic equations*, Cambridge University Press, 2010.

[134] U. Krause, *Positive dynamical systems in discrete time: Theory, Models, and Applications*, Walter de Gruyter, 2015.

[135] H. W. Kuhn, A. W. Tucker (eds.), *Linear inequalities and related systems*, Annal. Math. Stud. Princeton Univ. Press. 1985.

[136] A. Labra, M. Ladra, U. A. Rozikov, *An evolution algebra in population genetics*, Linear Algebra Appl. **457** (2014), 348–362.

[137] M. Ladra, B. A. Omirov, U. A. Rozikov, *Dibaric and evolution algebras in biology*, Lobachevskii Jour. Math. **35**(3) (2014), 198–210.

[138] M. Ladra, U. A. Rozikov, *Evolution algebra of a bisexual population*, Jour. Algebra. **378** (2013), 153–172.

[139] M. Ladra, U. A. Rozikov, *Algebras of cubic matrices*, Linear and Multilinear Alg. **65**(7) (2017), 1316–1328.

[140] M. Ladra, U. A. Rozikov, *Flow of finite-dimensional algebras*, Jour. Algebra. **470** (2017), 263–288.

[141] M. Ladra, U. A. Rozikov, *Construction of flows of finite-dimensional algebras*, Jour. Algebra. **492** (2017), 475–489.

[142] M. Ladra, U. A. Rozikov, *Evolution algebra of a "chicken" population*, arXiv:1307.4916v1.

[143] T. Y. Lam, *Lectures on modules and rings*, vol. 189 of Graduate Texts in Mathematics, Springer-Verlag, New York, 1999.

[144] H. Länger, *An interesting application of algebra to genetics*, Elem. Math. **60** (2005) 25–32.

[145] G. T. Lee, *Abstract algebra. An introductory course*, Springer Undergraduate Mathematics Series. Springer, Cham, 2018. xi+301 pp.

[146] S. P. Lee, M.-H. Chung, C. K. Kim, K. Nahm, *Fractional populations in sex-linked inheritance*, Phys. A **291**(1–4) (2001), 533–541.

[147] A. Levin, *Graphs, matrices, and populations: Linear algebraic techniques in theoretical computer science and population genetics*, Thesis (Ph.D.) Massachusetts Institute of Technology. 2013.

[148] Y. Li, X. Liu, *An impulsive model for Wolbachia infection control of mosquito-borne diseases with general birth and death rate functions*, Nonlinear Analysis: Real World Applications, **37** (2017), 412–432.

[149] J. Lindström, H. Kokko, *Sexual reproduction and population dynamics: the role of polygyny and demographic sex differences*, Proc. R. Soc. Lond. B **265** (1998) 483–488.

[150] H. Liu, *Population dynamics of different sex with different birth and death rate*, Chinese Journd of Ecology, **22** (2003), 63–65.

[151] S. Lovett, *Abstract algebra. Structures and applications*, CRC Press, Boca Raton, FL, 2016. xii+708 pp.

[152] Y. Lou, K. Liu, D. He, D. Gao, S. Ruan, *Modelling diapause in mosquito population growth*, J. Math. Biol. **78**(7) (2019), 2259–2288.

[153] J. López-Gómez, *Metasolutions of parabolic equations in population dynamics*, CRC Press, Boca Raton, FL, 2016.

[154] A. M. Lutambi, M. A. Penny, T. Smith, N. Chitnis, *Mathematical modelling of mosquito dispersal in a heterogeneous environment*, Math. Biosci. **241**(2) (2013), 198–216.

[155] Y. I. Lyubich, *Mathematical structures in population genetics*, Springer-Verlag, Berlin, 1992.

[156] Y. I. Lyubich, *Ultranormal case of the Bernstein problem*, Func. Anal. Appl. **31**(1) (1997), 60–62.

[157] Y. I. Lyubich, *Basic concepts and theorems of the evolutionary genetics of free populations*, Russian Math. Surveys, **26**(5) (1971), 51–123.

[158] V. M. Maksimov, *Cubic stochastic matrices and their probability interpretations*, Theory Probab. Appl. **41**(1) (1996) 55–69.

[159] A. Malevanets, R. Kapral, *Solute molecular dynamics in a mesoscale solvent*, J. Chem. Phys. **112** (2000), 7260–7269.

[160] B. J. Mamurov, U. A. Rozikov, *On cubic stochastic operators and processes*, J. Phys. Conf. Ser. **697** (2016), 012017, 15 pages.

[161] B. J. Mamurov, *On the averagings of cubic processes*, Uzbek. Mat. Zh. 4 (2010), 113–117.

[162] M. Martcheva, *Exponential growth in age-structured two-sex populations*, Mathematical Biosciences. **157** (1999), 1–22.

[163] P. Massatt, *Stability and fixed points of point-dissipative systems*, J. Diff. Equat. **40**(2) (1981), 217–231.

[164] K. Matsumoto, *Continuous orbit equivalence, flow equivalence of Markov shifts and circle actions on Cuntz-Krieger algebras*, Math. Z. **285**(1–2) (2017), 121–141.

[165] C. D. Mayer, *Matrix Analysis and Applied Linear Algebra*, SIAM, 2000.

[166] J. Maynard-Smith, G. R. Price, *The logic of animal conflict*, Nature **246**(5427) (1973), 15–18.

[167] C. McCaig, R. Norman, C. Shankland, *Process algebra models of population dynamics*, In: Horimoto K., Regensburger G., Rosenkranz M., Yoshida H. (eds.), Algebraic Biology. Lecture Notes in Computer Science, vol. 5147. Springer, Berlin, Heidelberg. AB 2008.

[168] C. McCaig, A. Fenton, A. Graham, C. Shankland, R. Norman, *Using process algebra to develop predator-prey models of within-host parasite dynamics*, J. Theoret. Biol. **329** (2013), 74–81.

[169] P. Mellon, M. V. Velasco, *Analytic aspects of evolution algebras*, Banach J. Math. Anal. **13**(1) (2019), 113–132.

[170] A. Micali, M. Ouattara, *Structure des algébres de Bernstein*, Linear Alg. Appl., **218** (1995), 77–88.

[171] A. Micali, O. Suzuki, *Algebraic methods for genetics (I), (Some useful algebras in genetics)*, RIMS, Math. Analysis. **1551** (2007), 117–122.

[172] A. Micali, O. Suzuki, *Algebraic methods for genetics (II) (Cuntz algebra and Cuntz-Krieger algebras)*, RIMS, Math. Analysis. **1597** (2008), 214–219.

[173] A. D. Mishkis, *Linear differensial equations with delay of argument*, Nauka, Moscow. 1972. (Russian)

[174] M. Möhle, M. Notohara, *An extension of a convergence theorem for Markov chains arising in population genetics*, J. Appl. Probab. **53**(3) (2016), 953–956.

[175] F. M. Mukhamedov, O. N. Khakimov, B. A. Omirov, I. Qaralleh, *Derivations and automorphisms of nilpotent evolution algebras with maximal nilindex*, J. Algebra Appl. **18**(12) (2019), 1950233, 23 pp.

[176] F. M. Mukhamedov, U. U. Jamilov, A. T. Pirnapasov, *On nonergodic uniform Lotka-Volterra operators*, Math. Notes **105**(1–2) (2019), 258–264.

[177] F. M. Mukhamedov, A. F. Embong, *On stable b-bistochastic quadratic stochastic operators and associated non-homogeneous Markov chains*, Linear Multilinear Algebra **66**(1) (2018), 1–21.

[178] F. M. Mukhamedov, A. F. Embong, C. H. Pah, *Orthogonal preserving quadratic stochastic operators: infinite dimensional case*, J. Phys. Conf. Ser. **819** (2017), 012010, 7 pages.

[179] F. M. Mukhamedov, A. F. Embong, A. Rosli, *Orthogonal-preserving and surjective cubic stochastic operators*, Ann. Funct. Anal. **8**(4) (2017), 490–501.

[180] F. M. Mukhamedov, *On circle preserving quadratic operators*, Bull. Malays. Math. Sci. Soc. **40**(2) (2017), 765–782.

[181] F. M. Mukhamedov, M. Saburov, *Stability and monotonicity of Lotka-Volterra type operators*, Qual. Theory Dyn. Syst. **16**(2) (2017), 249–267.

[182] F. M. Mukhamedov, M. T. Bin, H. Muhammad, *On Volterra and orthogonality preserving quadratic stochastic operators*, Miskolc Math. Notes **17**(1) (2016), 457–470.

[183] F. M. Mukhamedov, N. N. Ganikhodjaev, *Quantum quadratic operators and processes*, Lecture Notes in Mathematics, 2133. Springer, Cham, 2015.

[184] F. M. Mukhamedov, N. A. Supar, *On marginal processes of quadratic stochastic processes*, Bull. Malay. Math. Sci. Soc. **38** (2015), 1281–1296.

[185] F. M. Mukhamedov, N. A. Supar, Ch. H. Pah, *On quadratic stochastic processes and related differential equations*, J. Phys.: Conf. Ser. **435** (2013), 012013.

[186] J. Müller, C. Kuttler, *Methods and models in mathematical biology*, Deterministic and stochastic approaches. Lecture Notes on Mathematical Modelling in the Life Sciences. Springer, Heidelberg, 2015.

[187] L. Multerer, T. Smith, N. Chitnis, *Modeling the impact of sterile males on an Aedes aegypti population with optimal control*, Math. Biosci. **311** (2019), 91–102.

[188] Sh. N. Murodov, *Classification dynamics of two-dimensional chains of evolution algebras*, Internat. J. Math. **25**(2) (2014), 1450012.

[189] Sh. N. Murodov, *Time depending dynamics of chains of evolution algebras*, PhD thesis. University Santiago de Compostela (2019).

[190] S. Muthiah, *Some Statistical and Dynamical Models for the Analysis of Microbial Ecosystems and Their Genomic Data*, Thesis (Ph.D.) University of Maryland, College Park. 2019. 215 pp.

[191] B. A. Narkuziev, *Evolution algebras corresponding to permutations*, Uzbek. Mat. Zh. **4** (2014), 109–114.

[192] N. B. Narziev, S. N. Nishanov, *On a genetic algebra arising in biological population models*, (Russian) Uzbek. Mat. Zh. **4** (2011), 150–156.

[193] J. Newton, *Evolutionary game theory: A renaissance*, Games. **9**(2) (2018), paper 31, 67 pages.

[194] R. W. Ogden, *Non-Linear Elastic Deformations*, Dover Publications, Inc., New York, 1984.

[195] B. A. Omirov, U. A. Rozikov, M. V. Velasco, *A class of nilpotent evolution algebras*, Communications in Algebra. **47**(4) (2019), 1556–1567.

[196] B. A. Omirov, U. A. Rozikov, *On subalgebras of an evolution algebra of a "chicken" population*, J. Algebra Relat. Topics. **5**(2) (2017), 13–24.

[197] B. A. Omirov, U. A. Rozikov, K. N. Tulenbayev, *On real chains of evolution algebras*, Linear Multilinear Algebra. **63**(3) (2015) 586–600.

[198] J. M. Ortega, W. C. Rheinboldt, *Iterative solution of nonlinear equations in several variables*, Academic Press, New York, 1970.

[199] H. S. Oxley, *Genetics in terms of dynamical systems*, Thesis (D.A.) Idaho State University. 1994. 75 pp,

[200] L.-P. Pang, E. Spedicato, Z.-Q. Xia, W. Wang, *A method for solving the system of linear equations and linear inequalities*, Math. Comp. Model. **46**, (2007) 823–836.

[201] I. Paniello, *Genetic coalgebras and their cubic stochastic matrices*, J. Algebra Appl. **16**(12) (2017), 1750239, 22 pp.

[202] I. Paniello, *On actions on cubic stochastic matrices*, Markov Process. Related Fields **23**(2) (2017), 325–348.

[203] I. Paniello, *Backwards genetic inheritance through coalgebra-graphs*, Linear Multilinear Algebra **65**(5) (2017), 943–961.

[204] I. Paniello, *On evolution operators of genetic coalgebras*, J. Math. Biol. **74**(1–2) (2017), 149–168.

[205] I. Paniello, *Marginal distributions of genetic coalgebras*, J. Math. Biol. **68**(5) (2014), 1071–1087.

[206] L. A. Peresi, *The derivation algebra of gametic algebra for linked loci*, Math. Biosci. **91** (1988), 151–156.

[207] J. H. Pollard, *Mathematical models for the growth of human populations, the two sex problem*, Cambridge University Press. 1973.

[208] S. Pramanick, A. Chatterjee, S. Pal, *Interspecies competition between phytoplankton and zooplankton in marine ecosystem*, Nonlinear Stud. **25**(4) (2018), 867–882.

[209] I. Qaralleh, *On genetic and evolution algebras*, J. Phys. Conf. Ser. **819** (2017), 012011, 6 pages.

[210] W. Qin, X. Tan, X. Shi, J. Chen, X. Liu, *Dynamics and bifurcation analysis of a Filippov predator-prey ecosystem in a seasonally fluctuating environment*, Internat. J. Bifur. Chaos Appl. Sci. Engrg. **29**(2) (2019), 1950020, 16 pp.

[211] M. L. Reed, *Algebraic structure of genetic inheritance*, Bull. Amer. Math. Soc. (N.S.) **34** (2) (1997), 107–130.

[212] R. T. Rockafellar, *Convex Analysis*, Princeton Univ. Press. 1996.

[213] U. A. Rozikov, *An introduction to mathematical billiards*, World Scientific Publishing Co. Pte. Ltd., Hackensack, NJ, 2019.

[214] U. A. Rozikov, *Evolution operators and algebras of sex linked inheritance*, Asia Pac. Math. Newsl. **3**(1) (2013), 6–11.

[215] U. A. Rozikov, A. Yu. Khamraev, *On construction and a class of non-Volterra cubic stochastic operators*, Nonlinear Dyn. Syst. Theory. **14**(1) (2014), 92–100.

[216] U. A. Rozikov, A. Yu. Khamraev, *On cubic operators, defined on the finite-dimensional simplexes*, Ukraine Math. Jour. **56**(10) (2004), 1699–1711.

[217] U. A. Rozikov, Sh. N. Murodov, *Chain of evolution algebras of "chicken" population*, Linear Algebra Appl. **450** (2014), 186–201.

[218] U. A. Rozikov, Sh. N. Murodov, *Dynamics of two-dimensional evolution algebras*, Lobachevskii Jour. Math. **34**(4) (2013), 344–358.

[219] U. A. Rozikov, A. Zada, *ℓ-Volterra quadratic stochastic operators: Lyapunov functions, trajectories*, Appl. Math. Inf. Sci. **6**(2) 2012, 329–335.

[220] U. A. Rozikov, A. Zada, *On a class of separable quadratic stochastic operators*, Lobachevskii Jour. Math. **32**(4) (2011), 397–406.

[221] U. A. Rozikov, A. Zada, *On ℓ-Volterra quadratic stochastic operators*, Inter. Journal Biomath. **3**(2) (2010), 143–159.

[222] U. A. Rozikov, A. Zada, *On ℓ-Volterra quadratic stochastic operators*, Doklady Math. **79**(1) (2009), 32–34.

[223] U. A. Rozikov, J. P. Tian, *Evolution algebras generated by Gibbs measures*, Lobachevskii Jour. Math. **32**(4) (2011), 270–277.

[224] U. A. Rozikov, S. Nazir, *Separable quadratic stochastic operators*, Lobachevskii Jour. Math. **31**(3) (2010), 214–220.

[225] U. A. Rozikov, N. B. Shamsiddinov, *On Non-Volterra quadratic stochastic operators generated by a product measure*, Stoch. Anal. Appl. **27** (2) (2009), 353–362.

[226] U. A. Rozikov, S. K. Shoyimardonov, *On ocean ecosystem discrete time dynamics generated by ℓ-Volterra operators*, Inter. Jour. Biomath. **12**(2) (2019), 1950015, (24 pages).

[227] U. A. Rozikov, S. K. Shoyimardonov, *Discrete-time dynamics of an ocean ecosystem*, Doklady Acad. Nauk RUz. 3 (2018), 13–18.

[228] U. A. Rozikov, R. Varro, *Dynamical systems generated by a gonosomal evolution operator*, Jour. Discontinuity, Nonlinearity, and Complexity. **5**(2) (2016), 173–185.

[229] U. A. Rozikov, M. V. Velasco, *Discrete-time dynamical system and an evolution algebra of mosquito population*, Jour. Math. Biology. **78**(4) (2019), 1225–1244.

[230] U. A. Rozikov, U. U. Zhamilov, *Volterra quadratic stochastic operators of bisexual population*, Ukraine Math. Jour., **63**(7) (2011), 985–998.

[231] U. A. Rozikov, U. U. Zhamilov, *On F-quadratic stochastic operators*, Math. Notes., **83**(4) (2008), 554–559.

[232] U. A. Rozikov, U. U. Zhamilov, *On dynamics of strictly non-Volterra quadratic operators on two-dimensional simplex*, Sbornik: Math. **200**(9) (2009), 1339–1351.

[233] U. A. Rozikov, U. U. Zhamilov, *On trajectories of strongly non-Volterra quadratic operators*, Doklady Acad. Nauk RUz. 1 (2009), 3–6.

[234] U. A. Rozikov, U. U. Zhamilov (Jamilov), *On quadratic stochastic operators corresponding to cyclic groups*, Jour. Discontinuity, Nonlinearity, and Complexity. **6**(2) (2017), 147–164.

[235] K. Sabelfeld, *Stochastic algorithms in linear algebra-beyond the Markov chains and von Neumann-Ulam scheme*, Numerical methods and applications, 14–28, Lecture Notes in Comput. Sci., 6046, Springer, Heidelberg, 2011.

[236] R. Sanchez, E. Morgado, R. Grau, *Gene algebra from a genetic code algebraic structure*, J. Math. Biol. **51**(4) (2005), 431–457.

[237] T. A. Sarymsakov, N. N. Ganikhodjaev, *Analytic methods in the theory of quadric stochastic operators*, J. Theoret. Probab. **3**(1) (1990), 51–70.

[238] T. A. Sarymsakov, N. N. Ganikhodjaev, *On the ergodic principle for quadratic processes*, Soviet Math. Dokl. **43** (1991), 279–283.

[239] B. E. Schaffer, *The Stochastic Dynamics of Biomass and Soil Moisture in Water-Limited Ecosystems*, Thesis (Ph.D.) Princeton University. 2018. 197 pp.

[240] R. D. Schafer, *An introduction to nonassociative algebras*, Academic Press, New York, 1966.

[241] R. D. Schafer. *Structure of genetic algebras*, American J. Math., **71** (1949), 121–135.

[242] M. Schartl, *A comparative view on sex determination in medaka*, Mechanisms of Development **121** (7–8) (2004), 639–645.

[243] P. H. T. Schimit, *Evolutionary aspects of spatial prisoner's dilemma in a population modeled by continuous probabilistic cellular automata and genetic algorithm*, Appl. Math. Comput. **290** (2016), 178–188.

[244] H. Singh, J. Dhar, *Mathematical population dynamics and epidemiology in temporal and spatio-temporal domains*, Apple Academic Press, Oakville, ON, 2019.

[245] R. Singh, S. Ali, M. Jain, A. A. Raina, *Mathematical model for malaria with mosquito-dependent coefficient for human population with exposed class*, J. National Sci. Found. Sri Lanka, **47**(2) (2019), 185–108.

[246] C. H. Skiadas, C. Skiadas, *Charilaos Exploring the health state of a population by dynamic modeling methods*, The Springer Series on Demographic Methods and Population Analysis, 45. Springer, Cham, 2018.

[247] A. N. Sharkovskii, S. F. Kolyada, A. G. Sivak, V. V. Fedorenko, *Dynamics of one-dimensional mappings*, Naukova Dumka, Kiev, (1989) (Russian).

[248] A. N. Shiryaev, *Probability*, 2nd ed., Springer-Verlag, New York, 1996.

[249] J. P. Spence, *Probabilistic Models and Statistical Inference in Population Genetics*, Thesis (Ph.D.) University of California, Berkeley. 2019. 125 pp.

[250] J. S. Stonebraker, *Product-generation transition decision making for Bayer's hemophilia drugs: global capacity expansion under uncertainty with supply-demand imbalances*, Oper. Res. **61**(5) (2013), 1119–1133.

[251] Y. Suhov, M. Kelbert, *Probability and statistics by example*, vol. II, Markov chains: a primer in random processes and their applications, Cambridge Univ. Press, Cambridge, 2008.

[252] W. H. Sulis, *Modeling stochastic complexity in complex adaptive systems: non-Kolmogorov probability and the process algebra approach*, Nonlinear Dyn. Psychol. Life Sci. **21**(4) (2017), 407–440.

[253] G. Teschl, *Ordinary differential equations and dynamical systems*, Providence: American Mathematical Society. (2012).

[254] J. P. Tian, *Evolution algebras and their applications*, Lecture Notes in Mathematics, 1921, Springer-Verlag, Berlin, 2008.

[255] J. P. Tian, *Invitation to research of new mathematics from biology: Evolution algebras*, Topics in functional analysis and algebra, 257–272, Contemp. Math., 672, Amer. Math. Soc., Providence, RI, 2016.

[256] J. P. Tian, Y. M. Zou, *Finitely generated nil but not nilpotent evolution algebras*, J. Algebra Appl. **13**(1) (2014), 1350070, 10 pp.

[257] J. P. Tian, *The replicability of oncolytic virus: defining conditions in tumor virotherapy*, Math. Biosci. Engin. **8**(3) (2011) 841–860.

[258] J. P. Tian, *Algebraic model of non-Mendelian inheritance*, Discrete Contin. Dyn. Syst. Ser. S **4**(6) (2011), 1577–1586.

[259] Z. Tianran, W. Wang, *Mathematical models of two-sex population dynamics*, Kôkyûroku, **1432** (2005), 96–104.

[260] M. Toro, *A general overview of formal languages for individual-based modelling of ecosystems*, J. Log. Algebr. Methods Program. **104** (2019), 117–126.

[261] B. Traoré, B. Sangaré, S. Traoré, *Mathematical model of mosquito populations dynamics with logistic growth in a periodic environment*, An. Univ. Craiova Ser. Mat. Inform. **45**(1) (2018), 86–102.

[262] P. Turchin, *Complex population dynamics: a theoretical/empirical synthesis*, Monographs in Population Biology, 35. Princeton University Press, Princeton, NJ, 2003.

[263] R. Varro, *Gonosomal algebra*, Jour. Algebra, **447**(1) (2016), 1–30.

[264] M. V. Velasco, *The Jacobson radical of an evolution algebra*, Journal of Spectral Theory (EMS), **9**(2) (2019), 601–634.

[265] F. Verrilli, H. Kebriaei, L. Glielmo, M. Corless, C. Del Vecchio, *Carmen Effects of selection and mutation on epidemiology of X-linked genetic diseases*, Math. Biosci. Eng. **14**(3) (2017), 755–775.

[266] M. Walker, J. Blackwood, V. Brown, L. M. Childs, *Modelling Allee effects in a transgenic mosquito population during range expansion*, J. Biol. Dyn. **13** (2019), suppl. 1, 2–22.

[267] W. Wang, L. Chen. *A predator-prey system with stage structure for predator*, Computers Math Applic, **33** (1997), 83–91.

[268] E. B. Wilson, *Book Review: Mathematische Bevölkerungstheorie, auf Grund von G. H. Knibbs' "The Mathematical Theory of Population"*, Bull. Amer. Math. Soc. **30**(9–10) (1924), 561.

[269] A. Wörz-Busekros, *Algebras in genetics*, Lecture Notes in Biomathematics, 36, Springer-Verlag, Berlin-New York, 1980.

[270] A. Wörz-Busekros, *The zygotic algebra for sex linkage*, II. J. Math. Biol. **2**(4) (1975), 359–371.

[271] F. Xu, W. Gan, D. Tang, *Population dynamics and evolution in river ecosystems*, Nonlinear Anal. Real World Appl. 51 (2020), 102983, 16 pp.

[272] J. Yu, *Modeling mosquito population suppression based on delay differential equations*, SIAM J. Appl. Math. **78**(6) (2018), 3168–3187.

[273] M. I. Zakharevich, *The behavior of trajectories and the ergodic hypothesis for quadratic mappings of a simplex*, Russian Math. Surveys. **33** (1978), 207–208.

[274] J. Zhao, J. P. Tian, J. Wei, *Minimal model of plankton systems revisited with spatial diffusion and maturation delay*, Bull. Math. Biol. **78**(3) (2016), 381–412.

Index

CPSIA information can be obtained
at www.ICGtesting.com
Printed in the USA
LVHW082003280420
653775LV00001B/1